T0280680

HADRONIC MULTIPARTICLE PRODUCTION

ADVANCED SERIES ON DIRECTIONS IN HIGH ENERGY PHYSICS

Cover Artwork by courtesy of Los Alamos National Laboratory.
"This work was performed by the University of California, Los Alamos National Laboratory, under the auspices of the United States Department of Energy."

Advanced Series on
Directions in High Energy Physics-Vol.2

HADRONIC MULTIPARTICLE PRODUCTION

Editor
P. CARRUTHERS

World Scientific
Singapore • New Jersey • London • Hong Kong

Published by

World Scientific Publishing Co. Pte. Ltd.
P O Box 128, Farrer Road, Singapore 9128

USA office: World Scientific Publishing Co., Inc.
687 Hartwell Street, Teaneck, NJ 07666, USA

UK office: World Scientific Publishing Co. Pte. Ltd.
P O Box 379, London N12 7JS, England

HADRONIC MULTIPARTICLE PRODUCTION

Copyright © 1988 by World Scientific Publishing Co Pte Ltd.

All rights reserved. This book, or parts thereof, may not be reproduced in any form or by any means, electronic or mechanical, including photo-copying, recording or any information storage and retrieval system now known or to be invented, without written permission from the Publisher.

ISBN 9971-50-558-4
    9971-50-559-2 (pbk)

ISSN 0218-0324

Printed in Singapore by General Printing Services Pte. Ltd.

PREFACE

Much new experimental information on hadronic multiparticle production has accumulated in recent years. In consequence, several theoretical approaches have been developed in an attempt to understand or at least organize the data. The present volume contains insightful reviews of these developments by some of the most active physicists in this field. Apart from the paper by White on the Pomeron in Quantum Chromodynamics, the theoretical papers have a strongly phenomenological and modelistic flavor. As everybody knows, this is inevitable in view of the continuing primitive status of the theory of strong interactions for small momentum transfers. The consensus that QCD is the correct theory of strong interactions might be true but is actually based on rather few concrete results. We consider the current status of particle physics to be almost embarrassing because of our inability to truly understand the majority of collision events. It is not satisfactory to declare these processes uninteresting and to imagine that the physics of Born approximations and Clebsch-Gordan coefficients is the main intellectual content of high energy physics.

The experimental contributions present many interesting challenges for model builders. We hope that this volume will encourage predictions to be made before the next generation of accelerators come alive. The reader will be aware that during the preparation of this volume several very interesting nucleus-nucleus experiments have been completed at CERN. We had neither time nor space to include these developments. Nor have we been able to include multiparticle production data from the CERN collider. This defect will be remedied in part by a forthcoming review article by the UA5 group. In any event, there is today an increased appreciation of the interesting challenges posed by multihadronic processes.

Theoretical efforts range from deductive approaches (such as the Lund Model and the Dual Parton Model) to geometrical and statistical approaches. In fact there is considerable overlap among these approaches but the precise nature of this overlap remains to be understood. This situation motivated us in our recent work* to extract model independent results whenever possible; frequently the appearance of a particular counting distribution depends on probabilistic relations rather than detailed dynamics. (A famous example is the well known negative binomial distribution; by now there are many ways to "justify" this distribution.) As an example of this type of reasoning, we note that given an overall negative binomial and an ordinary binomial for conditional forward/backward probability, a linear forward backward correlation is predicted independently of any dynamical details. This type of result has value but is unfortunately rather dry and very incomplete.

* P. Carruthers and C. C. Shih, *Journal of Modern Physics A*, vol. 2, p. 1447 (1987).

I wish to express my gratitude to the authors for the quality of their contributions. I also wish to thank my collaborators, Minh Duong-Van, Ina Sarcevic, Chia Chang Shih, Fredrik Zachariasen and Richard Weiner for many educational discussions on hadronic physics over the past twenty years.

CONTENTS

Experiment

Hadron production in high energy collisions of leptons with nucleons and nuclei

NORBERT SCHMITZ

MAX-PLANCK-INSTITUT FÜR PHYSIK UND ASTROPHYSIK
MÜNCHEN, GERMANY

Abstract:

Recent experimental results on the production of hadrons in collisions of muons, neutrinos and antineutrinos with nucleons and nuclei at high energies are reviewed and discussed in the framework of the quark-parton model. Data are presented on the following topics: Multiplicities, fragmentation functions, transverse momenta and jet properties, azimuthal asymmetry, distribution of quantum numbers, production of strange particles, and single meson production.

1. Introduction

This review presents and discusses recent experimental results on the production of hadrons in lepton-nucleon (lN) scattering (leptoproduction) at high energies, i.e. in the reactions

$$\mu + N \rightarrow \mu \quad + \text{hadrons} \qquad \text{(muoproduction)} \qquad (1a)$$

$$\nu_\mu + N \rightarrow \mu^- + \text{hadrons} \qquad \text{(charged-current} \qquad (1b)$$

$$\bar{\nu}_\mu + N \rightarrow \mu^+ + \text{hadrons} \qquad \text{neutrinoproduction)} \qquad (1c)$$

The interaction takes place by the exchange of a virtual vector boson between the lepton and the hadron vertex (Fig. 1). The exchanged boson is a photon for μN and a $W^+(W^-)$ boson for charged-current (CC) $\nu N(\bar{\nu} N)$ scattering. Very little is known experimentally on hadron production in neutral-current (NC) νN and $\bar{\nu} N$ scattering,

$$\overset{(-)}{\nu}_\mu + N \rightarrow \overset{(-)}{\nu}_\mu + \text{hadrons} \qquad \text{(neutral-current neutrinoproduction)}, \qquad (2)$$

via Z^0 exchange. Such events are difficult to analyse, since the incident (anti)neutrino energy is unknown and the outgoing (anti)neutrino is not observed.

Early results on leptoproduction of hadrons were obtained in the 1970's with electron and muon beams at the electron accelerators of SLAC, DESY and Cornell, using bubble and streamer chambers, and with (anti)neutrino beams at the proton synchrotrons of Brookhaven (AGS), Argonne (ZGS) and CERN (PS). These experiments were however at lower lepton energies, and it is only with the advent of the large proton synchrotrons of CERN and Fermilab that leptoproduction could be studied over a wider range of the hadronic center of mass (cms) energy W up to the highest W values accessible today ($W \lesssim 20$ GeV for muoproduction, $W \lesssim 15$ GeV for neutrinoproduction). Nevertheless these W values are not yet high enough to observe leptoproduction events with a clear 3-jet structure due to hard QCD processes, as observed at even higher cms energies ($W \approx 40$ GeV) in e^+e^- annihilation at PETRA of DESY and PEP of SLAC. Only forthcoming experiments at the Tevatron of Fermilab will allow in the near future to study leptoproduction by muons and neutrinos at energies comparable to those in e^+e^- annihilation.

Electron, muon and (anti)neutrino beams have been used extensively over many years in deep-inelastic lepton-nucleon scattering experiments, mostly using electronic detectors, to probe the internal parton structure of the nucleons, i.e. to measure their structure functions and deduce from them, in the framework of the quark-parton model, the momentum distribution functions of the various quarks, antiquarks and of the gluon inside the proton and neutron (see e.g. refs. [1 –12]). On the other hand, a detailed investigation of hadron production favors the application of bubble or streamer chambers together with other detector components since, in the ideal case, a momentum measurement and particle identification (mass determination) for each final state hadron in an event is required.

Most features of deep-inelastic lepton-nucleon scattering are successfully described by the quark-parton model (QPM) (Fig. 2, chapt. 3) in which the lepton interacts, via boson exchange, with a quark or antiquark inside the nucleon. The model is extended by the effects of quantum chromodynamics (QCD), in particular the emission of soft and hard gluons. Subsequently the partons fragment into the observed final state hadrons. The QPM allows e.g. to understand the dynamics of deep-inelastic scattering and the kinematic properties of the produced hadrons, to relate the quantum numbers (charge, strangeness etc.) of the final state hadrons to the flavors of the scattered quarks, and to compare the various leptoproduction reactions with each other and to understand their common features and differences.

This review, covering the years since 1982, is organized as follows: Chapter 2 summarises the kinematic variables used to describe a leptoproduction event. Chapter 3 describes the main features of the QPM. In chapter 4 details are given on the experiments which have yielded information on hadron production in recent years. These experimental results are reviewed in chapter 5, the main chapter of this article. Chapter 6 gives the conclusions and an outlook into the future. Earlier results on hadron production in deep-inelastic lepton-nucleon scattering can be found in the reviews of refs. [13 –25].

2. Kinematics and variables

From the laboratory energies E, E' and momenta k, k' of the incident and outgoing lepton l and l', respectively, and the lepton scattering angle Θ (Fig. 1) the following *event variables* are obtained:

$$Q^2 = 2EE' - 2kk'cos\Theta - m_l^2 - m_{l'}^2 \tag{3}$$

$$= \text{four-momentum transfer squared} \approx 4EE'sin^2\frac{\Theta}{2} \,,$$

$$\nu = E - E' = \text{lab energy of the exchanged boson (energy transfer),} \tag{4}$$

$$W^2 = M^2 + 2M\nu - Q^2 \qquad (M = \text{nucleon mass}) \tag{5}$$

$$= \text{total energy } W \text{ (effective mass) squared of all final state hadrons in their cms.}$$

The two Bjorken scaling variables x and y are defined as

$$x = \frac{Q^2}{2M\nu} = \text{Bjorken } x \text{ with } 0 \le x \le 1 \,, \tag{6}$$

$$y = \frac{\nu}{E} = 1 - \frac{E'}{E} \text{ with } 0 \le y \le (1 + \frac{Mx}{2E})^{-1} \text{ for fixed } x. \tag{7}$$

From eqs. (5) and (6) the following relation is obtained between W, Q^2 and x:

$$W^2 = M^2 + Q^2(\frac{1}{x} - 1). \tag{8}$$

Thus, at fixed beam energy E, an event is determined by two independent event variables, e.g. $(E', \Theta), (\nu, Q^2), (W, Q^2), (x, Q^2)$ or (x, y).

A produced hadron h is characterized by the following *particle variables*:

$$x_F = \frac{p_L^*}{p_{Lmax}^*} \approx \frac{2p_L^*}{W} \quad \text{with} \quad -1 \leq x_F \leq 1$$

$$= \text{Feynman } x \text{ (fractional longitudinal cms momentum)} \tag{9}$$

where p_L^* is the longitudinal momentum component of the hadron along the virtual-boson direction ("current direction") in the center of mass of the virtual boson and target nucleon, i.e. in the hadronic cms; $p_L^* > 0$ ($p_L^* < 0$) is in (opposite to) the boson direction. $p_{Lmax}^* \approx W/2$ is the maximum kinematic value of p_L^*.

$$z = \frac{E_h}{\nu} = \text{fraction of the boson energy carried by the hadron } (z > 0), \tag{10}$$

$$y^* = \frac{1}{2} \ln \frac{E^* + p_L^*}{E^* - p_L^*} = \text{cms rapidity}, \tag{11}$$

where E_h and E^* are the hadron laboratory and cms energy respectively. A hadron with $x_F > 0$ (i.e. $y^* > 0$) is travelling into the forward cms hemisphere, a hadron with $x_F < 0$ ($y^* < 0$) into the backward hemisphere of boson-nucleon scattering. For high W, the kinematic correlation between x_F and z is such that backward hadrons have small z values whereas for large positive x_F the two variables are roughly equal. Therefore, if one wants to study hadron production in the two cms hemispheres separately, x_F or y^* and not z is the appropriate variable.

While x_F and y^* measure the longitudinal motion of the hadron, its transverse motion is measured by the momentum component p_T transverse to the boson direction and the azimuthal angle ϕ around this direction.

Two quantities, sphericity S and thrust T [26], are usually used to measure the collimation of the produced hadrons in an event in their overall cms, i.e. to characterize their jet properties. S and T are defined as

$$S = \frac{3}{2} \frac{\sum p_T^{*2}}{\sum p^{*2}} \quad \text{and} \quad T = \frac{\sum |p_L^*|}{\sum p^*} \tag{12}$$

where the sums extend over all final state hadrons in the event or, for an investigation of forward and backward jets separately, over the hadrons in one hemisphere only. Instead of measuring p_T^* or p_L^* with respect to the current direction, it is usually measured (as in e^+e^- annihilation) with respect to an event axis, "sphericity axis" or "thrust axis", determined such that, respectively, $\sum p_T^{*2}$ is a minimum or $\sum |p_L^*|$ is a maximum. Using the sphericity axis, $0 \leq S \leq 1$ with $S = 0$ for a pencil-like event and $S = 1$ for a spherical hadron distribution. With respect to the thrust axis, $\frac{1}{2} \leq T \leq 1$ with $T = 1$ for a pencil-like event and $T = \frac{1}{2}$ for a spherical event.

3. The quark-parton model

In the simple quark-parton model (QPM) (Fig. 2) the incoming lepton interacts, via virtual boson exchange, with one of the quarks or antiquarks inside the target nucleon. This (anti)quark absorbs the boson and is thereby knocked out of the nucleon. In the boson-nucleon center of mass, the interaction leads, on the parton level, to a final state with two colored objects, namely the forward going scattered (anti)quark and the backward going nucleon remnant system, which as a spectator is not involved in the interaction. In the simplest case of scattering on a valence quark the target remnant is a diquark.

Subsequently, both the scattered (anti)quark and the target remnant fragment (hadronize) into the final state hadrons, which can be classified, respectively, as current (quark) fragments, going predominantly into the forward cms hemisphere ($x_F > 0$), and target (diquark) fragments, travelling predominantly backward ($x_F < 0$). At sufficiently high cms energies W the current and target fragments appear as two hadron jets, the current (forward) jet and the target (backward) jet respectively, the jets getting more collimated and thus more pronounced with increasing W. However, the two jets are not independent due to conservation of quantum numbers and energy-momentum. Furthermore, since fragmentation produces a continuous chain of hadrons there is no clean separation between the two types of fragments, i.e. for some hadrons in the overlap region a unique allocation to one of the two jets is not possible. Thus a rapidity distribution shows at high W a smooth and continuous transition from the current fragmentation region via a central region to the target fragmentation region without sharp boundaries. At smaller W, the central region has not yet developed and the two fragmentation regions overlap. Kinematically the central rapidity region transforms into a narrow x_F region around $x_F = 0$.

In practice, when studying quark and diquark fragmentation in leptoproduction, an approximate separation of current and target fragments is achieved by selecting as current (target) fragments those hadrons which go forward ($x_F > 0$) (backward, $x_F < 0$) in the hadronic cms. This separation becomes better with increasing W. The fragmentation itself, e.g. of a quark or a diquark into hadrons of a particular type h (e.g. $\pi^\pm, K^\pm$, etc.) is described by the quark or diquark fragmentation functions, $D_q^h(z)$ and $D_{qq}^h(z)$ respectively.

More quantitatively, the cross section $d\sigma^h/dxdydz$ for the semi-inclusive production of a hadron h in the reaction

$$l + N \rightarrow l' + h + X, \tag{13}$$

at high energies has, in the QPM, three ingredients, assuming h to be a current fragment:

— The momentum distribution functions $q(x)$ and $\bar{q}(x)$ of the various (anti)quarks in the nucleon which participate in the interaction. Here x is the fraction of the nucleon four-momentum carried by the struck quark; it follows from kinematics that x is identical to Bjorken x, eq. (6).

— The cross sections $d\sigma/dy$ for the hard (quasi)elastic scattering processes $l + q \to l' + q'$ and $l + \bar{q} \to l' + \bar{q}'$, where y is related to the lq cms scattering angle by $y = \nu/E \approx (1 - cos\Theta^*)/2$.

— The fragmentation functions $D_{q'}^h(z)$ and $D_{\bar{q}'}^h(z)$ describing the fragmentation of q' and $\bar{q}'$ into h.

The first two ingredients yield the cross section $d\sigma/dxdy$ for the inclusive reaction $l + N \to l' + X$ as an incoherent sum, over all *participating* quark flavors, of the (quasi)elastic lq and $l\bar{q}$ cross sections, weighted with the quark and antiquark distribution functions in the nucleon. This is the basic idea of the QPM. The semi-inclusive single particle cross section is then obtained from the inclusive cross section by multiplying each term in the sum by the appropriate fragmentation function $D_{q'}^h(z)$ or $D_{\bar{q}'}^h(z)$. This prescription yields:

$$\frac{d\sigma^h}{dxdydz}(\mu N \to \mu h X) = \frac{2\pi\alpha^2}{MExy^2}(1 - y + \frac{y^2}{2}) \sum_q e_q^2[q(x)D_q^h(z) + \bar{q}(x)D_{\bar{q}}^h(z)]$$

$$\frac{d\sigma^h}{dxdydz}(\nu N \to \mu^- h X) = \frac{2G^2MEx}{\pi}\left[\sum_{q,q'} q(x)g_{qq'}^2 D_{q'}^h(z) + (1-y)^2 \sum_{\bar{q},\bar{q}'} \bar{q}(x)g_{qq'}^2 D_{\bar{q}'}^h(z)\right] \quad (14)$$

$$\frac{d\sigma^h}{dxdydz}(\bar{\nu} N \to \mu^+ h X) = \frac{2G^2MEx}{\pi}\left[\sum_{\bar{q},\bar{q}'} \bar{q}(x)g_{qq'}^2 D_{\bar{q}'}^h(z) + (1-y)^2 \sum_{q,q'} q(x)g_{qq'}^2 D_{q'}^h(z)\right]$$

The $g_{qq'}$'s are the relative coupling strengths (given by the Cabibbo angle Θ_c) for the flavor changing CC processes $W^\pm q \to q'$ and $W^\pm \bar{q} \to \bar{q}'$. They are listed in Table 1.*

If one neglects in the simplest case the sea quarks (i.e. at large x) and sets the small Cabibbo angle to zero (i.e. $g_{qq'}^2 = 1$ or 0 in Table 1) equations (14) simplify e.g. for a proton target to

$$\frac{d\sigma^h}{dxdydz}(\mu p \to \mu h X) = \frac{2\pi\alpha^2}{MExy^2}\left(1 - y + \frac{y^2}{2}\right)[\frac{4}{9}u(x)D_u^h(z) + \frac{1}{9}d(x)D_d^h(z)]$$

$$\frac{d\sigma^h}{dxdydz}(\nu p \to \mu^- h X) = \frac{2G^2MEx}{\pi} d(x)D_u^h(z) = \frac{d\sigma}{dxdy}(\nu p) \cdot D_u^h(z) \quad (15)$$

$$\frac{d\sigma^h}{dxdydz}(\bar{\nu} p \to \mu^+ h X) = \frac{2G^2MEx}{\pi} (1-y)^2 u(x)D_d^h(z) = \frac{d\sigma}{dxdy}(\bar{\nu} p) \cdot D_d^h(z)$$

where $u(x)$ and $d(x)$ are the u and d quark distribution functions in the proton. The corresponding formulae for a neutron target are obtained by the charge symmetry replacement $u(x) \leftrightarrow d(x)$. The contributing diagrams for this simplest case are sketched in Fig. 3. The cross section per nucleon of an isoscalar target is the average of the proton and neutron cross

---

*The energy is assumed to be sufficiently high to produce charm. Charm threshold effects and transitions involving b and t quarks are neglected.

Tab. 1: Coupling strengths $g_{qq'}^2$ for the flavor changing transitions $W^\pm q \to q'$ and $W^\pm \bar{q} \to \bar{q}'$ contributing to νN and $\bar{\nu} N$ scattering. The main transitions (valence quark and Cabibbo-favored) are underlined, the doubly suppressed transitions (sea quark and Cabibbo-suppressed) and transitions from $c, \bar{c}$ are in brackets.

ν N scattering		$\bar{\nu}$ N scattering	
transition	$g_{qq'}^2$	transition	$g_{qq'}^2$
$\underline{d \to u}$	$cos^2\Theta_c$ } 1	$\underline{u \to d}$	$cos^2\Theta_c$ } 1
$d \to c$	$sin^2\Theta_c$	$u \to s$	$sin^2\Theta_c$
$(s \to u)$	$sin^2\Theta_c$ } 1	$(c \to d)$	$sin^2\Theta_c$ } 1
$s \to c$	$cos^2\Theta_c$	$(c \to s)$	$cos^2\Theta_c$
$\bar{u} \to \bar{d}$	$cos^2\Theta_c$ } 1	$\bar{d} \to \bar{u}$	$cos^2\Theta_c$ } 1
$(\bar{u} \to \bar{s})$	$sin^2\Theta_c$	$(\bar{d} \to \bar{c})$	$sin^2\Theta_c$
$(\bar{c} \to \bar{d})$	$sin^2\Theta_c$ } 1	$(\bar{s} \to \bar{u})$	$sin^2\Theta_c$ } 1
$(\bar{c} \to \bar{s})$	$cos^2\Theta_c$	$\bar{s} \to \bar{c}$	$cos^2\Theta_c$

sections. If h is a diquark fragment, the quark fragmentation functions in eq. (15) have to be replaced by the appropriate diquark fragmentation functions according to Fig. 3.

Experimentally one measures the normalized z (or x_F, y^*) distribution $D^h(z, x, Q^2)$ of hadron h (using e.g. x and Q^2 as independent event variables), defined as

$$D^h(z, x, Q^2) = \frac{1}{N_{ev}(x, Q^2)} \cdot \frac{dN^h}{dz}(z, x, Q^2) \qquad (16)$$

where $N_{ev}(x, Q^2)$ is the number of events at (x, Q^2) and N^h the number of hadrons h in these events. (Corresponding definitions hold for x_F and y^* distributions). $D^h(z, x, Q^2)$ is thus the ratio of the semi-inclusive and inclusive cross sections:

$$D^h(z, x, Q^2) = \frac{d\sigma^h}{dxdQ^2dz}(x, Q^2, z) / \frac{d\sigma}{dxdQ^2}(x, Q^2) . \qquad (17)$$

It is often called the (generalized) fragmentation function of hadron h. For μp scattering for example one obtains in the forward hemisphere

$$D^h(z, x, Q^2) = \frac{\sum_q e_q^2 [q(x, Q^2)D_q^h(z, Q^2) + \bar{q}(x, Q^2)D_{\bar{q}}^h(z, Q^2)]}{\sum_q e_q^2 [q(x, Q^2) + \bar{q}(x, Q^2)]} \qquad (18)$$

where the QCD predicted Q^2 dependence of the distribution and fragmentation functions is explicitly written. The integral of the fragmentation function is the average multiplicity of hadron h:

$$<n_h>(x, Q^2) = \int_0^1 D^h(z, x, Q^2) \, dz = \int_{-1}^{+1} D^h(x_F, x, Q^2) \, dx_F \qquad (19)$$

Several comments can be made for the simplest case of valence quarks only and $\Theta_c = 0$ (Fig. 3):

— In νN and $\bar{\nu} N$ scattering one is dealing with a clean flavor situation: Only one flavor contributes and the flavor of the fragmenting quark and diquark is well defined. The semi-inclusive cross section factorises into the inclusive cross section and the fragmentation function (eq. (15)) such that the measured hadron distribution (17) in $\nu p(\bar{\nu} p)$ scattering is just the $u(d)$-quark fragmentation function, if h is a quark fragment ($x_F > 0$), and the $uu(ud)$-diquark fragmentation function, if h is a target fragment ($x_F < 0$). In μN scattering two quark flavors contribute, although the scattering on a u quark dominates.

— The simple QPM makes a number of predictions about the forward or backward hadron jets in the various reactions, thus allowing interesting comparisons and providing easy tests of the model. For instance, the forward jets in νp and νn as well as in $\bar{\nu} p$ and $\bar{\nu} n$ scattering should be the same:

$$D^h(\nu p, x_F > 0) = D^h(\nu n, x_F > 0) = D_u^h \tag{20}$$

$$D^h(\bar{\nu} p, x_F > 0) = D^h(\bar{\nu} n, x_F > 0) = D_d^h. \tag{21}$$

Thus, the forward jets in μp, νp and $\bar{\nu} p$ scattering are related by (assuming $u(x) \approx 2d(x)$):

$$D^h(\mu p, x_F > 0) = \frac{8}{9} D^h(\nu p, x_F > 0) + \frac{1}{9} D^h(\bar{\nu} p, x_F > 0) \approx D^h(\nu p, x_F > 0). \tag{22}$$

Similarly for the backward jets:

$$D^h(\bar{\nu} p, x_F < 0) = D^h(\nu n, x_F < 0) = D_{ud}^h \tag{23}$$

$$D^h(\mu p, x_F < 0) = \frac{8}{9} D^h(\bar{\nu} p, x_F < 0) + \frac{1}{9} D^h(\nu p, x_F < 0) \approx D^h(\bar{\nu} p, x_F < 0). \tag{24}$$

Finally the backward jet in νp and the charge symmetric jet in $\bar{\nu} n$ scattering should be the same, for example

$$D^{\pi^\pm}(\nu p, x_F < 0) = D_{uu}^{\pi^\pm} = D_{dd}^{\pi^\mp} = D^{\pi^\mp}(\bar{\nu} n, x_F < 0). \tag{25}$$

— Further predictions, for instance for charged pions, are obtained applying isospin symmetry:

$$D^{\pi^\pm}(\nu p, x_F > 0) = D_u^{\pi^\pm} = D_d^{\pi^\mp} = D^{\pi^\mp}(\bar{\nu} p, x_F > 0) \tag{26}$$

$$D^{\pi^+}(\bar{\nu} p, x_F < 0) = D_{ud}^{\pi^+} = D_{ud}^{\pi^-} = D^{\pi^-}(\bar{\nu} p, x_F < 0) \tag{27}$$

where the central equalities follow from isospin symmetry.

In reality the situation is of course more complicated, since at low x scattering occurs also on sea (anti)quarks, the Cabibbo angle is not exactly zero and $2d(x) < u(x)$ [27] at large x, favoring even more the u-quark. Nevertheless, comparisons of single particle distributions are still possible on the basis of the general equations (14), if one uses the known (anti)quark distribution functions (see chapt. 5.2).

The simple quark parton picture discussed so far has to be extended for two effects:

a.) Primordial transverse momentum: Due to their Fermi motion the confined (anti)quarks have also a transverse momentum k_T inside the nucleon. From the size of the nucleon and the uncertainty principle $< k_T > \approx 0.25$ GeV. This primordial (intrinsic) k_T becomes apparent (Fig. 4), by transverse momentum conservation, as a transverse momentum k_T of the struck quark and spectator diquark and thus as a tilt of the event axis relative to the current direction. Its influence on the transverse momentum and jet properties of the final state hadrons is discussed in chapt. 5.3.

b.) Quantum chromodynamics: QCD predicts the emission of hard and soft gluons by the struck quark before or after the boson absorption (Figs. 5a,b). In addition, boson-gluon fusion (Figs. 5c,d) is expected at low x, where the gluon distribution function is large. In this process a gluon inside the nucleon and the virtual boson materialize into a quark-antiquark pair, producing e.g. two forward jets at high energies. QCD processes lead to the well established Q^2 dependence of the nucleon structure functions (scaling violation), to a Q^2 dependence also of the fragmentation functions and to effects on the transverse momentum and jet properties of hadrons (Fig. 4) as discussed further in chapt. 5.3.

While the hard lepton-quark scattering, including gluon emission, can be calculated rigorously in the standard model of electroweak interactions and perturbative QCD, an exact theory for the soft transition of partons into hadrons is still lacking. Therefore various fragmentation models, based on the QPM and QCD, have been developed, some of them for e^+e^- annihilation, which can be classified into three main categories [28–31]: independent fragmentation (IF) models [32], cluster fragmentation (CF) models [33] and the Lund string fragmentation (SF) model [34, 35]. The most successful approach is the Lund string model in which a continuous chain of hadrons is produced from the color field between colored objects emerging from hard scattering or QCD processes with large momentum relative to each other, e.g. between the quark and the diquark in a leptoproduction event. This model describes hadron production in e^+e^- annihilation and in lepton-nucleon scattering in good agreement with experiments. Monte Carlo programs have been developed [35] which generate complete leptoproduction or e^+e^- annihilation events, incorporating, with adjustable parameters, also the effects of both primordial k_T and QCD. In particular the leptoproduction Lund Monte Carlo program has been extensively used by the experimenters, (a) to correct the raw data for experimental biases (chapt. 4), and (b) to analyse and interpret the experimental physics results. A disadvantage of the Lund model are its many adjustable parameters.

An alternative approach to hadron production, without using QCD, is provided by the fire-string (FS) model [36], in which the exchanged boson interacts with a $q\bar{q}$-pair from the nucleon yielding one or two fire-strings together with an excited baryon B^* (e.g. $\gamma^* + N \rightarrow B^* + FS$). Via a branching process the FS (as well as the B^*) then fragments into the final state hadrons.

4. Experiments and data analysis

The data for this review come primarily from the following leptoproduction experiments:

a.) The European Muon Collaboration (EMC) has studied muoproduction (1a) in two phases, NA2 [37 – 51] and NA9 [52 – 69], at the 450 GeV Super Proton Synchrotron (SPS) of CERN. The NA2 experiment uses a forward spectrometer system [37] measuring the scattered muon and the hadrons produced in the forward cms hemisphere. Data were taken at beam energies of E_μ = 120, 200 and 280 GeV, using long H_2 and D_2 targets and yielding a large data sample of $\sim$ 500 K events in total, although with limited particle identification.

In the NA9 phase later on, with E_μ = 280 GeV, the NA2 forward spectrometer was supplemented by a vertex detector [52] to register and identify particles with lower momenta at larger angles thus providing an almost 4π acceptance. The vertex detector consisted of a superconducting vertex spectrometer magnet with a streamer chamber positioned inside, and of Cerenkov and time-of-flight counters for particle identification. The streamer chamber contained a 1m long liquid H_2 or D_2 target. Both the vertex detector and the forward spectrometer were instrumented with proportional and drift chambers for particle tracking. The whole apparatus allowed $\sim$ 50 % of the final state particles to be identified. The maximum W with reasonable statistics is $\sim$ 20 GeV. In order to restrict the NA9 event sample to regions where the corrections were relatively small ($\lesssim$10 %) the following cuts were usually applied to the data: $Q^2 > 4$ GeV^2, $16 < W^2 < 400$ GeV^2, $20 < \nu < 260$ GeV, $E' > 20$ GeV, $y < 0.9$, and $\Theta > 0.5^0$ or 0.75^0. This yielded event samples of $\sim$ 29 K events on hydrogen and $\sim$ 22 K events on deuterium.

b.) Several collaborations [70 – 90] have used the 15′ bubble chamber, filled with H_2, D_2 or a $Ne - H_2$ mixture, to study neutrinoproduction in the wide-band ν and $\bar{\nu}$ beams at the 500 GeV PS of Fermilab. The (anti)neutrino energy spectrum ranged typically between 10 and 200 ÷ 250 GeV, peaking around $E_\nu \approx$ 25 GeV. The outgoing muon in a CC event was identified by an external muon identifier (EMI) behind the bubble chamber and/or by a kinematic algorithm for muon selection. Particle identification, essentially based on the track ionisation density, was poor, and so was, in case of H_2 and D_2 fillings, the detection of neutral particles, in particular of photons from π^0 decay. Therefore various methods, mostly based on transverse momentum balance, were applied to estimate the total outgoing hadronic energy and thus the unknown beam energy in each event.

c.) The three collaborations WA21 [91 – 102], WA25 [103 – 110] and WA59 [111 – 118] have exposed the Big European Bubble Chamber BEBC, filled respectively with H_2, D_2 and a $Ne - H_2$ mixture, to the ν and $\bar{\nu}$ wide-band beams at the CERN SPS with energy spectra extending to ~ 200 GeV. In each beam-target category there are typically between 10 K to 20 K events with $E_\mu > 3 \div 4$ GeV. The muon identification is achieved by a two-plane EMI [119] downstream of BEBC and, at a later stage of the WA21 experiment, by veto counters in front of BEBC and a cylindrical array of single wire proportional tubes (Internal Picket Fence IPF) [120] around the chamber. The IPF measures the time of an event thus allowing to separate NC events from background events due to neutral hadrons, to reduce the accidental background in the EMI, and to link a secondary vertex in the chamber (e.g. V^0 decay, neutron star) to the primary event vertex. Particle identification is poor also in the BEBC experiments, relying up to ~ 1 GeV/c mainly on bubble density, change of curvature, range in the chamber liquid and kinematic fits. The unknown neutrino energy of an event is again estimated from transverse momentum conservation. The maximum W with usable statistics is $\sim 12 \div 15$ GeV.

d.) The SKAT collaboration [121 – 123] has used the freon-filled bubble chamber SKAT in a wide-band $\nu, \bar{\nu}$ beam at the 70 GeV PS of Serpukhov. The beam energy ranges from 3 to 30 GeV. The event samples used in the analyses consist of ~ 8300 (650) $\nu A(\bar{\nu}A)$ events.

Corrections have been applied to the raw data for various experimental and instrumental effects. We briefly discuss the NA9 and WA21 corrections as examples. The NA9 data were corrected for limited detector acceptance, smearing introduced by the resolution of the apparatus, radiative effects, particle misidentification and inefficiencies of the off-line pattern recognition, track fitting and vertex finding procedures. The WA21 data were corrected for measurement errors, uncertainties in estimating the incident (anti)neutrino energy and for unidentified charged particles. In general, the correction proceeded as follows: Monte Carlo (MC) events ("original events") were generated by the appropriate leptoproduction versions [35] of the Lund MC program. Each generated event was then modified according to the experimental situation ("modified events") thus simulating the sample of real events with all their experimental imperfections. Any distribution D_{data}, obtained from the raw data, of a particular variable is then corrected by multiplying it by the ratio of the corresponding distributions obtained from the original and modified MC events, such that

$$D_{corr} = D_{data} \cdot \frac{D_{orig}^{MC}}{D_{mod}^{MC}} \,. \tag{28}$$

The event samples and ranges of variables in the recent leptoproduction experiments are sufficiently large to obtain interesting physics results on the various reactions which often complement each other such that a rather complete picture emerges.

5. Results

In this chapter an extensive selection of published physics results is presented, with emphasis on the main features and on those data that are relatively new, corrected for experimental effects (chapt. 4), and extending to the highest available energies. The list of references on the other hand is attempted to be rather complete. The majority of the data were obtained on protons using H_2 targets, with some data coming from D_2 targets, thus yielding also information on lepton-neutron scattering, and from heavy nuclear targets (neon, freon). The interpretation of the results will be mostly in the powerful and successful framework of the QPM (chapt. 3).

The topics discussed are:

— Multiplicities
— Fragmentation functions
— Transverse momenta and jet properties
— Azimuthal asymmetry
— Distributions of quantum numbers (charge, strangeness)
— Production of strange particles
— Single meson production

5.1. MULTIPLICITIES

a.) <u>Average multiplicities</u>

Average multiplicities $< n >$ of charged hadrons produced in $\mu p, \nu N$ and $\bar{\nu} N$ scattering have been measured in refs. [57, 58, 70, 72, 77, 96, 103, 107]. For all reactions, $< n >$ shows a strong W^2 dependence, which is observed when integrating over Q^2 as well as at fixed Q^2. In the available W ranges a parametrisation linear in lnW^2,

$$< n >= a + b \cdot ln(W^2/GeV^2) \,, \tag{29}$$

gives good fits to the data points. Fig. 6 shows a compilation of $< n >$ vs. W^2 for $\mu p, \nu p, \bar{\nu} p, \nu n$ and $\bar{\nu} n$ scattering [57, 96, 107]. The slopes for the various reactions are seen to be similar ($b \approx 1.2 \div 1.4$), except for the smaller slope for $\bar{\nu} n$ ($b \approx 1.1$).

Fig. 7 compiles, for the five reactions, the average charged multiplicities $< n_F >$ and $< n_B >$ separately in the forward and backward cms hemispheres, vs. W^2 [57, 96, 107]*. At fixed W^2, the following features are observed (*test of QPM relations*):

— For a given reaction, $< n_F > .> < n_B >$ (except for νp at lower W), $< n_F >$ showing a stronger increase with W than $< n_B >$. This implies that at higher W the quark jet has a larger average multiplicity than the diquark jet, which is plausible since the diquark jet contains in general a baryon as a leading particle consuming a large fraction of the available energy.

— In the forward hemisphere (quark fragmentation region) the QPM predictions resulting via eq. (19) from eqs. (20) $\div$ (22) (Fig. 3) are reasonably well fulfilled: the $< n_F >$ values for $\mu p, \nu p$ and νn are approximately equal (mainly u-quark fragmentation), as are the $< n_F >$ for $\bar{\nu} p$ and $\bar{\nu} n$ (mainly d-quark fragmentation). Actually the $< n_F >$ for *all five reactions* are approximately equal (although the μp points are somewhat lower at high W). This can be explained from the fact that most forward hadrons are pions for which

$$D_u^\pi \equiv D_u^{\pi^+} + D_u^{\pi^-} = D_d^{\pi^-} + D_d^{\pi^+} \equiv D_d^\pi \tag{30}$$

from isospin symmetry, eq. (26), implying similar forward jets in νN and $\bar{\nu} N$ scattering.

— In the backward hemisphere (diquark fragmentation region) the $< n_B >$ values for $\mu p, \bar{\nu} p$ and νn are roughly equal as expected from ud-diquark fragmentation, eqs. (23) and (24). Furthermore $< n_B > (\nu p) > < n_B > (\bar{\nu} p) > < n_B > (\bar{\nu} n)$ implying more charged hadrons in uu than in ud than in dd-diquark fragmentation.

The average multiplicities $< n_F >$ and $< n_B >$ as functions of W^2 are found [107] to be in rather good agreement with the predictions of the Lund model which is of course more realistic since it includes the scattering on sea quarks neglected in the simplified discussion above.

For the $\nu / \bar{\nu}$ reactions the average multiplicities $< n_{F,B}^\pm >$ in each hemisphere have been measured [72, 77, 96, 107] for each hadron charge separately, see e.g. Fig. 8. One finds:

— The QPM predictions following from eqs. (20), (21) (forward) and (23) (backward) are approximately satisfied for each hadron charge separately.

— In $\nu p, \nu n$ scattering $< n_F^+ > > < n_F^- >$, whereas in $\bar{\nu} p, \bar{\nu} n$ scattering $< n_F^- > > < n_F^+ >$. These inequalities follow from *quark charge retention* implying the $u(d)$ quark to fragment more readily into a positive (negative) hadron. Furthermore $< n_F^\pm > (\nu p) \approx < n_F^\mp > (\bar{\nu} p)$ as expected for pions from isospin symmetry, eq. (26).

---

*Here it is essential to correct for particle misidentification, since e.g. a backward going proton may fall, when misidentified as a π^+, into the forward hemisphere when Lorentz transformed from the laboratory into the hadronic cm system.

— In νp (uu-fragmentation) and $\bar{\nu}p, \nu n$ (ud-fragmentation) scattering $< n_B^+ > > < n_B^- >$, whereas in $\bar{\nu}n$ (dd-fragmentation) $< n_B^- > > < n_B^+ >$. These relations reflect diquark charge retention. The total hadronic charge $Q_H = < n_F^+ > + < n_B^+ > - < n_F^- > - < n_B^- >$ is of course fixed, namely $Q_H = 2, 1, 0, -1$ for $\nu p, \nu n, \bar{\nu}p$ and $\bar{\nu}n$ scattering respectively.

Average multiplicities as functions of W^2 have also been measured for specific types of hadrons, namely for $\pi^\pm$ [56, 60]; $K^\pm$ [56]; $K^0, \bar{K}^0$ [54, 58]; $\Lambda, \bar{\Lambda}$ [54]; $p, \bar{p}$ [56] and ρ^0 [63] in μp scattering, and for π^0 [96]; ρ^0 [92, 102, 110, 123] and Λ, K^0 [85, 91, 104, 121, 124] in νN and $\bar{\nu}N$ scattering*. Figs. $9 \div 11$ show, as examples, $< n >$ vs. W^2 for charged pions in μp (for fixed intervals of Q^2), for π^0 in νp and $\bar{\nu}p$ and for ρ^0 in $\mu p, \nu p$ and $\bar{\nu}p$ scattering, respectively. Again a linear dependence on lnW^2 is observed.

In a search for a possible $Q^2 dependence$ of $< n_h > (W, Q^2)$ one has to keep W fixed (which requires sufficient statistics). Otherwise an observed Q^2 dependence of $< n >$ could simply be a kinematic reflection of its strong W dependence via eq. (8) which implies in general a rise of the average W with increasing Q^2. Two possible origins of a Q^2 dependence of $< n_h >$, if it exists, are discussed in chapt. 5.2.

For μp scattering Fig. 12 shows the average multiplicity of charged pions (a) and of π^+(b) and π^-(c) separately vs. Q^2 for 8 narrow intervals of W [60]. At fixed W, a weak Q^2 dependence is observed in all three cases (a), (b) and (c). A simultaneous fit to the W^2 and Q^2 dependence of the form

$$< n_\pi > = a_1 + a_2 \cdot ln(W^2/\mathrm{GeV}^2) + a_3 \cdot ln(Q^2/\mathrm{GeV}^2) \qquad (31)$$

describes the data well yielding $a_1 = -0.66 \pm 0.17, a_2 = 1.04 \pm 0.03, a_3 = 0.10 \pm 0.05$ for all charged pions. Thus the lnQ^2 dependence is a 2σ effect and amounts to ~ 10 % of the lnW^2 dependence. Similarly, a weak Q^2 dependence of $< n >$ at fixed W has also been observed for all charged hadrons ($a_3 = 0.22 \pm 0.07$) and for K^0 mesons ($a_3 = 0.08 \pm 0.03$) produced in μp scattering [58]. For $\nu p, \bar{\nu}p$ and $\bar{\nu}n$ scattering, the average charged multiplicity is rather independent of Q^2 in fixed intervals of W [96, 103] with $a_3 = 0.04 \pm 0.02$ for νp and 0.05 ± 0.04 for $\bar{\nu}p$ from the data of ref. [96].

b.) <u>Higher moments</u>

The second order moments of multiplicity distributions, namely the dispersion (width) D and the correlation parameter f_2, have been investigated in refs. [57, 70, 72, 77, 96, 103]. They are defined as

$$D^2 = < n^2 > - < n >^2 \quad \text{and} \quad f_2 = < n(n-1) > - < n >^2 = D^2 - < n > \qquad (32)$$

---

*In some references the fraction F of CC events containing a Λ or K^0 is plotted instead of $< n >$. $F = < n >$ only in case that there are no events with $n > 1$.

with $f_2 = 0$ for a Poisson distribution. Plots of D^- vs. $<n^->$, where D^- is the dispersion of the multiplicity distribution of the negative hadrons and $<n^->$ their average multiplicity, have established a linear relation ("Wroblewski relation") $D^- = c^- + d^- \cdot <n^->$. This relation is very well fulfilled, with slopes $d^- \approx 0.30 \div 0.40$ for the various reactions. Similar slopes were found for e^+e^- and $\bar{p}p$ annihilation in the same energy ranges. The relation implies a linear dependence of D on $<n>$, with $c = 2c^- - Q_H d^-$ and $d = d^-$, for all charged hadrons, and a quadratic dependence of the correlation parameters f_2^{--} and f_2 on $<n^->$ and $<n>$, respectively, as is easily seen from eq. (32) and $n = Q_H + 2n^-$ where Q_H is the charge of all final state hadrons $(D = 2D^-, f_2 = 4f_2^{--} + 2 <n^-> -Q_H)$. Fig. 13 shows D^- and f_2^{--} vs. $<n^->$ for μp, νp and $\bar{\nu} p$ scattering. Furthermore, in μN and νN, $\bar{\nu} N$ scattering D depends linearly on lnW^2 such that at higher W $D/<n>$ becomes approximately W independent as required for KNO scaling to be valid (see below). Finally linear dependences of $D_{F,B}$ on $<n_{F,B}>$ have also been observed in the forward and backward hemispheres separately [57, 96].

c.) Multiplicity distributions

Charged multiplicity distributions $P(n)$ (with $\sum_n P(n)=1$) have been measured for the various reactions [57, 72, 77, 96, 103]. They are often plotted as KNO distributions [125], i.e. as $\psi(z) \equiv <n> \cdot P(n)$ vs. the reduced multiplicity $z = n/<n>$, for fixed intervals of W. KNO scaling implies that for a given reaction $\psi(z)$ is independent of W.

Fig. 14 shows as three more recent examples the charged hadron KNO distributions for μp [57], νp and $\bar{\nu} p$ [96] scattering. KNO scaling is seen to be roughly fulfilled in the available W ranges. The data points are well fitted by the one parameter scaling function [126]

$$\psi(z) = 2 \cdot \frac{e^{-c} c^{cz+1}}{\Gamma(cz+1)}, \tag{33}$$

which is based on a cluster model with a Poisson distribution for the cluster multiplicity. The fits are shown by the curves, with the fitted parameter values $c = 7.10 \pm 0.12$ for μp, 8.67 ± 0.13 for νp and 6.25 ± 0.17 for $\bar{\nu} p$ scattering. The distribution for μp falls between those for νp and $\bar{\nu} p$, but is closer to the latter. More generally the following systematics have been observed for the KNO distributions in νN and $\bar{\nu} N$ scattering [16, 77, 96, 103]:

— The νp and νn distributions are (approximately) equal and close to the distributions for $\bar{p}p$ and e^+e^- annihilation [127].[*]

— The $\bar{\nu} p$ and $\bar{\nu} n$ distributions are (approximately) equal and close to the pp and $\pi^- p$ distributions.

---

[*]It should be noticed that in ref. [57] the μp KNO distribution is compared with the distribution for $p\bar{p}$ annihilation at considerably lower W, namely $W < 3.9$ GeV where it is somewhat wider than at higher W [128].

— The $\bar{\nu}N$ distributions are considerably wider than the νN distributions.

Recently the EMC [67] has investigated the charged multiplicity distributions in limited cms rapidity intervals Δy^* for μp scattering. Such distributions were found to be very well fitted for a variety of different reactions [129 – 131] by the *negative binomial distribution* (NBD) $P(n; k, \bar{n})$ given by

$$P(n; k, \bar{n}) = \frac{k(k+1)....(k+n-1)}{n!} \, (\frac{\bar{n}}{\bar{n}+k})^n \, (\frac{k}{\bar{n}+k})^k \tag{34}$$

with $k > 0$ (or $1/k$) and $\bar{n}$ as free parameters in the fit. The average multiplicity $< n >$ and the dispersion D of the NBD are related to the two parameters by

$$< n >= \bar{n} \text{ and } D^2 = \bar{n} + \frac{\bar{n}^2}{k} . \tag{35}$$

The first two moments of the corresponding KNO distribution $\psi(z)$ thus are

$$< z >= 1 \text{ and } D_z^2 = \frac{1}{\bar{n}} + \frac{1}{k} . \tag{36}$$

In case of KNO scaling D_z is independent of W.

It is easy to show that for $k \to \infty$ the NBD (34) reduces to a Poissonian

$$P(n; \bar{n}) = \frac{\bar{n}^n e^{-\bar{n}}}{n!} \tag{37}$$

with $D^2 = \bar{n}$. For negative values of k ($k = -k'$ with $k' > 0$) the NBD becomes an ordinary binomial distribution (BD) (if $k' > n$):

$$P(n; -k', \bar{n}) \equiv P'(n; k', \bar{n}) = \binom{k'}{n} (\frac{\bar{n}}{k'})^n (1 - \frac{\bar{n}}{k'})^{k'-n} . \tag{38}$$

Thus it follows from eq. (35) that for fixed $\bar{n}$ a NBD ($k > 0$) is wider than the Poisson distribution whereas a BD ($k < 0$) is narrower.

In ref. [67] the EMC has fitted the charged multiplicity distributions for eight W intervals to the NBD in limited cms rapidity intervals of width $2 \cdot \Delta y^*$ centered around $y^* = 0$, i.e. extending from $-\Delta y^*$ in the backward to $+\Delta y^*$ in the forward hemisphere. The NBD fits the data very well as shown in Fig. 15 for the highest W interval ($18 < W < 20$ GeV). For each distribution the NBD parameters $1/k$ (W, Δy^*) and $\bar{n}(W, \Delta y^*)$ were determined from the fit. They are plotted vs. W in Fig. 16 for various Δy^* values, the largest Δy^* corresponding to the maximum rapidity range in which particles are produced.

The following main features are observed:

— For small Δy^*, i.e. in the central region around $y^* = 0$, the average multiplicity $\bar{n}$ is independent of W as expected from the well-known fact, that the particle density at zero rapidity is energy independent (see also chapt. 5.2). On the other hand $1/k$

increases significantly with W at small Δy^*. Consequently, the KNO distribution in the central region broadens with increasing W, implying a clear violation of KNO scaling.

— For the full rapidity range $1/k$ is rather small above $W \approx 8$ GeV. Thus for $W \gtrsim 8$ GeV the charged multiplicity distribution in μp scattering is close to Poisson, as is indeed directly observed in Fig. 17 from ref. [57].

— For the full rapidity range the W dependences of $1/k$ (increase) and $1/\bar{n}$ (decrease) roughly compensate each other in the width D_z of the KNO distribution, eq. (36), such that, as already discussed above (Fig. 14a), KNO scaling is roughly fulfilled in the W range of the experiment*. In this spirit the observed KNO scaling is accidental, due to a fortuitous cancellation in this particular W range, and need not hold at higher W. This point of view has first been stressed by the NA5 collaboration in ref. [129] for $\bar{p}p$ and pp scattering.

Charged multiplicity distributions in limited rapidity intervals were also measured in the forward and backward cms hemisphere of μp scattering separately. They also were found to be well described by the NBD parametrisation. From Fig. 18 which shows the parameters $\bar{n}$ and $1/k$ for the full forward and backward rapidity ranges as functions of W, the difference between quark and diquark fragmentation can be seen. Furthermore the $1/k$ data points for the forward μp hemisphere are close to the values for one $e^+ e^-$ hemisphere (straight line) which supports the concept of universality ("environmental independence") of quark fragmentation.

Numerous papers, not discussed here, have appeared on the interpretation of the observed NBD law in terms of various hadron production models. Some of them are listed in ref. [132] where further literature can be found. In addition, alternative parametrisations and related models for multiplicity (and in particular KNO) distributions have recently been proposed, see ref. [133] and the references cited therein.

d.) <u>Multiplicity correlations</u>

Correlations between the charged multiplicities n_F and n_B in the forward and backward hemispheres have been investigated in μp [57] and νp, $\bar{\nu}p$ [96] scattering, by plotting $< n_B >$ vs. n_F for fixed W. In νp and $\bar{\nu}p$ scattering (Fig. 19) a negative correlation is observed for $W \lesssim 5$ GeV, probably due to kinematic constraints. At higher W, n_F and n_B are rather uncorrelated in μp, νp and $\bar{\nu}p$, a fit of $< n_B >= a + b \cdot n_F$ yielding b values compatible with zero. This indicates that the quarks and diquarks fragment relatively independently of each other. In a compilation of b values for various types of reactions in ref. [134], b increases from

---

*Of course, since $< n >$ depends on W, a distribution cannot be Poissonian and obey KNO scaling simultaneously. So the above two observations hold only approximately in a limited W range.

negative values at low $W = \sqrt{s}$ to $b \approx 0.5$ for $p\bar{p}$ collisions at $W = 540$ GeV, indicating a strong positive correlation at high W.

In a search for a correlation between the π^0 and charged multiplicities, n_{π^0} and n, in νp and $\bar{\nu} p$ interactions [96], $< n_{\pi^0} >$ has been plotted vs. the negative multiplicity n^- ($n = 2n^- + 2$ for νp, $n = 2n^-$ for $\bar{\nu} p$), at fixed W. Such a correlation could arise e.g. from the production of $\rho^\pm \to \pi^\pm \pi^0$. At fixed $W \gtrsim 4$ GeV no significant $n_{\pi^0} - n^-$ correlation is observed. A plot of $< n_{\pi^0} >$ vs. n^- for all W shows a clear rise [72], which is a consequence of the increase of $< n_{\pi^0} >$ and $< n >$ with W, events with larger n^- having on average larger W and thus larger n_{π^0}. The rise cannot be interpreted as a direct $n_{\pi^0} - n^-$ correlation.

5.2. FRAGMENTATION FUNCTIONS

Single particle distributions in the longitudinal variables x_F, y^* or z (fragmentation functions) have been measured and discussed within the QPM for particular hadron types produced in the various semi-inclusive lN reactions (13), namely for the following particles: $h^\pm$ [46, 70]; $\pi^\pm$ [56, 60, 85, 95, 107]; π^0 [42]; $K^\pm$ [56, 65]; K^0, Λ, $\bar{\Lambda}$ [54, 56, 65, 70, 73, 81, 85, 91, 104, 121, 124]; $p, \bar{p}$ [56, 59, 68]; Δ^{++} [59]; ρ^0 [44, 63, 92, 102, 110, 123]; ω [63]; ϕ, f [25]; J/ψ [25, 40]; $K^{*\pm}$ [109]; K^{*0} [25] and D^0 [50].

Single particle distributions yield direct information on the fragmentation of quarks ($x_F > 0$) and diquarks ($x_F < 0$). As stressed in chapt. 2, x_F or y^* and not z are the appropriate variables for such studies. In this context the distributions of resonances are of particular interest, since they are the primary products of fragmentation thus yielding more information on the fragmenting partons than their decay products. In the following we discuss several interesting examples.

Fig. 20 shows the normalized x_F distributions $D^{\pi^\pm}(x_F)$ of π^+ and π^- produced in νp and $\bar{\nu} p$ interactions [95] with $W > 3$ GeV and $x > 0.1$, separately for the forward ($x_F > 0$) and backward ($x_F < 0$) hemisphere. The x cut suppresses the scattering on sea quarks in the proton, thus allowing to test the predictions (26) and (27) of the simple QPM. The following features are observed:

— In each hemisphere the distributions are well approximated by an exponential of the form $A \cdot \exp(-B|x_F|)$.

— In the forward hemisphere the QPM relation (26) is well fulfilled: $D^{\pi^\pm}(\nu p) = D^{\pi^\mp}(\bar{\nu} p)$. Furthermore the νp π^- and $\bar{\nu} p$ π^+ distributions fall off more steeply than the νp π^+ and $\bar{\nu} p$ π^- distributions. This is again expected from the QPM, since the π^+ in νp (π^- in $\bar{\nu} p$) scattering can be the leading particle containing the fragmenting u quark (d quark) (Fig. 3), whereas the oppositely charged pion contains only quarks created later on in the fragmentation process and is therefore unfavored. These considerations are independent of whether the $\pi^\pm$ is produced directly or comes from $\rho^\pm$ decay. In the latter case the pion is of course softer (smaller $|x_F|$) on average than a directly produced pion.

— In the backward hemisphere the QPM relation (27) is well fulfilled: $D^{\pi^+}(\bar{\nu}p) = D^{\pi^-}(\nu p)$. Furthermore the two distributions fall between the νp π^+ and π^- distributions as expected intuitively from the diquark flavors and quark contents of the pions when comparing $uu \to \pi^+$ (doubly favored), $ud \to \pi^{\pm}$ (favored) and $uu \to \pi^-$ (unfavored).

— In all four cases the distributions are steeper in the backward than in the forward hemisphere, as expected from the QPM since a pion from single quark fragmentation gets on average a higher momentum fraction than a pion from diquark fragmentation with a (possibly leading) baryon amongst its fragments.

Fig. 21 provides a precise test [135] of the QPM by comparing the π^+ and π^- invariant x_F distributions in μp (from EMC) and νp, $\bar{\nu}p$ [95] scattering. The figure shows the μp distributions (circles) and linear combinations (squares) of the νp and $\bar{\nu}p$ distributions with coefficients given in the figure for each hemisphere. These coefficients were obtained by numerically integrating the distribution functions $q(x)$ and $\bar{q}(x)$ (in the parametrisation of ref. [136]), which occur in the QPM formulae for the x_F distributions, over x and by eliminating from these formulae the quark or diquark fragmentation functions. (In the simple QPM without sea quarks the relations are given by eqs. (22) and (24)). As the figure shows the agreement is excellent and the QPM predictions are well satisfied for both charges and in both hemispheres.

In Fig. 22 the normalized x_F distributions of the (anti)baryons p, Δ^{++}, Λ and $\bar{\Lambda}$ produced in μp scattering [54, 59] are shown. Baryons can come from the target remnant as well as from $B\bar{B}$ pair production in the fragmentation chain. One notices immediately the pronounced forward-backward asymmetry of the three baryon distributions, favoring strongly the backward hemisphere, where predominantly ud-diquark fragmentation occurs (Fig. 3). This diquark fragments most easily into a proton (udu) or a Λ (uds) by picking up respectively a u or s quark from a $u\bar{u}$ or $s\bar{s}$ pair created in the fragmentation process. This leads in the backward hemisphere to the relatively large proton multiplicity and a Λ multiplicity, which is smaller because of the well-known suppression of $s\bar{s}$ over $u\bar{u}$, $d\bar{d}$ pair creation, see e.g. ref. [137] and chapt. 5.6. The $\Delta^{++}(uuu)$ is also backward-produced, but again suppressed relatively to the proton (with a ratio p/Δ^{++} of 6.2 ± 1.2), since it cannot come directly from a ud-diquark. It may originate either directly, with an additional u-quark, from the uu-diquark in the rarer case of muon scattering on the d-quark in the proton or from more complicated processes like scattering on a sea quark, or break-up of the diquark together with diquark-antidiquark pair creation. In the forward hemisphere it is nice to observe similar multiplicities for Λ and $\bar{\Lambda}$ consistent with diquark-antidiquark pair creation at a later stage in the fragmentation chain. In the diquark fragmentation region $\bar{\Lambda}$ ($\bar{u}\bar{d}\bar{s}$) production is strongly suppressed as expected. A behaviour similar in shape to that of $\bar{\Lambda}$ is observed for antiprotons [68]. The data points in Fig. 22 agree reasonably well with the curves predicted by the Lund model, except for the Δ^{++} at $x_F < 0$ where the Lund prediction is above the data points. This may be due to a (ud) bound state with $I = J = 0$ inside the proton [138], which is not included in the Lund model. Since its break-up is suppressed, Δ^{++} production would also be suppressed.

As a further example Fig. 23 (from ref. [24]) shows a compilation of x_F distributions of the ρ^0 produced in various reactions. While the data agree at large $|x_F|$ in both hemispheres, the μp points are higher than the νN, $\bar{\nu} N$ points around $x_F = 0$. This may be due to the higher W in the EMC experiment, indicating an increase of ρ^0 production with increasing W in the central region. As in Fig. 11, the Lund model prediction for μp, using V/PS $= 1$ for the vector to pseudoscalar meson production ratio, is above the μp data points [63].

We now discuss the question of a possible W and/or Q^2 dependence of fragmentation functions $D^h(x_F, W, Q^2)$. A W dependence is expected from the increase of particle production in the central rapidity region due to the kinematic widening of the available rapidity range according to $\Delta y^* \sim ln\, W^2$. If the height of the rapidity distribution is W independent, this increase leads directly, via eq. (19), to the observed rise, linear in $ln\, W^2$, of the average multiplicity of hadrons h. A Q^2 dependence could arise from a scaling violating Q^2 dependence of the (anti)quark fragmentation functions as predicted by QCD [139]. Finally, an x dependence is expected in the general case, that $D(x_F, W, Q^2)$ is a sum of quark or di-quark fragmentation functions with the x (and scaling violating Q^2) dependent distribution functions $q(x)$ and $\bar{q}(x)$ entering as weights, see e.g. eq. (18). Via eq. (8) this implies a Q^2 dependence at fixed W, and a W dependence at fixed Q^2. A Q^2 dependence of $D^h(x_F, W, Q^2)$ implies of course, via eq. (19), also a Q^2 dependence of $< n_h > (W, Q^2)$ (chapt. 5.1).

In earlier publications [22, 38, 105, 107, 140] a Q^2 dependence of fragmentation functions in leptoproduction has been observed and was analysed in some papers in terms of QCD by comparing the experimental moments of fragmentation functions with their QCD predicted Q^2 dependence. However, the Q^2 dependence showed up only if W was not restricted. It is therefore probably not a genuine Q^2 dependence, but rather a kinematic reflection of the W dependence via eq. (8). Thus in order to search for an explicit Q^2 dependence of $D^h(x_F, W, Q^2)$ one has to keep W fixed (and vice versa), as was already discussed in chapt.5.1 for a possible Q^2 dependence of average multiplicities.

The fragmentation function $D^\pi(x_F, W^2, Q^2)$ of charged pions produced in μp scattering has been investigated by the EMC [60]. Fig. 24 shows, for fixed intervals of x_F, $D^\pi(x_F, W^2, Q^2)$ vs. W^2 for fixed intervals of Q^2, and vs. Q^2 for fixed intervals of W^2, for forward going charged pions (see ref. [60] for the $x_F < 0$ plot). Two features are recognized:

— The distribution $D^\pi(x_F, W^2, Q^2)$ rises with W^2 at small $|x_F| \lesssim 0.1$, both in the forward

($x_F > 0$) and backward ($x_F < 0$) hemisphere. For larger $|x_F|$ no W dependence is observed. Thus the increase of pion production occurs in the central region.

— At fixed W and at any x_F, $D^\pi(x_F, W^2, Q^2)$ is independent of Q^2 within errors. The explicit Q^2 dependence of $< n_\pi >$ in Fig. 12 is too weak to be seen in this distribution.

A very similar behaviour of $D^\pi(x_F, W^2, Q^2)$ is observed for charged pions produced in νp and $\bar{\nu} p$ scattering [99].

The origin of the increase of pion production at small $|x_F|$ with increasing W^2 is directly seen from Fig. 25 where the cms rapidity distributions for charged pions, and for π^+ and π^- separately, are plotted for four W^2 intervals and all $Q^2 > 4$ GeV^2. The widths of the distributions increase with W^2 and for larger W^2 a central plateau starts to develop. As expected kinematically, this increase is proportional to $ln\,W^2$ as shown by a quantitative analysis in which a suitable fitting function was fitted through the data points (see curves). Furthermore, the central heights of the distributions are independent of W. Both features then result directly in the rise, linear in $ln\,W^2$, of the average pion multiplicity $< n_\pi >$ (Fig. 9). Finally the π^+ and π^- distributions in Fig. 25 are equal around $y^* = 0$ at higher W which is expected from $\pi^+\pi^-$ pair production in the central region.

5.3. TRANSVERSE MOMENTA AND JET PROPERTIES

Several physical effects, sketched in Fig. 4, contribute to the observed p_T of a hadron with respect to the boson direction:

a.) The primordial transverse momentum k_T of the struck quark inside the nucleon manifests itself as a k_T of the forward and backward jets and thus as a tilt of the event axis relative to the boson direction. A hadron in the jet acquires a contribution $z \cdot k_T$, z being the fraction of the jet energy carried by the hadron. In ref. [66] the average k_T^2 has been measured as

$$< k_T^2 >= [0.29 \begin{smallmatrix} +0.05 \\ -0.07 \end{smallmatrix} (stat.) \begin{smallmatrix} +0.14 \\ -0.18 \end{smallmatrix} (syst.) \; GeV]^2 . \tag{39}$$

b.) QCD predicts the emission of soft gluons by the struck quark. This gives a recoil p_T to the quark and thus to its hadronic fragments. The effect, which increases with W, thus broadens the hadronic p_T distributions in the forward hemisphere. This QCD induced p_T is balanced by the soft hadrons in the central region. The target remnant (diquark), as a spectator, does not radiate gluons and its fragments are unaffected.

c.) At present W values hard QCD effects, i.e. hard gluon emission or quark-antiquark pair production (Fig. 5), are sufficiently energetic only in a very small fraction of events (< 1 %) to produce a separate, identifiable gluon or antiquark jet in addition, i.e. a 2-jet structure in the forward hemisphere.

d.) Finally, the fragmentation of the partons contributes to p_T in both hemispheres. Describing the distribution of this contribution by a Gaussian $\exp(-p_{T frag}^2/\sigma^2)$ where $p_{T frag}$ is with respect to the parton direction, a value of $\sigma = 0.41 \pm 0.02$ GeV is obtained in ref. [66].

The average p_T^2 of a hadron is thus composed as

$$< p_T^2 >= z^2 \cdot < k_T^2 > + < p_T^2 >_{QCD} + < p_T^2 >_{frag} . \tag{40}$$

The p_T properties of charged hadrons have been investigated by the various leptoproduction experiments [39, 46, 55, 66, 69, 24, 25, 71, 97, 108]. From the p_T^2 distributions and plots of $< p_T^2 >$ vs. x_F ("seagull plot") for various W intervals, and from $< p_T^2 >$ vs. W the following features are observed in all reactions:

— In the forward hemisphere the p_T^2 distribution broadens, i.e. $< p_T^2 >$ rises, with increasing W, whereas backwards the rise is very weak (if any) at higher W. The p_T^2 distribution is thus wider in the forward than in the backward region at higher W.

— As kinematically expected, $< p_T^2 >$ rises with increasing $|x_F|$ in both hemispheres (seagull effect).

— No significant dependence of $< p_T^2 >$ on Q^2 is observed at fixed W.

As an illustration Fig. 26 shows the seagull plots at lower and higher W for $\nu p, \bar{\nu} p, \nu n$ and $\bar{\nu} n$ scattering [108]. The plots are rather forward-backward symmetric at low W; they become asymmetric at higher W due to the increase of $< p_T^2 >$ with W in the forward hemisphere.

In a quantitative analysis the EMC [55, 69, 25] has compared the seagull plot of charged hadrons in μp scattering (Fig. 27) with the predictions of four versions of the Lund model: The standard version (hard QCD effects, soft gluons, $< k_T^2 > = (0.44$ GeV$)^2$) describes the data reasonably well (full curve). Removing the hard QCD processes (dashed curve) or soft gluon radiation (dash-dotted curve) predicts too low a $< p_T^2 >$ in the forward hemisphere. The removal of soft gluons can be compensated in the forward region by increasing k_T (dotted curve for $< k_T^2 > = (0.88$ GeV$)^2$). This however introduces a sizable rotation of the event axis, as described above, causing too large $< p_T^2 >$ values in the backward hemisphere and thus destroying the observed asymmetry. Therefore, hard QCD and soft gluon radiation are needed, together with a rather small $< k_T^2 >$, to describe the data. Very similar results and conclusions were obtained for positive and negative hadrons separately, and by the WA21 [97] and WA25 [108] collaborations for neutrino scattering.

Additional insight is gained from studying the question as to how the p_T of the leading forward or backward hadron ("trigger particle"), carrying a large momentum fraction $|x_F|$ and containing (for $x_F > 0$) most likely the struck quark, is balanced by the p_T of the remaining particles in the event. In Fig. 28 of the EMC [55, 69] the rapidity distribution of this balancing p_T ("p_T flow") is plotted for a trigger particle with (a) $x_F > 0.5$ and (b) $-0.5 < x_F < -0.2$. The data points show a p_T balance mainly in the central region. They are compared with three versions of the Lund model, which all reproduce the p_T-flow distribution of the trigger particle itself (not shown) rather well: The standard version with hard QCD, soft gluons and $< k_T^2 > = (0.44 \text{ GeV})^2$ (full curves) is in reasonable agreement with the data. Removing the soft gluons destroys the agreement (dash-dotted curves). A large $< k_T^2 >$ alone without soft gluons ($< k_T^2 > = (0.88 \text{ GeV})^2$) (dashed curves) predicts, due to the rotation of the event axis, in the hemisphere opposite to the trigger particle a peak of the balancing p_T flow which is not observed. Hence soft gluons are needed and $< k_T^2 >$ cannot be large. Similar results were obtained from the neutrino experiments [97, 108].

In ref. [24, 69] the seagull plots and balancing-p_T-flow plots of the EMC are also compared with the predictions of two further models (besides the Lund model), the independent fragmentation (IF) model [32] and the fire-string (FS) model [36]. The IF model does not fit the data; the FS model does not describe well the seagull plot at larger x_F and the p_T-flow balancing a backward trigger particle; the Lund model fits best.

Hard QCD effects, although not yet causing at present W values an evident 3-jet structure (quark-gluon-diquark or quark-antiquark-remnant, Fig. 5) in a sizable fraction of events, nevertheless lead to a planar event structure. Determining an event plane by minimizing $\sum p_{Tout}^2$, the EMC [25, 69] shows in Fig. 29 for events with more than 3 forward charged hadrons the distribution of $\sum p_{Tin}^2$, where p_{Tout} and p_{Tin} are the components of a hadron's transverse momentum out of and in the event plane, respectively. The distribution has a long tail, due to planar events, which is fitted by the Lund model only if hard QCD effects are included, with little difference between the standard version and the version without soft gluons. A very similar result, although in a somewhat smaller range of $\sum p_{Tin}^2$, has been obtained by WA25 [108]. Hard QCD effects are also needed [69] to reproduce correctly the angular distribution of the energy flow in the event plane and its dependence on the scaling variable $\lambda = x_F/p_T = 2cot\Theta^*/W$ [141]. Comparing the data with the Lund model yields for the QCD coupling constant in leading order a value of $\alpha_s(Q^2 = 20 \text{ GeV}^2) = 0.29 \pm 0.01$ (stat.) ± 0.02 (syst.).

Guided by the Lund model predictions, a cluster algorithm has been used in ref. [69] to extract a small sample of 3-jet events with relatively little (~ 25 %) contamination by 2-jet events. In the forward hemisphere the 3-jet events have a considerably larger $< p_T^2 >$ which goes beyond the increase expected from the selection procedure, and a significantly higher production of (anti)protons.

Summarizing one concludes, that in the Lund model all three effects, namely hard QCD, soft gluon radiation and a small k_T, are needed to describe the p_T properties of final state hadrons at present energies.

p_T^2 distributions, seagull plots and plots of $< p_T^2 >$ $vs.$ W^2, Q^2 and x have been published by the various collaborations also for specific types of hadrons, namely $\pi^{\pm}$ [56]; π^0 [42]; $K^{\pm}$ [56]; K^0, Λ [54, 85, 91, 104, 121, 124]; $p, \bar{p}$ [45, 56]; Δ^{++} [59]; ρ^0 [44, 63, 92, 102, 110, 123]; ω [63]; J/ψ [40] and D^0 [50]. As an example Fig. 30 shows a compilation [63] of ρ^0 and ω p_T^2 distributions in $\mu p, \nu p$ and $\bar{\nu} p$ scattering. Within errors the distributions are seen to be the same.

Finally, the EMC [66] has compared the properties of (anti)quark jets in e^+e^- annihilation [127] and in the forward hemisphere of μp scattering, at similar energies ($W \approx 14$ GeV) and choosing the sphericity (or thrust) axis as event axis. With respect to this axis, a typical hadron is found to have on the average a smaller p_T and a larger p_L in μp scattering than in e^+e^- annihilation; in other words, a μp forward jet is narrower (less spherical) than an e^+e^- jet. This is also seen from Fig. 31 where the sphericity distributions in μp and e^+e^- are compared: the μp distribution is substantially narrower and centered around smaller S values than the e^+e^- distribution. The QPM offers several sources for these differences:

a) In e^+e^- annihilation heavy quark pairs $c\bar{c}$ (and $b\bar{b}$) are produced relatively frequently (in 45 % of the cases according to the quark charges below the top threshold), whereas in μp scattering the production of heavy quarks is rather rare (< 1 %). Due to its high mass and lower cms momentum, a heavy quark fragments into a much broader jet than a light quark.

b) In e^+e^- annihilation both the quark and antiquark can emit a hard gluon, whereas in μp only the struck quark can radiate. Thus an e^+e^- event has a higher chance to become broader by a radiated gluon than a μp event.

c) There is no analogue in e^+e^- to the photon-gluon fusion (Figs. 5c,d).

To test the first two sources more quantitatively four different versions of the Lund model for e^+e^- annihilation are shown by the curves in Fig. 31. The standard version (full curve) is reasonably close to the e^+e^- data points. If heavy quark production or hard gluon radiation is switched off, the S distribution gets narrower. If both are switched off simultaneously, the Lund curve (dotted) approaches the μp data points. Thus, production of heavy quarks and more frequent radiation of hard gluons in e^+e^- annihilation cause or at least strongly contribute to broadening the e^+e^- jet as compared to the μp forward jet.

5.4. AZIMUTHAL ASYMMETRY

An asymmetry is predicted by QCD [142], primordial k_T [143] and higher twist [144] in the distribution of the azimuthal angle ϕ of a hadron around the boson direction. ϕ is defined to be zero in the lepton plane along the direction of the incident and outgoing lepton, see

Fig. 32. In general the ϕ distribution can be written as

$$\frac{1}{N_{ev}} \frac{dN^h}{d\phi} = A + B\,cos\phi + C\,cos2\phi + D\,sin\phi + E\,sin2\phi \,. \tag{41}$$

The ϕ asymmetries are calculated and occur most clearly on the parton level. They are preserved from the partons to the (leading) hadrons, although smeared out by the fragmentation. The theoretical expressions [142] for the coefficients in the ϕ distribution (41) contain the parton distribution and fragmentation functions.

In first order perturbative QCD for example, a gluon in lepton-quark scattering (Figs. 4, 5b) is not emitted isotropically around the boson direction, but rather prefers to travel, in the p_T plane (Fig. 32), along with the lepton ($cos\phi > 0$) and to be closer to the lepton plane ($cos2\phi > 0$). This asymmetry originates from the various polarisation states of the intermediate boson and their interferences. The recoiling quark is opposite to the gluon in the p_T plane by p_T conservation. Since a hadron from quark fragmentation is in general harder than from gluon fragmentation, $< cos\phi >= B/2A < 0$ (left-right asymmetry) and $< cos2\phi >= C/2A > 0$ (flatness towards the lepton plane) is predicted for large $x_F(A > 0)$. $D = E = 0$ in $O(\alpha_s)$ QCD. The primordial k_T also leads to $B < 0$ (compensated by $B > 0$ in the backward hemisphere) and $C > 0$ for large x_F, although with a different Q^2 dependence of the coefficients ($1/lnQ^2$ in QCD, power of $1/Q$ in case of k_T). Higher twist predicts $B > 0$ for large x_F, i.e. a sign of $< cos\phi >$ opposite to that due to QCD and k_T.

The ϕ distribution of charged hadrons has recently been investigated in several publications [43, 49, 64, 75, 80, 101, 113]. Fig. 33 shows $< cos\phi >$ vs. x_F (up to a known positive kinematic y dependent factor) from the EMC [43, 64]. $< cos\phi >$ is slightly positive for $x_F < 0$, indicating only a small k_T contribution, and drops to negative values below -0.1 with increasing $x_F > 0$. The data points are compared with the predictions of a model by König and Kroll [145] in which $O(\alpha_s)$ QCD and k_T (but not higher twist) are included. The model uses the Lund string model rather than fragmentation functions to simulate fragmentation. The curves in Fig. 33 are for two sets of values for $< k_T^2 >$ and the QCD scale parameter Λ. The version with $k_T \neq 0$ reproduces the data points reasonably well, although not perfectly at larger x_F. No evidence is observed for a higher twist contribution at large x_F [49, 64]. $< cos2\phi >$ is slightly positive, but close to zero.

Negative values of $< cos\phi >$ and small positive values of $< cos2\phi >$ are also found for hadrons in semi-inclusive neutrino reactions [24, 75, 113]. On the other hand positive $< cos\phi >$ values are obtained for the pion in *exclusive* (coherent) reactions with a single diffractively produced pion (chapt. 5.7): $< cos\phi >= 0.20 \pm 0.06$ in $\nu p \to \mu^- p \pi^+$ and 0.18 ± 0.07 in $\bar{\nu}p \to \mu^+ p\pi^-$ [101] for $W > 2$ GeV; $< cos\phi >= 0.17 \pm 0.08$ in $\nu A \to \mu^- A\pi^+$ and 0.29 ± 0.18 in $\bar{\nu}A \to \mu^+ A\pi^-$ [122], where A is a nucleus in freon.

A small time-reversal-odd $sin\phi$ term (and an even smaller $sin2\phi$ term) in eq. (41) is predicted in higher-order QCD (one-loop order) [146]. This term is sensitive to the gluon self-coupling (3-gluon vertex) and causes a small (~1 %) up–down asymmetry $\tilde{A} = (N_u - N_d)/(N_u + N_d) = 4 < sin\phi > /\pi$ with respect to the lepton plane ($< sin\phi >= D/2A$). For this term to occur the incoming lepton has to be longitudinally polarized ($D \propto P$) which is naturally the case for (anti)neutrinos, but also for muons from parity violating π and K decays (e.g. $P_{\mu^+} \approx -0.8$ in NA9 at 280 GeV [43]). In refs. [43, 64, 75, 101] $< sin\phi >$ is compatible with zero, whereas in the νNe experiment of ref. [80] an asymmetry of $\tilde{A} = -0.054 \pm 0.017$ is obtained for fast charged hadrons ($z > 0.3$) which is probably associated with reinteractions in the Ne nucleus. ($\tilde{A}$ is -0.111 ± 0.028 for events showing evidence for reinteraction, and -0.001 ± 0.026 for no reinteraction.)

5.5. DISTRIBUTIONS OF QUANTUM NUMBERS

From the QPM one expects the final state hadrons to partially retain the quantum numbers, like electric charge Q, of the fragmenting (anti)quark and target remnant, thus carrying information on the flavors of the parent partons. In particular the fastest (leading) hadron with the largest x_F will most likely contain the fragmenting quark as a constituent. Therefore, parton-charge retention should become apparent as two (positive or negative) charge peaks in the cms rapidity distribution of the hadronic net charge, defined as

$$\frac{d<Q>}{dy^*} = \frac{1}{N_{ev}} \left(\frac{dN^+}{dy^*} - \frac{dN^-}{dy^*} \right) , \qquad (42)$$

$N^+(N^-)$ being the number of positive (negative) hadrons in N_{ev} events. The situation is simplest at larger x where the scattering is predominantly on valence quarks, leaving a quark and a diquark in the final state (Fig. 3), and at high W where the overlap of the current and target fragments is small. Unfortunately both requirements reduce the statistics considerably, since large W corresponds preferentially to small x via eq. (8).

Net charge distributions have been investigated in various experiments [51, 53, 94, 103]. Fig. 34 shows the net charge rapidity distributions in μp scattering [53, 135] for three W^2 intervals; no x cut is applied. At small W, corresponding to larger $< x >$ via eq. (8) such that the final state consists most likely of a u-quark and ud-diquark, considerable overlap of the two charge peaks is observed. At higher W a valley around $y^* = 0$ develops between the peaks, due to h^+h^- (mainly $\pi^+\pi^-$) pair creation ("local charge compensation") in the central region of the fragmentation chain. In the highest W interval (i.e. at small $< x >$) the forward peak gets smaller because at small x also the sea quarks, some of them having negative charge, contribute. The integral of $d<Q>/dy^*$ is of course one, the total hadronic charge in a μp event.

Integrating $d<Q>/dy^*$ in each hemisphere gives the forward and backward hadronic net charges $< Q_{F,B} >$. In a simple picture of fragmentation by creating a chain of $q\bar{q}$ pairs, the hadronic net charge $< Q >_q$ from the fragmentation of a quark q is related to the quark charges e_q by [32, 13, 16]

$$< Q >_q = e_q - \sum_a e_a \gamma_a . \qquad (43)$$

Here γ_a is the probability to create a neutral quark-antiquark pair $a\bar{a}$ in the fragmentation and the sum extends over all participating quark flavors. The second term ("leakage term") accounts for the fact that in the observed meson fragments the quark of the last pair $a\bar{a}$ in the fragmentation chain, created with probability γ_a, is not included and its charge must therefore be subtracted. Neglecting heavy quarks ($\gamma_c = \gamma_b = 0$) such that $\gamma_u + \gamma_d + \gamma_s = 1$ with $\gamma_u = \gamma_d$ (isospin invariance), and using $e_u - e_d = 1(\pi^+)$, $e_u - e_s = 1(K^+)$ one obtains from eq. (43):

$$< Q >_u = 1 - \gamma_u \quad \text{and} \quad < Q >_d = < Q >_s = -\gamma_u . \qquad (44)$$

The forward net charge $< Q_F >$ has been plotted as a function of $1/W$ in neutrino experiments where at large x only one quark flavor is involved such that $< Q_F >$ is equal to $< Q >_u$ or $< Q >_d$ in νN or $\bar{\nu} N$ interactions respectively (Fig. 3). The plots were linearly extrapolated to infinite W where the quark and diquark fragments do not overlap. Using eq. (44), WA25 [103] obtains $\gamma_u = 0.38 \pm 0.09$ from $\bar{\nu} D$ interactions with $x > 0.15$; WA21 [94] obtains $\gamma_u = 0.39 \pm 0.10$ (0.50 ± 0.17) from νp $(\bar{\nu} p)$ interactions with $x > 0.1$. A compilation of earlier γ_u measurements is in ref. [16].

In μp scattering two quark flavors, u and d, contribute to $< Q_F >$ at large x. Under the simplifying assumptions that sea quarks are neglected and $u(x) \approx 2d(x)$, one obtains, applying eqs. (43) and (44):

$$< Q_F > = \frac{8}{9} \cdot \left(\frac{2}{3} - 0.1\right) + \frac{1}{9} \cdot \left(-\frac{1}{3} - 0.1\right) = \frac{8}{9} \cdot (1 - 0.43) - \frac{1}{9} \cdot 0.43 = 0.45 . \qquad (45)$$

Here a value of 0.1 has been inserted for the leakage term as obtained with $\gamma_s/\gamma_u = 0.3$ (see chapt. 5.6), corresponding to $\gamma_u = 0.43$. In Fig. 35 from the EMC [53, 135] $< Q_F >$ is plotted vs. x. At large x the data points approach the simple prediction of 0.45. At small x, $< Q_F >$ is considerably lower due to the sea quark contributions. The curve shows the more rigorous QPM prediction

$$< Q_F > = \frac{4 u_v(1 - \gamma_u) - d_v \gamma_u}{4(u + \bar{u} + c + \bar{c}) + (d + \bar{d} + s + \bar{s})} \qquad (46)$$

using $\gamma_u = 0.44$ and the quark distribution functions in the parametrisation of ref. [136]. (The contributions from the sea quarks and antiquarks cancel in the numerator). The agreement with the data points is excellent. Also the Lund model nicely describes $< Q_F >$ vs. x [53].

The EMC [65] has also measured the rapidity distribution of strangeness, defined analogously to eq. (42), in μp events. Fig. 36a shows the distributions of the strangeness contained in charged kaons and in Λ hyperons, Fig. 36b the sum of these two distributions, i.e. the distribution $d < S > /dy^*$ of the total observable strangeness. (The other hyperons are undetected, the strangeness of the observed neutral kaons is undetermined). The forward hemisphere is dominated by $K^+(u\bar{s})$ from u-quark, the backward hemisphere by the Λ (uds) from ud-diquark fragmentation (see also Fig. 22b). The Lund model curves are in qualitative agreement with the data points, although the model overestimates negative and underestimates positive strangeness.

5.6. PRODUCTION OF STRANGE PARTICLES

In strangeness conserving μN scattering strange particles can be produced in pairs ("associated production") either by scattering on an s or $\bar{s}$ sea quark, leaving a strange target remnant, or by $s\bar{s}$ (or $ss\bar{s}\bar{s}$) pair creation in the fragmentation process. In νN and $\bar{\nu}N$ scattering (Table 1) the following CC processes lead to single or associated strange particle production, neglecting doubly unfavored transitions (on a sea quark and Cabibbo-suppressed) and transitions from $c, \bar{c}$ quarks:

$$
\begin{array}{llll}
\text{In } \nu\text{N:} & d \to c & \text{In } \bar{\nu}\text{N:} & u \to s \\
& \hookrightarrow s & & \\
& s(\bar{s}) \to c(\bar{s}) & & \bar{s}(s) \to \bar{c}(s) \\
& \hookrightarrow s & & \hookrightarrow \bar{s}
\end{array} \tag{47}
$$

The backward going spectator s or $\bar{s}$ quark is written in brackets. Furthermore, $s\bar{s}$ (or $ss\bar{s}\bar{s}$) pairs can also be created in the fragmentation chain of events involving u and d quarks.

Strange particle production has been investigated in μN [54, 56, 58, 65], CC νN and $\bar{\nu}N$ [70, 73, 74, 78, 79, 85, 89, 91, 98, 100, 101, 104, 109, 121, 124, 147], and NC νN and $\bar{\nu}N$ [81, 82] interactions. Neutral strange particles $(K^0, \bar{K}^0, \Lambda, \bar{\Lambda})$ are relatively easily recognized by their decay whereas for $K^{\pm}$ mesons particle identification is required to distinguish them from $\pi^{\pm}$ or (anti)protons. Relevant quantities have been measured such as production cross sections and rates (as functions of E, W, Q^2, x), average multiplicities, single particle distributions (in z, x_F, y^*, p_T^2), seagull plots, strangeness distributions etc. In addition the production of strange particle pairs and of strange resonances (K^*, Σ^*) has been studied. The experimental results have been compared, in general successfully, with the predictions of the simple QPM and the Lund or other fragmentation models. For instance in μN scattering [65], the K^+ $(u\bar{s})$ is produced more abundantly than the K^- $(s\bar{u})$ (Fig. 36) since the scattering is mainly on a u-quark, which may be contained in K^+ as the first-rank particle, but not in K^-. Similarly, $K^{*+}(u\bar{s})$ is favored over $K^{*-}(s\bar{u})$ in the forward region of νp and νn scattering [109], again involving predominantly a fragmenting u-quark.

Some of the aspects of strange particle production were already discussed in the preceding chapters. Here we concentrate on the following four particular topics: Λ polarisation, strangeness suppression, diquark break-up and charm-production.

a.) <u>Λ polarisation</u>

The polarisation vector $\vec{P}_\Lambda$ of Λ hyperons produced in the target fragmentation region of νp and $\bar{\nu} p$ scattering has been measured by WA21 [100]. A polarisation is expected from the simple QPM since the weak V-A interaction is on a left-handed quark (d in νp, u in $\bar{\nu} p$), leaving the spectator diquark from the unpolarized target proton longitudinally polarized in the proton direction for compensation. This polarisation may be transferred in the diquark fragmentation to the Λ hyperon. The situation is however complicated by the fact, that the Λ may come indirectly from the decay of a Σ^0 or the $\Sigma(1385)$ resonance. More quantitative QPM predictions for P_Λ, based on SU(6) symmetry, have been made by Bigi [148].

The main result of the analysis of ref. [100] is shown in Fig. 37. In the production plane, defined by the directions of the incident boson and proton in the Λ rest frame, a significant Λ polarisation of $\sim$ 70 % is observed for $x_F < 0$ in both reactions; it is in the proton direction in $\bar{\nu} p$ and opposite to the boson direction in νp scattering, as seen in the Λ rest frame. The polarisation is largest at small W and Q^2 and large x. In νp scattering a strong contribution to P_Λ comes from the production of a highly polarized $\Sigma^+(1385)$. Perpendicular to the production plane the polarisation is consistent with zero. The results qualitatively agree with the QPM predictions of ref. [148].

A Λ polarisation perpendicular to the production plane has also been searched for by WA25 [104] in νD and $\bar{\nu} D$ interactions with $W^2 > 5$ GeV^2, where the production plane is defined, differently from the previous definition, by the directions of the incident (anti)neutrino and deuteron in the Λ rest frame. Within the large statistical errors the perpendicular polarisation is consistent with zero for νD, whereas it could be different from zero (-0.57 ± 0.22) at small x for $\bar{\nu} D$ scattering.

b.) <u>Strangeness suppression</u>

The ratio $D^K(z)/D^\pi(z)$ of kaon and pion fragmentation functions ("K over π ratio") is related to the ratio $\lambda = \gamma_s/\gamma_u$ of the probabilities γ_u and γ_s defined in chapt. 5.5. As a simple illustration Fig. 38 shows directly that in case of a fragmenting u-quark the ratio K^+/π^+ for $z \to 1$ (where the meson most likely contains the u-quark) is equal to λ ($\gamma_u = \gamma_d$). More generally, in the Lund MC program the parameter λ controls the rate of strange meson production.

The WA21 collaboration [98] has measured the ratio K^0/π^- in νp and $\bar{\nu} p$ interactions with $W > 2$ GeV as a function of z in the current fragmentation region ($x_F > 0$). A comparison with the Lund model predictions for various λ yields a value of $\lambda = 0.203 \pm 0.014$ (stat.) ± 0.010 (syst.). The values obtained from νp and $\bar{\nu} p$ interactions separately are the same within errors. The WA25 collaboration [109] also needs a value of $\lambda \approx 0.2$ to reproduce their measured K^0 and Λ rates in νD and $\bar{\nu} D$ interactions by the Lund model.

The EMC [65] has measured the ratio $< n_K > / < n_{h\pm} >$ of average kaon and charged hadron multiplicities as a function of W in μp scattering (Fig. 39). From a comparison with the Lund predictions for various λ (see curves) a value of $\lambda = 0.30 \pm 0.01$ (*stat.*) ± 0.07 (*syst.*) is obtained by linear interpolation. Within the errors, the two λ values from WA21 and EMC are compatible. Using $\gamma_u = 1/(\lambda + 2)$ (chapt. 5.5), the two values yield $\gamma_u = 0.454 \pm 0.004$ and 0.435 ± 0.013 respectively, compatible with the γ_u values as determined from the forward hadronic net charge in neutrino reactions, chapt. 5.5.

Recent compilations of λ values from various experiments are given in ref. [137]. Simple SU(3) symmetry predicts $\lambda = 1$. The fact that λ is found considerably smaller ("strangeness suppression") may be due to the mass difference of strange and non-strange quarks. This explanation is supported by the observed slight increase of λ with energy W, although most present λ values are still far away from unity. From the observed number of $\Lambda\bar{\Lambda}$ pairs and the Lund model the EMC [65] concludes that the production of strange diquarks is even more strongly suppressed ($\gamma_{us}/\gamma_{ud} \approx 0.06$).

c.) A production and diquark break-up

Λ hyperons are either produced directly or may come from the decay of Σ^0 or strange baryon resonances, e.g. $\Sigma(1385) \rightarrow \Lambda\pi$. Assuming (anti)neutrino scattering on a valence quark (Fig. 3), direct $\Lambda(uds)$ production can proceed in a simple way, by the ud-diquark picking up an s-quark, in $\bar{\nu}p$ and νn, whereas in νp and $\bar{\nu}n$ interactions it is possible only by a break-up of the uu or dd diquark, its two quarks fragmenting independently. On the other hand $\Sigma^+(1385)(uus)$, $\Sigma^0(1385)(uds)$ and $\Sigma^-(1385)(dds)$ can be produced respectively in νp, $\bar{\nu}p$ or νn, and $\bar{\nu}n$ scattering without break-up. Furthermore Λ's can also come from charmed baryon decay (small), from pair production ($\Lambda\bar{\Lambda}$) deeper in the fragmentation chain and from scattering on sea quarks, leaving a more complicated target remnant.

In ref. [109] the WA25 collaboration has estimated the fractions of Λ's from pair production (using the number of $\bar{\Lambda}$'s), from $\Sigma^\pm(1385)$ decays (using the signals in the $\Lambda\pi^\pm$ mass plots) and from scattering on sea quarks (using the Lund model). Subtracting these Λ sources, the authors then arrive at upper limits of 2.3 % (4.0 %) of all νp ($\bar{\nu}n$) CC events with $W^2 > 5$ GeV^2, in which Λ's were produced via diquark break-up. From this they deduce a diquark break-up probability f of less than $\sim$ 75 % for both uu and dd-diquarks.

In a similar analysis the $15'$ νD_2 collaboration [78] obtains $f = 0.53 \pm 0.33$ for uu-diquark break-up, using νp and νn events with $W > 4$ GeV and $x > 0.2$, thereby reducing the overlap of the current and target regions, and the scattering on sea quarks. In ref. [149] $f \approx 0.25$ is obtained for the uu-diquark in νp scattering.

d.) Charm production

In νN and $\bar{\nu}N$ interactions a substantial portion of strange particles originates from the production and subsequent decay of charmed mesons and baryons according to (47) [74, 76, 79, 83, 85, 88, 91, 93, 121, 124]. Charmed particle production is established by the complete kinematic reconstruction of the decay chain in single events in a bubble chamber, from peaks in effective mass distributions or indirectly from the occurrence in νN interactions of singly produced strange particles, which according to (47)(see also Table 1) is a signature for charm production. Using the QPM predicted relative contributions of the various strangeness production channels [91, 121, 124] roughly $7 \div 10$ % of the CC cross section above the charm threshold is estimated to be due to charm production. In ref. [85] a rate of (5.0 ± 1.5) % is obtained for single s-quark production via charm in νN interactions.

A different signature for charm production in CC neutrino scattering is the occurrence of dilepton events where, in addition to the muon from the scattering, a second lepton (e or μ) of opposite charge is observed from a semileptonic charm decay ($c \rightarrow sl^+\nu$ in νN, $\bar{c} \rightarrow \bar{s}l^-\bar{\nu}$ in $\bar{\nu}N$), associated with a strange particle. Dilepton events have been investigated in refs. [84, 112, 114, 150, 151, 152] and are successfully compared with Monte Carlo predictions based on the QPM for the production and subsequent decays of charmed particles. Alternatively they have been used to determine the ratio $\kappa = 2S/(\bar{U} + \bar{D})$ of the strange and non-strange sea in the nucleon, yielding $\kappa = 0.52 \pm 0.09$ [150] and $0.52^{+0.17}_{-0.15}$ [152].

In μN scattering charm may be produced either by scattering on a rare c or $\bar{c}$ sea quark in the nucleon or by the process of photon-gluon fusion (PGF, Figs. 5c,d) into a $c\bar{c}$ pair. In the latter case the c and $\bar{c}$ may either form a bound charmonium state, e.g. a J/ψ decaying into $\mu^+\mu^-$, or fragment independently into two charmed particles, possibly with semileptonic decays. Thus using events with dimuons or trimuons as experimental signature, the EMC [40, 41] has studied the production of charmed particles and of charmonium states in μFe interactions at 250 GeV. For open charm production the data were found in good agreement with the predictions of the PGF model. This is essentially confirmed in a comparison of the model with distributions of D^0 mesons observed directly (in $K\pi$ mass plots) in ref. [50]. For the J/ψ, the first-order PGF model does not describe the z and p_T^2 distributions, and higher order QCD effects have to be included to explain the data.

5.7. SINGLE MESON PRODUCTION

Several groups have observed the coherent production of a single $\rho^\pm$ [87, 111, 118, 153] or $\pi^\pm$ [90, 111, 116, 122, 153] meson in CC (anti)neutrino interactions with nuclei, mostly Ne:

$$\overset{(-)}{\nu}Ne \rightarrow \mu^\mp \rho^\pm Ne \qquad (48a)$$
$$\overset{(-)}{\nu}Ne \rightarrow \mu^\mp \pi^\pm Ne \qquad (48b)$$

In these reactions the nucleus interacts as a whole without break-up and is undetected due to its small recoil energy. Experimentally such events are therefore characterized by the absence of nuclear fragments or low momentum protons, and by a small momentum transfer t to the nucleus (Fig. 40, $t \lesssim 0.1$ GeV^2 for most events).

The relevant diagram for coherent $\rho^\pm$ production is shown in Fig. 40a. According to the vector meson dominance model (VDM) the $W^\pm$ couples (by $W^+ \rightarrow u\bar{d}$, $W^- \rightarrow d\bar{u}$) to a $\rho^\pm$ which scatters off the nucleus by elastic diffraction scattering (Pomeron exchange). This scattering is characterized by a steep t distribution, parametrized by an exponential e^{-bt} where the slope b is proportional to the lateral dimension of the nucleus, $b = \alpha A^{2/3}$ with $\alpha \approx 11$ GeV^{-2}. The differential cross section $d\sigma/dQ^2 d\nu dt$ and its dependence on the neutrino energy, predicted by the VDM, is in good agreement with the experimental distributions, Fig. 41 [118, 153]. The ρ decay angular distribution indicates a substantial production of ρ mesons with helicity zero.

The dominant contribution to coherent $\pi^\pm$ production comes from the meson dominance coupling of the axial-vector current to the A_1 meson with $J^{PC} = 1^{++}$ (Fig. 40b) which is diffractively scattered by Pomeron exchange in the t channel. According to the partially conserved axial current (PCAC) hypothesis which relates the divergence of the axial current to the pion field $(\partial_\mu A^\mu = f_\pi \phi_\pi)$, the longitudinal part of A_1 behaves like a pion at $Q^2 = 0$. Hence, as Adler [154] has shown, the cross section for the process $\nu N \rightarrow \mu X$ is proportional to that of $\pi N \rightarrow X$ where N is a nucleon or nucleus and X any hadronic final state. In our case of reaction (48b) the pion reaction is elastic diffractive $\pi^\pm Ne$ scattering by Pomeron exchange in the t channel with an exponential t distribution. The Q^2 dependence is described by the A_1 propagator. In addition there is a small contribution, proportional to the lepton mass squared, from the direct coupling of the axial current to the pion (as in π decay) (Fig. 40b) which again is diffractively scattered off the Ne nucleus. The Q^2 dependence of this contribution is given by the pion propagator. A third process in which the $W^\pm$ couples to the $\rho^\pm$ and an ω meson is exchanged in the t channel, vanishes by CVC (conserved vector current) at $Q^2 = 0$ and is negligible for $Q^2 \neq 0$. The final formula for $d\sigma/dQ^2 d\nu dt$ of reaction (48b) describes the experimental distributions (Fig. 42) and the dependence on the neutrino energy very well [90, 116, 122]. The analogous coherent diffractive production of a single π^0 in NC (anti)neutrino reactions has been investigated in refs. [86, 122, 155, 156, 157].

Exclusive single pion production by (anti)neutrinos has also been studied on (quasi) free nucleons [101, 106, 158] in the reactions

$$\nu p \rightarrow \mu^- p \pi^+ \qquad (49a)$$

$$\bar{\nu} p \rightarrow \mu^+ p \pi^- \qquad (49b)$$

$$\bar{\nu} n \rightarrow \mu^+ n \pi^- \qquad (49c)$$

for which W is the $N\pi$ effective mass $M(N\pi)$. In the resonance region ($W < 2$ GeV) the reactions (49a) and (49c) are dominated by the production of the $\Delta^{++}(1232)$ and $\Delta^-(1232)$

resonances, respectively ($\nu p \to \mu^- \Delta^{++}$, $\bar{\nu} n \to \mu^+ \Delta^-$), showing up as peaks in the W distributions, whereas in addition to the $\Delta^0(1232)$ also $I = \frac{1}{2}$ resonances contribute to reaction (49b). A model of Rein and Sehgal [159], including the resonances and a non-resonant background, describes the $N\pi$ mass distributions, the cross sections as functions of the neutrino energy, and the steeply falling Q^2 distributions at low $W \lesssim 1.4$ GeV reasonably well. The model is however in disagreement with the moments of the "decay" angular distribution of the $p\pi^-$ system in reaction (49b).

At higher $W > 2$ GeV, a peripheral mechanism is responsible for single pion production, with a diagram of the same structure as in Fig. 40. The processes $W^\pm p \to \pi^\pm p$ are well described [101] by Reggeized π^0 exchange in the t channel, with a possible admixture of some vector meson exchange. Pomeron exchange is excluded, i.e. there is no evidence for diffractive production. This Regge model reproduces the observed W^2 dependence of the cross section. The Treiman-Yang test confirms that the t-channel object is spinless.

Finally, the EMC [48] and CHIO [160] collaborations have investigated exclusive single ρ^0 production in μp scattering, $\mu p \to \mu p \rho^0$, i.e. ρ^0 photoproduction $\gamma^* p \to p \rho^0$ by virtual photons. At low Q^2 the photon is expected to behave hadron-like ($\gamma^* \to q\bar{q}$); the VDM, with the γ^* coupling to the ρ^0 and subsequent diffractive $\rho^0 p$ scattering via Pomeron exchange in the t-channel (in analogy to Fig. 40a), should be an appropriate model, as in real photoproduction ($Q^2 = 0$). At high Q^2, the γ^* should act as a pointlike probe, the ρ^0 being produced in a hard scattering process like photon-gluon fusion ($\gamma^* g \to \rho^0$). Such a process is characterized by a small slope b of the t distribution. The EMC [48], including data from ref. [160], finds the slope b indeed to decrease with increasing Q^2, whereas the ρ^0 density element r_{00} increases such that at high Q^2 the ρ^0 is almost purely longitudinal (helicity $\lambda = 0$, i.e. r_{00} close to one). Both results cannot be reconciled with the VDM at higher Q^2. Furthermore, the observed ρ production by transverse photons ($\lambda = \pm 1$), together with the longitudinal ρ polarisation, indicates that the s-channel helicity is not conserved in $\gamma^* p \to p \rho^0$. The data are compared with a Pomeron model calculation in ref. [161].

6. Conclusions and outlook

As this review shows, a vast amount of experimental results has been accumulated in recent years on hadron production in lepton-nucleon scattering at high energies. The main features like multiplicities, longitudinal and transverse single particle distributions, and jet properties vary with W and show only little (if any) Q^2 dependence, W thus being the relevant variable. In general the quark-parton model, extended by QCD effects and incorporating the Lund string model of fragmentation, is a powerful framework in which the production and hadronisation of quarks and diquarks is successfully described, the data yielding overwhelming evidence for its validity. Thus a rather complete picture for the fragmentation of partons into hadrons has emerged from the data. However, also other models of hadron production such as the fire-string model should attract more attention in the future. For specific (exclusive) reactions, like single meson production, and in particular kinematic regions of phase space, specific models like the vector meson dominance model turn out to be appropriate.

For further progress higher energies are needed. They would allow an even better separation of the current and target fragmentation regions and open even wider ranges of the relevant variables permitting a further study, with longer lever arms, of e.g. W and Q^2 dependences. Higher energies are also required to enter a regime where QCD effects, such as hard gluon radiation, show up more clearly and directly, e.g. in the form of 3-jet events as in e^+e^- annihilation.

At the end we list a few examples for specific open questions of interest. For multiplicities: Does the average hadron multiplicity $< n >$ continue to rise linearly with lnW^2 or is the rise steeper at higher W, as in e^+e^- annihilation? Does $< n >$ depend on Q^2 or not? Is KNO scaling violated at higher W? Is the negative binomial distribution still a good parametrisation of multiplicity distributions at higher W? Is there a positive forward-backward charged multiplicity correlation at higher W as in hadron-hadron collisions? For fragmentation functions: Is there a central plateau with constant height in the pion rapidity distribution also at higher W? Does the pion fragmentation function show a Q^2 dependence at fixed W? If yes, what is its origin? (A possible scaling violation of fragmentation functions and its interpretation in terms of QCD is still a rather open question). What is the W and Q^2 dependence of the fragmentation functions of other hadrons (kaon, proton, Λ etc.)? More work is also needed on the fragmentation (and break-up) of diquarks, which can be studied in a clean way only in leptoproduction, not in e^+e^- annihilation and only badly in hadron-hadron collisions. For transverse momentum and jets: Search and investigation of 3-jet events (angular distribution of the jet axes, of energy flow, etc.) and gluon jets, determination of the running QCD coupling constant α_s. For strange particles: Does the strangeness suppression parameter γ_s/γ_u increase with W? More information is also desirable e.g. on Λ polarisation, charm production, flavor retention in fragmentation, and on short and long range correlations of particles and quantum numbers.

There is hope, that answers to some of these problems, and further results, will come from high energy μN and νN experiments at the Tevatron of Fermilab in the near future.

Acknowledgements

I am grateful to Mrs. E. Thurner for her diligent and careful work in typing the manuscript and to H. Kühlwein for his continuous help with the TEX system. Thanks are due to P. Breitenlohner for adopting this system for special purposes, to M. Vidal for producing the compilation plots, and to I. Derado for a critical reading of the manuscript.

REFERENCES

1. H.E. Fisk, F. Sciulli: Ann. Rev. Nucl. Part. Sci. **32**, 499 (1982)
2. J. Drees, H.E. Montgomery: Ann. Rev. Nucl. Part. Sci. **33**, 383 (1983)
3. F. Dydak: Proc. 1983 Intern. Symp. on Lepton and Photon Interactions at High Energies, Cornell, USA, p. 634 (1983)
4. H. Abramowicz et al: Z. Phys. **C17**, 283 (1983); **C25**, 29 (1984)
5. D. B. Mac Farlane et al.: Z. Phys. **C26**, 1 (1984)
6. F. Sciulli: Proc. 1985 Intern. Symp. on Lepton and Photon Interactions at High Energies, Kyoto, Japan, p. 8 (1985)
7. A. Bodek: Proc. 11th Intern. Conf. on Neutrino Physics and Astrophysics, Nordkirchen, Fed. Rep. Germany, p. 643 (1985)
8. J.J. Aubert et al.: Nucl. Phys. **B259**, 189 (1985); **B272**, 158 (1986)
9. W.D. Nowak: Fortschr. Phys. **34**, 57 (1986)
10. F. Eisele: Rep. Prog. Phys. **49**, 233 (1986)
11. M. Diemoz et al.: Phys. Rep. **130**, 293 (1986)
12. T. Sloan: Proc. 17th Intern. Symp. on Multiparticle Dynamics, Seewinkel, Austria (1986)
13. N. Schmitz: Proc. 1979 Intern. Symp. on Lepton and Photon Interactions at High Energies, Fermilab, USA, p. 359 (1979)
14. B. Saitta: Proc. 10th Intern. Conf. on Neutrino Physics and Astrophysics, Erice, Italy, p. 107 (1980)
15. H.E. Montgomery: Proc. 1981 Intern. Symp. on Lepton and Photon Interactions at High Energies, Bonn, Fed. Rep. Germany, p. 508 (1981)
16. N. Schmitz: ibidem, p. 527 (1981)
17. D. Haidt: ibidem, p. 558 (1981)
18. P.B. Renton: Proc. Intern. Conf. on High Energy Physics, Lisbon, Portugal, p. 236 (1981)
19. P.B. Renton, W.S.C. Williams: Ann. Rev. Nucl. Part. Sci. **31**, 193 (1981)
20. P.B. Renton: Proc. 13th Intern. Symp. on Multiparticle Dynamics, Volendam, Netherlands, p. 394 (1982)
21. F. Janata: Proc. 14th Intern. Symp. on Multiparticle Dynamics, Lake Tahoe, USA, p. 42 (1983)
22. N. Schmitz: Proc. 7th Warsaw Symp. on Elem. Part. Phys., Kazimierz, Poland, p. 143 (1984)
23. W. Wittek: Proc. 11th Intern. Conf. on Neutrino Physics and Astrophysics, Nordkirchen, Fed. Rep. Germany, p. 455 (1984)
24. P.B. Renton: Proc. 16th Intern. Symp. on Multiparticle Dynamics, Kiryat-Anavim, Israel, p. 17 (1985)
25. P.B. Renton: Proc. 9th Warsaw Symp. on Elem. Part. Phys., Kazimierz, Poland, p. 149 (1986)
26. J.D. Bjorken, S.J. Brodsky: Phys. Rev. **D1**, 1416 (1970); E. Farhi: Phys. Rev. Lett. **39**, 1587 (1977); S. Brandt et al.: Phys. Lett. **12**, 57 (1964); A. De Rujula et al.: Nucl. Phys. **B138**, 387 (1978); S. Brandt, H.D. Dahmen: Z. Phys. **C1**, 61 (1979)
27. F. Bobisut: Proc. 11th Intern. Conf. on Neutrino Physics and Astrophysics, Nordkirchen, Fed. Rep. Germany, p. 422 (1984)
28. H. Yamamoto: Proc. 1985 Intern. Symp. on Lepton and Photon Interactions at High Energies, Kyoto, Japan, p. 50 (1985)
29. B.R. Webber: Proc. 16th Intern. Symp. on Multiparticle Dynamics, Kiryat-Anavim, Israel, p. 41 (1985); Ann. Rev. Nucl. Part. Sci. **36**, 253 (1986)

38

30. T. Sjöstrand: Proc. 23rd Intern. Conf. on High Energy Physics, Berkeley, USA (1986)
31. T.D. Gottschalk: Lectures at 19th Intern. School of Elem.Part.Physics, Kupari-Dubrovnik, Yugoslavia (1983), CALT-68-1075; preprint CALT-68-1413
32. R.D. Field, R.P. Feynman: Phys. Rev. **D15**, 2590 (1977); Nucl. Phys. **B136**, 1 (1978); P. Hoyer et al.: Nucl. Phys. **B161**, 349 (1979); A. Ali et al.: Phys. Lett. **B93**, 155 (1980)
33. G.C. Fox, S. Wolfram: Nucl. Phys. **B168**, 285 (1980); R.D. Field, S. Wolfram: Nucl. Phys. **B213**, 65 (1983); G. Marchesini, B.R. Webber: Nucl. Phys. **B238**, 1 (1984); B.R. Webber: Nucl. Phys. **B238**, 492 (1984); T.D. Gottschalk: Nucl. Phys. **B214**, 201 (1983); **B239**, 325 (1984); **B239**, 349 (1984)
34. B. Andersson et al.: Phys. Rep. **97**, 31 (1983); G. Gustafson: Proc. 16th Intern. Symp. on Multiparticle Dynamics, Kiryat-Anavim, Israel, p. 61 (1985); B. Andersson et al.: Nucl. Phys. **B264**, 29 (1986)
35. T. Sjöstrand: Comp. Phys. Comm. **27**, 243 (1982); **28**, 229 (1983); **39**, 347 (1986); CERN "The LUND Monte Carlo Programs" (long writeup), June 1986
36. G. Preparata: Proc. SLAC Summer Institute on Particle Physics 1983, Stanford, USA, p. 395 (1983); A. Giannelli et al.: Phys. Lett. **150B**, 214 (1985)
37. O.C. Allkofer et al.: Nucl. Instr. Meth. **179**, 445 (1981)
38. J.J. Aubert et al.: Phys. Lett. **114B**, 373 (1982)
39. J.J. Aubert et al.: Phys. Lett. **119B**, 233 (1982)
40. J.J. Aubert et al.: Nucl. Phys. **B213**, 1 (1983)
41. J.J. Aubert et al.: Nucl. Phys. **B213**, 31 (1983)
42. J.J. Aubert et al.: Z. Phys. **C18**, 189 (1983)
43. J.J. Aubert et al.: Phys. Lett. **130B**, 118 (1983)
44. J.J. Aubert et al.: Phys. Lett. **133B**, 370 (1983)
45. J.J. Aubert et al.: Phys. Lett. **135B**, 225 (1984)
46. A. Arvidson et al.: Nucl. Phys. **B246**, 381 (1984)
47. J.J. Aubert et al.: Phys. Lett. **160B**, 417 (1985)
48. J.J. Aubert et al.: Phys. Lett. **161B**, 203 (1985)
49. J.J. Aubert et al.: Z. Phys. **C30**, 23 (1986)
50. J.J. Aubert et al.: Phys. Lett. **167B**, 127 (1986)
51. J.J. Aubert et al.: Z. Phys. **C31**, 175 (1986)
52. J.P. Albanese et al.: Nucl. Instr. Meth. **212**, 111 (1983)
53. J.P. Albanese et al.: Phys. Lett. **144B**, 302 (1984)
54. M. Arneodo et al.: Phys. Lett. **145B**, 156 (1984)
55. M. Arneodo et al.: Phys. Lett. **149B**, 415 (1984)
56. M. Arneodo et al.: Phys. Lett. **150B**, 458 (1985)
57. M. Arneodo et al.: Nucl. Phys. **B258**, 249 (1985)
58. M. Arneodo et al.: Phys. Lett. **165B**, 222 (1985)
59. M. Arneodo et al.: Nucl. Phys. **B264**, 739 (1986)
60. M. Arneodo et al.: Z. Phys. **C31**, 1 (1986)
61. M. Arneodo et al.: Z. Phys. **C31**, 333 (1986)
62. M. Arneodo et al.: Z. Phys. **C32**, 1 (1986)
63. M. Arneodo et al.: Z. Phys. **C33**, 167 (1986)
64. M. Arneodo et al.: Z. Phys. **C34**, 277 (1987)
65. M. Arneodo et al.: Z. Phys. **C34**, 283 (1987)
66. M. Arneodo et al.: preprint CERN-EP/86-119 (1986)
67. M. Arneodo et al.: preprint CERN-EP/87-35 (1987)
68. M. Arneodo et al.: preprint CERN-EP/87-74 (1987)

69. M. Arneodo et al.: CERN-preprint (1987)
70. J.P. Berge et al.: Nucl. Phys. **B203**, 1 (1982)
71. J.P. Berge et al.: Nucl. Phys. **B203**, 16 (1982)
72. M. Derrick et al.: Phys. Rev. **D25**, 624 (1982)
73. R. Brock et al.: Phys. Rev. **D25**, 1753 (1982)
74. D. Son et al.: Phys. Rev. Lett. **49**, 1128 (1982); **49**, 1800 (1982) erratum
75. V.V. Ammosov et al.: JETP Lett. **38**, 248 (1983)
76. A.E. Asratyan et al.: Phys. Lett. **132B**, 246 (1983)
77. D. Zieminska et al.: Phys. Rev. **D27**, 47 (1983)
78. C.C. Chang et al.: Phys. Rev. **D27**, 2776 (1983)
79. D. Son et al.: Phys. Rev. **D28**, 2129 (1983)
80. H.C. Ballagh et al.: Phys. Rev. **D30**, 1130 (1984)
81. A.E. Asratyan et al.: Phys. Lett. **140B**, 127 (1984)
82. V.V. Ammosov et al.: JETP Lett. **39**, 209 (1984)
83. A.E. Asratyan et al.: Phys. Lett. **156B**, 441 (1985)
84. N.J. Baker et al.: Phys. Rev. **D32**, 531 (1985)
85. N.J. Baker et al.: Phys. Rev. **D34**, 1251 (1986)
86. C. Baltay et al.: Phys. Rev. Lett. **57**, 2629 (1986)
87. H.C. Ballagh et al.: Proc. 12th Intern. Conf. on Neutrino Physics and Astrophysics, Sendai, Japan, p. 555 (1986)
88. W. Smart et al.: Acta Phys. Polonica, **B17**, 41 (1986)
89. V.V. Ammosov et al.: Preprint FNAL-E-180 (1986)
90. V.V. Ammosov et al.: Serpukhov preprint 86-160 (1986)
91. H. Grässler et al.: Nucl. Phys. **B194**, 1 (1982)
92. P. Allen et al.: Nucl. Phys. **B194**, 373 (1982)
93. P.C. Bosetti et al.: Phys. Lett. **109B**, 234 (1982)
94. P. Allen et al.: Phys. Lett. **112B**, 88 (1982)
95. P. Allen et al.: Nucl. Phys. **B214**, 369 (1983)
96. H. Grässler et al.: Nucl. Phys. **B223**, 269 (1983)
97. G.T. Jones et al.: Z. Phys. **C25**, 121 (1984)
98. G.T. Jones et al.: Z. Phys. **C27**, 43 (1985)
99. N. Schmitz: Proc. 15th Intern. Symp. on Multiparticle Dynamics, Lund, Sweden, p. 764 (1984)
100. G.T. Jones et al.: Z. Phys. **C28**, 23 (1985)
101. P. Allen et al.: Nucl. Phys. **B264**, 221 (1986)
102. H. Grässler et al.: Nucl. Phys. **B272**, 253 (1986)
103. S. Barlag et al.: Z. Phys. **C11**, 283 (1982); **C14**, 281 (1982) erratum
104. D. Allasia et al.: Nucl. Phys. **B224**, 1 (1983)
105. D. Allasia et al.: Phys. Lett. **124B**, 543 (1983)
106. D. Allasia et al.: Z. Phys. **C20**, 95 (1983)
107. D. Allasia et al.: Z. Phys. **C24**, 119 (1984)
108. D. Allasia et al.: Z. Phys. **C27**, 239 (1985)
109. D. Allasia et al.: Phys. Lett. **154B**, 231 (1985)
110. D. Allasia et al.: Nucl. Phys. **B268**, 1 (1986)
111. P. Marage et al.: Phys. Lett. **140B**, 137 (1984)
112. P. Marage et al.: Z. Phys. **C21**, 307 (1984)
113. P. Kasper: Proc. 11th Intern. Conf. on Neutrino Physics and Astrophysics, Nordkirchen, Fed. Rep. Germany, p. 473 (1984)
114. G. Gerbier et al.: Z. Phys. **C29**, 15 (1985)
115. P.J. Fitch et al.: Z. Phys. **C31**, 51 (1986)

116. P. Marage et al.: Z. Phys. C31, 191 (1986)
117. W. Wittek et al.: Phys. Lett. 187B, 179 (1987)
118. P. Marage et al.: Preprint (1987)
119. R. Beuselinck et al.: Nucl. Instr. Meth. 154, 445 (1978)
120. H. Foeth et al.: Nucl. Instr. Meth. A253, 245 (1987)
121. V.V. Ammosov et al.: Z. Phys. C30, 183 (1986)
122. H.J. Grabosch et al.: Z. Phys. C31, 203 (1986)
123. V.V. Ammosov et al.: Berlin-Zeuthen preprint PHE/86-10 (1986)
124. P. Bosetti et al.: Nucl. Phys. B209, 29 (1982)
125. Z. Koba, H.B. Nielsen, P. Olesen: Nucl. Phys. B40, 317 (1972)
126. D. Levy: Nucl. Phys. B59, 583 (1973); F. Hayot, G. Sterman: Phys. Lett. 121B, 419 (1983)
127. M. Althoff et al.: Z. Phys. C22, 307 (1984)
128. J.G. Rushbrooke et al.: Phys. Lett. 59B, 303 (1975)
129. G.J. Alner et al.: Phys. Lett. 160B, 193 (1985); 160B, 199 (1985); 167B, 476 (1986)
130. M. Derrick et al.: Phys. Lett. 168B, 299 (1986); M. Adamus et al.: Phys. Lett. 177B, 239 (1986); F. Dengler et al.: Z. Phys. C33, 187 (1986)
131. M. Adamus et al.: Z. Phys. C32, 475 (1986)
132. P. Carruthers, C.C. Shih: Phys. Lett. 127B, 242 (1983); A. Giovannini, L. Van Hove: Z. Phys. C30, 391 (1986); L. Van Hove, A. Giovannini: Proc. 17th Intern. Symp. on Multiparticle Dynamics, Seewinkel, Austria (1986); A. Bialas, A. Szczerba: Acta Phys. Polonica B17, 1085 (1986); K. Fialkowski: Phys. Lett. 169B, 436 (1986); 173B, 197 (1986); E.D. Malaza, B.R. Webber: Nucl. Phys. B267, 702 (1986); C.S. Lam: Proc. 21st Rencontre de Moriond, Les Arcs, France, Vol.2, p. 241 (1986); R.C. Hwa, C.S. Lam: Phys. Lett. 173B, 346 (1986); W.R. Chen, R.C. Hwa: Oregon-preprint OITS-347 (1986); A. Bassetto: Padova-preprint DFPD 9/87 (1987)
133. P. Carruthers, C.C. Shih: Phys. Lett. 137B, 425 (1984); Mod. Phys. Lett. A2, 89 (1987); A. Capella et al.: Phys. Rev. D32, 2933 (1985); P. Carruthers: Proc. 21st Rencontre de Moriond, Les Arcs, France, Vol.2, p. 229 (1986); Chao Wei-qin et al.: Phys. Lett. 176B, 211 (1986); Phys. Rev. D35, 152 (1987); Berlin-preprint FUB-HEP/87-1 (1987); C.C. Shih, P. Carruthers: Proc. 17th Intern. Symp. on Multiparticle Dynamics, Seewinkel, Austria (1986); M. Blazek: Phys. Rev. D35, 102 (1987); A. Capella, A.V. Ramallo: preprint LPTHE Orsay 87/08 (1987)
134. A. Wroblewski: Proc. 14th Intern. Symp. on Multiparticle Dynamics, Lake Tahoe, USA, p. 573 (1983); Proc. 15th Intern. Symp. on Multiparticle Dynamics, Lund, Sweden, p. 30 (1984)
135. F. Dengler: Ph.D. Thesis, München (1985)
136. M. Glück et al.: Z. Phys. C13, 119 (1982)
137. A. Wroblewski: Acta Phys. Polonica B16, 379 (1985)
138. S. Fredriksson et al.: Z. Phys. C14, 35 (1982); C19, 53 (1983); Phys. Rev. D28, 255 (1983)
139. J.F. Owens: Phys. Lett. 76B, 85 (1978); T. Uematsu: Phys. Lett. 79B, 97 (1978); N. Sakai: Phys. Lett. 85B, 67 (1979); G. Altarelli et al.: Nucl. Phys. B160, 301 (1979); R. Baier, K. Fey: Z.Phys. C2, 339 (1979); G. Altarelli: Phys. Rep. 81, 1 (1982); G. Ingelman: Z. Phys. C26, 483 (1984)
140. H. Deden et al.: Nucl. Phys. B198, 365 (1982)
141. W. Ochs, L. Stodolsky: Phys. Lett. 69B, 225 (1977); H. Fesefeldt et al.: Phys. Lett. 74B, 389 (1978); W. Ochs, T. Shimada: Z. Phys. C4, 141 (1980)

142. H. Georgi, H.D. Politzer: Phys. Rev. Lett. **40**, 3 (1978); A. Mendez: Nucl. Phys. **B145**, 199 (1978); A. Mendez et al.: Nucl. Phys. **B148**, 499 (1979)
143. R.N. Cahn: Phys. Lett. **78B**, 269 (1978)
144. E.L. Berger: Z. Phys. **C4**, 289 (1980); Phys. Lett. **89B**, 241 (1980)
145. A. König, P. Kroll: Z. Phys. **C16**, 89 (1982)
146. K. Hagiwara et al.: Phys. Rev. Lett. **47**, 983 (1981); Phys. Rev. **D27**, 84 (1983)
147. W.A. Mann et al.: Phys. Rev. **D34**, 2545 (1986)
148. I.I.Y. Bigi: Nuovo Cim. **41A**, 43 (1977); **41A**, 581 (1977)
149. K. Moriyasu, E. Wolin: Z. Phys. **C20**, 151 (1983)
150. H. Abramowicz et al.: Z. Phys. **C15**, 19 (1982)
151. A. Haatuft et al.: Nucl. Phys. **B222**, 365 (1983)
152. K. Lang et al.: Z. Phys. **C33**, 483 (1987)
153. P. Marage: Proc. 12th Intern. Conf. on Neutrino Physics and Astrophysics, Sendai, Japan, p. 540 (1986)
154. S.L. Adler: Phys. Rev. **135B**, 963 (1964)
155. H. Faissner et al.: Phys. Lett. **125B**, 230 (1983)
156. E. Isiksal et al.: Phys. Rev. Lett. **52**, 1096 (1984)
157. F. Bergsma et al.: Phys. Lett. **157B**, 469 (1985)
158. T. Kitagaki et al.: Phys. Rev. **D34**, 2554 (1986); N.J. Baker et al.: Phys. Rev. **D28**, 2900 (1983)
159. D. Rein, L.M. Sehgal: Ann. Phys. **133**, 79 (1981)
160. W.D. Shambroom et al.: Phys. Rev. **D26**, 1 (1982)
161. A. Donnachie, P.V. Landshoff: Phys. Lett. **185B**, 403 (1987)

42

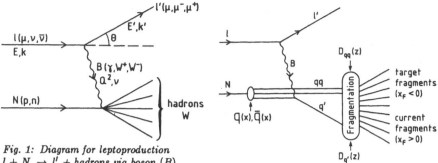

Fig. 1: Diagram for leptoproduction
$l + N \rightarrow l' + hadrons$ via boson (B)
exchange.

Fig. 2: QPM diagram for leptoproduction.

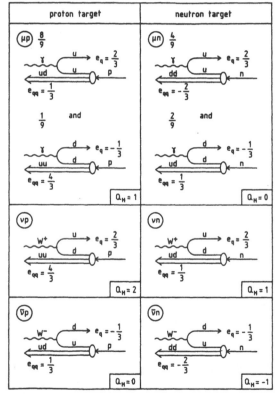

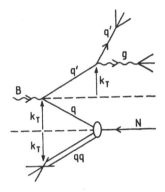

Fig. 4: The effects of primordial k_T, gluon radiation and
fragmentation on the transverse momenta of the final
state hadrons.

Fig. 3: The main QPM diagrams (only valence quarks) for $\mu N, \nu N$ and $\bar{\nu} N$ scattering in
the boson-nucleon cms. The quark and diquark charges e_q and e_{qq}, and the total hadronic
charge Q_H are indicated. For the μN diagrams also the weight factors $e_q^2 \cdot N_q$ are given (N_q
= number of quarks with flavor q inside the nucleon).

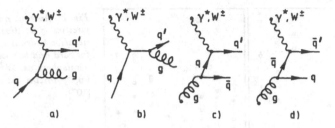

Fig. 5: *First-order QCD diagrams contributing to leptoproduction: gluon radiation (a,b), boson-gluon fusion (c,d).*

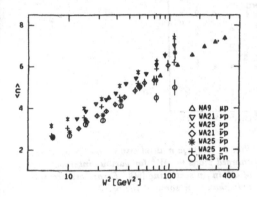

Fig. 6: *Compilation of average charged multiplicities $< n >$ vs. W^2 from various experiments: NA9 [57] μp; WA21 [96] $\nu p, \bar{\nu} p$; WA25 [107] $\nu p, \bar{\nu} p, \nu n, \bar{\nu} n$. The data points of WA25 were obtained from adding the average multiplicities in the forward and backward hemisphere in ref. [107].*

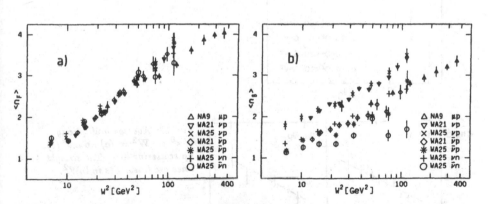

Fig. 7: *Compilation of average charged multiplicities $< n_{F,B} >$ in the (a) forward and (b) backward hemisphere vs. W^2 from various experiments: NA9 [57] μp; WA21 [96] $\nu p, \bar{\nu} p$; WA25 [107] $\nu p, \bar{\nu} p, \nu n, \bar{\nu} n$.*

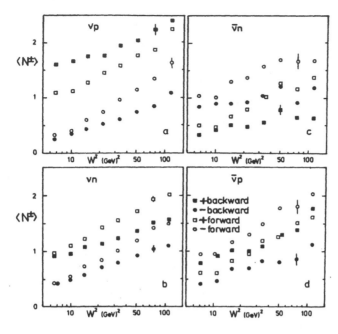

Fig. 8: Average multiplicities of positive and negative hadrons in the forward and backward hemisphere vs. W^2, for $\nu p, \bar{\nu} p, \nu n, \bar{\nu} n$ scattering [107].

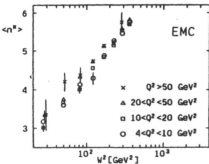

Fig. 9: Average multiplicity $< n_\pi >$ of charged pions vs. W^2 for various intervals of Q^2 in μp scattering [60]. Error bars are only shown for some of the points.

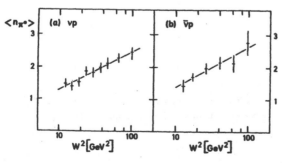

Fig. 10: Average multiplicity of π^0 vs. W^2 in (a) νp and (b) $\bar{\nu} p$ scattering [96]. The straight lines are linear fits in $\ln W^2$.

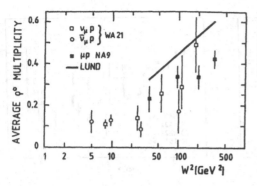

Fig. 11: Average multiplicity of ρ^0 mesons vs. W^2 in μp [63], νp [92] and $\bar{\nu} p$ [102] scattering. The curve is the prediction of the Lund model for μp. From ref. [63].

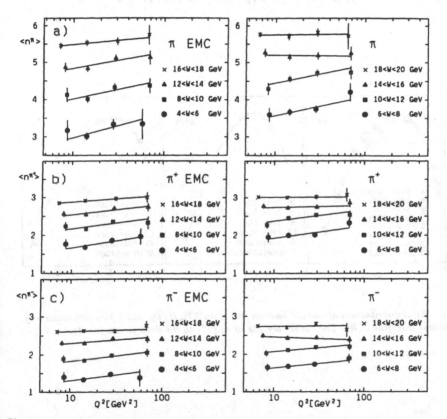

Fig. 12: Average multiplicity $< n_\pi >$ of (a) charged pions and of (b) π^+ and (c) π^- separately vs. Q^2 for fixed intervals of W in μp scattering [60]. The straight lines are linear fits in $\ln Q^2$.

46

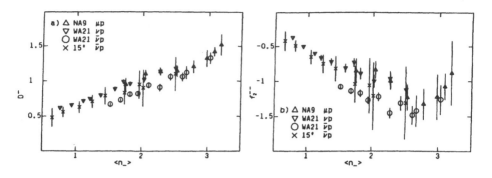

Fig. 13: Dispersion D^- (a) and correlation parameter f_2^{--} (b) of the multiplicity distribution of negative hadrons in μp [57], νp [96] and $\bar{\nu} p$ [72,96] scattering vs. $< n^- >$.

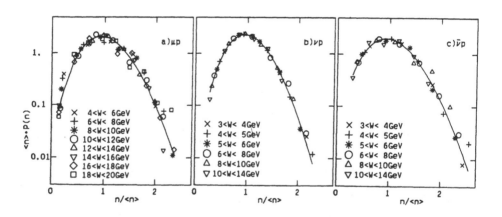

Fig. 14: KNO distributions of charged hadrons in (a) μp [57], (b) νp and (c) $\bar{\nu} p$ [96] scattering for various intervals of W. The curves are fits of eq.(33) to the data points (see text).

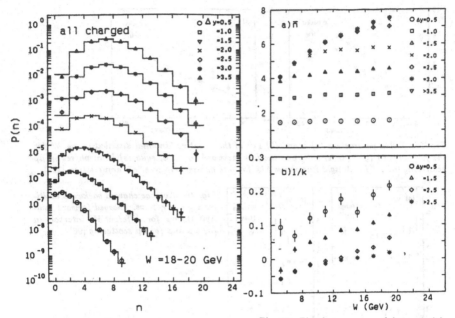

Fig. 15: Charged multiplicity distributions in μp scattering [67] for various cms rapidity intervals $2 \cdot \Delta y^*$ centered around $y^* = 0$, for $18 < W < 20$ GeV. The histograms show the fitted negative binomial distributions. The distribution for the largest Δy^* is in ordinary scale, each consecutive one is shifted down by a factor of ten.

Fig. 16: Fitted parameters (a) $\bar{n}$ and (b) $1/k$ of the negative binomial distribution in various intervals $2 \cdot \Delta y^*$, centered around $y^* = 0$, vs. W for μp scattering [67].

Fig. 17: Charged multiplicity distributions $P(n)$ for μp scattering [57] in four intervals of W (in GeV). The dashed curve is a Poisson distribution for the charged multiplicity n, the full curve is $P(n)$ under the assumption that the multiplicity of negative hadrons, $n^- = (n-1)/2$, obeys a Poisson distribution.

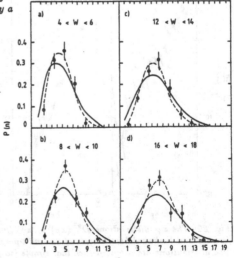

48

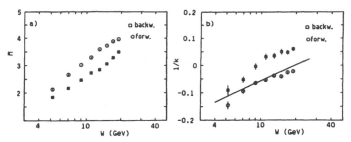

Fig. 18: Fitted parameters (a) $\bar{n}$ and (b) $1/k$ of the negative binomial distribution vs. W for the full forward ($y^* > 0$, circles) and the full backward ($y^* < 0$, squares) hemispheres in μp scattering [67]. The straight line shows $1/k$ for one hemisphere in e^+e^- annihilation [131].

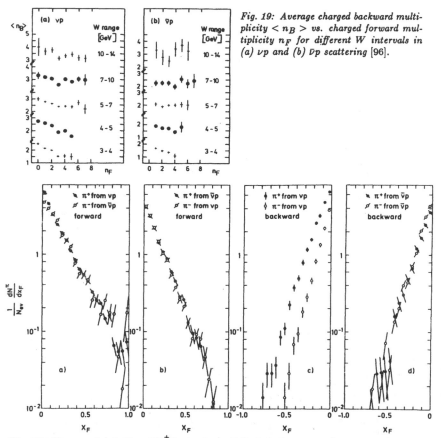

Fig. 19: Average charged backward multiplicity $< n_B >$ vs. charged forward multiplicity n_F for different W intervals in (a) νp and (b) $\bar{\nu} p$ scattering [96].

Fig. 20: The x_F distributions $D^{\pi^\pm}(x_F)$ of π^+ (full circles) and π^- (open circles) in the forward and backward hemispheres of νp and $\bar{\nu} p$ scattering [95] with $W > 3$ GeV, $x > 0.1$. The distributions are arranged such that simple isospin relations of the QPM can be tested.

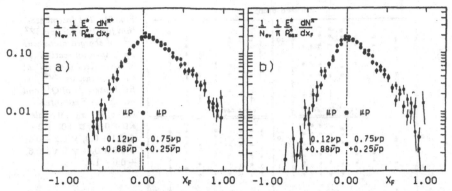

Fig. 21: Invariant x_F distributions of (a) π^+ and (b) π^- in μp scattering (circles)(from EMC), compared with the QPM predicted linear combinations (squares) of the corresponding νp and $\bar{\nu} p$ distributions [95] in each hemisphere, from ref. [135].

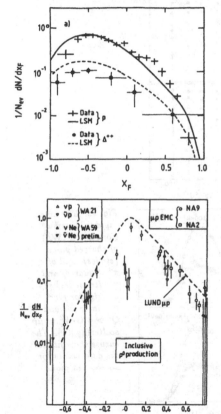

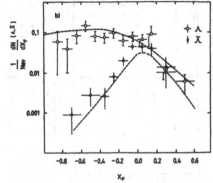

Fig. 22: The x_F distributions of (a) protons (including those from Δ^{++} decay) and $\Delta^{++}(1232)$, (b) Λ and $\bar{\Lambda}$ in μp scattering [54, 59]. The curves show the Lund model predictions.

Fig. 23: Compilation of x_F distributions of ρ^0 mesons in μp and $\nu N, \bar{\nu} N$ interactions, from ref. [24]. The curve shows the Lund model prediction for μp.

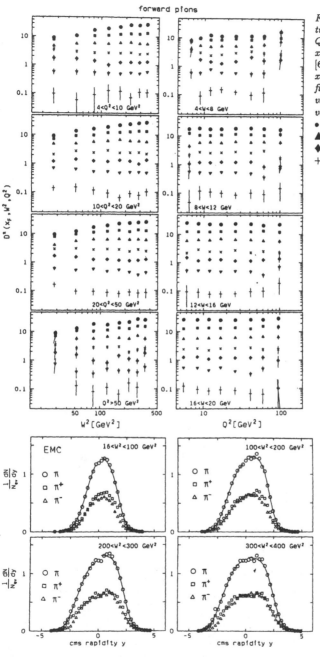

forward pions

Fig. 24: The fragmentation function $D^\pi(x_F, W^2, Q^2)$ of charged pions with $x_F > 0$ in μp scattering [60], for fixed intervals of x_F, is shown vs. W^2 for fixed intervals of Q^2, and vs. Q^2 for fixed intervals of W. x_F intervals: $\bullet\ 0 \div 0.05$, $\blacksquare\ 0.05 \div 0.1$, $\blacktriangle\ 0.1 \div 0.2$, $\times\ 0.2 \div 0.3$, $\blacklozenge\ 0.3 \div 0.4$, $\blacktriangledown\ 0.4 \div 0.6$, $+\ 0.6 \div 1.0$.

Fig. 25: The cms rapidity distributions $D^\pi(y^*, W^2)$ of charged pions (circles), and of π^+ (squares) and π^- (triangles) separately, for four W^2 intervals and $Q^2 > 4\ GeV^2$ in μp scattering [60]. The curves show fits to the data points for all charged pions.

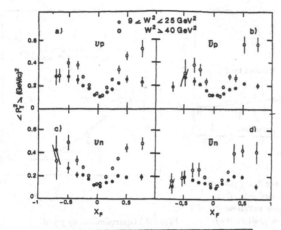

Fig. 26: Seagull plots ($< p_T^2 >$ vs. x_F) of charged hadrons in $\nu p, \bar{\nu} p, \nu n, \bar{\nu} n$ scattering [108] at low (dots) and high (circles) W.

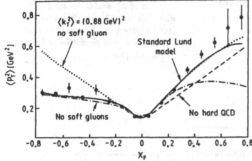

Fig. 27: Seagull plot of charged hadrons in μp scattering [69, 25] with $W > 4$ GeV, compared with four versions of the Lund model: standard version with hard QCD effects, soft gluons and $< k_T^2 > = (0.44 \text{ GeV})^2$: full curve; no hard QCD effects: dashed curve; no soft gluons: dash-dotted curve; no soft gluons and $< k_T^2 > = (0.88 \text{ GeV})^2$: dotted curve.

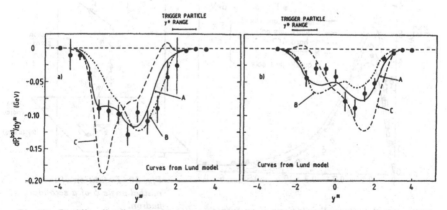

Fig. 28: Rapidity distribution of transverse momentum balancing the p_T of a trigger particle with (a) $x_F > 0.5$ and (b) $-0.5 < x_F < -0.2$ in μp scattering [69] with $W > 4$ GeV. The curves show three versions of the Lund model: standard version with hard QCD, soft gluons and $< k_T^2 > = (0.44 \text{ GeV})^2$: A; no soft gluons: B; no soft gluons and $< k_T^2 > = (0.88 \text{ GeV})^2$: C.

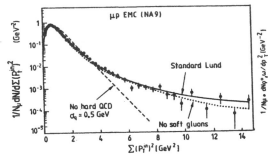

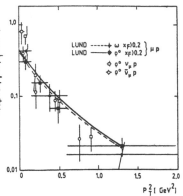

Fig. 29: Distribution of Σp_{Tin}^2 (see text) for μp events with more than 3 charged hadrons, from ref. [69, 25]. The curves show three versions of the Lund model: standard version (full curve), no hard QCD (dashed curve), no soft gluons (dotted curve).

Fig. 30: Compilation of p_T^2 distributions of ρ^0 and ω mesons produced in μp [63], νp [92] and $\bar{\nu} p$ [102] scattering. The curves show the Lund model predictions for μp. From ref. [63].

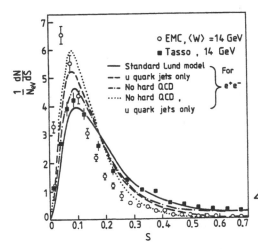

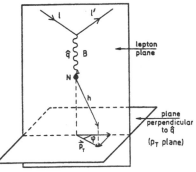

Fig. 32: Definition of the azimuthal angle ϕ of a secondary hadron.

Fig. 31: Sphericity distributions of μp events (forward hemisphere, circles) [66] and e^+e^- events (full squares) [127]. The curves show the predictions of four versions of the Lund model for e^+e^-. From ref. [66].

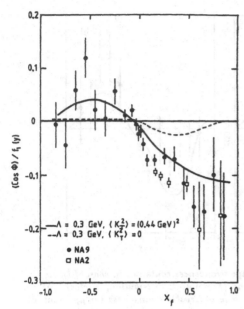

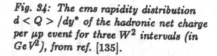

Fig. 34: The cms rapidity distribution $d<Q>/dy^*$ of the hadronic net charge per μp event for three W^2 intervals (in GeV^2), from ref. [135].

Fig. 33: Azimuthal asymmetry $<\cos\phi>/f_1(y)$ of charged hadrons vs. x_F in μp scattering [43,64], where $f_1(y)$ is a known kinematical factor. The curves are the predictions of ref. [145] for two sets of parameters.

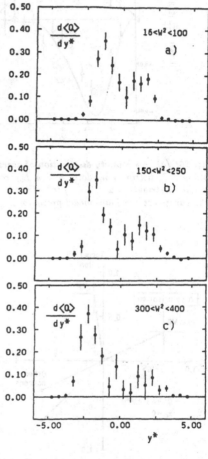

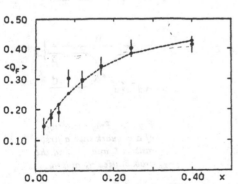

Fig. 35: Average hadronic net charge $<Q_F>$ in the forward hemisphere of μp scattering vs. x, from ref. [135]. The curve shows the prediction of the QPM, eq.(46).

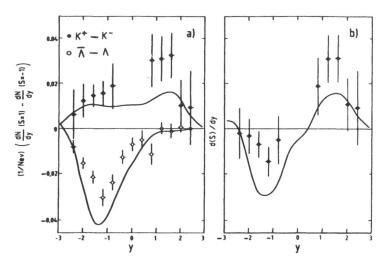

Fig. 36: (a) cms rapidity distribution of average strangeness contained in charged kaons (K^+ minus K^- distribution, dots) and Λ hyperons ($\bar{\Lambda}$ minus Λ distribution, circles); (b) cms rapidity distribution $d<S>/dy^*$ of the average observable strangeness per μp event [65]. The curves are the Lund model predictions.

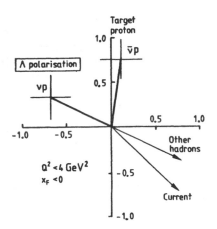

Fig. 37: Λ polarization vectors in the production plane for $x_F < 0$, $Q^2 < 4$ GeV^2 in νp and $\bar{\nu}p$ scattering [100]. Also shown are the momentum vectors of the target proton, intermediate boson and hadron system in the Λ rest frame, averaged over the νp and $\bar{\nu}p$ event sample. From ref. [24].

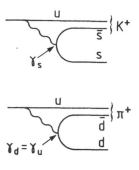

Fig. 38: Fragmentation of a u-quark into a first-rank K^+ and π^+ and the probabilities γ_s and γ_u.

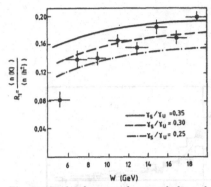

Fig. 39: Ratio of average kaon and charged hadron multiplicities in μp scattering [65]. The curves show the Lund model predictions for three values of the strangeness suppression factor $\lambda = \gamma_s/\gamma_u$.

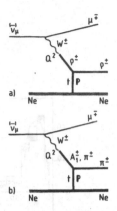

Fig. 40: Diagrams for coherent single meson production by (anti)neutrinos on Ne, (a) $\rho^\pm$ and (b) $\pi^\pm$.

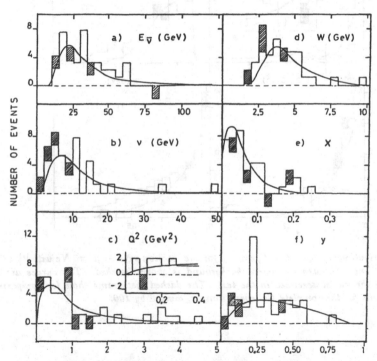

Fig. 41: Distributions of E_ν, ν, Q^2, W, x and y for the reaction $\bar\nu Ne \to \mu^+ \rho^- Ne$ with $|t| < 0.1$ GeV^2 [118]. The estimated incoherent background is shown hatched. The curves are the predictions of the vector dominance model (see text).

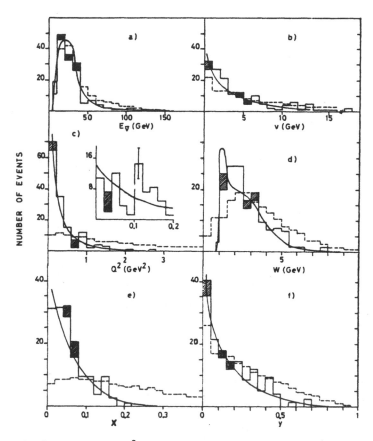

Fig. 42: Distributions of E_ν, ν, Q^2, W, x and y for the reaction $\bar{\nu}Ne \to \mu^+\pi^-Ne$ with $|t| < 0.05$ GeV^2 [116]. The estimated incoherent background is shown hatched. The curves are the predictions of the model described in the text. The dashed histograms show for comparison the distributions for the complete CC event sample, divided by 100.

PARTICLE PRODUCTION IN CONTINUUM e^+e^- ANNIHILATION AT HIGH ENERGY

M. Derrick and S. Abachi
Argonne National Laboratory

Introduction

High energy e^+e^- annihilation provides the simplest laboratory for studying the transformation of quarks and gluons into hadrons. The center-of-mass energy is known and it is shared a few partons, either by a $q\bar{q}$ pair or a $q\bar{q}$ pair plus a small number of gluons. In addition, the jets of particles resulting from the parton-hadron transitions come from quarks of different flavors in a known ratio. Because there is no hadronic target, there is no additional debris so that it is possible to make unique measurements of the behavior of both the low energy and the high energy hadrons.

The results from e^+e^- studies are prototypical of the situation to be expected in more complex reactions such as deep inelastic lepton scattering and hadronic collisions. In the simplest picture, given by the parton model, the former gives a jet of particles originating from the struck quark plus the remaining nucleon fragments involving a diquark. Hard hadronic collisions, on the other hand, give two high energy partons that subsequently fragment. The relative mixture of quark and gluon parents depends on the fractional energies of the interacting partons but, in general, gluons predominate. The parton energies are also a priori unknown in that the longitudinal momentum must be reconstructed from the observed hadrons in the jet. Soft hadronic collisions are even more complex at the parton level so that the results are still discussed in terms of phenomenological models. The interplay of data from

these different reactions is illuminating of the nature of the complex
phenomena involved.

In this review, we present the results on particle production coming from
high energy e^+e^- annihilation in the continuum. There is a significant body
of data at center-of-mass energies near 30 GeV resulting from the operation of
several detectors at the PEP and PETRA storage rings. The latter has now
ceased operation and the program at the former has been in suspension for over
two years so this phase of experimentation is almost complete, and now is an
appropriate time to overview the data. The PEP ring was operated at $\sqrt{s}$ = 29
GeV and data samples corresponding to integrated luminosities of up to 300 pb^{-1}
per detector have been analyzed. Most of the PEP results in this subject come
from the High Resolution Spectrometer (HRS), the MarkII detector, and the Time
Projection Chamber (TPC). The PETRA ring was operated at many energies but
the main data taking was at $\sqrt{s}$ = 34 GeV, and results from ~ 70 pb^{-1} have been
published by the JADE and TASSO collaborations. Results are also available
from the CELLO and PLUTO detectors. High statistics data samples have been
taken with the ARGUS detector at the DORIS storage ring operated at DESY and
with the CESR detector at Cornell. These data are at center-of-mass energies
of about 10 GeV.

Several reviews have been written about this subject. Some of them
summarize the results from a particular detector,[1,2] but others are more
general.[3,4] However, more precise measurements have recently been published
and so, in this article, we concentrate on the new data, including earlier
contributions only as pertinent. In general, only published results are
quoted using conference contributions and Theses only when the data is not
otherwise available.

Experimental Details

The cross section for e^+e^- annihilation into hadrons via one photon
exchange is:

$$\sigma = \frac{4\pi\alpha^2 R}{3s} = \frac{86.8R}{s} \text{ nb} \quad,$$

where R is given by the sum of the squares of the quark charges times a color factor of three.

For the PEP and PETRA data, the u, d, s, c, and b quarks are all produced so that:

$$R = 3\left[(\tfrac{2}{3})^2 + (\tfrac{1}{3})^2 + (\tfrac{1}{3})^2 + (\tfrac{2}{3})^2 + (\tfrac{1}{3})^2\right]$$

$$= \frac{11}{3} .$$

The continuum data from CESR and ARGUS is below threshold for the $e^+e^- \rightarrow b\bar{b}$ reaction, so R = 10/3 for these data sets. There is a very small correction for the Z^0 exchange contribution, but the second-order QCD correction to the cross section, $(1 + \frac{\alpha_s}{\pi})$, is more significant. Experimental values for R above $b\bar{b}$ threshold are about 4. For example, the JADE measurements[4] give R = 4.02 ± 0.04 ± 0.10 for $\sqrt{s}$ > 14 GeV.

All of the detectors contributing data to this review use a solenoidal magnet. A multilayer open-geometry drift-chamber is usually used to provide identification and momentum measurements of the charged particles. Surrounding the tracking is a system of shower counters to identify photons from, say, π^0 decay. The segmentation and energy resolution of all such electromagnetic calorimeters is poor so that neutral particles are not well measured as compared to the charged particles. Neutral hadrons such as K^0 mesons and Λ hyperons are identified through their decays to charged particles, although the resulting detection efficiency is low.

The separation of $\pi^\pm$ from $K^\pm$ and $p^\pm$ is done by time of flight for low momenta ($\sim$ 1(2) GeV for π-K (π-p) separation) or by a dE/dx system. The best such measurements come from the TPC. In this detector, the tracking is also done by the time projection chamber itself rather than with a conventional drift chamber.

As an example of a modern detector, the cross section of the ARGUS spectrometer is shown in Fig. 1.

The data samples per detector are about 400-500 K events at 10 GeV, 50-100 K events at 29 GeV, and 10-30 K events at 34 GeV. A typical event at

60

ARGUS

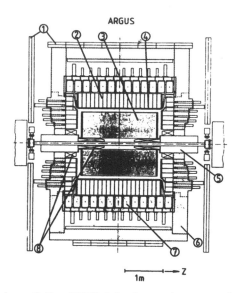

1m ⊢————⊢→ Z

Fig. 1. Schematic view of the ARGUS detector. 1 = muon chambers, 2 = shower
counters, 3 = drift chamber, 4 = time-of-flight counters, 5 = mini-
beta quadrupoles, 6 = iron yoke, 7 = solenoid coils, 8 = compensation
coils.

29 GeV, observed in the HRS, is shown in Fig. 2. The two-jet nature of the
$e^+e^- \rightarrow q\bar{q}$ process is evident.

Although the events are clean, all of the experiments use Monte Carlo
simulations to correct for the selection cuts that are needed to reject tau-
pair events, two-photon interactions, and higher-order QED processes. In
making the corrections, it is assumed that the correction factorizes, i.e., it
depends only on the variable in question. In generating the Monte Carlo
events, some parton level model and a fragmentation algorithm is assumed. The
corrections depend only weakly on these assumptions.

Definition of Variables

The global properties of the events are studied in terms of the jet
variables thrust (T), aplanarity (A), and sphericity (S). The thrust is

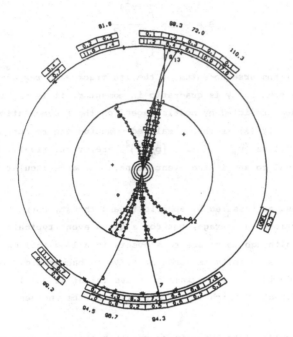

Fig. 2. Typical $e^+e^- \to q\bar{q}$ event at 29 GeV observed in the HRS.

selected by maximizing the longitudinal momentum:

$$T = \max \frac{\sum_{i}^{n} |p_L^i|}{\sum_{i}^{n} |p^i|}$$

where p^i is the momentum of the ith track, p_L^i is the component of p^i along the T axis, and the sum extends over all charged tracks of the event. The jet axis may also be taken as that direction which minimizes the transverse momentum squared. In this case, the sphericity (S) of the event is defined as

$$S = \frac{3}{2} \min \frac{\sum\limits_{i}^{n} (p_T^i)^2}{\sum\limits_{i}^{n} (p^i)^2} \quad ,$$

where p_T is the transverse momentum of the ith track with respect to a trial axis. Since the sphericity is quadratic in momentum, it is not infrared safe so that the values predicted by models depend on the fragmentation scheme used. The aplanarity (A) is the normalized momentum square out of the event plane. The quantities $(p_{T_{out}})^2$ and $(p_{T_{in}})^2$, are, respectively, the square of the momentum normal to and in the event plane, but perpendicular to the S or T axes.

The jet axes, as measured by sphericity and thrust, correlate well with the parton directions on average, but for a given event typically differ by a few degrees;[5] although there are some events in a long tail extending as far as 20°. Since the sphericity is more sensitive to background, the thrust axis is best used to define the jet axis: positive thrust is usually chosen along the positive z direction which is the direction of the incident positron beam.

The quarks produced in the primary reaction $e^+e^- \to q\bar{q}$ fragment into hadrons, each of which takes a fraction of the quark energy. As a measure of this fraction, several variables are used: (i) $X = 2p/\sqrt{s}$, where p is the particle momentum. This variable is simple experimentally since it can be measured without knowing the particle species. Other related measures of the longitudinal distribution of the hadrons are Feynman X (X_F), where $X_F = 2p_L/\sqrt{s}$ and rapidity (Y),

$$Y = \frac{1}{2} \ln(E + p_L)/(E - p_L) \quad ,$$

where p_L is the component of momentum parallel to the jet axis and E is the particle energy. (ii) $Z = 2E/\sqrt{s}$ where E is the energy of the particle in question. For the 30 GeV data, there is little difference between X and Z except near $X = 0$ where the particle mass introduces a threshold in the Z variable. (iii) The natural variable to use when comparing results to models is the light cone variable:

$$x^+ = \frac{E + P_L}{E_{max} + P_{max}} \; .$$

Since the jet direction is not known, one usually substitutes p for p_L so that the variable becomes

$$x^+ = \frac{E + p}{(E + p)_{max}} \; .$$

The normalized two-body particle rapidity correlation is defined as:

$$R = \frac{\rho_2(Y_1 Y_2)}{f \rho_1(Y) \rho_1(Y_2)} - 1 \; ,$$

where $\rho_2(Y_1, Y_2)$ is the two-particle density in rapidity

$$\rho_2(Y_1, Y_2) = \frac{1}{\sigma} \frac{d^2 \sigma}{dY_1 dY_2} \; ,$$

ρ_1 is the equivalent single particle density, and f is a normaliztion factor which is equal to $\langle n(n-1) \rangle / \langle n \rangle^2$ if particle types are not distinquished. Since R is a ratio, it is relatively free of experimental biases.

The fragmentation function for a quark of type q to produce a hadron of type h is defined as the hadron density per unit of Z:

$$D^h(Z) = \frac{1}{N_{ev}} \frac{dN^h}{dZ} = \frac{1}{\sigma} \frac{d\sigma^h}{dZ} \; .$$

In terms of the cross section, the invariant differential energy distribution is given by $\frac{s}{\beta} \frac{d\sigma}{dZ}$ μb Gev^2 or in dimensionless units $\frac{1}{\beta \sigma} \frac{d\sigma}{dZ}$.

For particles with spin, an additional degree of freedom is available and one can write[6]

$$D_{\lambda \lambda'}(Z) = D(Z) \; \rho_{\lambda \lambda'}(Z)$$

where $\rho_{\lambda\lambda'}$ is the helicity density matrix. The latter is measured by analyzing the decay angular distribution. For example, for the decay of a $J^P = 1^-$ particle such as $\rho^0 \to \pi^+\pi^-$, the decay angular distribution may be written as

$$W(\cos\theta,\phi) = (3/4\pi)\Big[\tfrac{1}{2}(1 - \rho_{00}) + \tfrac{1}{2}(3\rho_{00} - 1)\cos^2\theta$$
$$- \rho_{1-1}\sin^2\theta\cos 2\phi - \sqrt{2}\,\mathrm{Re}\rho_{10}\sin 2\theta\cos\phi\Big] \quad,$$

where the z-axis is in the direction of the momentum of the vector particle and the y-axis is along the production plane normal. The azimuthal angle, ϕ, is chosen to be zero in the production plane. The polar angle, θ, is the helicity angle of the vector particle.

In measuring these variables, it is normal to correct for the effects of initial and final state electromagnetic radiation. This can be done at different levels of precision. The simplest correction is merely to reduce the effective center-of-mass energy by a fraction of a GeV to account for the mean energy radiated by an initial state photon. The second level is to make a bin-by-bin correction that is estimated by running the Monte Carlo simulation both with and without radiative corrections.

Although, in principle, one could correct for the effects on the quark energy of final state QCD radiation, such a correction would be so model-dependent as to render comparison between experiments impossible. Indeed, there is a choice in QCD about the very definition of the quark decay functions.

The multiplicity distribution for the charged particles is specified by several variables. The number of charged particles in any event is denoted by n. Charge conservation requires n to be even. The mean value of the multiplicity distribution is denoted by $< n >$, while the shape of the distribution is characterized, in general, by the moments $C_q = < n^q >/< n >^q$, but is often parametrized by the dispersion D_2, where $D_2^2 = < n^2 > - < n >^2$ and by the f_2 moment defined as

$$f_2 = \langle n(n-1) \rangle - \langle n \rangle^2 = D_2^2 - \langle n \rangle \quad .$$

The f_2 moment is zero for a Poisson distribution. The events can be divided into two individual jets by a plane perpendicular to the thrust axis, and the variables describing the properties of the individual jets are denoted by n, $\mathscr{D}_2$, etc.

The multiplicity distributions for e^+e^- annihilation at different center-of-mass energies are often compared using the form first suggested by Koba Nielsen and Olesen[7] (KNO). If z is the scaled multiplicity $n/\langle n \rangle$, then KNO scaling says that all of the data can be represented by a universal function $\psi(z)$ given by

$$\psi(z) = \langle n \rangle \; \frac{\sigma_n}{\Sigma \sigma_n} \quad ,$$

where σ_n is the cross section for producing an event with n charged particles.

The multiplicity distributions may also be compared to the negative-binomial (NB) distribution in the variable n:

$$P(n, \langle n \rangle, k) = \frac{k(k+1) \cdots (k+n-1)}{n!}$$

$$\times \left[\frac{\langle n \rangle/k}{1 + \langle n \rangle/k} \right]^n \left[1 + \frac{\langle n \rangle}{k} \right]^{-k} \quad .$$

The position of this distribution is fixed by the mean value $\langle n \rangle$ and its shape is determined by the parameter k. When k is an integer, the resulting shape is known as the generalized Bose-Einstein distribution and gives the distribution resulting from the superposition of k identical emitters. The variable k is related to $\langle n \rangle$ and D_2 by

$$\frac{D_2^2}{\langle n \rangle^2} = \frac{1}{\langle n \rangle} + \frac{1}{k} \quad .$$

In the limit $k \to \infty$, $D_2^2 = <n>$ and the NB expression become a Poisson distribution:

$$P(n) = \frac{<n>^n}{n!} e^{-<n>} .$$

Since the multiplicity distributions in KNO form have a sharp rise at low z, followed by an exponential fall, fits to data have often used the gamma distribution (GD)

$$\psi(z) = \frac{K^K}{(K - 1)!} z^{K-1} e^{-Kz} .$$

When $K = 1$, the KNO distribution in this form is a simple exponential, $\psi(z) = e^{-z}$, and for $K = 2$ it becomes $\psi(z) = 4Ze^{-2z}$. The GD follows from the NB in the limit when $<n> >> k$.

If KNO scaling holds, then the $k(K)$ values and the C_q moments of the multiplicity distribution will be energy independent.

Fits to the multiplicity distributions with the NB and GD give different numerical values for the $k(K)$ parameters. They are approximately related by $1/K = 1/k + 1/<n>$, so that for a Poisson distribution, with $1/k = 0$, $K \approx <n>$.

Fragmentation Models

Although, in principle, the parton level processes can be calculated within QCD, the transformation of the partons into hadrons is both non-perturbative and complex. Much of the effort in this subject has been directed at using the data to parametrize these transitions in terms of some model. In the first attempt to do this, by Feynman and Field[8], the initially produced quark was iteratively connected to a $\bar{q}q$ pair forming a meson and another quark. This process was continued until all the energy was used up. This simple scheme with the probability that the first rank meson takes a fraction $\sim a + b(1 - Z)^2$ of the initial energy agreed with the basic characteristics of the data available at that time. Even with further refinements[9], the model was ad hoc and has now been superseded.

In the Lund[10] string picture, a color flux tube joins the q and $\bar{q}$. The particles are formed as the string tension exceeds the particles masses at a value of ~ 1 GeV/fm. The suppression of $s\bar{s}$ pairs and other heavier particles comes about because of the greater length of string that must be used up. The fragmentation is parametrized as:

$$f(Z) = \frac{(1 - Z)^a}{Z} e^{-b\, M_T^2/Z}$$

where $M_T = (m^2 + p_T^2)^{1/2}$ is the transverse mass, and a and b are constants to be fixed by comparison with data.

This approach still has many free parameters, such as the vector to pseudoscalar ratio, the strangeness suppression factor, and the diquark-to-quark ratio. Baryons are a difficulty that is circumvented by introducing diquark pairs, even though no evidence for point-like diquarks has been found.

In the final class of models[11], a parton level QCD cascade is followed up to some cutoff. At this point, color neutral clusters are formed which then decay into the observed particles. The suppression of heavier particles in these models comes about because of phase space limitations in the cluster decay. One important element, first introduced in the Webber model, was to incorporate coherence effects between successive soft gluon emissions. This depletes the particle spectrum at low momentum in the region between the q and the $\bar{q}$.

A pictorial representation of an $e^+e^- \rightarrow q\bar{q}g$ event, according to the pictures given by three classes of model, is shown in Fig. 3.

Display of 30 GeV Data

a. Event Shape Measurements

Several groups have reported measurements of event shapes using global event variables. Such data were historically crucial in establishing the

68

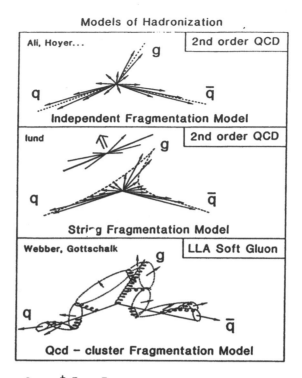

Fig. 3. Pictures of an $e^+e^- \to q\bar{q}g$ event according to the three classes of models.

existence of the gluon. The sphericity and thrust distributions published by the HRS[5], MarkII[12], and TASSO[13] detector collaborations at ~ 30 GeV are shown in Fig. 4. In general, the data agree well, although there are some differences between the results at low and high values of the variables, which are an indication of the systematic uncertainties of such measurements.

The energy variation of the mean values of 1–T are shown[5,12–14] in Fig. 5, compared to the predictions of four model[10,11] and extrapolated to $\sqrt{s}$ = 90 GeV, where new data will become available over the next few years. Since there is little difference between the predictions of the models, these measures will be a sensitive indication of new physics, such as possible quark

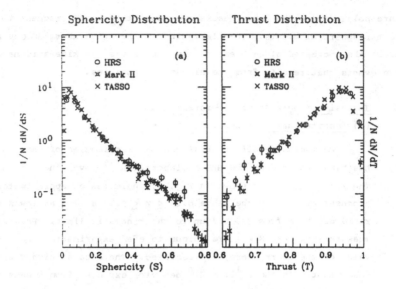

Fig. 4. Sphericity and thrust distributions near 30 Gev.

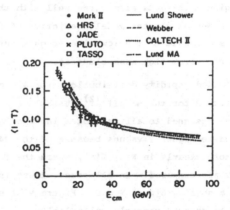

Fig. 5 Mean value of 1-T as a function of cm energy compared to several models.

thresholds, since the heavy quarks are then at rest and fragment to give rather spherical events. The flattening of the curve above 30 GeV arises since the increased gluon bremsstrahlung balances the kinematic narrowing of the events that results from the higher jet energies.

b. Inclusive Single Particle Distributions

(i) Longitudinal Distributions

The inclusive differential energy distribution of charged particles has been measured by the HRS collaboration[15] over the full Z range. The result, which is shown in Fig. 6, indicates an approximately exponential fall in the range $0.25 \leq Z \leq 0.7$, a rise at low Z and a more rapid decrease from $Z \sim 0.7$ up to the kinematic limit. Precision measurements near $Z = 1$ are unique to e^+e^- experiments since the maximum value of the quark energy is specified. The data at high Z are consistent with the $Z^{-1}(1 - Z)^n$ behavior expected from dimensional counting arguments[16] for a first-order quark-meson transition. The fitted value of n for $Z > 0.5$ is 2.08 ± 0.21, as shown in Fig. 7, suggesting that the particles containing the leading quark dominate in this Z region. The data also agree well with the Lund fragmentation, whereas the Webber cluster model prediction of that time fell far below the data as a result of the low cluster mass cutoff: high momentum particles come from high mass clusters.

The folded rapidity distributions for all charged particles are shown in Fig. 8 for the MarkII[12], TASSO[13], and HRS[17] data. Pion masses were assigned to all particles in calculating rapidity. The TASSO data extend to higher Y values because of the 34 GeV energy. This effect is shown more clearly in Fig. 9(a), where the TASSO results at 14, 22, and 34 GeV are compared. In all cases, there is a broad maximum around Y = 1.5 and a shallow dip at Y = 0. Figure 9(b) shows the HRS data divided according to charged particle multiplicity n. The dip at Y = 0 seems to be associated with the higher values of n.

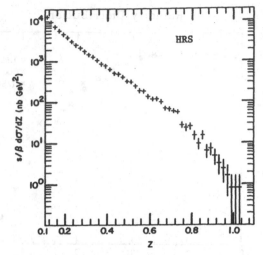

Fig. 6. Inclusive fragmentation function for all charged particles.

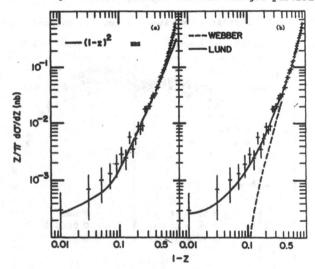

Fig. 7. (a) Invariant differential cross section of charged particles as a function of 1 - z together with the prediction from dimensional counting. (b) Invariant differential cross section of charged particles as a function of 1 - z, together with the predictions from the Lund Monte Carlo (full line) and the Webber Monte Carlo (dashed line).

Rapidity Distribution

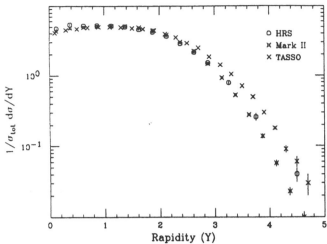

Fig. 8. Folded rapidity distribution for all charged particles.

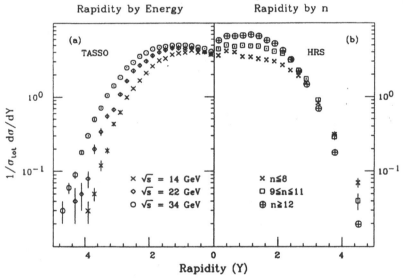

Fig. 9. (a) Variation of rapidity distribution with cm energy $\sqrt{s}$.
(b) Variation of rapidity distribution with charged particle
multiplicity n.

Similar measurements have also been made for many individual stable particles. The TPC detector has a good capability of doing such experiments[18] as the masses of the charged particles are well identified by the dE/dx determinations. Cerenkov counters and time-of-flight systems have been used in other detectors to identify the particle species in limited momentum ranges. Figure 10 shows a compilation of cross sections at 29 GeV up to $Z \sim 0.5$. The kaon and proton data have been shifted downwards by one and two orders of magnitude, respectively, for clarity in the display. The new TPC data are also shown as particle fractions in Fig. 11. The proton fraction seems to be constant at about 10%, whereas the K meson percentage rises to about 30% at $Z \simeq 0.35$ and then levels off.

The corresponding rapidity distributions for π, K, and p of Fig. 12 show that the peak near $Y = 1.5$ is associated with the K mesons, a point to which we shall return.

The shape of the fragmentation function at very low Z is of interest since such particles are, by definition, very soft. It has been suggested[19] that these particles may, in fact, reflect the characteristics of the soft partons - a kind of parton-hadron duality. The HRS[5] and the TPC[18] distributions in $Z/\sigma \, d\sigma/dZ$ are shown with a logarithmic Z scale in Fig. 13. The kinematic thresholds require that the data turn over at the appropriate mass values, but the low Z region is also depleted by the coherence of the low energy gluon radiation, which leads to a destructive interference. The position of the observed maximum, $Z \sim 0.05$ for the inclusive and the pion data, should be given by $(Q_0/\sqrt{s})^{1/2}$, where Q_0 is the cluster mass cutoff. Natural values of Q_0 of Λ_{QCD} or m_π lead to a maximum at $Z \sim 0.07$, close to that observed. The kaon and proton data must peak higher (0.13 for K, 0.18 for p) because of the larger masses. Calculations based on these ideas[19] give good qualitative representations of the data as seen by the lines in Fig. 13.

74

Charged π, K, p Cross Sections at 29 GeV

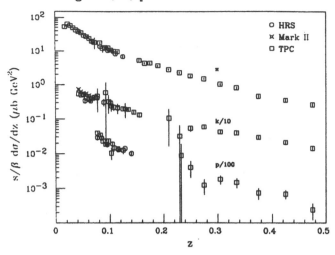

Fig. 10. Cross sections for production of charged pions, kaons, and protons at 29 GeV. The kaon (proton) data have been shifted down by a factor of 10(100) for clarity.

Particle Fractions %

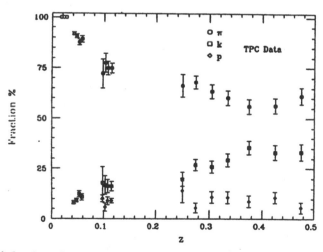

Fig. 11. Particle fractions as a function of Z at $\sqrt{s}$ = 29 GeV.

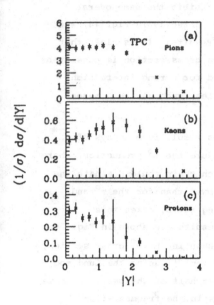

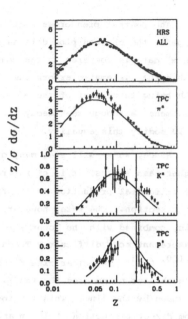

Fig. 12. Rapidity distributions for (a) pions, (b) kaons, and (c) protons at $\sqrt{s} = 29$ GeV.

Fig. 13. Behavior of the fragmentation functions at low Z compared to model predictions.

π°, $\pi^{\pm}$ Cross Sections

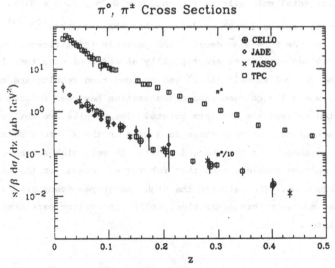

Fig. 14. Comparison of charged and neutral pion cross sections. The π° data have been shifted down by a factor of 10.

The neutral pion cross sections[20,21] exhibit the same overall shape as the charged pion data as can be seen directly in Fig. 14, where the π^o data is compared to the charged pion results. It is clear that within the systematic errors the charged pion cross section is consistent with being twice that of the π^o - as required for strong interaction phenomena, although weak decays of heavy states, such as the D mesons, could modify this equality.

Since, as we will see, many of the pions result from decays of higher mass states, it is of interest to measure the η^o production, where the primordial production in the fragmentation chain could dominate the cross section. The rate, however, is much lower than for the π^o and this, combined with the lower $\eta \rightarrow \gamma\gamma$ branching ratio, makes this measurement more difficult. The available results are shown in Fig. 15.[20,22,23] The clear rise for Z < 0.15, seen in the π^o cross section, is much less pronounced for the η. A prediction[22] using the Lund model is shown by the line. This fit indicates that half of the particles come from direct production of the η and η' mesons in the fragmentation chain. The latter has been directly measured by the MarkII collaboration with the results shown by the filled symbols in Fig. 15. The corresponding total multiplicity is $< n_{\eta'} > = 0.26 \pm 0.09 \pm 0.05$.

The cross section for K^o production has been measured[24-28] by reconstructing the $K^o_s = \pi^+ \pi^-$ decay. The particle identification is not in doubt, but the statistics are typically an order of magnitude lower than for the charged kaons. The K^o and charged kaon results are compared in Fig. 16, where the charged kaon cross section has been reduced by a factor of ten to separate the data points. The results are obviously in excellent agreement, although there is a rise in the K^o cross section at low Z that is absent for the charged kaons. At very high Z, the charged kaon cross section should exceed that for the K^o because of the stronger u quark production. The data for the high energy pseudoscalars is insufficient to check this prediction, unlike the vector particles discussed later.

$\eta,\ \eta'$ Cross Sections

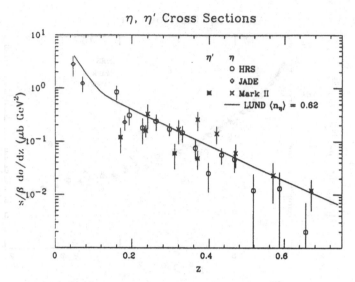

Fig. 15. Eta and eta prime meson cross sections near $\sqrt{s}$ = 30 GeV. The line shows the Lund prediction with $< n_\eta > = 0.62$. The best fit to the data corresponds to $< n_\eta > = 0.58 \pm 0.10$.

$K^0,\ K^\pm$ Cross Sections

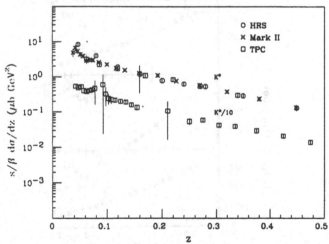

Fig. 16. Comparison of K^0 and $K^\pm$ cross sections. The latter have been shifted down by a factor of 10.

p_T^2 Distribution

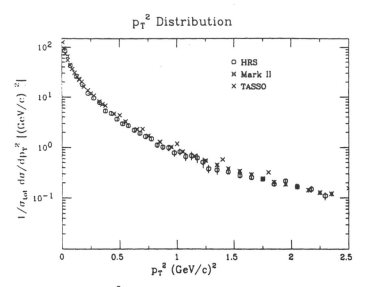

Fig. 17. Distribution of p_T^2 for all charged particles near $\sqrt{s}$ = 30 GeV.

Transverse Distributions

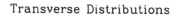

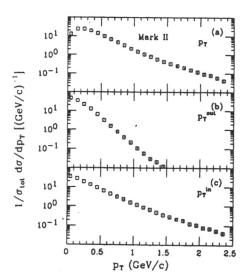

Fig. 18. Transverse momentum distributions: (a) total, (b) out of event plane, and (c) in the event plane.

(ii) <u>Transverse Distributions</u>

The inclusive transverse momentum squared distributions relative to the jet axes are shown in Fig. 17. After a sharp fall, there is a long tail extending to $p_T^2 \sim 2.5$ $(\text{GeV/c})^2$. Again, the TASSO data points at 34 GeV, are a little above the 29 GeV results but, although the HRS uses the thrust axis and MarkII and TASSO the sphericity axis, the agreement between the three data sets is remarkable. The overall P_T distribution for the MarkII data is shown in Fig. 18(a), and divided into two components: out of the plane of the event in Fig. 18(b) and in the plane of the event in Fig. 18(c). The effects of gluon emission that broadens the transverse distribution in the plane is evident.

The heavier particles show a wider p_T distribution, as is also the case with other reactions. For example, the TPC group[26] report mean p_T^2 values of 0.30 ± 0.02 GeV/c^2 for $\pi^\pm$, 0.51 ± 0.20 $(\text{GeV/c})^2$ for K^0 and 0.57 ± 0.08 $(\text{GeV/c})^2$ for the K^{*0}.

c.   <u>Meson Resonance Production</u>

The production of the vector mesons ρ^0, K^{*0}, $K^{*\pm}$, and ϕ has been measured by several detectors. The peak corresponding to the charged K^* mesons can be observed over a simple background, as shown in Fig. 19, which gives the HRS mass distribution[29]; the resulting fragmentation function is compared to the JADE[30] and MarkII results[31] in the upper data set of Fig. 20.

The neutral K^* and ρ^0 cross sections are harder to measure, although the statistics are much higher, since, in the absence of K:π separation, the two decays $K^* \rightarrow K\pi$ and $\rho \rightarrow \pi\pi$ have significant kinematic overlaps under the K:π interchange. There are also serious reflections from the decays of other resonances, not all of which are well determined. However, several groups[26,30-33] have reported analyses of such data. Figure 21 shows the ρ^0 data: the K^{*0} results (divided by 10) are the lower data set on Fig. 20. There is a good agreement between the overall charged and neutral K^* cross sections, although the cross section for $K^{*\pm}$ systematically exceeds that for the K^{*0} in the higher Z range.

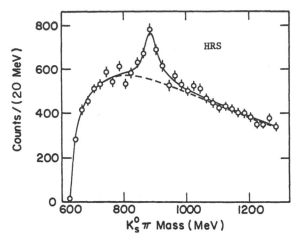

Fig. 19. $K^0\pi^{\pm}$ mass distribution at $\sqrt{s}$ = 29 GeV.

K* Cross Sections

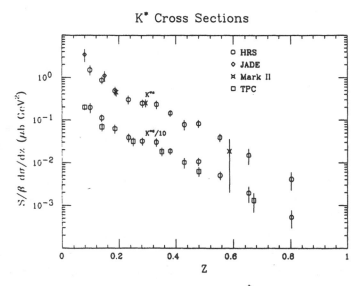

Fig. 20. Cross sections for charged and neutral $K^*(890)$ production. The K^{*0} data has been shifted down by a factor of 10.

ρ° Cross Section

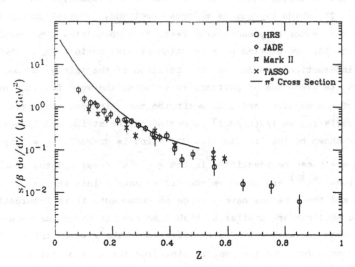

Fig. 21. Cross sections for ρ° production. The line shows the Z variation of the π° cross section.

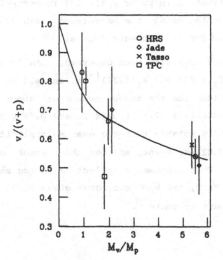

Fig. 22. The V/(V + P) ratio as a function of the mass ratio M_V/M_P for charm, strange, and non-strange mesons.

The vector meson cross sections lack the peak at low Z that characterizes the pseudoscalar data, but at high Z, the cross sections for vector meson are comparable to, or larger than, those for the pseudoscalars, as seen by the line in Fig. 21, which represents π^0 cross sections. Several groups have determined the vector to pseudoscalar ratio from such data. The results, shown in Fig. 22, are plotted as a function of mass ratio M_V/M_P. The quark spin-spin interaction spreads the wave function of the triplet meson, as compared to its pseudoscalar partner, so reducing the wave function overlap with that of the $q\bar{q}$ pair, and thus modifying the spin counting ratio of V/V+P = 3/4. A variation as $1/3(M_V/M_P)^\gamma$, expected on the string model, agrees with the data as shown by the line in Fig. 22, which is drawn for $\gamma = 0.55$.

The ϕ meson can be identified in the $\phi \rightarrow K^+K^-$ decay and the resulting cross sections[34,35] are shown as the filled data points in Fig. 23. Since both the ϕ and the η mesons have a large $s\bar{s}$ component, it is interesting that their cross sections are similar at high Z as seen in the comparison of Fig. 23, even though the masses are quite different. The rise of the η cross section at low Z reflects the contribution from the $\eta' \rightarrow \eta$ decay.

No evidence for vector meson alignment has been reported and all groups assume a flat distribution in polar angle (θ) to correct for the decays that are removed by selection cuts. The background peaks along the jet direction, which corresponds to forward-backward decays.

Other resonances have also been observed, notably the scalar $a_0(980)$ and the tensor mesons $f_2(1270)$ and $K_2(1430)$[36], as well as the $\eta'(958)$.[23] The fragmentation functions for the tensor mesons are shown in Fig. 24. The mean multiplicities are similar: 0.14 ± 0.04 f_2 and 0.12 ± 0.06 $K_2^*(1430)$ per event at $\sqrt{s} = 29$ GeV. It is notable that the mean multiplicity of the scalar $a_0(980)$ is 0.05 ± 0.02 per event, smaller than tensor meson values even though the a_0 is lighter. This presumably reflects a strong $s\bar{s}$ component in the a_0 wave function. In fact, the Webber program gives 0.033 $\eta'(985)$ particles per event, assuming a pure $s\bar{s}$ content.

ϕ, η Cross Sections

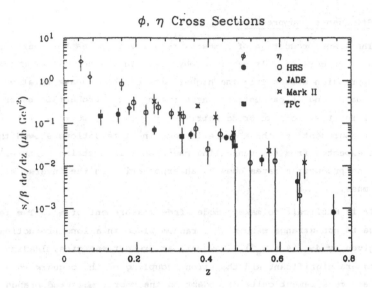

Fig. 23. Cross sections for ϕ and η production at $\sqrt{s}$ = 29 GeV.

Tensor Meson Cross Sections

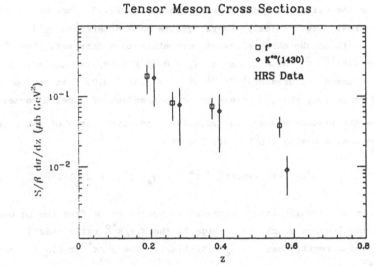

Fig. 24. Cross sections for production of the tensor mesons ρ^0 and $K^*(1430)$ at $\sqrt{s}$ = 29 GeV.

d. Strangeness Suppression

The lower production of K mesons relative to pions has been known since the earliest days of multiparticle physics. In the context of current ideas, the suppression arises from the higher mass of the s quark relative to the u and d quarks and is parametrized as a ratio of the probabilities of making an $s\bar{s}$ pair relative to a $u\bar{u}$ or $d\bar{d}$ pair: $\gamma_s = \frac{P_s}{P_u} = \frac{P_s}{P_d}$, where $P_s + P_u + P_d = 1$. Early measurements of the K^o rate in neutrino interactions showed that $\gamma_s \sim$ 1/3 as expected from the tunnelling phenomenon in a string picture,[37] in which the production rates goes as an exponential in the inverse square of the quark mass.

It is difficult to make a model-free measurement of γ_s. The ratio of strange to non-strange mesons at a random place in a long production chain is just given by $4\gamma_s/(4 + \gamma_s^2)$. However, at present energies, leading particle effects are significant and the strong coupling of the c quark to strangeness makes any measurement quite dependent on the poorly measured branching ratios for the heavy-quark states. In addition, the different vector-to-pseudoscalar ratios for the different quark flavors, shown in Fig. 22, brings in other uncertainties. Nevertheless, several groups have determined γ_s in the context of fitting all of the other relevant parameters to a data set. The K^o inclusive data[25,27] gave a mean of $\gamma_s = 0.33 \pm 0.03$, in good agreement with the vector meson measurements[26,28] of $\gamma_s = 0.34 \pm 0.03$. At the ten percent level of precision, then, γ_s seems to be independent of model uncertainties.

From the leading quarks, the relative production rates of the ρ and K^* vector mesons, assuming $\sigma(\omega^o) = \sigma(\rho^o)$ are:

$$\rho^o : K^{+o} : K^{*\pm} :: 2.5 f_V^{u,d} : f_V^s (1 + \gamma_s) : f_V^s (1 + 4\gamma_s) \ .$$

where f_V is the probability of producing a vector meson from the $q\bar{q}$ pair that includes the leading quark. So at high Z, the $K^{*\pm}/K^{*o}$ ratio should be greater than one and directly measures γ_s. Similarly, the ρ^o/K^{*o} ratio is sensitive to $f_V^{u,d}/f_V^s$.

The updated HRS data[33] for ρ^o and K^{*o} production are shown on a logarithmic (1-Z) scale in Fig. 25. Above Z ~ 0.4 to 0.5, the data are

K*⁰ ρ⁰ Comparison

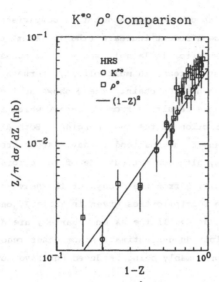

Fig. 25. Comparison of ρ⁰ and K*⁰ production at high Z.

consistent with the $(1 - Z)^2$ behavior expected for first-order quark:meson transitions. The ρ⁰ (and K*±) cross sections are also systematically larger than the K*⁰ as expected. However, data of higher statistics are required to measure γ_s and the vector fractions with higher precision.

Mean Multiplicities: Comparison with Data at 10 GeV

The total mean multiplicities for the mesons, as measured by the CLEO[38] and ARGUS[39] experiments in the continuum near 10 GeV and by the PEP and PETRA groups near 30 GeV, are given in Table I and shown in Fig. 26(a). The 30 GeV results in this table are an update of those given by Saxon.[2] In order to calculate the total multiplicity from the observed fragmentation data, it is necessary to extrapolate to threshold using a fit to the energy distribution given by, say, the Lund Monte Carlo model. The correction is usually at the 10-20% level but can be larger if the low Z region is not measured, or if the shape of the fragmentation function is not known.[22] The errors shown in Fig. 26 include an estimate of the systematic errors arising from these uncertainties.

Several comments should be made about this comparison: The multiplicity of pions and kaons show a modest increase (the 10 GeV π^o cross section seems to be low, although the error is large), whereas the non-strange resonances ρ^o and f^o show the largest increase in multiplicity, perhaps indicating dominant production in the fragmentation chain. The ϕ shows no increase, consistent with many of these mesons resulting from D_s decay, which is produced with an approximately equal fraction at the two energies. Both at 10 GeV and 30 GeV, the charged K mesons and K^* mesons tend to have larger cross sections than their neutral partners, although not outside of the errors.

If the pions resulting from the decays of the heavier mesons and baryons are estimated from the multiplicities given in Table I, one finds that fewer than one half and, perhaps as little as one quarter, are directly produced. The one half is certainly an overestimate since other resonances such as the 1^+ and 2^+ mesons are presumably being produced but have not yet been observed.

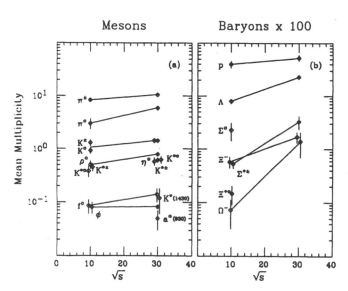

Fig. 26. Comparison of mean multiplicities at 10 GeV and 30 GeV: (a) mesons, (b) baryons.

Table I. Comparison of Mean Multiplicities

Particle	Mean Multiplicity in Particles Per Event		
	$\sqrt{s}$ = 10 GeV		$\sqrt{s}$ = 30 GeV
Charged	10.0	± 0.5	12.87 ± 0.03 ± 0.30
$\pi^{\pm}$	8.3	± 0.4	10.86 ± 0.2
$K^{\pm}$	1.3	± 0.2	1.47 ± 0.06
$p^{\pm}$	0.4	± 0.06	0.54 ± 0.07
π^{o}	3.0	± 0.7	5.9 ± 0.3
η^{o}			0.58 ± 0.10
η'			0.26 ± 0.10
K^{o}	0.92	± 0.12	1.44 ± 0.06
Λ	0.08	± 0.008	0.23 ± 0.01
Σ^{o}	0.0229	± 0.0085	
Ξ^{-}	0.0058	± 0.0014	0.017 ± 0.004
Ω^{-}	0.00072	± 0.0004	0.014 ± 0.007
ρ^{o}	0.5	± 0.09	0.81 ± 0.04
$K^{*\pm}$	0.45	± 0.08	0.62 ± 0.06
K^{*o}	0.38	± 0.09	0.54 ± 0.03
ϕ	0.08	± 0.02	0.082 ± 0.006
$\Sigma^{*}(1385)$	0.0053	± 0.001	0.033 ± 0.008
$\Xi^{*}(1530)$	0.00146	± 0.00056	
$f_2(1270)$	0.085	± 0.025	0.14 ± 0.04
$K_2^{*}(1430)$			0.12 ± 0.06
$a_o(980)$			0.05 ± 0.02

Heavy Quark Fragmentation

In contrast to the rapidly falling energy distribution that characterizes all of the fragmentation functions so far presented, mesons containing the heavy quarks, c and b, show a hard fragmentation. The situation here is unique in that one can select the meson containing the leading quark since the rate for additional $c\bar{c}$ or $b\bar{b}$ pair production is negligible. Most of the detectors[40] taking data near 30 GeV have measured the production of the $D^{*+}(2010)$ in the $D^{*+} \rightarrow D^{o}\pi^{+}$ decay with the D^{o} subsequently decaying to $K^{-}\pi^{+}$, $K^{-}\pi^{+}\pi^{+}\pi^{-}$ or $K^{-}\pi^{+}\pi^{o}$. The low Q value of the decay of 5.8 MeV leads to a well measured peak in the $D^{*+}-D^{o}$ mass difference distribution. The $D^{*o} \rightarrow \gamma D^{o}$ decay has also been observed[41] using this technique.

Similar results have been obtained near 10 GeV by both the ARGUS[42] and CLEO[43] detector collaborations. At this lower energy, the individual D^{+} and D^{o} mesons can be reconstructed in several decay modes, whereas at 29 GeV the high momentum precision that characterizes the HRS is necessary for such a reconstruction.

A second unique feature of the charm sector is that it allows measurement of the absolute cross section for $e^{+}e^{-} \rightarrow c\bar{c}$ since all such charm quarks eventually give a D^{+}, D^{o}, or D_{s}^{+} meson or a Λ_{c} baryon.

To measure the total cross section, some absolute branching ratio must be known and these have only been measured for the D^{+} and D^{o} mesons and, even here, recent measurements of the crucial $D^{o} \rightarrow K^{-}\pi^{+}$ branching ratio have varied by a large factor. Using the latest values[44] of $B(D^{o} \rightarrow K^{-}\pi^{+})$ of (4.2 ± 0.56)% and $B(D^{+} \rightarrow K^{-}\pi^{+}\pi^{+})$ of (9.1 ± 1.4)%, the HRS collaboration[45] measured $\sigma(D^{o} + D^{+}) = 0.28 \pm 0.04$ nb, which, when compared to the expected value of $\sigma_{charm} = 0.42$ nb, gives 67 ± 10% for the fraction of charm quarks that lead to D mesons. A similar determination by the CLEO collaboration[43] at 10 GeV gave $\sigma(D^{o} + D^{+}) = 1.76 \pm 0.23$ nb, compared to an expected total charm cross section of 2.46 ± 0.20 nb or a D meson fraction of (71 ± 12)%. These results leave (31 ± 8)% of the charm quarks to be accounted for in D_{s} mesons and Λ_{c} baryons, in agreement with 25% expected from analyses of the non-charm sector with a strange quark suppression factor of $\gamma_{s} \sim 0.3$ and a diquark/quark ratio of ~ 8%.

Figure 27 compares the D^* fragmentation function in terms of the light cone variable x^+ as measured at 10 GeV and at 30 GeV. The cross sections reported by all of the experiments have been corrected to the new MarkIII[42] decay branching ratios, although the errors shown do not include the branching ratio uncertainties.

The hard fragmentation observed in these data is in strong contrast to the results for light quarks, although in the latter case, no unambiguous separation of the particles containing the leading quark has so far proved possible. For the 30 GeV data, charm mesons from the decay of b quark states are included and populate the low x^+ region.

The simplest understanding of these data comes from noting that the high-mass of the charm quark means that a light antiquark can be picked up from the

D^* Cross Sections

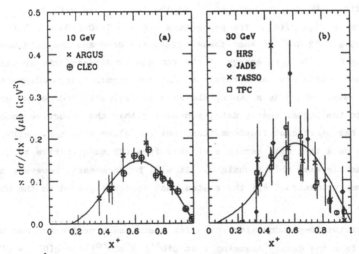

Fig. 27. (a) D^* cross sections at 10 GeV fitted to the Lund form shown as the line, (b) D^* cross sections near 30 GeV. The line shows the evolution up to 30 GeV of the fit to the 10 GeV data.

vacuum without slowing down the heavier quark. A parametrization, based on the application of the uncertainty relation to the $Q \rightarrow Mq$ transition,[46] leads to the prediction that:

$$D(Z) \sim \frac{1}{Z} \left(\frac{1}{1 - \frac{1}{Z} - \frac{\varepsilon}{1 - Z}} \right)^2 ,$$

where the parameter ε is related to the quark mass ratio via $\varepsilon \sim (m_q/m_Q)^2$. This form, which goes as $ZD(Z) \sim (1 - Z)^2$ as $Z \rightarrow 1$, represented the early data quite well. However, as discussed by Bethke[46], effects of QED and QCD radiation must be understood before drawing hard conclusions from comparing predictions with observed data.

Several other forms have also been suggested. The parametrization used in the Lund model:

$$D(x^+) \sim \frac{(1-x^+)^a}{x^+} \exp \frac{-bM_T^2}{x^+} ,$$

has been fitted by the CLEO group[43] to the 10 GeV data with the result shown by the line in Fig. 27(a). The parameters are a = 0.60 ± 0.10 ± 0.04 and b = 0.52 ± 0.05 ± 0.02 GeV^{-2}. When these values are used and the Lund model is run at 30.4 GeV, the mean energy of the PEP and PETRA data, the spectrum broadens and shifts down as a result of the additional gluon radiation. The result, however, which is shown by the line in Fig. 27(b), does not give a good fit to the higher energy data, suggesting that the structive function evolution has other contributions than just QCD gluon emission. With the high statistics data that will become available from LEP quantitative tests of QCD may be possible with this technique. It will be necessary, however, to have complete measurements of all the states that eventually lead to the observed D^* mesons.

The vector-to-pseudoscalar ratio for charm meson production can be estimated from the data. Assuming that $\sigma(D^{*+}) = \sigma(D^{*0})$ and $\sigma(D^+) = \sigma(D^0)$, the CLEO group[43] measure V/(V + P) = 0.85 ± 0.11 ± 0.17: the HRS collaboration[45] found a similar result of 0.83 ± 0.17: values in agreement with the simple spin counting result of 3/4.

The ARGUS group[48] have reported a state at a mass of 2420 MeV that decays to $D^*\pi$. About 20% of the observed D^*'s come from this source. The spin parity of the state is unknown; 0^+, 1^+, and 2^+ are obvious possibilities: the 0^+ state will only decay to $D\pi$. In any case, a correction to the V/(V + P) ratio will be necessary. It is likely to be small, but no hard estimate can be made at this time.

Another way of investigating such details of the quark-meson transition comes from exploiting the spin state of the vector mesons. The HRS collaboration has searched for alignment of the D^* mesons.[49] Adding all of the data for the three D meson decay modes, $K^-\pi^+$, $K^-\pi^+\pi^+\pi^-$, and $K^-\pi^+\pi^0$ and using the region with $Z > 0.4$, where there is little background, density matrix elements of $\rho_{00} = 0.37 \pm 0.04$, $\rho_{1-1} = 0.04 \pm 0.03$, and $Re\rho_{10} = 0.00 \pm 0.01$ were measured. The alignment parameter η, defined as $\eta = \frac{1}{2}(3\rho_{00}-1)$, has the value 0.06 ± 0.06. From simple statistical arguments[6], η is related to the V/P ratio by the relationship $\eta = \frac{1}{4}\left(1 - \frac{3P}{V}\right)$, so that the alignment measurement corresponds to V/(V + P) = 0.80 ± 0.05, a more precise result than the direct determination. This value is also shown in Fig. 22.

Several groups[50] have observed the $c\bar{s}$ mesons D_s^+ in the $\phi\pi^+$, $\phi\pi^+\pi^+\pi^-$, and K^*K decay modes, in addition to the vector state D_s^{*+}. The data is weaker than for the D mesons, but generally in accord with expectations, although no absolute decay branching ratios are known.

No B mesons have been observed in the continuum so the energy distribution of these 5 GeV objects has not been directly measured. Indirect observations of the B mesons using semileptonic decays have been reported.[2] The e or μ leptons from B decay populate the high transverse momentum region. The results indicate an even harder fragmentation than for the D mesons, as expected from the large b quark mass.

Properties of Jets with Identified Leading Flavor

Since the basic reaction in e^+e^- annihilation to hadrons is $e^+e^- \rightarrow q_f\bar{q}_f$, where q_f is a quark of definite flavor, it is possible to study the fragmentation of quarks of a known flavor by tagging the flavor of one jet and looking at the properties of the second jet.[51]

Several techniques have been used: a D^* of high momentum selects charm quark jets and a single charged particle with high fractional energy ($Z \gtrsim 0.7$) selects light quark jets (u, d, s). Although, as we have discussed, there is a hard fragmentation for particles containing the c and b quarks, the high decay multiplicity of the resulting c and b mesons and baryons means that the decay products have a relatively low momentum. The light quark by contrast can lead directly to a high momentum π or K meson. Finally, direct leptons in certain ranges of p and p_T give jets enriched in c and b quarks. In all cases, there is a tradeoff between selection efficiency and purity, and the final samples are typically a few hundred events compared to total of $\sim 10^5$ events recorded in the experiment. This means that only the most general characteristics of the events have been measured.

The TASSO collaboration isolated a sample of 82 charm jets, with an estimated background of 16 ± 6 events. The events were chosen to have a D^* with $Z > 0.5$ to reduce background and to exclude b quark decays. The characteristics of the jet opposite to the D^* were compared to the properties of the average jets. Within errors, no differences were found between the two samples.

The particles in the trigger D^* jet, excluding the D^* decay products themselves, had, an average energy of 6.2 GeV. The characteristics of these particles were found to be similar to the average jet in e^+e^- annihilation at $\sqrt{s} = 14$ GeV, i.e., a 7 GeV jet. The mean charged multiplicities were $< n > = 4.45 \pm 0.03 \pm 0.2$ for the 7 GeV jets as compared to $4.4 \pm 0.6 \pm 0.4$ for the residual particles in the high energy D^* jet. Figure 28 shows the rapidity distribution of the residual particles compared to that of the average jet at 34.4 GeV (histogram) and at 14 GeV (dashed line). The latter passes through the data points.

The HRS collaboration selected 111 charm events using the D^* technique and 287 events coming from light quark fragmentation by requiring $Z > 0.7$. The latter had an estimated background of 11%. Some differences were observed in the charged multiplicity and the mean longitudinal momentum: $< n > = 6.6 \pm$

0.2 and $< p_L > = 1.38 \pm 0.06$ GeV/c for c jets as compared to $< n > = 5.8 \pm 0.1$ and $< p_L > = 1.52 \pm 0.04$ GeV/c for (u,d,s) jets. The rapidity distributions are also a little different as shown in Figs. 29(a) and 29(b). The broad peak near Y = 1.5 in the inclusive Y distribution of Fig. 8 is clearly associated with the peaking seen in the dashed–dotted curves in Fig. 29, which show the prediction for the rapidity distribution for the decay products of the primary meson, as is also the association with the higher multiplicities seen in Fig. 29(b), and the peak in the K meson data of Fig. 12(b).

The DELCO group used a high transverse momentum electron to select 146 b-quark events, including a background of $17 \pm 6\%$. The mean charged multiplicity for b quark jets was 7.16 ± 0.46 and the mean longitudinal momentum was 1.06 ± 0.04 GeV/c, again different than for c or (u,d,s) jets.

Figure 29(c) shows the rapidity distribution. Because of the higher decay multiplicity, the peaking near Y = 1.5 is even stronger in this case than for the charm jet case. These characteristics of the rapidity distributions of heavy quark jets may be used in the future in searching for higher mass quarks.[52]

Baryon Production

Baryons are copiously produced in hard processes, including high energy e^+e^- annihilation. The production mechanism is a mystery that continues to provides an important challenge to model builders. Perhaps the most striking result is the report of the ARGUS collaboration[53] of antideuteron production in a limited momentum range, but with a rate of $(1.6 {+1.0 \atop -0.7})10^{-5}$ per event. By contrast, the proton production is at the five percent level, as presented in Table I. When interpreted in the context of the Lund model, the earlier TPC data[54] on proton production corresponded to a diquark/quark ratio of 0.078 ± 0.005.

The Λ hyperon is experimentally the easiest baryon to measure, and data are available[27,38,55-58] from most of the detectors at both 10 GeV and 30 GeV. The fragmentation functions for the Λ, as well as the Ξ^- hyperon,[59,60]

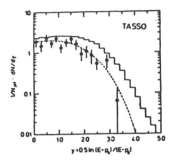

Fig. 28. Rapidity distribution of residual particles in charm jets (data points) compared to $e^+e^- \rightarrow$ anything at the appropriate lower energy of 14 GeV (dashed line) and at the full energy of 34.4 GeV (histogram).

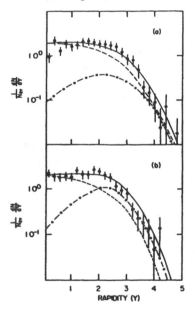

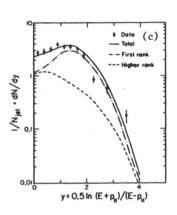

Fig. 29. Rapidity distributions for (a) u,d,s quark jets, (b) c-quark jets, (c) b-quark jets. The dashed-dotted lines show the prediction for the Y distribution of particles from the decay of the heavy hadrons.

shown in Fig. 30, are quite similar to those of the mesons in that they fall strongly as Z is increased, and at 30 GeV there is a rise of the Λ cross section at small Z. The mean multiplicities of the Λ and Ξ^- increase by the same factor of 2.9 in going from 10 GeV to 30 GeV, whereas for the proton, the multiplicity increase is only about 35%.

One obvious measure of the nature of the diquarks that must be involved in baryon production is to look at the fragmentation of the Λ_c. This state has not been observed above $\sqrt{s}$ = 10 GeV: the ARGUS[61] and CLEO[43] data are shown in Fig. 31. It is clear that the fragmentation shape is very similar to that of the charmed mesons. In fact, a fit to the Peterson form gives ϵ = 0.24 ± 0.05, as compared to $\epsilon \simeq$ 0.17 for fits to the charm meson data at the same energy. This implies a similar effective mass for the diquark as the u or d quark. The model of DeGrand[62], in which there are two independent quark vertices, implies a softer fragmentation, shown by the dashed line, which is in disagreement with the data.

The Λ and Ξ^- hyperons can be produced by attaching a strange diquark, say sd to a u or to an s quark, respectively. The Λ also has a contribution from ud, plus s. Any additional suppression of strange diquarks is specified in the Lund model by the parameter δ = (us/ud)/(s/d), which can be measured by comparing the Λ and Ξ multiplicities with model predictions. However, the Λ cross section has a major contribution from the $\Lambda_c \to \Lambda$ decay, which is not well known. The HRS collaboration[58] have analyzed the Λ rapidity distribution, shown in Fig. 32, to separate the direct production and the Λ_c decay contribution. The latter is shown (multiplied by 10) in the figure as the dashed-dotted line for a 23% branching ratio of $\Lambda_c \to \Lambda$. The solid curve corresponds to δ = 0.89 ± 0.10; there is, however, a large systematic uncertainty connected with the 50% error on the Λ_c branching ratio. The 29 GeV Ξ^- rate[54,59] of 0.017 ± 0.004 corresponds to δ = 0.60 ± 0.15. This is a less model-dependent result.

There is some suggestion from the MarkII analysis that there is an enhanced Λ production in three-jet events. There is known to be about twice as much baryon production from the γ bound states[60], as compared to the

96

Λ, Ξ^- Cross Sections

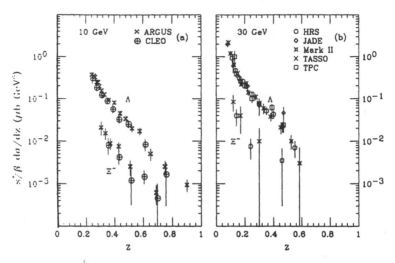

Fig. 30. Cross sections for Λ and Ξ^- production at (a) 10 GeV and (b) 30 GeV.

Λ_c Fragmentation Function

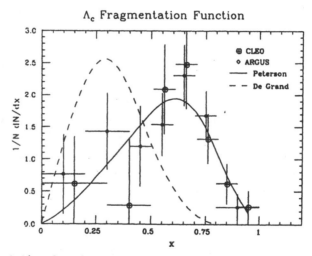

Fig. 31. Fragmentation function for Λ_c production at 10 GeV compared to the Peterson function (full line) and the DeGrand prediction (dashed line).

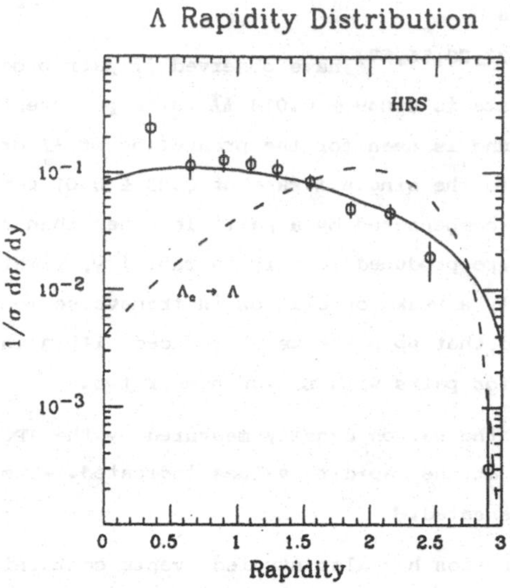

Λ Rapidity Distribution

Fig. 32. Rapidity distribution for Λ production at $\sqrt{s}$ = 29 GeV compared
to Lund model (full line). The dashed line shows ten times
the $\Lambda_c \to \Lambda$ contribution.

continuum near 10 GeV, as expected for a $\gamma \to 3g$ decay, each gluon in turn
giving a $q\bar{q}$ pair.

Measurements of the $\Sigma^*(1385)$ are sensitive to the question of spin one
diquarks. Table I shows that the cross section for this state increases
strongly from 10 to 30 GeV. With $\delta = 0.6$ and a spin one diquark suppression
factor of $\alpha = 0.05$, the Lund model predicts 0.023 $\Sigma^*(1385)$ hyperons per events
with $0.1 < X < 0.8$, as compared to 0.024 ± 0.007 measured by the HRS
collaboration at 29 GeV.

The Ω^- production measures the ss spin one diquarks. The high value of
0.014 ± 0.007 Ω^- per event reported by the MarkII collaboration[63] is
anomalous in that it is equal, within errors, to the Ξ^- rate. The result is
presumably connected with an upward statistical fluctuation, particularly
since the ARGUS measurement[61] shows no such effect.

98

Rapidity Correlations

Several groups[27,28,56,57] have observed $\Lambda\bar\Lambda$ pair production. The mean of the HRS and TPC rate is 0.049 ± 0.014 $\Lambda\bar\Lambda$ pairs per event at 29 GeV. No signal above background is seen for the production of $\Lambda\Lambda$ or $\bar\Lambda\bar\Lambda$ pairs. This $\Lambda\bar\Lambda$ rate, when compared to the single Λ rate of 0.23 ± 0.01 per event, shows that the baryon number is compensated by a particle other than a Λ one half of the time. The $\Lambda\bar\Lambda$ pairs are produced locally in rapidity, within 1 to 2 units, although there is only a weak correlation in transverse momentum. The TASSO group[65] also showed that $p\bar p$ pairs were produced within the same jet: no signal was observed for pairs with baryon number two.

Figure 33 shows the baryon density measured by the TPC group[64], relative to a $\bar p$ or $\bar\Lambda$ at the rapidity values indicated. The baryon number is clearly locally compensated.

The TPC collaboration has also studied events containing an identified proton as well as a Λ. For $p\bar p$ pairs, they find the distribution in the angle between the baryon momentum vector and the sphericity axis (θ^*) to peak near $\cos\theta^* = 1$. If the baryon pairs come from the decay of a color neutral cluster, this distribution should be flat, whereas a peaking along the jet axis is expected for a local qq-$\bar q\bar q$ color flux tube. The updated TPC data[64], shown in Fig. 34, clearly favors the string over the cluster model.

The baryon production data also suggests a significant contribution from a so-called popcorn mechanism in which a meson is produced between the $B\bar B$ pair.

Short-range rapidity correlations have been seen in the compensation probabilities for charge[66] and flavor.[67] In addition, since pairs of quarks of opposite flavor are produced in the primary annihilation, a long-range correlation is also found. This is particularly strong for K^+:K^- pairs as observed by the TPC group.[67]

Measurements of correlations in rapidity are a standard technique for investigating the details of multiparticle reactions. The two-particle rapidity correlation, R, measured by the TASSO collaboration is compared to pp

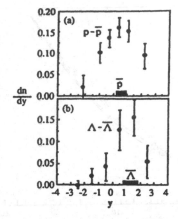

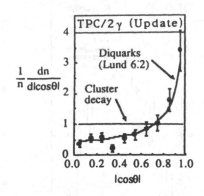

Fig. 33. (a) Net baryon number density observed in e^+e^- annihilation events with an antiproton near y ≈ 0.6. The baryon number density is defined as the number of protons per event and rapidity minus the number of antiprotons. (b) As (a) but using lambdas, and for a "trigger"-antilambda.

Fig. 34. Distribution in |cosθ| of proton-antiproton pairs produced in e^+e^- annihilation at √s = 29 GeV, after background subtraction. θ is the angle between the proton momentum and the jet axis, measured in the proton-antiproton rest frame. Lines give model predictions for proton production via a diquark mechanism and via cluster decay.

inelastic data in Fig. 35. The overall shapes are similar although the e^+e^- correlation is stonger. Such comparisons are difficult to quantify because of the strong leading particle effects of the hadronic data. In the future, a comparison of correlations in gluon jets from hard hadronic collisions and quark jets from e^+e^- annihilation could be illuminating.

Multiplicity Distributions

The mean charged particle multiplicities have been measured by many experiments. The results are shown in Fig. 36. The solid line corresponds to < n > = 3.33 - 0.4 lnS + 0.26 $\ln^2$S. The simple function < n > = 2.3 $S^{1/4}$, which would be expected from phase space alone, also represents the data quite well as shown by the dashed line in Fig. 36.

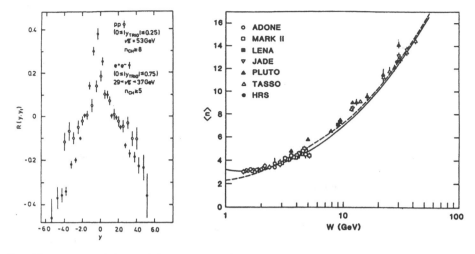

Fig. 35. Two-particle rapidity correlations at $\sqrt{s}$ = 34 GeV
compared to pp interactions.

Fig. 36. Mean charged multiplicity
$< n >$ as a function of center-
of-mass energy W. The full
(dashed) line shows an a +
$b\ln(W^2) + c\, pn^2(W^2)\, (\alpha W^{1/2})$
variation with the values of
the constants given in the
text.

The multiplicity distribution for the TASSO [13] and HRS [17] data is
shown in KNO form in Fig. 37. There is approximate scaling in the variable
$z = \dfrac{n}{< n >}$, although the small energy span available and the differences
between the data from the two experiments precludes any definitive
conclusion. If the events are divided by a plane perpendicular to the thrust
axis, similar distributions can be measured for each of the two jets that make
up almost all of the events.

The shape of the multiplicity distribution can be characterized by the
moments. Some values for the 29 GeV HRS data, such as the dispersion and f_2
moment are given in Table II. The values are close to those expected for a
Poisson: $f_2 = 0$, $D_2 = \sqrt{n}$ and $1/k$, for the negative binomial fit, equal to
zero. The variation of the dispersion with the e^+e^- center-of-mass energy is

given in Fig. 38. The line shows a $\sqrt{n}$ variation expected for a Poisson at all energies. It is clear that the tendency of the higher energy data is to be above the line. Precision multiplicity measurements at the highest energies are needed to clarify this important point.[68]

Table II. Charged Particle Multiplicity Parameters at 29 GeV

Variable	Whole Event	Single Jet
$< n >$	$12.87 \pm 0.03 \pm 0.03$	$6.43 \pm 0.02 \pm 0.15$
D_2	$3.67 \pm 0.02 \pm 0.18$	$2.55 \pm 0.02 \pm 0.13$
f_2	$0.60 \pm 0.02 \pm 0.18$	$-0.07 \pm 0.02 + 0.13$
$< n >/D_2$	3.51 ± 0.18	2.52 ± 0.13
K GD fit	12.3	7.11
$1/k$ NB fit	3.9×10^{-3}	1.4×10^{-4}

The widths of the single jet and complete event multiplicities measured at 29 GeV are in $\sqrt{2}$ ratio, expected if the total multiplicity, is just the folding of those of the two jets. This can be checked by measuring the forward:backward multiplicity correlations; the measurements[17,69] show no correlation within errors . For the linear relationship $< n_B > = a + b\, n_F$, b values of -0.001 ± 0.015 for the unselected HRS data and $+0.002 \pm 0.006$ for the data with $|Y| > 1$ are obtained. A small positive correlation is expected since, as we have discussed, $< n >$ is somewhat larger for c and b quark jets than for the light quarks (u,d,s) and the quarks are produced in pairs. The ratio of the dispersion for the whole event D_2 and that for a single jet $\mathcal{D}_2$ is related to b by the equation

$$\mathcal{D}_2 = \frac{D_2}{\sqrt{2(1+b)}} \quad .$$

The data of Table II yield b = $0.04 \pm 0.02 \pm 0.10$, in agreement with the direct measurements.

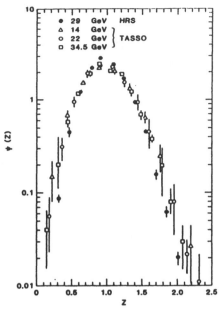

Fig. 37. Charged particle multiplicity distribution for e^+e^- annihilation in KNO form.

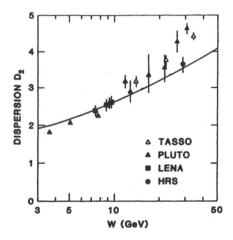

Fig. 38. Variation of the dispersion D_2 with center-of-mass energy W. The line shows a $D_2 \propto \sqrt{n}$ variation.

The multiplicity distributions in selected rapidity intervals has been measured by the HRS collaboration[70] with the result shown in Fig. 39. The negative binomial distribution represents this data quite well as seen by the histograms. The C_q moments, shown in Fig. 40, indicate a smooth variation as the rapidity span is changed. The higher moments grow significantly as $|Y|$ is decreased.

These and related data from other reactions have been interpreted by many authors from several different points of view.[71] One appealing suggestion of Giovannini and Van Hove[71] is that the wide occurrence of the negative binomial expression in high energy multiplicity distributions suggests a cascading process. They call such hadron cascades that come from a single parent, a clan. They are perhaps bremsstrahlung gluon jets, at least in the central regions of rapidity. The analysis of the e^+e^- data with this idea gives a mean multiplicity per clan $< n >_c$ of just over one charged particle. The value is controlled by a parameter $\beta = \frac{< n >}{< n > + k}$ being given by $< n >_c = \frac{\beta}{\beta - 1} \frac{1}{\ln(1 - \beta)}$. This value of $< n >_c$ is close to independent emission of single particles in agreement with the Poisson multiplicity distribution observed at 29 GeV, but in disagreement with the observed strong resonance production and the rapidity correlations of Fig. 35.

Monte Carlo studies of 200 GeV and 2 TeV e^+e^- annihilations[72], using version 6.3 of the Lund Monte Carlo program, predict that the multiplicity distributions will continue to follow the NB distribution, as shown in Fig. 41. The results for $< n >_c$, shown in Fig. 42, suggest a modest increase in the parton multiplicity per clan to $< n >_c \sim 1.5$ for $\sqrt{s} = 200$ GeV and $< n >_c \sim 2$ to 3 for $\sqrt{s} = 2$ TeV. The average number of clans, $\bar{N}$, is almost the same at all three energies for $|Y| < Y_0$ up to 3 units and only diverges for the higher Y_0 selections as seen in Fig. 43. Such studies also show that $< n >$ for the hadrons is twice that for the partons and the k values in the NB fit are equal, supporting the idea of local parton-hadron duality.

104

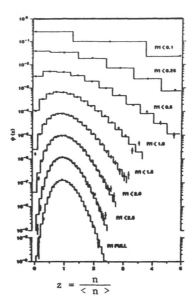

$$z = \frac{n}{\langle n \rangle}$$

Fig. 39. Multiplicity distributions for two-jet events of e^+e^- annihilation at 29 GeV as a function of the rapidity span selection. The histogram shows the best fit to the negative binomial of Eq. (2) expressed in KNO form. Each of the selected distributions has been shifted down by a factor of ten relative to the $|Y| < 0.1$ data. The ordinate scale for the data set with no rapidity selection is shown separately.

Multiplicity Moments

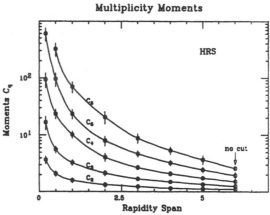

Fig. 40. Moments of the two-jet multiplicity distributions as a function of rapidity span.

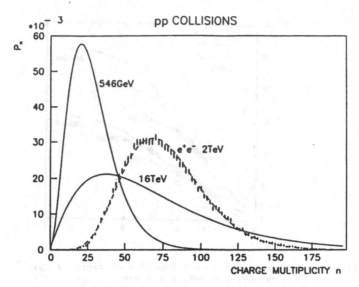

Fig. 41. Charged multiplicity distribution at $\sqrt{s}$ = 2 TeV compared to $p\bar{p}$ interactions at 546 GeV and 16 TeV. The dashed line shows the fit to the negative binomial distribution.

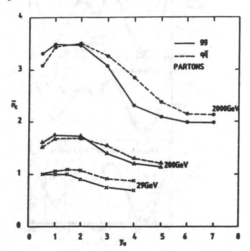

Fig. 42. Mean number of partons per clan for e^+e^- annihilation at 29 GeV, 200 GeV, and 2 TeV as a function of rapidity span Y_0 according to the Lund Monte Carlo generator.

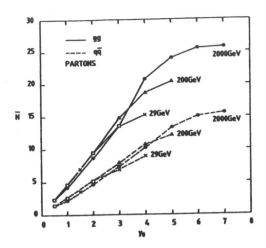

Fig. 43. Mean number of clans as a function of rapidity span $|Y| < Y_o$ for $\sqrt{s}$ = 29 GeV, 200 GeV, and 2 TeV.

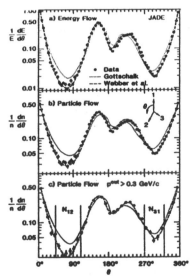

Fig. 44. Energy and particle flow in three-jet events. The depletion near 80° comes from the coherent gluon interference effect.

Particle Flow in Three-Jet Events

The wider transverse momentum distribution in the jet plane seen in Fig. 18(c) is a clear indication of gluon bremsstrahlung, and so some information on the properties of gluon jets can be obtained. However, it is, in principle, not possible to decouple the gluon jet from the total event, particularly at present energies. Nevertheless, a number of interesting studies have been done.

The JADE group[73] made the first comparisons of the particle flow in the event plane of three-jet events with fragmentation models. Fig. 44 shows such data: the zero of the angle is chosen in the center of the highest energy jet (that opposite to the smallest opening angle) and rotating to the jet with the second highest energy. The region between jet 1 and jet 2, which are preferentially the quark jets, is depleted and that between jet 2 and jet 3 enhanced, as compared to the predictions of the independent fragmentation model. The effect is more pronounced if particles with somewhat higher p_T, relative to the event plane, are selected as shown in Fig. 44(c). Both the string and cluster models can be arranged to fit the data. The Webber model result is shown in the figure.

A more dramatic comparison can be made between $q\bar{q}\gamma$ events and the $q\bar{q}g$ three-jet events.[74] The MarkII data on the particle flow for these two types of event is compared in Fig. 45(a) for all tracks, in Fig. 45(b) for tracks with $p_T^{out} > 0.3$ GeV/c. The absence of particles near 240° for the $q\bar{q}\gamma$ events is clear but, in addition, there are fewer particles near 80° in the $q\bar{q}g$ case than for the $q\bar{q}\gamma$ events. This is ascribed to the drag effect of the high energy gluon or to the destructive interference between the soft gluons from the quarks in the region between the q and $\bar{q}$.

The HRS collaboration[75] has measured the charged particle multiplicities associated with the three jets. The results are shown in Fig. 46 and compared to the line which gives the variation of < n > with jet energy for inclusive e^+e^- annihilations. The data points for all jets follow this curve. Even though Jet 3 is enriched in gluons, the charged multiplicity is the same. Because of the color charge of the gluon, the simplest expectation

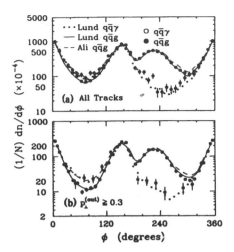

Fig. 45. The charged-track density as a function of the event plane angle ϕ. The angular region between $\phi = 0°$ and $\phi = 150°$ separates the q and $\bar{q}$ for all of the $q\bar{q}\gamma$ events and for 65% of the $q\bar{q}g$ events.

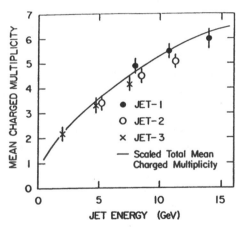

Fig. 46. Mean charged particle multiplicity as a function of jet energy for three-jet events. Jet 3 is enriched in gluons. The line shows the variation of single jet multiplicity at a jet energy of $\sqrt{s}/2$.

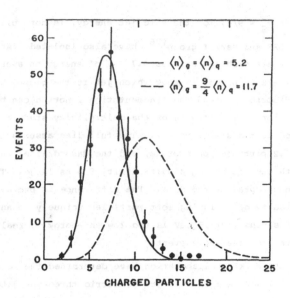

Fig. 47. Charged particle multiplicity distribution of symmetric three-jet events compared to two assumptions about the gluon jet multiplicity.

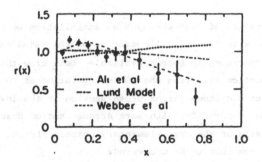

Fig. 48. Ratio of gluon-to-quark fragmentation.

that $< n >_g / < n >_q = 9/4$, for the same jet energy, is not followed.

Both the HRS and MarkII group[76] have also isolated samples of symmetric 3-jet events, where each jet carries 9.7 GeV of energy on average. Although in this case there is no way to know which jet is the gluon, both the multiplicity distribution and the fragmentation function can be compared to simple expectations. The results of the multiplicity study are shown in Fig. 47 and compared to two simple models. The full line assumes three jets each with $< n > = 5.2$ particles on average, and the dashed line, which corresponds to a model with one jet having a multiplicity 9/4 as large. The second case clearly does not represent the data. The difference is presumably connected with the difficulty of assigning soft particles uniquely to any jet, which is another way of saying that 29 GeV is too low an energy to really separate the individual jets from the whole event.

However, the MarkII collaboration have determined the ratio D_g/D_q by comparing the 9.7 GeV jets from 29 GeV symmetric three-jet data with inclusive $e^+e^- \rightarrow q\bar{q}$ annihilations at lower energies. The results, shown in Fig. 48, indicate a softer gluon than quark fragmentation in agreement with simple expectations.

Conclusions

The characteristics of high energy e^+e^- annihilation were first studied about ten years ago with the initial operation of the PETRA storage ring. High quality data continues to become available, both from the PEP experiments, but particularly from the 10 GeV operation of the CESR and DORIS rings. The order of magnitude larger cross section at the lower energy means that the production of particles with more strange and/or charm quarks can be studied. More details of the most elementary characteristics of the parton-hadron transitions continue to be uncovered.

The Lund string model provides a good, although incomplete, parametrization of the fragmentation process. Since a realistic representation of the fragmentation of parton jets is an essential building block for understanding the physics that may be uncovered by the highest energy hadron-hadron and lepton-hadron collisions, it is important that the e^+e^- studies continue.

The cluster models also do a surprisingly good job at describing the overall features of the data and, with a more detailed description of cluster decay, could do even better.

A convergence between these two approaches is occurring with the coherent gluon effect being included in the QCD parton generation in the latest version of the Lund model, and the use of string algorithms to fragment heavy clusters in the cluster models.

Finally, phenomenological fits to things like the rapidity distribution in e^+e^- annihilation have illuminated similar data from soft hadronic collisions.

In a few years, very large data samples of $e^+e^- \rightarrow Z^0 \rightarrow$ hadrons will become available from the LEP detectors. The combination of high energy, high rates, and more sophisticated detectors will make our present efforts look no more than a prelude.

We wish to thank Paula Waschbusch for an invaluable contribution in preparing the figures for this article.

112

References

1.   S. L. Wu, Physics Reports $\underline{107}$, 59 (1984).

2.   D. H. Saxon, "Quark and Gluon Fragmentation in High Energy e^+e^- Annihilation," RAL 86-057.

3.   B. Adeva et al. (MARKJ collaboration), Physics Reports $\underline{109}$, 131 (1984).

4.   B. Naroska (JADE collaboration), Physics Reports $\underline{148}$, 68 (1987).

5.   HRS - M. Derrick et al., Phys. Rev. $\underline{D31}$, 31 (1985).

6.   J. F. Donoghue, Phys. Rev. $\underline{D19}$, 3806 (1979).

7.   Z. Koba, H. B. Nielsen and P. Olesen, Nucl. Phys. $\underline{B40}$, 317 (1972).

8.   R. D. Field and R. P. Feynman, Nucl. Phys. $\underline{B136}$, 1 (1978).

9.   P. Hoyer et al., Nucl. Phys. $\underline{B161}$, 349 (1979);
A. Ali et al., Phys. Lett. $\underline{93B}$, 155 (1980).

10. B. Andersson, G. Gustafson and T. Sjostrand, Phys. Rep. $\underline{97}$, 32 (1983).

11. R. D. Field and S. Wolfram, Nucl. Phys. $\underline{B213}$, 65 (1983);
B. R. Webber, Nucl. Phys. $\underline{B238}$, 492 (1984);
T. D. Gottschalk, ibid., $\underline{B239}$, 325 (1984).
T. D. Gottschalk and D. Morris, CALT 68-1365 (1986).

12. MarkII - A. Petersen et al., Phys. Rev. $\underline{D37}$, 1 (1988).

13. TASSO - M. Althoff et al., Z. Phys. $\underline{C22}$, 307 (1984).

14. JADE - W. Bartel et al., Z. Phys. C20, 187 (1983); CELLO - M. J. Behrend et al., Z. Phys. C14, 45 (1982); PLUTO - C. Berger et al., Z. Phys. C12, 297 (1982); ibid., C28, 365 (1985).

15. HRS - M. Derrick et al., Phys. Lett. 164B, 199 (1985).

16. D. Sivers, Ann. Rev. Nuc. Part. Sci. 32, 149 (1982).

17. HRS - M. Derrick et al., Phys. Rev. D34, 3304 (1986).

18. TPC - H. Aihara et al., Phys. Rev. Lett. 52, 577(1984). The data presented here came from the upgraded detector: Z. Wolf, LBL-23738 (1987).

19. L. V. Gribov, E. M. Levin, and M. G. Ryskin, Phys. Rep. 100, 1 (1983); Ya. I. Azimov et al., Z. Phys. C27, 65 (1985).

20. JADE - W. Bartel et al., Z. Phys.C31, 213 (1980); C28, 343 (1985).

21. CELLO - H. J. Behrend et al., Z. Phys. C20, 207 (1983); TPC - H. Aihara et al. Z. Phys. C27, 187 (1985). TASSO - W. Braunschweig et al., DESY 86-078;

22. HRS - S. Abachi et al., Phys. Lett. 205B, 111 (1988). A satisfactory fit could only be obtained after changing the D decay branching ratios to agree with recent measurements (Ref. 44).

23. Mark II - G. Wormser et al., submitted to Phys. Rev. Lett.

24. Mark II - H. Schellman et al., Phys. Rev. D31, 3013 (1985).

25. JADE - W. Bartel et al., Z. Phys. C20, 187 (1983).

26. TPC - H. Aihara et al., Phys. Rev. Lett. 53, 2378 (1984).

27. TASSO - M. Althoff et al., Z. Phys. C27, 27 (1985).

28. HRS - M. Derrick et al., Phys. Rev. D35, 2639 (1987).

29. HRS - S. Abachi et al., Phys. Lett. 199B, 151 (1987).

30. JADE - W. Bartel et al., Phys. Lett. 145B, 441 (1985).

31. MarkII - H. Schellman et al., LBL 18699.

32. TASSO - W. Brandelik et al., Phys. Lett. 101B, 129 (1981).

33. HRS - M. Derrick et al., Phys. Lett. 158B, 519 (1985). The results shown
     in Fig. 20 and Fig. 21 have been updated to the full 300 pb^{-1} data
     sample.

34. TPC - H. Aihara et al., Phys. Rev. Lett. 52, 2201 (1984).

35. HRS - M. Derrick et al., Phys. Rev. Lett. 54, 2568 (1985).

36. HRS - S. Abachi et al., Phys. Rev. Lett. 57, 1990 (1987).

37. A. Casher, H. Neuberger and S. Nussinov, Phys. Rev. D20, 179 (1979).

38. CLEO - S. Behrends et al., Phys. Rev. D31, 2161 (1985).

39. ARGUS - H. Albrecht et al., Phys. Lett. 183B, 419 (1983); DESY 87-141.

40. MarkII - J. M. Yelton et al., Phys. Rev. Lett. 49, 430 (1982);
     HRS   - S. Ahlen et al., Phys. Rev. Lett. 51, 1147 (1983),
               M. Derrick et al., Phys. Lett. 146B, 261 (1984);

TASSO   - M. Althoff et al., Phys. Lett. 136B, 139 (1984),
              ibid. 138B, 317 (1984);
JADE   - W. Bartel et al., Phys. Lett. 146B, 121 (1984);
DELCO   - H. Yamamoto et al., Phys. Rev. Lett. 54, 522 (1985);
TPC   - H. Aihara et al., Phys. Rev. D34, 1145 (1986).

41. JADE - W. Bartel et al., Phys. Lett. B161, 197 (1985);
     HRS - E. Low et al., Phys. Lett. B183, 232 (1987).

42. ARGUS - H. Albrecht et al., Phys. Lett. 150B, 235 (1985).

43. CLEO - D. Bortoletto et al., Phys. Rev. D37, 1719 (1988).

44. Mark III - J. Adler et al., Phys. Rev. Lett. 60, 89 (1988);
     See also SLAC PUB-4518 (1988);
     HRS - S. Abachi et al., ANL-HEP-PR-87-100, Phys. Lett. (to be published).

45. HRS - B. Baringer et al., PU-88-613, Phys. Lett. (to be published).

46. C. Peterson et al., Phys. Rev. D27, 105 (1983).

47. S. Bethke, Z. Phys. C29, 175 (1985).

48. ARGUS - H. Albrecht et al., Phys. Rev. Lett. 56, 549 (1986).

49. HRS - S. Abachi et al., Phys. Lett. 119B, 585 (1987).

50. CLEO - A. Chen et al., Phys. Rev. Lett. 51, 634 (1983);
     TASSO - M. Althoff et al., Phys. Lett. 136B, 139 (1984);
     ARGUS - H. Albrecht et al., Phys. Lett. 153B, 343 (1985);
     HRS - M. Derrick et al., Phys. Rev. Lett. 54, 2568 (1985).

51. TASSO - M. Althoff et al., Phys. Lett. 135B, 243 (1984).
    DELCO - M. Sakuda et al., Phys. Lett. 152B, 399 (1985).
    HRS - P. Kesten et al., Phys. Lett. 161B, 412 (1985).

52. M. Derrick and T. Gottschalk, Proc. of the 1984 Summer Study on the
    Design and Utilization of the SSC, p. 49, Eds. R. Donaldson and J. G.
    Morfin, Snowmass, CO.

53. ARGUS - H. Albrecht et al., Phys. Lett. 157B, 326 (1985).

54. TPC - H. Yamamoto, Proc. of the 1985 Rencontre de Moriond, Ed. J. Tran
    Than, p. 91.

55. ARGUS - H. Albrecht et al, Phys. Lett. 183B, 419 (1987).

56. TPC - H. Aihara et al., Phys. Rev. Lett. 54, 274 (1985).

57. MarkII - C. de la Vaissiere et al., Phys. Rev. Lett. 54, 2071 (1985).

58. HRS - P. Baringer et al., Phys. Rev. Lett. 56, 1346 (1986). The results
    presented here are an update to the full 300 pb^{-1} data sample.

59. TASSO - M. Althoff, Phys. Lett. 130B, 340 (1983).
    MarkII - S. R. Klein et al., Phys. Rev. Lett. 58, 644 (1987).
    HRS - S. Abachi et al., Phys. Rev. Lett. 58, 2627 (1987).

60. ARGUS - H. Albrecht et al., DESY 87-141.

61. ARGUS - H. Albrecht et al., DESY 88-011.

62. T. DeGrand, Phys. Rev. D25, 3928 (1982).

63. MarkII - S. R. Klein et al., Phys. Rev. Lett. 59, 2412 (1987).

64. TPC - H. Aihara et al., Phys. Rev. Lett. 55, 1007 (1985); ibid., 57, 3140
    (1986). W. Hoffmann, LBL-24086.

65. TASSO - M. Althoff et al., Phys. Lett. 139B, 126 (1984).

66. TASSO - R. Brandelik et al., Phys. Lett. 100B 357 (1981),
        M. Althoff et al., Z. Phys. 29, 347 (1986).
    PLUTO - Ch. Berger et al., Z. Phys. C12, 297 (1982);

67. TPC - H. Aihara et al., Phys. Rev. Lett. 53, 2199 (1984).

68. A. Capella and A. V. Ramallo, Phys. Rev. D37, 1763 (1988).

69. TPC - H. Aihara et al., Z. Phys. C27, 495 (1985);
    TASSO - M. Althoff et al., ibid., C29, 347 (1985).

70. HRS - M. Derrick et al., Phys. Lett. 168B, 299 (1986).

71. A detailed discussion of the different approaches to understanding
    multiplicity distributions is given by P. Carruthers and C. C. Shih,
    Intl. Journal of Mod. Phys. A2, 1447 (1987). See also A. Giovannini and
    L. Van Hove, Z. Phys. C90, 391 (1986), A. Giovannini, Proc. of Physics in
    Collision VI, p. 39, Ed. M. Derrick, World Scientific.

72. W. Kittel, HEN-282, contribution to Workshop on Physics with Future
    Accelerators, La Thuile and CERN(1987); A. Giovannini and L. Van Hove,
    CERN-TH-4885/87.

73. JADE - W. Bartel, Z. Phys. C21, 37 (1983);
    Phys. Lett. 134B, 275 (1984); ibid., 157B, 340 (1985).

74. TPC – H. Aihara et al., Phys. Rev. Lett. 57, 945 (1986);
    MarkII – P. D. Sheldon et al., ibid., 57, 1398 (1986).

75. HRS – M. Derrick et al., Phys. Lett. 165B, 449 (1985).

76. MarkII – A. Petersen et al., Phys. Rev. Lett. 55, 1954 (1985).

SELECTED RESULTS ON MULTIHADRON PRODUCTION
AT ISR ENERGIES

R. Campanini

Dipartimento di Fisica, Università di Bologna
and INFN,Sezione di Bologna
Italy

Experiments at ISR runned from 1972 to 1983, studyng pp, $\bar{p}p$, $p\alpha$, $\alpha\alpha$ and dd interactions at c.m.s energies of 22,26,30,45,53 and 63 Gev per nucleon pair. In the following we will review some aspects of reactions with many particles in the final state. Previous excellent reviews are listed in [1].

1.   MULTIPLICITIES DISTRIBUTIONS

We can sumarize results as follows:

1.1. Multiplicities in the central pseudorapidity region in pp.

In [2] the mean charged value $<n_{ch}>$ and distribution moments have been studied for pseudorapidity $|\eta| < 1.5$ at different energies, for events with at least one particle in the given range (fig. 1). The observed multiplicity distribution moments are much higher than Poisson moments for the same mean multiplicity: this means strong correlations in the central region. A clusters model was used to fit the data: generated low mass clusters with $<m> = 1.2$ Gev decay isotropically in their individual rest frame with Poisson like decay multiplicity k, with $<k> = 3$. The charged particle multiplicity was forced to agree with the measured distribution: this requirement introduced correlations between clusters in addition to those arising from 4 momentum conservation.

1.2. Multiplicity in the full pseudorapidity range in pp.

In the full angular range results are given in [2] and in [3], see fig. 2,3. In [3] kinematical cuts are applied to separate non single diffractive events. About 15% of the events were determined to be diffractive. In both experiments [2] and [3] the dependence of $<n>$ to squared energy s is found to be of the form:

$$<n> = A + B \cdot \ln s + C \ln^2 s$$

The energy dependence of the dispersion $D_2/<n>$ is in good agreement with KNO scaling for the non single diffractive data [2] whereas the scaling is broken for the sample of inelastic events without single diffraction subtraction [2,3].

120

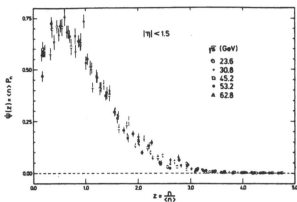

Fig. 1 – Normalized multiplicity distributions of charged
particles in the central region ($|\eta|<1.5$) as a function of
the KNO variable $z = n/\langle n\rangle$.

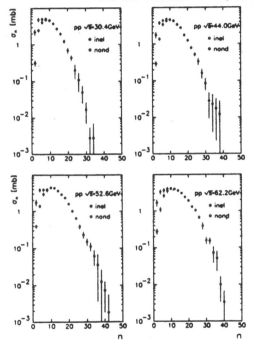

Fig. 2 – Topological cross sections as a function of the charged
particle multiplicity for inelastic
and non-single diffractive events
at different ISR energies.

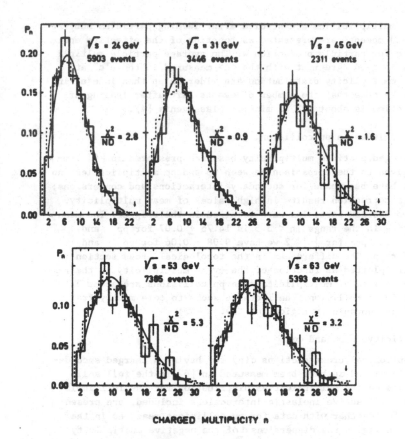

Fig. 3 – Total charged particle multiplicity distributions (normalized) for the five ISR energies. The dashed histograms indicate the distributions for raw data.

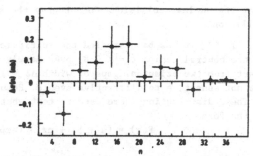

Fig. 4 – The difference in the topological cross sections obtained in pp and p̄p plotted as a function of the event multiplicity n.

In [4] multiplicities are studied after leading particles subtrac
tion; <n> and moments are presented as function of the effective energy available for particle production w. The average particle multipli
city <n> is in good agreement with the <n> measured in the e^+e^- at
$\sqrt{s}$ = w; the multiplicity distribution are wider in pp than in e^+e^- interactions. We note that the number of events left after leading effect subtraction is about 10% of minimum bias events [4].

1.3. Multiplicities in p̄p collision.

The charged particle multiplicity has been presented in [5]. Emphasis is given to the comparison between pp and p̄p multiplicities; no
corrections have been made for secondary interactions and conversions;
this lack of correction results in high values of mean multiplicity.
At least two charged particles were required in the trigger. The mean
multiplicities in the range $|\eta|< 3.5$ is 14.75 ± 0.07 for pp   and
15.10 ± 0.10 for p̄p; for $|\eta|< 2$ we have 9.38 ± 0.06 for pp   and
9.7 ± 0.1 for p̄p. The differences in the topological cross section in
pp and p̄p are plotted as function of the event multiplicity in the ran
ge $|\eta|< 3.5$ in fig. 4. Multiplicities in p̄p collisions are about 3%
higher than in pp collision. The increase seems to come mostly from
the central pseudorapidity region($|\eta|$ <2).

1.4. Multiplicity in pα and αα.

The topological cross sections $\sigma(n)$ for having n charged secondary hadrons in αα and αp have been measured in [6] in the full solid
angle. Results on the multiplicities distribution for the full inelastic sample, diffractive inelastic interactions included, are presented in fig. 5, together with data for pp collisions measured in the
experiment. In fig. 6 the dispersions d of the negative multiplicity
distributions are plotted vs <n > and compared with data at various
energies. All pp data agree with a linear relation between d_ and <n >;
the line in the fig. 5 represents the original relation by Wrobleskwi.
The αα and αp data points do not fall on the same line. This indicates
that the nuclear collisions do not obey the same KNO scaling as   pp
collisions.

In [7] it has been studied the multiplicity of charged particles
in the central region ($|\eta|< 0.8$) over nearly the full azimuth range,
apart from two azimuthal gaps of width $\Delta\phi$ = 16. The requested trigger
was not sensitive to diffractive events. Results are presented in fig.
7. These distributions were used to test whether the same KNO scaling
of the form

$$F(z) = (a + bz + cz^2) (\exp(-dz)),$$

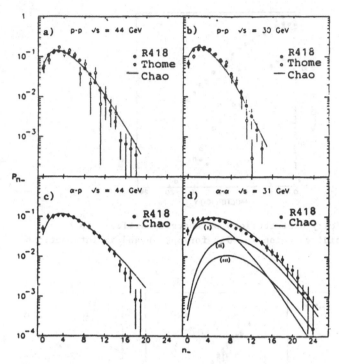

Fig. 5 – Negative multiplicity distributions $P_{n-} = \sigma_{n-}/\sigma^{prod}$ (a) and (b) for pp (c) for αp and (d) for αα collisions.

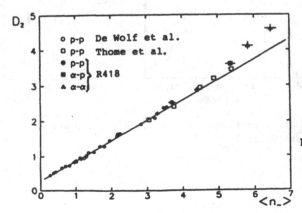

Fig. 6 – Dispersion of negative multiplicity distributions versus average multiplicity. The line indicates the original Wroblewski fit adapted to negative particles.

124

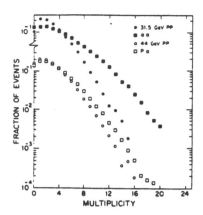

Fig. 7 - Multiplicity distributions of charged particles in the
central pseudorapidity region $|\eta|<0.8$ for pp, αp and $\alpha\alpha$ interactions.

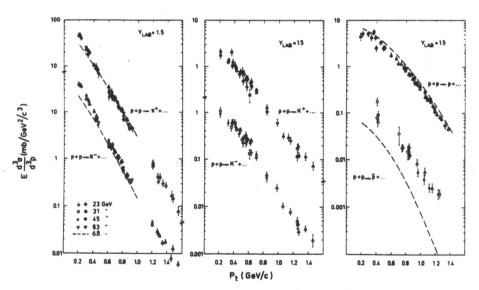

Fig. 8 - The invariant cross sections for π^+, π^-, K^+, K^-, p and $\bar{p}$
production plotted versus P_t at $Y_{LAB} = 1.5$.

with z=n/<n>, described pp, αp and αα data. The constants b,c and d were determined from pp data. Only the mean <n> and the overall normalization a were fitted. The fit to αp data is rather good, but the αα show a systematic departure from pp scaling function.

2. INCLUSIVE DISTRIBUTIONS

2.1. P_T distributions

Fig. 8 gives the transverse momentum distribution for $\pi^{\pm}$, $k^{\pm}$, $p^{\pm}$ at fixed values of rapidity. The dependence on p_T is approximately exponential at values lower than 1 Gev/c; however a better descrip tion of the data is obtained by including a quadratic term in p_T in the exponent. The p_T spectra depend only weakly on y [8]. The observed particle production distributions indicate the existence of two different regions, a low p_T and high p_T region, with transitions interval at 1 to 2 Gev/c. The regions differ with respect to the form of p_T distributions, the dependence of the production cross sections on the collision energy and the composition of the charged particle flux. The low p_T interval is characterized by:

i) approximatively exponentially decreasing p_T spectra. Typical values for the slopes are:

$$\pi^+ \ 6.4 \qquad \pi^- \ 6.8$$
$$K^+ \ 4.8 \qquad \underline{K}^- \ 5.5$$
$$p \ 3.4 \qquad \bar{p} \ 3.8 \ , \text{ almost independent of the collision}$$
energy.

ii) weak dependence on rapidity y.

iii) the particle flux is dominated by pions; the fractions of heavier particles is growing with p_T independently of s. The high p_T region is characterized by:

i) a p_T dependence approaching a power law with increasing p_T

ii) a strong s dependence, increasing with p_T

iii) $\pi/k/N$ ratios close to 2/1/1 which depends only weakly on p_T
or s.

The same behaviour in seen in pp, αp,αα interactions. The average $<p_T>$ is equal in all of these reactions [9].

The ratio of p_T spectra obtained in pp to those obtained in $p\bar{p}$

for identified p, $\bar{p}$, K^+, K^-, is showed in fig. 9. [10] .
The ratios of inclusive cross section:

$$R(\alpha\alpha/pp) = [d\sigma(\alpha\alpha)/dp_T]/d\sigma(pp)/dp_T$$

are considered as function of p_T in fig. 10. In particular for
π^0 $r(\alpha\alpha/pp) = 23.8 \pm 0.4$ for $p_T > 3$ Gev/c, showing that high p_T yield
is increasing faster than A^2, where A is the atomic number.

2.2. Rapidity, pseudorapidity and longitudinal momentum distribution

In fig. 11 we present for pp collisions the inclusive distributions as function of the rapidity in the rest frame of one of the incident protons, for fixed p_T of 0.4 Gev/c [8]. All y distributions,
with the proton exception, exhibit a decrease for $y_{LAB} < 2$ (fragmen
tation region) while for $y_{LAB} > 2$ the distribution is slowly increasing
and reachs a maximum at Y_{LAB} corresponding to $y_{cm} = 0$. The height of
this maximum is increasing with the c.m. energy (fig. 12) [12]. In
fig. 13 pseudorapidity distributions of charged particles with equal
and unequal beam energies are shown: from η above 2 the inclusive distribution is the same wheter a proton of 26 Gev has been hit by a
proton of 26 Gev or of 15 Gev (limiting fragmentation).

- The value of particle density in pp at y_{cm} around 0 is increasing
with the event multiplicity (fig. 14,15) [14,15].

- The particle density $\rho(y)$ and the charge density $\rho(y)=\rho^+(y)-\rho^-(y)$
for events with an observed multiplicity greater than 6 are shown
in fig. 16.

- The pseudorapidity distribution in pp is shown in fig. 17 [5].
The ratio of pseudorapidity distributions obtained by the same experiment in pp and $\bar{p}p$ collisions [5] is close to unity through the pseu
dorapidity range, except for $\eta > 3$ where there is a relative depletion
in $\bar{p}p$ collisions (fig. 18).

- In fig. 19 y distributions for negative hadrons produced in $\alpha\alpha$
and αp collisions are shown and compared with pp collisions at corresponding s [15]. The ratio of the rapidity distribution R ($\alpha\alpha$ /pp =
$[dn(\alpha\alpha + h^-)/dy]/[dn(pp + h^-)/dy]$ and R ($\alpha p/pp$) are presented in fig.
20. The ratio rises very sharply at the highest rapidities on both
sides for $\alpha\alpha$ and in the α side for αp collisions. The particle ratios
at y = 0 have the value R ($\alpha p/pp$) = 1.10 $\pm$ 0.05 and R($\alpha\alpha$/pp)= 1.67 $\pm$
0.05.

- For $\alpha\alpha$, αp and pp collisions, the central rapidity density is in-

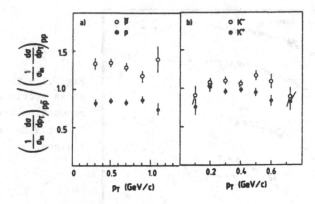

Fig. 9 – Ratio of the p_T spectra obtained in $\bar{p}p$ to that obtained in pp for identified a) $p, \bar{p}$ b) K^+, K^-.

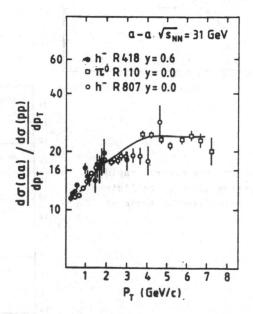

Fig. 10 – Ratios of $\alpha\alpha$/pp cross sections vs P_T for negative hadrons and π° production.

128

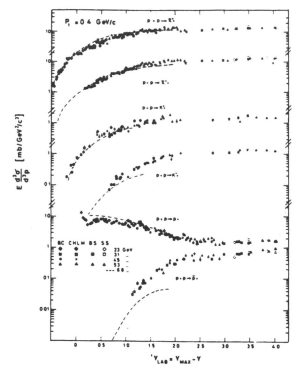

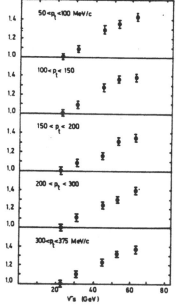

Fig. 11 - The invariant cross sections for $\pi^{\pm}, K^{\pm}, p^{\pm}$ production versus Y_{LAB} at P_T = 0.4 Gev/c in pp collisions.

Fig. 12 - Height of the central rapidity distribution versus $\sqrt{s}$ in pp collisions. The values are normalized to those at $\sqrt{s}$ = 23 Gev.

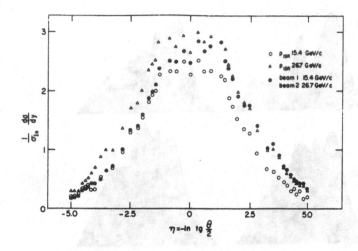

Fig. 13 - Inclusive pseudorapidity distribution for charged particles observed in collisions with identical and different beam energies.

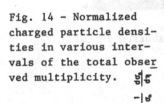

Fig. 14 - Normalized
charged particle densi-
ties in various inter-
vals of the total obser_
ved multiplicity.

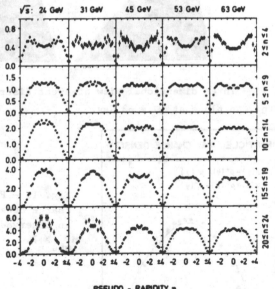

PSEUDO - RAPIDITY η

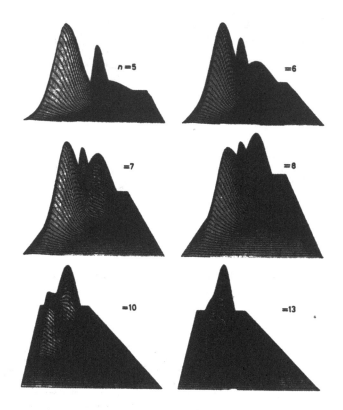

Fig. 15 - Three dimensional view of event distribution in the $\bar{\eta} = \langle \eta \rangle$, $\Delta\eta$ plane at $\sqrt{s}$ = 63 Gev.

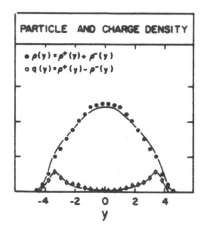

Fig. 16 - Particle density $\rho(y)$ and charge density $q(y)$ for events with an observed charged multiplicity $n_{obs} \gtrsim 7$.

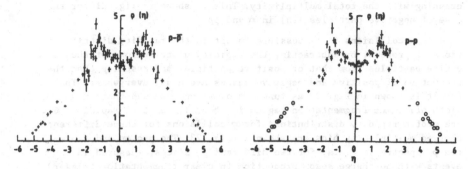

Fig. 17 - The pseudorapidity η distributions measured in p̄p and pp interactions.

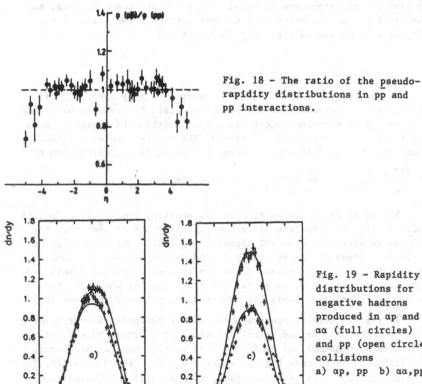

Fig. 18 - The ratio of the pseudorapidity distributions in p̄p and pp interactions.

Fig. 19 - Rapidity distributions for negative hadrons produced in αp and αα (full circles) and pp (open circles) collisions
a) αp, pp b) αα,pp

creasing with the total multiplicity. This is shown in fig. 21 for the case of negative particles [15] in αα and pp.

- In αα collisions it is possible to obtain the rapidity distributions of protons, by subtracting the rapidity distributions of the negative particles from that of positive particles and assuming that the production of positive and negative mesons are equal everywhere. The result is shown in fig. 22 as function of y; one can see that the width of nuclear fragmentation region is about 1.5 unit of rapidity [6]. The proton rapidity distributions for αα collisions for three different charged multiplicity bins are shown in fig. 23. The proton and deuteron p_L spectra in c.m.s, for two different p_T slices, measured in events with no charge meson production (nuclear fragmentation trigger) are shown in fig. 24. If one subtracts spectator protons from fig. 23, one obtains the rapidity density distributions of protons partecipant to the collision. The results are presented in fig. 25, where also data from dd collisions are included. For both the αα and dd data, the maxima of the proton distributions move toward larger value of rapidity lost Δy as the multiplicity increases [17].

3. TRANSVERSE ENERGY DISTRIBUTIONS

In fig. 26 the neutral total transverse energy spectra E_T° for pp, dd and αα interactions are presented [18]. The results have been obtained by an electromagnetic calorimeter within the rapidity interval $\pm$ 1.2 units over the full azimuth. The spectra are well fitted by gamma distribution [18]. A new form of scaling is indicated: the plot

of $\dfrac{\langle E_T^\circ \rangle}{N}$ dN/dE_T° vs $E_T^\circ / \langle E_T^\circ \rangle$ is very similar in pp, dd, αα.

- The ratio of pp and p̄p E_T° spectra, starting from E_T° at 6 Gev, is shown in fig. 27. The rate of production of events in the range $6 < E_T^\circ < 20$ Gev is observed to be 10% higher in p̄p collisions than in pp collisions. A study of the structure of the events through the sphericity variable shows this excess to be due to more isotropic events produced in p̄p collisions, as we can see from fig. 28, where the ratio p̄p/pp is plotted separately for events with spericity <0.2 and >0.2.

- In fig. 29 results are presented on the charged transverse energy E_T for the αα, αp and pp collisions, in the range $|y| < 0.8$ [20]. The curves are normalized at $E_T = 0.5$ to better compare the spectra for different reactions. Fig. 30 shows the ratios $r(p\alpha) = (d\sigma/dE_T(p\alpha))/(d\sigma/dE_T(pp))$ and $r(\alpha\alpha) = (d\sigma/dE_T(\alpha\alpha)/ (d\sigma/dE_T(pp))$ indicating that below 4 Gev they are lower than the value $r=A$ for pα and $r = A^2$ for αα. Above this energy they are enhanced.

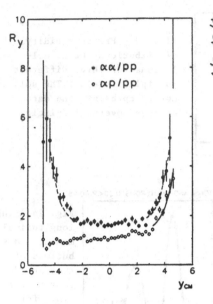

Fig. 20 - Ratio of the rapidity
distributions for negative hadrons
produced in αα and pp (full circles)
and αp and pp (open circles) interactions.

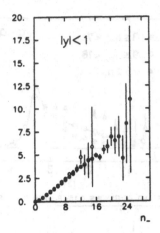

Fig. 21 - The dependence of
the rapidity density for negative hadrons in the region
|y|<1 on the total multiplicity;
(full circles for αα, open circles for pp).

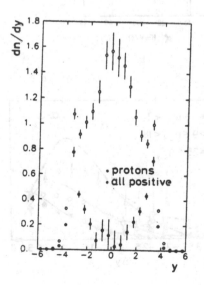

Fig. 22 - Rapidity distributions of
positive particles produced in αα
interactions.

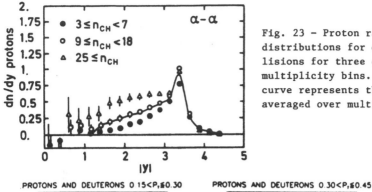

Fig. 23 – Proton rapidity distributions for αα collisions for three different multiplicity bins. The solid curve represents the data averaged over multiplicity.

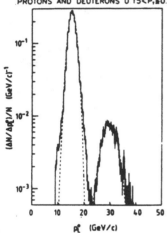

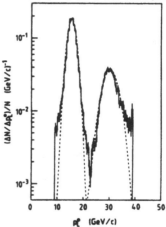

Fig. 24 – The longitudinal momentum distributions of protons and deuterons in the αα c.m.s for two different p_T slices.

Fig. 25 – Proton rapidity distribution for αα collisions for three different multiplicity bins with quasi elastic peak subtracted from the data.

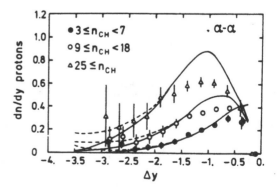

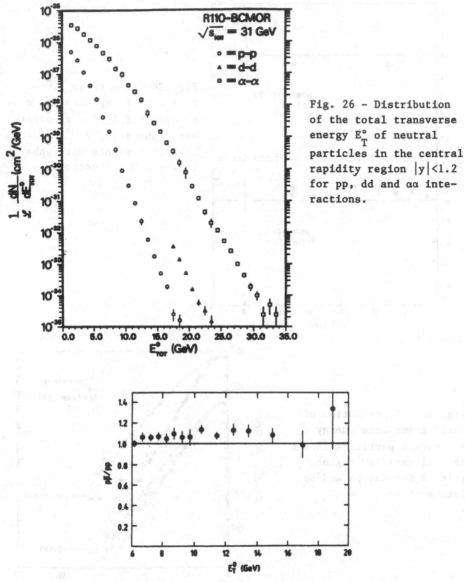

Fig. 26 – Distribution of the total transverse energy E_T^o of neutral particles in the central rapidity region $|y|<1.2$ for pp, dd and αα interactions.

Fig. 27 – The ratio of the p$\bar{\text{p}}$ and pp E_T^o spectra as a function of E_T^o.

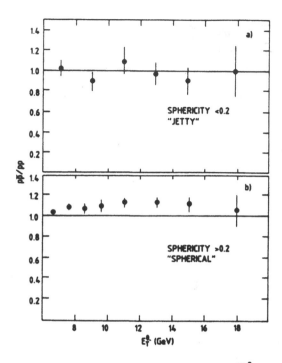

Fig. 28 - The ratio of the pp and pp̄ E_T^o spectra as a function of E_T^o for a) events with sphericity < 0.2 (Jet-like), b) events with sphericity > 0.2 (Isotropic).

Fig. 29 - Distribution of total transverse energy E_T of charged particles in the central rapidity region $|y|<0.8$ for $\alpha\alpha$, $p\alpha$ and pp interactions.

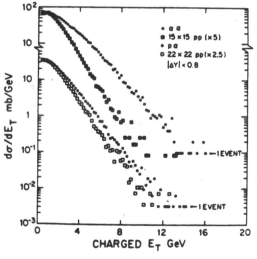

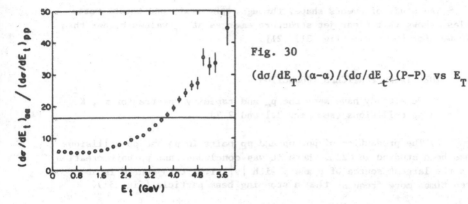

Fig. 30

$(d\sigma/dE_T)(\alpha-\alpha)/(d\sigma/dE_t)(P-P)$ vs E_T

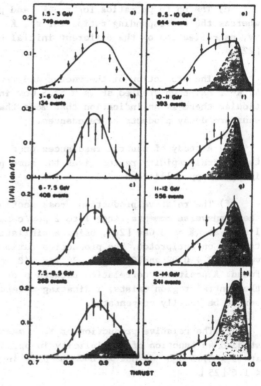

Fig. 31 – Distribution of
the thrust in the one wall of
the calorimeter for different
E_T bands. The solid curves
are the fits to a two component model, the shaded area
being the high P_T Jet component predicted by a QCD
Monte Carlo (ISAJET) .

- The study of events shape, through sphericity and thrust variables, shows that clear jet structure emerges at E_T values higher than
20 Gev for $|y| < 0.8$ (fig. 31) [21].

4. IDENTIFIED PARTICLES PRODUCTION

a) We already have seen the p_T and rapidity spectra for $\pi^{\pm}$, $K^{\pm}$,
p, $\bar{p}$ in pp collisions (see par. 2.1 and 2.2).

b) The production of pp, $\bar{p}\bar{p}$ and $p\bar{p}$ pairs in pp and $p\bar{p}$ collisions
has been studied in [22], where it was concluded that $p\bar{p}$ pair creation
is the largest source of p and $\bar{p}$ with $|y| <1.0$; this source is about
ten times more frequent than a stopping beam particle (fig. 32).

c) The $p\bar{p}$ to pp ratios for π^{+}, K^{+} and p are lower than unity,
whereas the corresponding ratios for π^{-}, K^{-} and $\bar{p}$ are larger than uni
ty, as reflection of the different initial charge state (fig. 33)
[22].

d) The production of the vector and tensor mesons $\rho°$, $f°$, $g°$, $\bar{K}*°$,
ϕ, $K*°$, has been studied at $\sqrt{s}$ = 52.5 Gev in pp collision [23]. In par
ticular there is an indication that more than 55% of all π^{-} and K^{-} mesons are decay products of resonances.

e) A study of neutral resonances production $\rho°$, f, k*,ϕ,$\Lambda°$,$\bar{\Lambda}°$ in
the central rapidity region gives the same production cross section
in pp and $p\bar{p}$ collisions [23].

f) The relative production cross section for $\bar{p}$, $\bar{\Lambda}$, $\bar{B}$ at y=0 in
the transverse momenta range 1 to 2 gev/c has been measured in pp col
lision at $\sqrt{s}$ = 63 Gev [24], using annihilation in a calorimeter as a
trigger on antiproton. The production ratios $\bar{\Lambda}/\bar{p}$, $\bar{B}/\bar{\Lambda}$ are respectively 0.27 $\pm$ 0.02 and 0.06 $\pm$ 0.02. For $\bar{\Omega}/\bar{B}$ an upper limit of 0.15 is
found. A positive correlation between $\bar{p}$ or $\bar{\Lambda}$ and associate proton in
the central region exists, indicating a tendency for the baryon num-
ber to be locally compensated.

g) The relative production of π, k meson, p, $\bar{p}$ and ϕ have been
studied as function of multiplicity in pp, dd, αp and $\alpha\alpha$ interactions
at $\sqrt{s}$ = 31.5 and 44 Gev per nucleon pair in the central region $|\eta|$
< 1.6 [25].

The rise of k/all with multiplicity is of the order of 15% going
from lowest to highest multiplicity in the given kinematical region

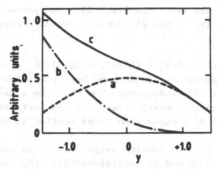

Fig. 32 – Origin of slow protons and antiprotons. Curves a and
b represent the probabilities of production through pair production and beam fragmentation respectively.

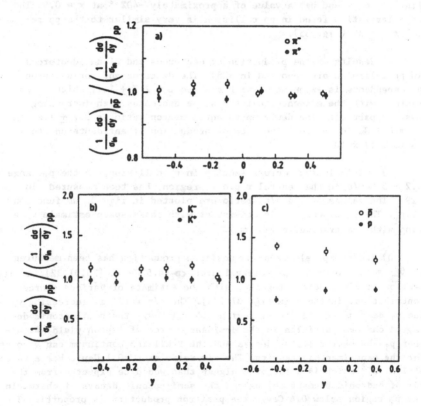

Fig. 33 – Ratio of the rapidity distributions in p̄p to the one in
pp for identified a) π⁺, π⁻, b) K⁺, K⁻ c) p, p̄.

and is similar for all reactions (fig. 34). The percentage of ϕ per $K^+ K^-$ pairs is similar in the given experimental errors for all the reactions.

h) The production of Λ, Λ (1520), $\phi(1020)$, $\bar{\Lambda}$, $\bar{p}$ in pp collisions in the proton fragmentation region have been presented in [26]. The data on Λ and on Λ (1520) productions show that the x dependence for baryons with same quark content depends on the particle mass; the x dependence of the $\bar{p}$ and $\bar{\Lambda}$ production cross section is very similar.

i) The Λ° and $\bar{\Lambda}^\circ$ polarization perpendicular to the reaction plane has been measured in pp and pp collisions [27]. The results show that Λ° polarization P_{Λ° doesn't depend on the $\sqrt{s}$ between 31 and 63 Gev, at given x, p_T values. At a given energy, P_{Λ° is growing with x and with transverse momentum. At an average P_T of 1.1 Gev/c P_{Λ° depends linearly on x and has a value of approximately -40% $^\circ$ at x = 0.8. The Λ° polarization found in pp collision is very similar to the pp results for Λ° (fig. 35).

l) Results on the production of deuterons and of antideuterons in pp collision are reported in [28]. The deuteron and antideuteron p_T dependence is exponential with a slope b = 2.7 $\pm$ 1.1, which is consistent with the evidence that the slope decreases with increasing mass of particle. The deuteron to antideuteron ratio is 3.7 $\pm$ 1.2. At p_T = 0.7 Gev/c the relative rate of production of antideuteron to π^- is (5 $\pm$ 1) x 10^{-5}.

m) The inclusive η cross section in pp collision, in the p_T range 0.2 - 5 Gev/c in the central rapidity region, has been measured in [29]. The values of the η/π° ratio are plotted in fig. 36 as function of p_T. The p_T dependence is consistent with phase space estimate (scaling with the transverse mass m_T).

n) The prompt electron and positron production has been measured at θ_{CM} = 90° over the p_T interval 0.08 $< p_T < 4.5$ Gev/c [30,31,32]. A collection of e/π results, together with the estimate of various sources contribution, is shown in fig. 37 [32]. The e/π ratio is increasing with the p_T decreasing. In the p_T region 0.5 - 1 Gev/c the semileptonic decay of charmed particles is the dominant source of lepton yield. At higher p_T the vector mesons decays and the Drell-Yan continuum can account for the experimental results. The e/π ratio at p_T=0.1 Gev/c has a value of 1 x 10^{-3} which is a factor 6 higher than would be expected from the sum of hadronic bremsstrahlung and the semileptonic decays of charm. In the p_T region below 0.4 Gev/c the positron production is proportional to the square of the mean multiplicity of the events (fig. 38); such

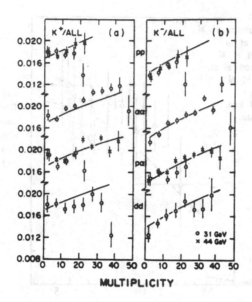

Fig. 34 - Fractions of tracks identified as kaons for pp, αα, pα and dd reactions. (a) K$^+$, b) K$^-$, the curves are fitted to pα data.

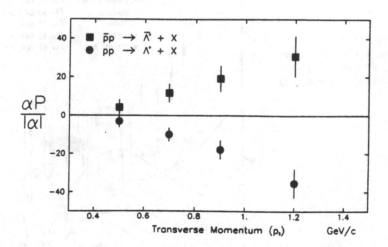

Fig. 35 - The polarization of $\bar{\Lambda}^°$ and $\Lambda^°$ versus transverse momentum.

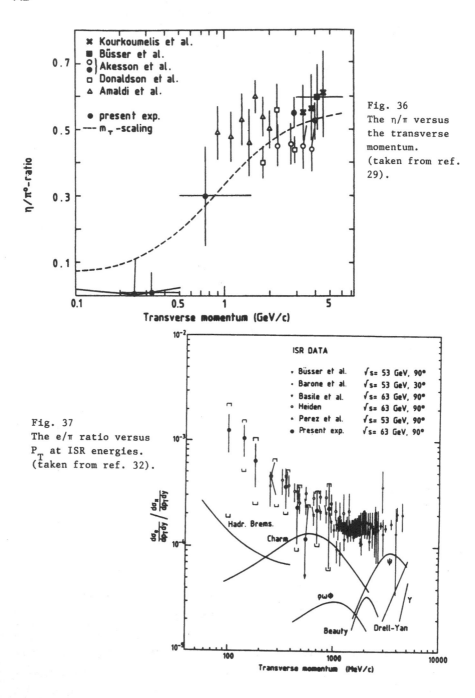

Fig. 36
The η/π versus
the transverse
momentum.
(taken from ref.
29).

Fig. 37
The e/π ratio versus
P$_T$ at ISR energies.
(taken from ref. 32).

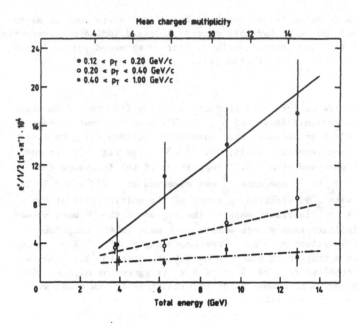

Fig. 38 – The e$^+$/π ratio as a function of total energy and mean charged multiplicity.

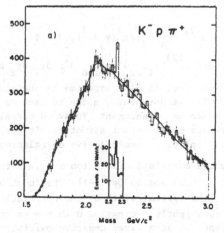

Fig. 39
The Λ_c^+ signal in the $K^- P\pi^{\mp}$ invariant mass plot.

a quadratic dependence is not expected from final state sources as hadronic bremsstrahlung or hadronic decays; it could indicate a production
mechanism of the soft lepton continuum over an extended volume during
the early stages of the collision [33].

   o) The evidence for charm production at the ISR comes from obser
vation of the hadronic decays of Λ_c^+, D^+, D°. Some information came
from prompt electron measurements. Mass peaks corresponding to Λ_c^+ -->
$K^- p \pi^+$ have been reported in [34, 35, 37, 38] (see fig. 39). In most
cases Λ_c^+ was observed at X > 0.3, suggestive of the fragmentation
process p-- > Λ_c^+ $\bar{D}$. In one case Λ_c^+ was observed at 0.03< x < 0.3 in
association with a 90° electron. A study of the multiplicity of the
events with $K^- p \pi^+$ invariant mass in the region of the Λ mass showed
higher multiplicity than events with $K^- p \pi^+$ mass in the neighbour
region [39]. Invariant mass peaks corresponding to D° --> $K^- \pi^+$ in con
junction with a triggering electron have been reported [51] . In addition [36] reported forward D^+ --> $K^- \pi^+ \pi^+$ triggered by high x and high
transverse momentum K^- meson. Large uncertainties in the cross section values are present [40].

4.   CORRELATIONS

a) Rapidity correlations

   The two particles normalized inclusive correlation function is
defined as:

$$R(y_1, y_2) = \rho^{(2)}(y_1,y_2)/\rho^1(y_1)\rho^1(y_2) - 1$$

where $\rho^{(1)}(y) = \sigma_{prod}^{-1} d\sigma/dy$ and $\rho^{(2)}(y_1,y_2) = \sigma_{prod}^{-1} d^2\sigma/dy_1 dy_2$ are the
single and two particle densities. In many experiments instead of ra
pidity y it has been used the pseudorapidity η due to lack of particle identification or of momentum measurement. To avoid correlations
originating from mixing events of different structure, it has been
studied R also at fixed multiplicity (seminclusive correlations).

   In fig. 40 the inclusive correlation function $R(\eta_1,\eta_2)$ is plotted vs η_2 at fixed η_1, at $\sqrt{s}$ = 23 and 53 Gev [41]. The function R
has a maximum when $\eta_1 = \eta_2$, the typical width of the correlation cur
ve is about two units in η, slightly increasing with the energy.In
fig. 41 the correlation function at a fixed negative multiplicity
$n_- = 5$ is shown for $\alpha\alpha$, αp, pp at various energies [42]; correlation

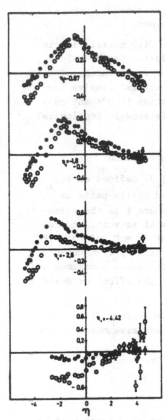

Fig. 40 – Correlation function $R(\eta_1, \eta_2)$ for fixed η_1 vs η_2; full circles $\sqrt{s}$ = 63 Gev, open circles $\sqrt{s}$ = 23 Gev.

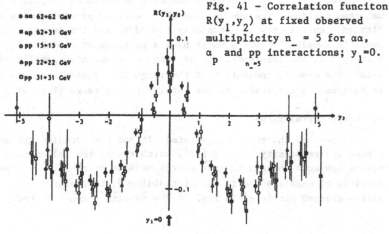

● αα 62+62 GeV
■ αp 62+31 GeV
○ pp 15+15 GeV
▲ pp 22+22 GeV
□ pp 31+31 GeV

Fig. 41 – Correlation funciton $R(y_1, y_2)$ at fixed observed multiplicity $n_- = 5$ for αα, α$_p$ and pp interactions; y_1=0.

analysis in terms of cluster production and decay showed no systematic difference between αα, αp and pp collisions.

In fig. 42 the difference between pseudorapidity correlation in pp and p̄p collisions, integrated over η_2, are plotted for η_1 values in the range -1, 1 [5]. In such an interval there is an excess of correlation in p̄p respect to pp. From the study of seminclusive correlation it has been found that the extracorrelation is present only in events at those multiplicities where the difference in topological cross section between pp and p̄p is higher.

b) Azimuthal correlation; transverse momentum compensation

In fig. 43 the correlation function $R(\vec{p}_1, \vec{p}_2)$, defined as $R(y_1, y_2)$ with y replaced by $\vec{p}$, is plotted for like and unlike pairs as function of $\Delta y = |y_1 - y_2|$ and $\Delta\phi = |\phi_1 - \phi_2|$, where ϕ is the azimuthal angle [43]. The picture shows the existence of azimuthal correlation, which is most evident for unlike pairs at $\Delta y = 0$ and $\Delta\phi = 180°$. Most part of this correlation is interpreted as arising from momentum conservation within one cluster. The $R(\vec{p}_1, \vec{p}_2)$ shows a maximum at $\Delta\phi = 0$ and $\Delta y = 0$ for like charge pairs: this is the effect of Bose/Einstein correlation.

To investigate the transverse momentum compensation, it has been studied the so called associated compensating transverse momentum density:

$$\pi_t(y_1, y_2) = -(1/N) < \sum_{i=1}^{n} \vec{p}_{Ti} \cdot (\frac{\vec{p}_{T2}}{|\vec{p}_{T2}|}) > .$$

Here, p_{T2} and y_2 are the transverse momentum and rapidity of a selected particle; the sum is over all n charged particles with rapidity in a unit interval of rapidity around y_1 and the average is taken over the sample of N events which have a particle at rapidity y_2; $\pi_T(y_1,y_2)$ is the mean component of transverse momentum locally compensated. The results indicate that the range in rapidity over which p_T is balanced is comparable to the total rapidity range (fig. 44).

c) Charge correlations

To see how charge is compensated, it has been studied the associated charge density balance [43], which answer the following question: how much changes the net charge density at rapidity y_1 when a particle of negative charge and at rapidity y_2 is replaced by a positive charged particle? In fig. 44 the results (dashed dotted line)

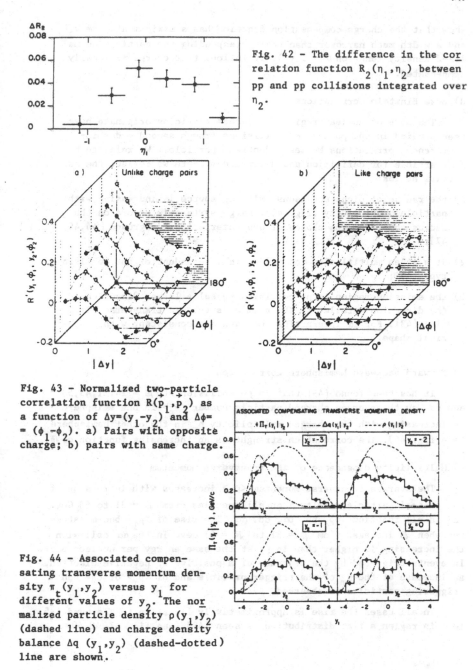

Fig. 42 - The difference in the correlation function $R_2(\eta_1,\eta_2)$ between $\bar{p}p$ and pp collisions integrated over η_2.

Fig. 43 - Normalized two-particle correlation function $R(\vec{p}_1,\vec{p}_2)$ as a function of $\Delta y=(y_1-y_2)$ and $\Delta\phi= (\phi_1-\phi_2)$. a) Pairs with opposite charge; b) pairs with same charge.

Fig. 44 - Associated compensating transverse momentum density $\pi_t(y_1,y_2)$ versus y_1 for different values of y_2. The normalized particle density $\rho(y_1,y_2)$ (dashed line) and charge density balance $\Delta q(y_1,y_2)$ (dashed-dotted) line are shown.

show that the charge compensation function has a maximum at $y_1 = y_2$ and a width much narrower than the corresponding two particle normalized density function $\rho(y_1, y_2)$. It follows that charge is locally compensated.

d) Bose Einstein correlations

The size of nuclear region from where particles originate have been studied in pp, p̄p and $\alpha\alpha$ collisions trough second order interference correlations between identical particles. We refer to W. Zajc article for definition and discussions. Here we collect the main results at ISR:

1) the radius of pions and kaons emitting source is increasing with particle multiplicity (fig. 45). Large systematic errors on the radius come from the choice of non interfering background [44,45, 47];

2) at a given multiplicity the radius is independent of the collision type [44];

3) the emission region in pp collision reveal a larger dimension in the direction parallel to the beam axis than in transverse direction [48]. For $\alpha\alpha$ interaction the data are consistent with a spherical shape.

e) Forward backward hemisphere correlation

It has been found [49] that in pp collision at $\sqrt{s}$ = 24,31,45,53 and 63 Gev the average charged particle number $\langle n_B \rangle$ in one emisphere is increasing with the charged multiplicity n_F in the other hemisphere (see fig. 46). The correlation strenght is an increasing function of $\sqrt{s}$.

f) Multiplicity dependence of the transverse momentum

The average transverse momentum $\langle p_T \rangle$ increases with increasing of the central particle density ρ at all energies from $\sqrt{s}$ = 31 to 63 Gev and in all reactions pp, p̄p, pα, dd, $\alpha\alpha$. The rise of $\langle p_T \rangle$ become steeper when $\sqrt{s}$ increase from $\sqrt{s}$ = 31 to $\sqrt{s}$ = 63 Gev. In the $\alpha\alpha$ collision the increasing is higher than in pp at the same energy per nucleon pair. In events triggered by the presence of a positive or negative electron, an increase of the p_T of the trigger particle and of $\langle p_T \rangle$ of the non triggering particles is observed.

In all cases the rise is approximatively linear up to ρ around 4-5. In this region a flat distribution is seen [48] (see fig. 47,48,49).

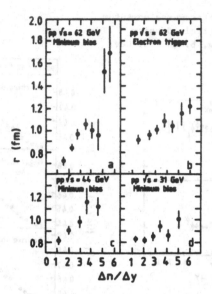

Fig. 45 - Radius of the particle emitting source as function of the charged particle density ρ = Δn/Δy.

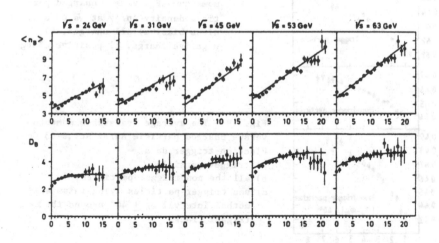

Fig. 46 - Relation between the mean multiplicity $\langle n_B \rangle$ and the dispersion D_B in one hemisphere to the multiplicity in the other hemisphere.

Fig. 47 - The average transverse momentum
$\langle P_T \rangle$ versus charged particle density
$\Delta n/\Delta y$ at a) $\sqrt{s}$ = 44 Gev, b) $\sqrt{s}$ = 31 Gev.

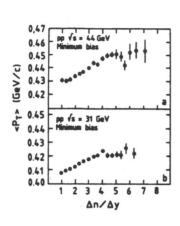

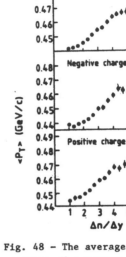

Fig. 48 - The average transverse
momentum $\langle P_t \rangle$ versus charged particle density $\Delta n/\Delta y$ at $\sqrt{s}$ 63 Gev. Mi
nimum bias. a) all charges, b)
negative charge, c) positive charge

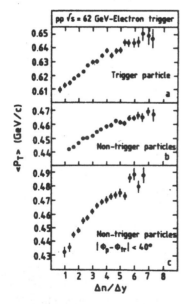

Fig. 49 - The average transverse momentum
versus charged particle $\Delta n/\Delta y$ at $\sqrt{s}$= 63 Gev,
electron trigger data
a) trigger particle,
b) all the non trigger particles,
c) non trigger particles lying in an azimuthal interval of $\pm$ 40° around the
   trigger track.

REFERENCES

[1] L. Camilleri, Phys. Rep. 144 (1987) 52.
    G. Giacomelli and M. Jacob, Phys. Rep. 55 (1979) 1.
    M. Jacob, CERN TH 3807 (1984).
    M. Faessler, Phys. Rep. 115 (1984) 1.

[2] W. Thome et al. Nucl. Phys. B 129 (1977) 365.

[3] A. Breakstone et al., Phys. Rev. D 30 (1984) 528.

[4] M. Basile et al., Lett. Nuovo Cimento 41 (1984) 293.

[5] Contribution from experiment R210 in proton-antiproton collisions
    in the ISR, Discussion meeting between experimentalist and theo-
    rist, eds. M.G. Albrow and M. Jacob, series 2, number 5, p. 29.

[6] W. Bell et al. Phys. Lett. 128B (1983) 349.

[7] T. Akesson et al. Phys. Lett. 119B (1982) 464.

[8] B. Alper et al. Nucl. Phys. B100 (1975) 237.

[9] W. Bell et al., Phys. Lett. 112B (1982) 271.

[10] T. Akesson et al., Phys. Lett. 108B (1982) 58.

[11] A.L.S. Angelis et al., Phys. Lett. 185B (1987) 213.

[12] K. Guetler et al., Nucl. Phys. B116 (1976) 17.

[13] G. Bellettini et al., Phys. Lett. 45B (1973) 69.

[14] T. Del Prete, Riv. Nuovo Cimento 5 (1975) 532.

[15] W. Bell et al. Z. Phys. C 27 (1985) 191.

[16] W. Bell et al. CERN EP/86-15.

[17] I. Otterlund, University of Lund Preprint, LUIP 8606/1986.

[18] A.L.S. Angelis et al., Phys. Lett. 168B (1986) 158.

[19] A.L.S. Angelis et al., Nucl. Phys. B263 (1986) 228.

[20] H. Gordon et al., Phys. Rev. D28 (1983) 2736.

[21] T. Akesson et al., Phys. Lett. 123B (1983) 133.

[22] T. Akesson et al., Nucl. Phys. B228 (1983) 409.

[23] D. Drijard et al., Z. Phys. C 9 (1981) 293.

[24] T. Akesson et al., Nucl. Phys. B246 (1984) 1.

[25] T. Akesson et al., Phys. Rev. Lett. 55 (1985) 2535.

[26] G. Bobbink et al., Nucl. Phys. B 217 (1983) 11.

[27] A.M. Smith et al., Phys. Lett. B185 (1987) 209.

[28] B. Alper et al., Phys. Lett. B46 (1973) 265.

[29] T. Akesson et al., Phys. Lett. B178 (1986) 447.

[30] M. Barone et al., Nucl. Phys. B132 (1978) 29.

[31] M. Heiden, Ph. D. Thesis, University of Heidelberg, 1981.

[32] T. Akesson et al., Phys. Lett. B152 (1985) 411.

[33] T. Akesson et al. CERN-EP/87-16 (1987).

[34] K.L. Giboni et al., Phys. Lett. 85B (1979) 437.

[35] W. Lockmann et al., Phys. Lett. 85B (1979) 443.

[36] D. Drijard et al., Phys. Lett. 81B (1979) 250.

[37] D. Drijard et al., Phys. Lett. 85B (1979) 452.

[38] M. Basile et al., Nuovo Cim., 63A (1981) 230.

[39] K.L. Giboni, Ph. D. Thesis, Aachen 1980.

[40] A. Kernan and G. Van Dalen, Phys. Rep. 106 (1984) 297.

[41] S.R. Amendolia et al., Phys. Lett. 48B (1974) 359.

[42] W. Bell et al., Z. Phys. C 22 (1984) 109.

[43] D. Drijard et al., Nucl. Phys. B155 (1979) 269.

[44] T. Akesson et al., Phys. Lett. 129B (1983) 269.

[45] T. Akesson et al., Phys. Lett. 155B (1985) 128.

[46] A. Breakstone et al., Phys. Lett. B132 (1983) 458.

[47] A. Breakstone et al., Z. Phys. C 33 (1987) 333.

[48] T. Akesson et al., CERN-EP/86-215 (Submitted to Phys. Lett. B).

[49] S. Uhlig et al., Nucl. Phys. B132 (1978) 15.

[50] A. Breakstone et al., Phys. Lett. 183B (1987) 229.

[51] W. Geist, Proc. Moriond workshop on new flavours. Les Arcs, 1982.
     M. Basile et al., Charm production at the CERN ISR, CERN/EP Internal Report 81-02.

HADRON-NUCLEAR INTERACTIONS

W. D. WALKER and P. C. BHAT
Duke University, Durham NC 27706

ABSTRACT

This paper relies mainly on data taken with 100 GeV/c hadrons incident on targets of Mg, Ag and Au. We present arguments and data which strongly indicate that it is possible to analyze hadron-nuclear collisions on the basis of initial impact parameter. We have observed the pion multiplicity for central and peripheral collisions with Ag and Au nuclei. We find evidence for the absorption of the secondary particles as they traverse the struck nucleus. We also find evidence that a considerable amount of the energy of the hadronic cascade is transferred to struck nucleons. We see very little evidence for either boson or nucleon resonance production among the produced particles. In other respects the Lund String Model program, FRITIOF, agrees well with the data. The impact parameter division also seems to be confirmed by the program. The bombarding energy dependence for multiplicity seems to confirm the Limiting Fragmentation Hypothesis.

I. INTRODUCTION

The study of hadron-nuclear collisions dates back to the early days of cosmic ray experiments on nuclear interactions.[C1] One of the first questions that arose as to whether there would be differences between hadron-nucleon and hadron-nuclear collisions since experimenters used light nuclei as a substitute for a nucleon target. Collisions involving light nuclei were the closest thing available to collisions with a hydrogen target. The advent of liquid hydrogen targets and multi-GeV accelerators such as the Cosmotron ended the use of nuclear targets at least as substitutes for nucleon targets.

The situation today is quite different from the previous period. We have available extensive data on both hadron-nucleon and hadron-nuclear interactions and can make useful comparisons between these collisions. This is to be the purpose of the present paper.

We will attempt to make deductions concerning the hadronic cascade that occurs in the matter of the nuclear target. The data together with specific models should allow one to draw at least qualitative conclusions concerning the gestation period for particle production.

Figure 1 shows a fairly typical π^- gold collision with 14 or 15 identifiable protons emerging along with 11 positive and 12 negative minimum tracks. This type of interaction is characteristic of hadron-nuclear collision in that they are

155

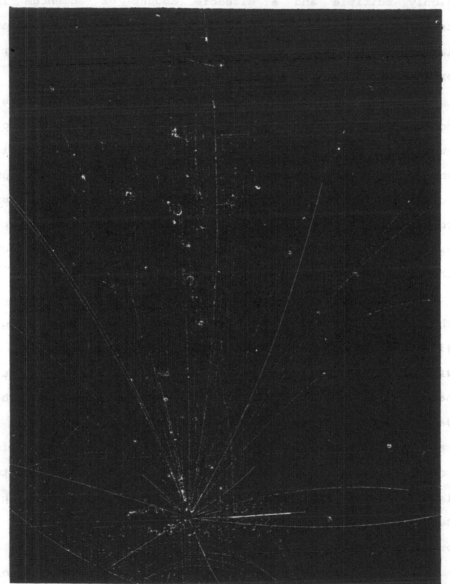

Figure 1: A π⁻ gold collision occurring in a .3 mm thick
gold foil. There are 14 or 15 identifiable protons emerging
from the collision as well as 11 positive and 12 negative
minimum ionizing particles.

highly multiple and a sizeable amount of momentum is transferred to recoiling nucleons. The pions are sprayed at large angles with respect to the direction of the incident projectile.

A very direct and important use of hadron-nuclear collisions has to do with our understanding of nucleus-nucleus collisions. The important question is whether one can account for the data on nucleus-nucleus collisions as a linear superposition of nucleon-nucleus collisions. It is very conceivable that nonlinear modes are excited which would be of interest and would possibly reflect the existence of a quark-gluon plasma.

We have utilized several sources of data in our review. The data of Barton et al[B1] is the most complete data on the differential inclusive cross section for the production of a wide range of particles coming from collisions of 100 GeV hadrons with a selection of nuclei. We also have used the data coming from experiments on π-neon collisions over a range of energies which have been done at Duke University in collaboration with SUNY-Albany.[D1,D2] The most recent data is that of E597 which is a collaboration involving Fermilab, Duke, Notre Dame and Penn State Universities. We give a brief discussion of this experiment.

Since we are interested in the spatial development of hadronic showers we then continue with a discussion of the production of black and gray protons which emerge from the

collisions. These particles are important signatures and we
believe they are strongly correlated with the impact
parameter of the collision.

II. DESCRIPTION OF THE EXPERIMENT E597

The experiment E597 was done using the Fermilab 30"
hydrogen bubble chamber. The bubble chamber was fitted with
three pairs of small foils, one pair each for targets silver,
gold and magnesium. The thicknesses were such as to be a
small fraction of a radiation length in order to minimize the
effect of photon conversion. The thicknesses and areas are
such as to facilitate a correction to zero target thickness.
The bubble chamber was used in order to determine the
particle ionization and momentum and thus obtain good π-p
discrimination up to a momenta of about 1 GeV/c. The fact
that one can readily identify protons is clear from Figure 1.
The chamber was exposed to 100 GeV beams of π^+'s, protons,
and a mixture of π^-'s and $\bar{p}$'s. A short portion of the run
was devoted to 320 GeV π^-'s. The incident particles were
tagged by Cerenkov counters and wire chambers upstream of the
bubble chamber. The experimental set-up is shown in Figure
2. In the present work, the downstream chambers were used
only for sharpening the momentum determinations.

III. IMPACT PARAMETER ANALYSIS

It has been known for nearly forty years that a strong

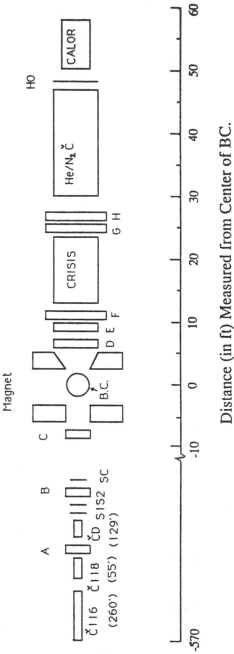

Distance (in ft) Measured from Center of BC.

Figure 2: The experimental arrangement for experiment E597. In this report we mainly utilize the upstream Cerenkov counters for identification as well as the downstream drift chambers.

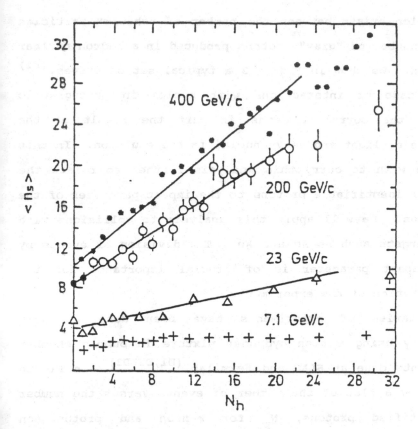

Figure 3: The mean shower multiplicity versus the number of heavily ionizing tracks observed in proton-emulsion experiments. The bombarding energies are noted on the figure. This figure was taken from the review of Frederickson et al.

correlation exists between the number of "shower" particles and the number of "gray" protons produced in a hadron nuclear collision. We show in Figure 3 a typical set of curves.[F1] In the case of interactions with nuclei in photographic emulsion the correlation was in part the result of the existence of light and heavy nuclei in the emulsion. In this paper we wish to carry this analysis further to relate the number of identifiable protons to the impact parameter of the collisions. We will apply this analysis to collisions with heavy targets such as Ag and Au. The division of events by their impact parameter is of crucial importance for the interpretation of the experiments.

A series of experiments have been done at Duke University using a neon hydrogen mixture in bubble chamber experiments done at SLAC and Fermilab.[D1,D2,D3] In Figure 4, we show a plot of the number of events versus the number of identified protons, N_p for π-neon and proton-neon collisions. The bombarding energies range from 10.5 GeV to 300 GeV for the proton projectiles.[A1] There are small differences in the number of identified protons, 1.79 at 200 Gev[D3] and 1.95 at 10.5 Gev[D2] The difference which is considerably larger than can be accounted for by statistical uncertainties may in part be due to slightly different operating conditions for the bubble chamber. It is likely however that there is a small difference since slightly fewer protons are easily identified with the higher energy

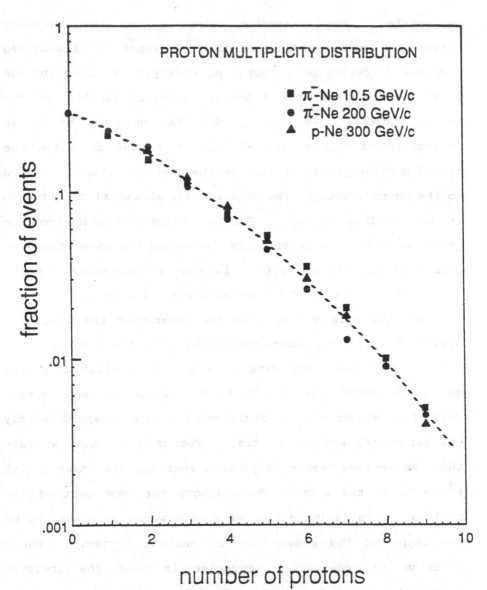

Figure 4: The fraction of events which have a given number of identified protons from hadron-neon collisions. The curves seem to be nearly independent of bombarding energy and projectile.

projectiles. The remarkable thing is that the frequency curves are nearly identical over a range of bombarding energies differing by a factor of thirty. The distribution of events in N_p is thus a nearly universal function of the target nucleus. We believe that the quantity of N_p is indicative of the volume of nuclear matter in which the reaction takes place. It is, on the average, closely related to the chord length of the axis of this shower or flight path in the nuclear matter. Further evidence comes from an experiment[B3] done at the CERN SPS which has shown that the number of identified protons is nearly independent of the projectile and energy for heavier nuclei than neon.

For the data rising from the experiment E597, we have divided the heavy element collisions into two groups: those with $N_p \leq 4$ and those with $N_p \geq 5$. This division placed comparable numbers of events in the peripheral and central collision categories. The division of events is approximately 60% peripheral and 40% central. For this division we note that the average number of protons emerging goes nearly with ℓ^3, ℓ being the average chord-length for that part of the nucleus. This is indicative of a cone which we assume to be the shape of the shower in the nuclear matter. Such a division is necessarily imprecise in that the incident projectile can strike the target with small impact parameter and yet travel several mean free paths before interacting. Such a collision will produce only a small number of recoil protons and be classified as a peripheral collision, but fortunately this type of collision is rather improbable. The physics we wish to address, however, has to do with the path length of the secondary particles in the nuclear matter.

We have divided up the cross sections for collisions with Ag and Au geometrically. By this we mean

$$\Delta\sigma = \int 2\pi b db (1 - e^{-\ell/\lambda})$$

where ℓ is the chord length for the impact parameter. We show in Figure 5 a plot of N_p against the average value of ℓ. In this calculation we assume a uniform distribuion of nuclear matter. The line through the points is given by $<N_p> = A\ell^3$. This suggests that N_p for a given interaction is proportional to the volume of a cone swept out by the reaction products of the interaction.

In Figures 6 and 7 we show the distribution of the number of gray protons produced in collisions with the projectile with the nuclei Ne, Ag, Au. These data were obtained from experiment E597 and the earlier Ne experiments. The curves were obtained using two models of the matter distributions in the nuclei. Model A assumes a uniform density up to a radius $R = 1.15 A^{1/3}$. Model B is done using the well known Woods-Saxon model of nuclear density. Both models show discrepancies for small numbers of identified protons. We believe this to be the result of diffractive collisions which are not considered in our simple model. The

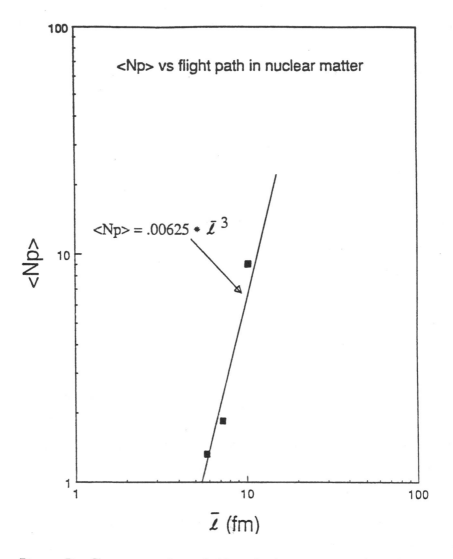

Figure 5: The mean number of identified protons is plotted against the mean chord length for peripheral and central collisions with silver and gold nuclei and with central collisions with magnesium. We have corrected the magnesium data to account for the different fraction of neutrons and protons in magnesium as compared to the heavy elements.

calculations are done by assuming that all of the nucleons along the path of the shower are swept out by the emerging hadronic shower. The volume swept out is a truncated cone of initial radius of about 1 fermi and half-angle of 0.1 radian. The curve was fitted for the Ne data and then the same parameters were used for the heavier nuclei.

The curves developed from our simple geometrical models give a modestly good description of the observed distributions in N_p. If we included the non-identified protons in the distributions the agreement would be better.

Table I gives a comparison of our magnesium data with the peripheral collisions involving Ag and Au based on the Np selection. The agreement of the numbers from the central Mg collision and peripheral collisions from Ag and Au are also good. We also give the values of ℓ, the average path-length of the projectile in nuclear matter measured in Fermis. We believe the divisions of the collisions with nuclei into peripheral and central are rather accurate.

IV. MULTIPLICITY

Table II and Table III compare the number of shower particles produced in hadron collisions with protons, magnesium, silver and gold nuclei for different numbers of recoiling protons. These numbers have been corrected for the finite thickness of the targets. The tables are broken into different momentum ranges of the secondaries and into central

166

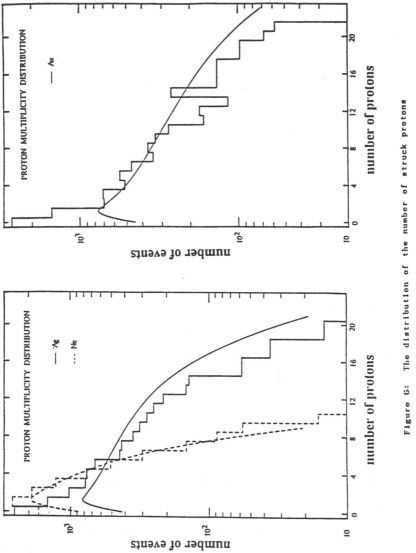

Figure G: The distribution of the number of struck protons emerging from a hadron-nuclear encounter. The nucleus is assumed to have a uniform density of nucleons and a radius

R = (1.15)A1/3.

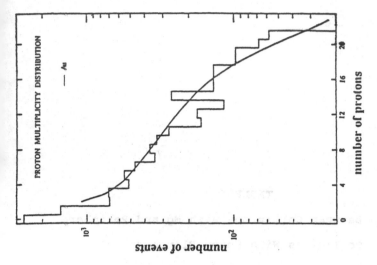

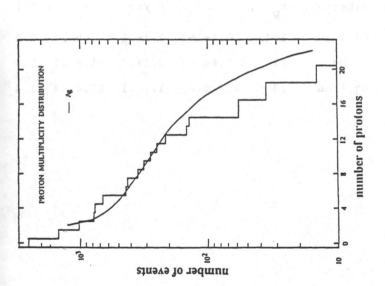

Figure 7: The distribution of the number of struck protons emerging from a hadron-nuclear encounter. The nucleus is assumed to have a nucleonic density given by the Woods-Saxon distribution. This distribution seems to agree better with the data.

TABLE I

Comparison Between Collisions with Mg and Peripheral

Collisions With Heavy Elements

% of Interactions	Material	N_p	$<n\pm>$	$<y\pm>$	$<N_p>$	$\bar{\ell}$ (fm)
60	Ag + Au	1-4	10.4±.5	2.1±.1	1.9±.1	7.2
68	Mg	≥1	9.6±.8	2.1±.1	1.6±.1	5.9
40	Ag + Au	≥5	17.7±.8	1.7±.1	11.4±.4	11.4

TABLE II

p(p̄) Collisions

Range of P_L		$N_p > 1$		--- $N_p < 4$ ---		--- $N_p > 5$ ---	
		p(p̄)-p	p(p̄)-Mg	p(p̄)-Ag	p(p̄)-Au	p(p̄)-Ag	p(p̄)-Au
>20 GeV/c	No. of +(−)	0.62(0.56)	0.48(0.60)	0.59(0.48)	0.76(0.60)	0.44(0.33)	0.26(0.28)
	No. of −(+)	0.10(0.12)	0.22(0.18)	0.11(0.18)	0.12(0.10)	0.10(0.12)	0.05(0.17)
	$\Sigma P^+_{(-)}$ (GeV/c)	36.4(28.2)	16.3(22.8)	24.8(20.9)	32.2(26.9)	15.0(12.9)	9.7(9.3)
	$\Sigma P^-_{(+)}$ (GeV/c)	6.2(3.7)	7.3(6.6)	4.0(6.2)	4.4(4.4)	4.7(4.3)	2.4(5.5)
10-20 GeV/c	No. of +(−)	0.39(0.37)	0.44(0.61)	0.58(0.47)	0.49(0.43)	0.60(0.46)	0.47(0.51)
	No. of −(+)	0.19(0.20)	0.41(0.31)	0.32(0.33)	0.23(0.37)	0.35(0.40)	0.34(0.38)
	$\Sigma P^+_{(-)}$ (GeV/c)	4.9(6.3)	7.1(9.1)	8.8(7.4)	7.8(6.6)	8.7(7.0)	6.8(8.1)
	$\Sigma P^-_{(+)}$ (GeV/c)	3.5(2.9)	6.7(5.1)	5.1(5.4)	4.2(5.6)	5.2(5.8)	5.3(6.2)
<10 GeV/c	No. of +(−)	2.9(2.8)	5.1(4.2)	4.2(4.0)	3.6(3.8)	7.6(7.3)	9.2(8.0)
	No. of −(+)	1.9(3.0)	4.2(4.5)	3.2(4.7)	3.1(3.8)	6.4(8.1)	7.1(9.2)
	$\Sigma P^+_{(-)}$ (GeV/c)	7.4(6.1)	13.1(10.9)	9.9(9.2)	10.1(9.1)	15.6(12.4)	18.3(14.2)
	$\Sigma P^-_{(+)}$ (GeV/c)	3.9(6.1)	10.3(10.3)	6.8(10.2)	7.6(10.5)	10.6(15.4)	10.9(16.0)
Total Number of Events			27(67)	202(144)	86(125)	86(95)	58(78)

TABLE III

π⁺(π⁻) Collisions

Range of P_L		$\pi^+(\pi^-)$-p	$N_p >1$ $\pi^+(\pi^-)$-Mg	--- $N_p <4$ --- $\pi^+(\pi^-)$-Ag*	$\pi^+(\pi^-)$-Au*	--- $N_p >5$ --- $\pi^+(\pi^-)$-Ag	$\pi^+(\pi^-)$-Au
>20 GeV/c	No. of +(−)	0.59(0.49)	0.59(0.43)	0.59(0.48)	0.57(0.53)	0.31(0.28)	0.24(0.31)
	No. of −(+)	0.20(0.21)	0.18(0.23)	0.23(0.26)	0.27(0.33)	0.19(0.25)	0.16(0.22)
	$\Sigma P_{+(-)}$ (GeV/c)	26.9(23.3)	22.7(17.4)	23.4(18.9)	24.0(20.8)	10.6(11.6)	8.9(11.9)
	$\Sigma P_{-(+)}$ (GeV/c)	7.3(7.1)	6.8(8.5)	9.0(9.1)	10.1(12.2)	7.5(9.4)	7.2(6.8)
10-20 GeV/c	No. of +(−)	0.66(0.33)	0.56(0.54)	0.56(0.55)	0.51(0.56)	0.50(0.40)	0.54(0.62)
	No. of −(+)	0.27(0.24)	0.38(0.44)	0.30(0.44)	0.31(0.33)	0.38(0.38)	0.36(0.51)
	$\Sigma P_{+(-)}$ (GeV/c)	6.7(5.8)	8.8(8.3)	8.6(8.3)	7.8(8.5)	7.6(6.4)	8.5(9.2)
	$\Sigma P_{-(+)}$ (GeV/c)	3.5(3.5)	6.0(7.0)	5.0(6.9)	5.2(5.2)	6.1(5.9)	5.6(7.8)
<10 GeV/c	No. of +(−)	3.0(2.5)	4.1(3.8)	4.0(3.5)	4.0(3.6)	7.1(5.9)	8.4(6.3)
	No. of −(+)	1.9(2.6)	3.6(3.6)	2.9(3.6)	2.9(3.6)	5.4(6.5)	6.5(7.4)
	$\Sigma P_{+(-)}$ (GeV/c)	5.6(5.6)	10.8(10.2)	9.7(8.3)	11.2(8.6)	14.5(11.7)	16.3(11.9)
	$\Sigma P_{-(+)}$ (GeV/c)	3.8(5.3)	8.5(8.3)	7.1(9.4)	7.4(9.2)	8.5(13.1)	11.7(13.9)
Total Number of Events			92(81)	230(203)	173(178)	108(103)	94(111)

* In the analysis of π⁻-Ag and π⁻-Au events with $N_p <4$, the criteria for accepting events are (i) that there be at least four shower particles and (ii) that for events with either 4 or 5 shower particles, the momenta of each of the tracks should be greater than 1 GeV/c.

TABLE IV

Multiplicity Ratios for Secondary
Particles of Momentum Less than 10 GeV/c

Projectile	Target	v	$R = \dfrac{(No.-)nuc^{*}}{(No.\pm)p}$
π	Ag, Np$\leq$4	1.94	1.30±.06
π	Au, Np$\leq$4	2.10	1.30±.06
P	Ag, Np$\leq$4	2.51	1.52±.08
P	Ag, Np$\leq$4	2.77	1.47±.11
π	Ag, Np$\geq$4	3.15	2.41±.07
π	Au, Np$\geq$4	3.65	2.57±.07
P	Ag, Np$\geq$4	4.74	2.91±.08
P	Au, Np$\geq$4	5.60	3.21±.10
π^{-}	Ne, Np$\geq$2	1.78	1.45±.03
P	Ne, Np$\geq$2	2.28	1.80±.04
π	Mg, Np$\geq$1	1.94	1.50±.15
P	Mg, Np$\geq$1	2.59	1.80±.20

* We have tabulated the number of negative particles only for the data on Ag, Au and Mg. In this way we avoid counting fast recoil protons. For π^{-} incident we have tabulated the number of π^{-} particles and for π^{+} we have tabulated the π^{-}-particles produced. The sum of these numbers is taken as the multiplicity for $\pi^{\pm}$.

and peripheral collisions in the case of the nuclei.   The hydrogen collisions have had the recoil protons removed from the low energy secondaries.   In the tables we also include the momentum carried by these produced particles.   A number of conclusions can be drawn from the results given in the tables.   The number of forward moving particles is diminished very little for collisions with the light nucleus, Mg, and for the peripheral collisions with Ag and Au.   In the case of the central collisions the number of particles of momentum 20 GeV/c or greater is decreased by a factor of the order of two.   The averge momentum carried by these particles is decreased by over a factor of two.   If we look at the numbers of particles of momentum of less than 10 GeV/c we see dramatic differences in the number of produced particles which are mainly pions.   For the central collisions with Ag and Au we find three times as many particles produced as in the case of hadron-proton collisions.   In Table IV we give the ratio of these multiplicities.

Table IV gives a comparison of the multiplicity of the produced π's and K's for the central and peripheral collisions with Ag and Au. We also include the data obtained from π^--Ne[D3] collisions and P-Ne[A1] collisions at projectile energies of 200 and 300 GeV respectively.   In all of these data we have excluded only particles with momentum less than 10 GeV/c as explained in the table. We have calculated υ by the formula: $\upsilon = (A_{in}\, \sigma_p)/\sigma_{in}$, where A_{in} is the atomic weight of the part of the nucleus considered and σ_{in} is the cross section for that part. In our procedure the nucleus is divided up in cross sections by the fraction of the cross section with $N_p \leq 4$ and $N_p > 4$ for Ag and Au. Using the Woods-Saxon distribution of matter we calculate what impact parameter corresponds to the observed division of cross section. Using this impact parameter we calculate the atomic number, A_{in}, included inside and outside the impact parameter. The same procedure is used for the data with Ne as a target. Examination of the table shows the usual correlation between R and υ. There is another effect, namely that for comparable values of υ the value of R is greater for the central collisions in the light nuclei than for peripheral collisions with a heavy nucleus. If one considers the geometry of the two situations then the results seem plausible. The exit path for the secondary particles from a light nucleus will on average traverse less nuclear matter than for secondaries from a peripheral collision with a heavy nucleus. The absorption of the secondary particles must be an important consideration.

We know that the ratio R is less than one from projectile incident on modestly heavy nuclei for bombarding energies less than 10 GeV. This means that the absorption of produced pions of momentum less than 1 GeV/c must be appreciable. We would also believe that objects like ρ, ω^o, Λ's etc. are strongly absorbed in nuclear matter. For example, the reaction $\rho^+ + d \rightarrow p + p$ would be a strongly exothermic reaction and would be expected to have a large cross section.

We have discussed the average multiplicities of particles produced in collisions with various nuclei. We next consider the distribution of multiplicities of the produced particles. The standard distribution is the usual KNO distribution which seems to be valid for hadronic projectile energies ranging from a few GeV to 800 GeV.[S1,K1] We show in Figure 8 the distributions plotted in the standard fashion. For the nuclear collisions we have the distribution in the number of negative particles produced in π^-, $\bar{p}$ collisions with Ag and Au as well as π^--Ne collisions. The heavy element collisions are divided into peripheral and central collisions. The scaling curve for the peripheral collisions seems to be consistent with the Ne and hydrogen data. In Figure 9 we plot the data for central collisions with Ag and Au. There seems to be a considerable difference between these data and the rest of the sample. There is a lack of events at the lowest and highest multiplicities and more in the region of $n/\langle n \rangle \cong 1$. The central collisions probably suppress the low multiplicity side because of a lack of diffractive events. The high multiplicity side may be suppressed because of the depletion of the energy available.

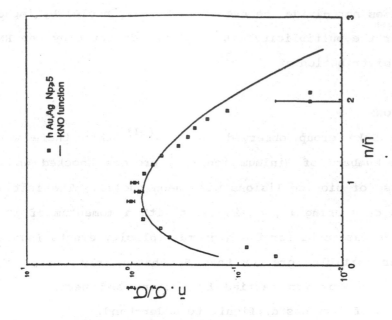

Figure 9: The scaled multiplicity distribution for negative particles emerging from central collisions of hadrons with silver and gold nuclei.

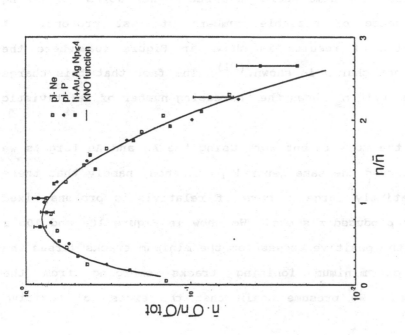

Figure 8: A scaled multiplicity distribution for negative particles from π⁻p, π⁻Ne and peripheral collisions with silver and gold nuclei.

If one does not divide the events from heavy nuclei by impact parameter the multiplicity distributions do not obey the KNO scaling distribution at all.[B2]

V. BARYONS

The Duke group observed in 1975[D1] that there were sizeable numbers of minimum ionizing protons knocked on in the course of pion collisions with neon nuclei. The initial work was done using π^+, π^- incident with a momentum of 10.5 GeV/c. It was found for the high multiplicity events (number of shower particles greater than 5) that nearly 40% of the outgoing momentum was carried by protons and neutrons.[D2] This piece of data was difficult to understand.

Since that time there have been other works confirming the presence of sizeable numbers of fast protons. A compilation of results is shown in Figure 10a where the average net charge is shown.[F1] The fact that this charge increases with n_s shows the increasing number of relativistic protons.

In the more recent work using the Ag and Au targets we have observed the same general phenomenon, namely that there are relatively large numbers of relativistic protons mixed with the produced π's etc. We show in Figure 10a and 10b a plot of the positive excess for the minimum tracks versus the number of minimum ionizing tracks emerging from the collision. We presume again that the excess of positive

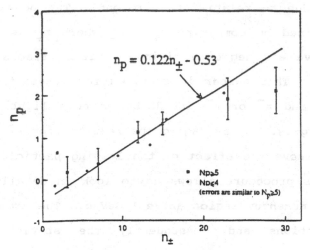

Figure 10b: The number of minimum ionizing protons emerging from central and peripheral collisions with silver and gold nuclei.

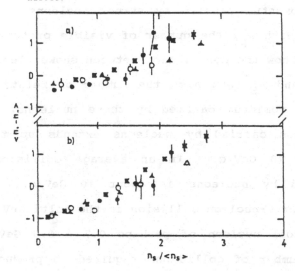

Fig.10a The average net charge $\langle n_+ - n_- \rangle$ versus the scaled number of shower particles ($n_s = n_+ + n_-$) in 40 GeV/c $\pi^- C$ collisions (squares) and in 10.5 GeV/c (triangles), 25 GeV/c (open circles) and 50 GeV/c (solid circles) $\pi^- Ne$ collisions. (a) Even values of n_s (b) Odd values of n_s.

particles to be due to relativistic protons. The positive excess is determined by comparing $n_+ - n_-$ where n_+ is the number of positive and negative mimimum ionizing tracks or identified pions. This number is compiled for interactions initiated by π^+ and π^- or π^- incident for both light and heavy nuclear targets. By averaging the results from π^+ and π^- incident we remove the effect of the leading particle in our sample. This procedure allows us to look in detail at particles in the momentum region above 1 GeV/c. The process requires subtractions and consequently the statistical accuracy is not great.

In our Figure 11 we note that both the number of shower particles and momentum carried by struck nucleons increase nearly linearly with N_p, the number of visible protons. We believe this to show the correlation between shower length in nuclear matter and $n_\pm$, and also the number of relativistic protons and the momentum carried by these nucleons. Note that the momentum carried by nucleons extends up to the neighborhood of 20 GeV/c. In an average collision the momentum carried by nucleons is about 10 GeV/c. In a nucleon-nucleon or π-nucleon collision in the multi GeV range we expect a recoil nucleon of perhaps 0.7 - 0.9 GeV/c on average. The number of collisions required to produce the nucleons observed seems prohibitively large if we require this to be done by the incident hadron. To understand these results, we need an efficient mechanism to transfer the

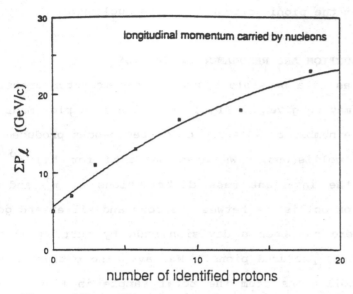

Figure 11b: The amount of momentum carried by nucleons emerging from collisions with silver and gold nuclei.

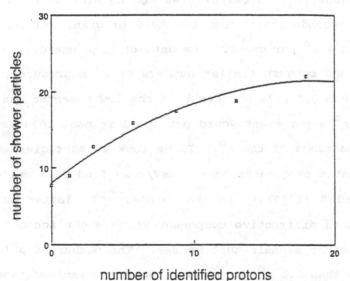

Figure 11a: A plot of the number of shower (minimum ionizing tracks) against the number of visible protons emerging from collisions with silver and gold nuclei.

momentum from the pionic products to the nucleons.

VI. ρ^o PRODUCTION AND RESONANCE PRODUCTION

A hint as to a possible mechanism for momentum transfer to nucleons may be given in Figure 12 which is a plot made to determine the number of ρ^o's and other resonances produced in the nuclear collisions. We have searched for the ρ^o by looking at the invariant mass distributions of π^+ and π^- resulting from collisions between hadrons and silver and gold nuclei. There has been a division made by cutting on the momentum of the produced pions. We have also extracted the peripheral collisions from the total sample in Figure 12b. In Figure 13 we look for the ρ^o among the high energy secondary pions ($p_\pi > 3$ GeV/c). We see no hint of the ρ^o in the slower secondaries. From the Lund program, FRITIOF, we expect about 1 ρ^o per event. The data on high energy p-p and π-p collisions suggest similar numbers of ρ^o's produced. We find less than 0.1 ρ^o's per event in the low momentum sample. One half a ρ^o's per event would put an additional 1000 counts in this mass range of the ρ^o. If we look at particles which have a momentum of greater than 3 GeV/c we find some evidence for ρ^o's and S (f^o(975)) in the sample. The latter sample must have some diffractive component which might account for the ρ^o and other signals that we see. The number of ρ^o's we see is less than 0.1 ρ^o per event. We have examined results from an experiment on π-Ne collisions at 10.5 GeV.[D3] These

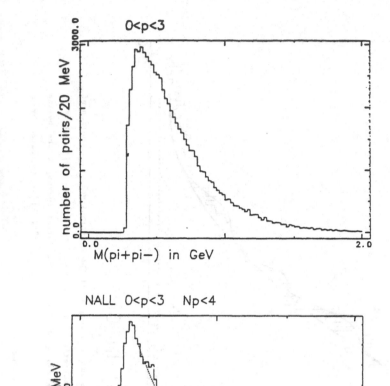

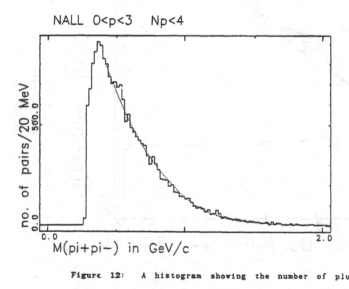

Figure 12: A histogram showing the number of plus-minus pairs of the low momentum pions emerging from hadron collisions with gold and silver nuclei. The lower figure shows the sample from peripheral collisions. There is some evidence of K^0 decays which were not identified but very little if any evidence of ρ^0 production. The upper histogram shows the result of 2000 collisions.

182

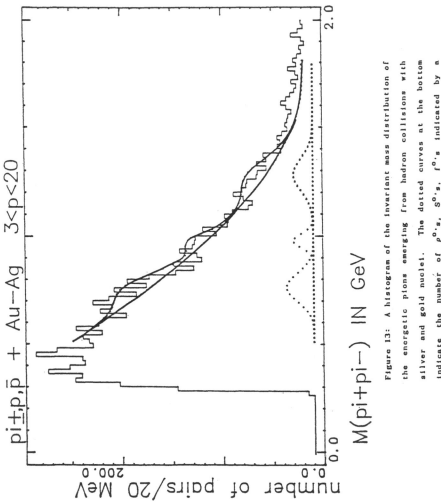

Figure 13: A histogram of the invariant mass distribution of the energetic pions emerging from hadron collisions with silver and gold nuclei. The dotted curves at the bottom indicate the number of ρ^0's, S^0's, f^0's indicated by a fitting program.

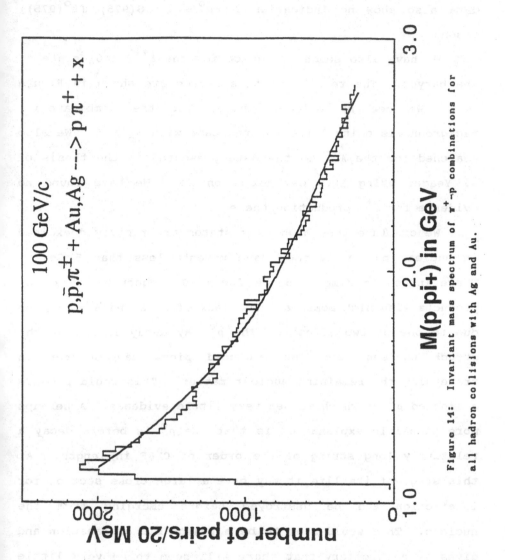

Figure 14: Invariant mass spectrum of π^+-p combinations for all hadron collisions with Ag and Au.

data also show no indication of ρ^o's or S(975) (f^o(975)) mesons.

We have also sought evidence for the Δ^{++}(1240) state of the baryon. The results of this search are shown in Figure 14. We see no evidence here, but the combinatorial background is quite large for the case with $N_p \geq 5$. We also searched for the Δ^{++} in the data presented in the thesis of W. Yeager using 10.5 GeV pions on Ne. We have found no evidence for Δ^{++} production there.

We conclude that high mass states are rapidly deexcited in nuclear matter if they have momenta less than 5 GeV/c. There indeed is some evidence for ρ, S production among the products with high momenta. The lack of ρ^o's and S^o's may be due to one of two effects. The ρ^o may decay inside of the struck nucleus, and the produced pions may scatter in traversing the remaining nuclear matter. This would produce a smeared ρ^o of which we see very little evidence. A perhaps more plausible explanation is that the ρ^o is before decay a relatively long string of the order of 0.8F in length. At this stage of its life it may have a large cross section for interaction and be destroyed before emerging from the nucleus. This would seem to be a reasonable explanation and gives as a corollary that there will seem to be very little resonance production in hadron-nuclear collisions. Indeed the ρ^o is dominantly produced in the soft or peripheral interactions at low energies. For pions incident on nucleons it is dominantly produced in one-pion-exchange or diffractive interactions. In photoproduction the ρ^o is primarily produced diffractively. Both of these processes tend to be very peripheral and are consistent with a physically large ρ^o. In any event, the deexcitation process is likely to be a very efficient means of transferring momentum to the target nucleons from pions.

VII. HADRONIC CASCADE

By breaking the events up into central and peripheral collisions for the events in Ag and Au we can begin to see how the cascade develops as the hadrons propagate through the nuclear matter. We examine the different energy domains of the secondary particle to gain insight as to what happens as the incident hadron traverses the nucleus. For the upper energies of the secondaries we use the data of Barton et al. We show in Figure 15 the sum[B1] of the energy carried by particles of momenta greater than 50 GeV/c as a function of average flight-path through nuclear matter. If one looks at the data with the eye of the wounded quark model [B5] one sees an interesting result.

Using the definition of "Radiation Length" in the same way as in electromagnetic shower theory, $\frac{dE}{dx} = E/L$.[D4] One finds an L = 7.6 F for nucleons incident and L = 10 F for pions incident. These results are dependent on where the energy cut is made. The idea of "Radiation Length" for a

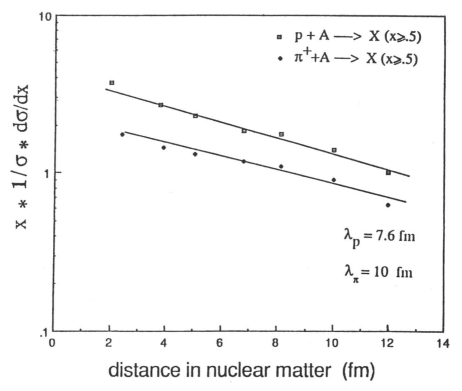

Figure 15: A plot of the energy content of particles of momentum more than 50 GeV/c plotted against the average path-length of the incident particle in nuclear matter. These data are from the experiment of Barton et al.

quark-gluon shower must be a qualitative rather than quantitative concept since not only is there a cascade shower going on but the fragmentation functions must play a large role in estimating the energy of the "spectator" quark or diquark. However, it seems clear that the diquark has a larger cross section for energy loss than the quark or anti-quark spectator.

The lower energy domain can be studied by an examination of Tables II and III which shows that the energy content of particles with momenta between 10 and 20 GeV/c varies little as the shower develops in nuclear matter. The "flow in" from higher energy particles is balanced by "flow out" into lower energy particles.

The energy or momentum content of the low energy component of the shower gives the most insight as to the development of the cascade. We show in Figure 16 the result of plotting these quantities as a function of distances in nuclear matter. The energy content of the pions seem to grow up to a distance of about 12 F. The momentum content of the nucleonic content of the shower is increasing rapidly with the flight path in nuclear matter. It could be that the momentum content of the nucleonic component increase as ℓ^3, ℓ being the flight path of the cascade products in nuclear matter. It is interesting that the multiplicity of pions seems to parallel the growth in the nucleonic component of the showers. The action seems to take place in a cone inside

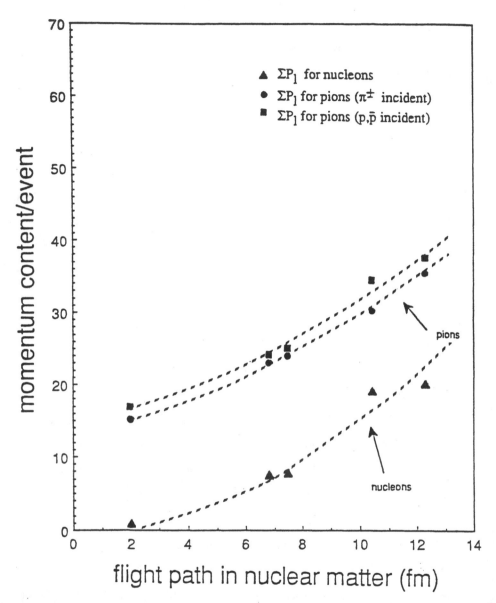

Figure 16: A plot of the longitudinal momentum content of struck nucleons and pions of momentum less than 10 GeV/c as a function of average path length in nuclear matter.

of the struck nucleus. After the initial interaction the strings that are produced begin to stretch. The slower objects may indeed fragment in the struck nucleus. As the produced strings stretch and fragment they develop larger color dipoles. The products begin to develop their cross section as they develop color dipoles. For very short time the color charges of the produced objects are very close together and have very little dipole strength. As the quarks separate with passing time their ultimate dipole strength is achieved and their ability to interact reaches its assymptotic value. It is also likely that as the strings stretch the gluons composing the string interact which contributes to the cascade.

The struck target quarks probably materialize into observable particles inside the nucleus. In the scale of distances between 5 and 10 Fermis, quarks and gluons in the few GeV range are likely to be materializing. These particles will have developed something like their assymptotic cross section (i.e. their ability to interact) inside the nucleus. Thus we believe that we see the products of the interactions of 1-2 GeV pions, ρ's etc, inside of the struck nucleus. This statement is supported by the fact that we see some ρ^{o}'s etc, when they have momenta above 5-6 GeV/c but essentially none below that momentum. The other evidence indicating that pions in the 1-2 GeV region indeed materialize (or develop their assymptotic cross section) comes from the enhancement in the $y = 1, 2$ region for the Np ≥ 5 events. These rapidity distributions are shown in Figures 17 and 18. From these results we say that the materialization distance is $0.7\tau_\pi\beta_\pi$ for pions. The constant (0.7F) is accurate to perhaps 50%.

VIII. ENERGY DEPOSITION

It was pointed out some years ago that a new form of hadronic matter might be formed at high energies. The name given this new form of matter was a quark-gluon plasma.[M1,B4] At sufficiently large energy densities the quarks are deconfined and consequently no longer team up in pairs or triplets of colorless objects. According to the estimates made by various authors a transition between ordinary hadronic matter and the plasma might be achieved when the energy density reaches the order of 1 GeV per (Fermi)^3. This means that such a transition might occur if the energy density had this value in a particular rest frame.

Several authors have analyzed the data on the forward going secondary particles derived from the experiment of Barton et al.[B1] Using these data Goldhaber and Busza[G1] found that a proton striking the center of a lead nucleus would lose 3 or 4 units of rapidity in the average collision. Data from this experiment confirm this result.[T1]

Since we have extensive data on the less energetic secondaries from hadron-nuclear collisions we have analyzed the data in a different way. Ideally one would like to be able to trace the diffusion of the projectile energy down through the struck nucleus. One of the points of interest would be to see if there is an accumulation of energy deposition which approaches the 1 GeV/F^3 necessary for a phase transition to the quark-gluon state.

In our considerations we sum the energy deposited by the projectile in particles which have a momentum of less than 10 GeV/c, between 10 and 20 GeV/c and finally greater than 20 GeV/c. The sums of the momenta carried by the lowest energy of particles are plotted in Figure 16. Examining Figure 16 we see that the momentum content of the low energy ($p_L \leq$ 10 GeV/c) pion group increases by about 20 GeV/c in traversing the order of 10 Fermis of nuclear matter. As the particles exit from the nucleus they are in a disc which has a diameter of the order of 3 Fermis and is perhaps 0.5 Fermi thick. The volume of such a disc is 4.5 F^3. The energy content of this disc is about 30 GeV. This would seem to be plenty of energy to produce a quark-gluon plasma. The products contained in this disc are spread over 4 units of rapidity. The question of the formation of a quark-gluon plasma is dependent on whether there is much mixing or interaction of these products. We find none of the expected signatures of the quark-gluon plasma except high multiplicity.

IX. MODEL COMPARISON

We have looked at the comparison of hadron-nuclear collisions with several models. The programs were those that were outgrowth of Paige and Protopopescu's ISAJET[P1] as extended by T. Ludlam, HIJET.[L1] We obtained the results of a program from CERN which was formulated by J. Ranft[A4] and collaborators. The third program was that which was an outgrowth of the Lund String Model.[A2] We did not run the Dual Parton Model[C2] since it was not readily available.

The CERN program was rejected since it did not take into account nucleon recoils in a realistic way and did not give a correlation between nucleon recoil and mutiplicity.

Most of our effort at comparison was centered on the Lund String Model[A2] as applied to hadron nuclear collisions (FRITIOF).[A3] FRITIOF agreed much better with our data than did HIJET. In our view the shortcomings of HIJET was probably due to the basic program ISAJET which is likely not applicable at this low an energy. The Lund model has been tuned to this energy range.

In the Lund model the elementary particle is viewed as a string with quarks (or antiquarks at the end). The string can be excited by stretching it or kinking it. The result of either of these modes of excitation result in the fragmentation of the string into hadrons.

In the FRITIOF program the incident hadron is allowed to interact more than once. Each interaction produces a new string in the target and can increase the excitation of the projectile string. The results of these calculations are shown in Table V. We find good agreement in pion multiplicities for both the central collisions and the peripheral ones. If one examines the comparison of the FRITIOF predictions of rapidity distributions with the data shown in Figures 17 and 18 we also find remarkably good agreement. There seems to be an excess of particles produced in the rapidity range of .5 to 1.5 units. This could be the result of conventional cascade processes. The only other deficiencies of the model has to do with the number of ρ^o's observed and the momentum distribution of the nucleons emergent from the collisions. These deficiencies would not be reflected in either the multiplicities nor very much in the rapidity distributions. There are less pions predicted at low rapidities than the data indicate. The possibility of cascading of the products has not been considered in the program. The other deficiency has to do with the small number of ρ^o's per event that we find. FRITIOF predicts at least five times more ρ^o's than we find.

The Lund program is well tuned by including lots of experimental information. The comparison with the data is quite useful in giving us insight into the excitation and hadronization of the string.

194

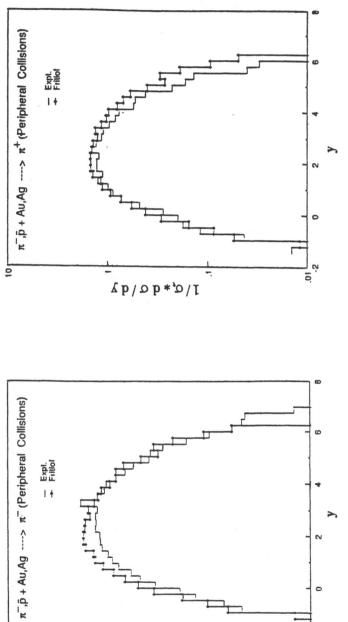

Figure 17: The rapidity distributions of + and - shower particles are compared with the predictions of the FRITIOF program. These data are for peripheral collisions with silver and gold nuclei.

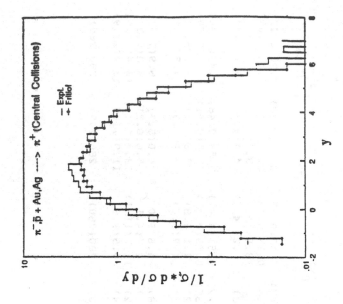

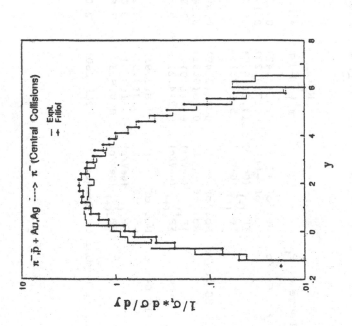

Figure 18: The rapidity distributions of + and - shower particles are compared with the predictions of the FRITIOF program. These data are for central collisions with silver and gold nuclei.

TABLE V

$\pi^+(\pi^-)$ Collisions Calculated in FRITIOF

Range of P_L		$N_p \geq 1$ $\pi^+(\pi^-)$-Mg	$N_p < 4$ $\pi^+(\pi^-)$-Ag	$N_p < 4$ $\pi^+(\pi^-)$-Au	$N_p > 5$ $\pi^+(\pi^-)$-Ag	$N_p > 5$ $\pi^+(\pi^-)$-Au
≥ 20 GeV/c	No. of +(−)	0.45(0.42)	0.47(0.47)	0.37(0.37)	0.26(0.28)	0.21(0.23)
	No. of −(+)	0.17(0.20)	0.17(0.18)	0.17(0.15)	0.15(0.13)	0.12(0.09)
	$\Sigma P_{+(-)}$ (GeV/c)	17.2(16.0)	19.0(19.0)	13.5(14.1)	8.5(9.3)	6.4(7.1)
	$\Sigma P_{-(+)}$ (GeV/c)	5.6(7.1)	5.9(5.8)	5.2(4.5)	4.4(3.7)	3.7(2.8)
10–20 GeV/c	No. of +(−)	0.42(0.49)	0.49(0.52)	0.51(0.49)	0.53(0.48)	0.49(0.44)
	No. of −(+)	0.38(0.31)	0.38(0.35)	0.35(0.37)	0.36(0.41)	0.33(0.39)
	$\Sigma P_{+(-)}$ (GeV/c)	6.0(6.9)	7.0(7.2)	7.1(6.8)	7.3(6.7)	6.9(6.0)
	$\Sigma P_{-(+)}$ (GeV/c)	5.2(4.2)	5.3(5.1)	4.7(5.0)	4.9(5.5)	4.5(5.5)
≤ 10 GeV/c	No. of +(−)	3.8(3.9)	3.6(3.8)	4.6(5.0)	6.0(5.3)	6.9(7.5)
	No. of −(+)	3.3(3.3)	3.2(3.1)	4.4(4.1)	5.6(6.3)	6.8(6.3)
	$\Sigma P_{+(-)}$ (GeV/c)	10.1(9.8)	9.0(9.2)	11.2(11.1)	13.4(13.4)	14.4(14.9)
	$\Sigma P_{-(+)}$ (GeV/c)	8.0(8.3)	7.6(7.4)	9.5(9.6)	11.9(11.9)	13.4(12.9)
Total Number of Events		280(260)	280(260)	280(260)	280(260)	280(260)

X. EFFECT OF INCREASE OF BOMBARDING ENERGY

In experiment E597 a small part of the experiment was run with 320 GeV/c π^- incident on the chamber. The effect of raising the energy is small if one centers their attention on the production of low energy (< 10 GeV) pions and has a considerable effect in the production of higher energy pions. This is certainly expected from Yang and collaborator's Limiting Fragmentation Hypothesis.[B6] This principle seems to be true. We show a comparison of the rapidity distributions of 100 GeV and 320 GeV projectiles in Figure 19. The general result of raising the bombarding energy is to increase the numbers of secondary particles with momenta above 10 GeV/c. We have found that there are useful scaling laws which are shown in Figure 20. If one plots the fraction of the bombarding energy contained among particles (mainly pions) of momentum less than P_L then a general scaling law seems to hold. Figure 20 shows a plot of $\Sigma E_\ell / P_B$ versus P_{max}/P_B. P_{max} is the maximum momentum of particles in a given interval and P_B is the bombarding momentum. For small values of x the distributions are linear with equal slopes. This result is the amount of energy carried by particles with longitudinal momenta less than P_{max} is the same.

The scaling law seems to be more rigorous for the peripheral collisions. The linearity of the ΣE versus x plot would seem to be consistent with a dx/x momentum spectrum of produced particles. For the $N_p \geq 5$ events the deviation from

198

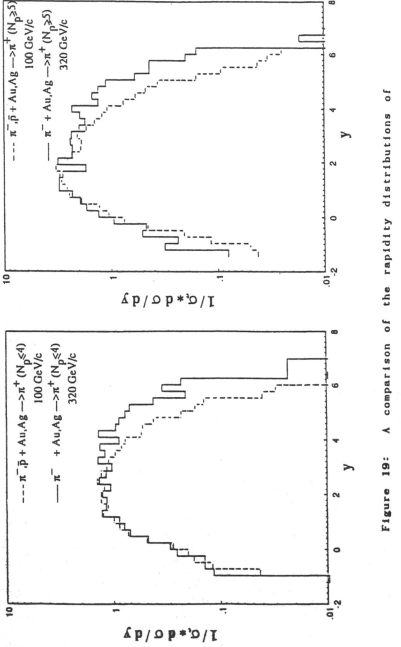

Figure 19: A comparison of the rapidity distributions of negative particles produced by negative particles ($\pi^-, \bar{p}$) at 100 GeV with those produced by 320 GeV π^-.

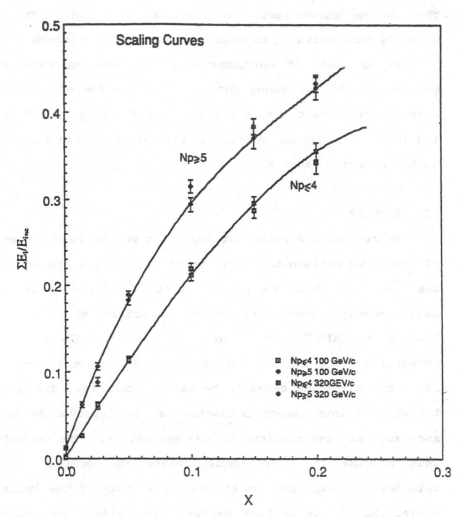

Figure 20: A plot showing the comparison of 100 GeV and 320 GeV data using an integral scaling law.

$\frac{\Sigma \epsilon_i}{E_i}$ for particles $P < x \, P_i$.

the scaling curve seems to start at x = .05. This is probably indicative of cascading in the struck nucleus.

If one looks at electromagnetic cascades one would find good x scaling for young showers. As the showers traverse more material the critical energy becomes important and the x scaling law no longer works. A similar phenomena may occur in the hadronic cascade.

XI. SUMMARY

We have made a number of important points in our summary of data. We believe the division of events by N_p to be among the most important. We believe that the division is very nearly energy independent. We show a scaling curve given by Elliott et al[D1] in Figure 21. The differences in bombarding energy in the two pieces of data were a factor of 20, yet the scaling seems to be rather accurate. The actual translation into impact parameter may be improved by using more accurate descriptions of nuclear matter. The method is less precise in light nuclei where the nuclear target thickness is comparable to the mean free path of the incoming projectile in the nuclear matter. We believe our insight that $N_p = A \ell^3$ is a valid and useful rule. It should make possible direct comparisons of events in different nuclei with the same value of N_p. If one compares events in Pb with those in carbon with equal N_p there are differences in ℓ for a given N_p. These differences are mainly due to the

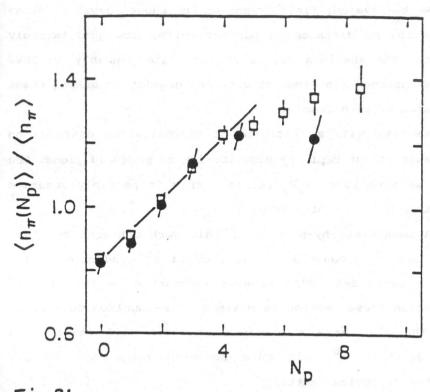

Fig. 21   Scaled values of the average pion multiplicity versus N_p, the number of protons per event, for π^--Ne interactions at 200 GeV/c (solid circles) and 10.5 GeV/c (open squares). N_p includes the energetic protons in addition to the visible protons.

different A/Z ratio for the two nuclei.

We believe our final graphs on the shower development as a function of distance in nuclear matter are qualitatively correct. The abscissa values are accurate probably to 20%. We have assumed a nucleus of constant density in making these estimates of path length.

We have made an estimate of hadronization distance on the basis of our rapidity distribution of produced pions. The value we give is $0.7\tau_\pi\beta_\pi$ Fermis. This is probably accurate to within a factor of 2 or 3.

A remarkable by-product of this work concerns the lack of ρ^0 and S^0 production. The lack of ρ^0 production with momenta below 5-6 GeV/c is very remarkable in that the ρ^0 production cross section is maximum in π- nucleon collisions when the ρ^0's have a momentum of 1.5 - 3.0 GeV/c. We conclude that ρ^0's in this momentum range are rapidly absorbed in nuclear matter.

Our analysis of energy deposition in pions of momenta less than 10 GeV/c is quite large (7 GeV/(Fermi)^3) as the shower propagates through the nuclear matter. We find no evidence for the production of a quark-gluon plasma other than high mutiplicity of pion produced.

We find that the predictions of FRITIOF to be remarkably accurate in many respects. The program fails to predict the large transfer of momentum to struck nucleons and also the lack of ρ^0 production.

We find the limiting fragmentation hypothesis to be quite accurate. As a by-product of this we find a simple scaling law which seems accurate.

ACKNOWLEDGEMENTS

We wish to acknowledge help and encouragement from a number of people. We are grateful to Prof. Peter Carruthers for the invitation to contribute to this volume and for useful conversations. We also wish to thank Dr. G. Stevenson for generating events with the FLUKA82 program. We acknowledge the help of Dr. T. Ludlam with HIJET.

BIBLIOGRAPHY

A1. S.A. Azimov et al, Phys. Rev. D23 (1981) 2512.

A2. B. Anderson, G. Gustafson, G. Ingelman, T. Sjostradd, Phys. Reports 97 (1983) 33.

A3. B. Anderson, G. Gustafson, B. Nilsson, B. Almquist, Nucl. Phys. B281 (1987) 289.

A4. P. Aurenche, F.W. Bopp, J. Ranft, Z. Fur Physik C23 (1987) 67.

B1. D. Barton et al, Phys. Rev. D27 (1983) 2580.

B2. N. Biswas et al, Phys Rev. D33 (1986) 3167.

B3. K. Braune et al, Z. fur Physik 13 (1982) 191.

B4. J.D. Bjorken, Phys. Rev. D27, (1983) 140.

B5. A. Bialas, E. Bialas, Phys. Rev. D20 (1979) 2854.

B6. J. Benecke, T. T. Chou, C. N. Yang, E. Yen, Phys. Rev. 188 (1969) 2159.

C1. U. Camerini, W. Lock, D. Perkins, Progress in Cosmic Ray Physics Vol. I, (1952) North Holland Publishing Co.

C2. A. Capella, U. Sukhatme, J. Tran Than Van, Z. fur Physik C3 (1980) 329.

D1. J.R. Elliott et al, Phys. Rev. Lett. 34 (1975) 607.

D2. W.M. Yeager et al, Phys. Rev. D16 (1977) 1294.

D3. H. Band, Ph.D. Thesis, Duke University (1980).

D4. S. Date', M. Gyulassy, H. Sumiyoshi, Phys. Rev. D32 (1985) 619.

F1.   S. Fredrikson, G. Eilam, G. Berlad, L. Bergstrom, Physics Reports **144** (1987)187.

G1.   A. Goldhaber, W. Busza, Phys. Lett. **139B** (1984) 235.

K1.   Z. Koba, H. Nielsen, P. Oleson, Nuc. Phys. **B40** (1972) 317.

L1.   T.W. Ludlam, BNL Report 51921 (1985) 373.

M1.   L. McLerran, B. Svetitsky, Phys. Lett. **98B** (1981) 195.

P1.   F.E. Paige, S.D. Protopopescu, Proceedings. of the Oregon Workshop on Super High Energy Physics (1985) 41.

S1.   P. Slattery, Phys. Rev. Lett **29** (1972) 1624.

T1.   W. Toothacker et al, Phys. Lett. **20B** (1987) 295.

TABLE I

Comparison Between Collisions with Mg and Peripheral Collisions With Heavy Elements

% of Interactions	Material	N_p	$<n\pm>$	$<y\pm>$	$<N_p>$	$\bar{\ell}$(fm)
60	Ag + Au	1-4	10.4±.5	2.1±.1	1.9±.1	7.2
68	Mg	≥1	9.6±.8	2.1±.1	1.6±.1	5.9
40	Ag + Au	≥5	17.7±.8	1.7±.1	11.4±.4	11.4

FIGURE CAPTIONS

Figure 1: A π^- gold collision occurring in a .3 mm thick gold foil. There are 14 or 15 identifiable protons emerging from the collision as well as 11 positive and 12 negative minimum ionizing particles.

Figure 2: The experimental arrangement for experiment E597. In this report we mainly utilize the upstream Cerenkov counters for identification as well as the downstream drift chambers.

Figure 3: The mean shower multiplicity versus the number of heavily ionizing tracks observed in proton-emulsion experiments. The bombarding energies are noted on the figure. This figure was taken from the review of Frederickson et al.

Figure 4: The fraction of events which have a given number of identified protons from hadron-neon collisions. The curves seem to be nearly independent of bombarding energy and projectile.

Figure 5: The mean number of identified protons is plotted against the mean chord length for peripheral and central collisions with silver and gold nuclei and with central collisions with magnesium. We have corrected the magnesium data to account for the different fraction of neutrons and protons in magnesium as compared to the heavy elements.

Figure 6: The distribution of the number of struck protons emerging from a hadron-nuclear encounter. The nucleus is assumed to have a uniform density of nucleons and a radius $R = (1.15)A1/3$.

Figure 7: The distribution of the number of struck protons emerging from a hadron-nuclear encounter. The nucleus is assumed to have a nucleonic density given by the Woods-Saxon distribution. This distribution seems to agree better with the data.

208

Figure 8: A scaled multiplicity distribution for negative particles from π^-p, π^-Ne and peripheral collisions with silver and gold nuclei.

Figure 9: The scaled multiplicity distribution for negative particles emerging from central collisions of hadrons with silver and gold nuclei.

Figure 10a: The average net charge $<n_+-n_->$ versus the scaled number of shower particles ($n_s=n_++n_-$) in 40 Gev/c π^-C collsions (squares) and in 10.5 GeV/c (triangles), 25 GeV/c (open circles) and 50 GeV/c (solid circles) π^-Ne collisions. (a) Even values of n_s, (b) Odd values of n_s.

Figure 10b: The number of minimum ionizing protons emerging from central and peripheral collisions with silver and gold nuclei.

Figure 11a: A plot of the number of shower (minimum ionizing tracks) against the number of visible protons emerging from collisions with silver and gold nuclei.

Figure 11b: The amount of momentum carried by nucleons emerging from collisions with silver and gold nuclei.

Figure 12: A YHhistogramshowing the number of plus-minus pairs of the low momentum pions emerging from hadron collisions with gold and silver nuclei. The lower figure shows the sample from peripheral collisions. There is some evidence of K^o decays which were not identified but very little if any evidence of ρ^o production. The upper histogram shows the result of 2000 collisions.

Figure 13: A histogram of the invariant mass distribution of the energetic pions emerging from hadron collisions with silver and gold nuclei. The dotted curves at the bottom indicate the number of ρ^o's, S^o's, f^o's indicated by a fitting program.

Figure 14: Invariant mass spectrum of π^+-p combinations for all hadron collisions with Ag and Au.

Figure 15: A plot of the energy content of particles of momentum more than 50 GeV plotted against the average path-length of the incident particle in nuclear matter. These data are from the experiment of Barton et al.

Figure 16: A plot of the longitudinal momentum content of struck nucleons and pions of momentum less than 10 GeV/c as a function of average path length in nuclear matter.

Figure 17: The rapidity distribution of + and - shower particles are compared with the predictions of the FRITIOF program. These data are for peripheral collisions with silver and gold nuclei.

Figure 18: The rapidity distribution of + and - shower particles are compared with the predictions of the FRITIOF program. These data are for central collisions with silver and gold nuclei.

Figure 19: A comparison of the rapidity distributions of negative particles produced by negative particles $(\pi^-,\bar{p})$ at 100 GeV with those produced by 320 GeV π^-.

Figure 20: A plot showing the comparison of 100 GeV and 320 GeV data using an integral scaling law.
$(\Sigma\epsilon_i)/E_i$ for particles P < x P_i.

Figure 21: Scaled values of the average pion multiplicity versus N_p, the number of protons per event, for π^--Ne interactions at 200 GeV/c (solid circles) and 10.5 GeV/c (open squares). N_p includes the energetic protons in addition to visible protons

DIFFRACTIVE MULTIPARTICLE PRODUCTION

(An Experimental and Pheonomenological Review)

K. Goulianos
The Rockefeller University
New York, New York   10021
U.S.A.

Abstract

We review the properties of multiparticle clusters produced in hadron diffraction dissociation at high energies, emphasizing the phenomenological description of experimental data.

We review the experimental data available on multiparticle clusters produced in hadron diffraction dissociation at high energies and outline the phenomenalogical description which can be used to predict diffractive multiparticle production in the multi-TeV energy region of present and future Colliders.

The processes we discuss are single diffraction dissociation (SD), double diffraction dissociation (DD) and double Pomeron exchange (DP):

$$p + p \rightarrow X + p \qquad\qquad SD \qquad\qquad (1)$$
$$\quad\;\; \big|$$
$$\quad\;\; \downarrow\rightarrow hadrons$$

$$p + p \rightarrow X_1 + X_2 \qquad\qquad DD \qquad\qquad (2)$$

$$p + p \rightarrow p + p + X \qquad\qquad DP \qquad\qquad (3)$$

Although only protons are shown above in the initial states, data for SD also exist for other particles dissociating on protons, namely $\bar{p}$, $\pi^{\pm}$, $K^{\pm}$ or γ, and for protons dissociating on other hadrons, such as $\bar{p}$ or d (deuteron). For DD and DP, data are available only for proton-proton interactions at ISR energies, $20 < \sqrt{s} < 60$ GeV, with the exception of some data at the S$\bar{p}$pS Collider by the UA5 Collaboration. The ISR DD data involve only exclusive or semi-inclusive final states. Nevertheless, these data are useful in checking the factorization rules on the basis of which predictions for inclusive data at ISR and higher energies can be made.

The goal of this article is to weave the existing experimental data on diffractive multiparticle production into a phenomenological framework, which can then be used to predict the diffractive behavior at the currently operating Fermilab Collider (Tevatron: $\bar{p}$p at $\sqrt{s} = 2$ TeV) and at the proposed Superconducting Super Collider (SSC: pp at $\sqrt{s} = 40$ TeV).

Our review is organized in sections, as follows:

(i)   Introduction to diffraction dissociation.
(ii)   Single Diffraction cross sections.
(iii) Diffractive multiplicity distributions.
(iv)   Diffractive pseudorapidity distributions.
(v)   Double diffraction dissociation.
(vi)   Double Pomeron exchange.
(vii) Conclusion.

In each section we discuss the relevant data along with the phenomenology that ties them together and allows extrapolations to higher energies. Although an effort has been made to keep the various sections independent, some references to later sections could not be avoided.

The review draws heavily from work done in whole or in part by the author (Ref.s 1, 3, 4, 8, 11, 12, 15, 16, 17).

I. INTRODUCTION TO DIFFRACTION DISSOCIATION

The phenomenology of diffraction dissociation is discussed in detail in a Physics Reports article [1] written by the author. For completeness, we present here a brief introduction to diffraction dissociation, sketching the features which are relevant to the sections that follow.

The amplitude for single diffraction dissociation of a hadron "h" on a proton "p" is presented schematically in Fig. 1.

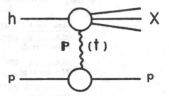

FIG.1: Amplitude for hadron dissociation.

The excitation mass, M_x, is related to the forward component of the four-momentum transfer, q_{11}, and to the Feynmann x variable, $x = p_{11}^*/p_{11,max}^*$ of the recoil proton, as follows:

$$|q_{11}| = m_p \frac{M_x^2 - M_h^2}{s} \approx m_p(1-x) \qquad (4)$$

As the wave length associated with q_{11}, $\lambda = 1/|q_{11}|$, becomes larger than the interaction radius, $R \approx 1/m_\pi$, one expects coherence or diffraction phenomena to appear. Thus, the condition for inelastic diffraction is

$$1/|q_{11}| \geq R \approx \frac{1}{m_\pi} \qquad (5a)$$

$$1 - x \approx \frac{M_x^2 - M_h^2}{s} \leq \frac{m_\pi}{m_p} \approx 0.15 \qquad (5b)$$

As seen in Fig. 2, the differential cross section, $d^2\sigma/dtdx$, which is fairly flat in the central x region (not shown entirely in the figure), increases rapidly with decreasing 1-x in the region 1-x ≤ 0.15. This increase is attributed to the setting in of the coherence condition and therefore the peak at small 1-x is identified as "diffractive." The higher the s-value, the larger the M_x^2 that fulfills the coherence condition. At the Fermilab $\bar{p}p$ Collider, $\sqrt{s} = 2000$ GeV, the mass corresponding to $M_x^2/s = 0.15$ is $M_x = 780$ GeV! As larger mass states are produced diffractively, the overall diffraction dissociation cross section increases. At $\sqrt{s} = 2000$ GeV as much as one-half of the inelastic cross section may be due to single and double diffraction dissociation (see section II).

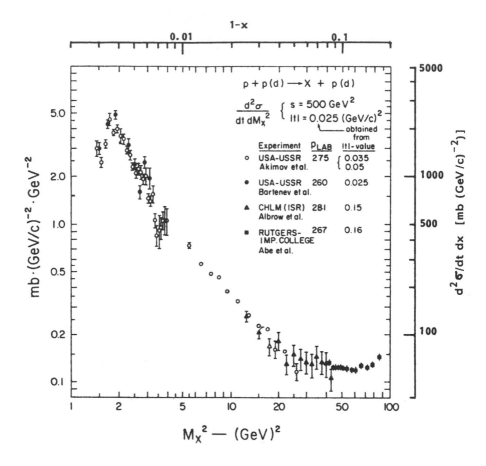

FIG.2: The differential cross section versus M_x^2 for pp→Xp at t = -0.025 $(GeV/c)^2$ and s = 500 GeV^2 [1]. The axes on the right and the top of the figure give the invariant cross section $d^2\sigma/dtdx$ versus of 1-x.

The differential cross section for the dissociation of a hadron "h" on a target "a",

$$\frac{d^2\sigma^{ha}}{dtdM_x^2} = f_{ha}(s,t,M_x^2)$$
(6)

depends weakly on s, is exponential in t, varies as $1/M_x^2$ and obeys factorization and the finite mass sum rule. This behavior is consistent with the hypothesis that diffraction dissociation is dominated by a triple-Pomeron Regge type amplitude.

In detail, in the mass region 5 GeV$^2 < M_x^2 < 0.15s$ and fot $|t| < 0.1$ (GeV/c)^2, the differential cross section for single diffraction dissociation of protons on protons is given by

$$\frac{d^2\sigma}{dtdM_x^2} = \frac{A}{M_x^2} (1 + \frac{B}{s}) (b_{SD} e^{b_{SD}t})$$
(7)

where $A = 0.68 \pm 0.05$ mb, $B = 36 \pm 8$ GeV^2 and $b_{SD}/b_{e\ell} \approx 2/3$ to $1/2$. At ISR energies, the slope for single diffraction dissociation is $b_{SD} \approx 7$ (GeV/c)$^{-2}$. The average $|t|$ of the diffractive states is thus $1/b_{SD} \approx 0.14$ (GeV/c)^2, corresponding to $p_T \approx 380$ MeV/c. At a beam momentum of 30 GeV/c (ISR), the diffractive states are produced at an average angle of ~ 13 mrad. At the Fermilab Collider, the average production angle will be ~ 0.3 mrad. Thus, the common characterization of the particle clusters arising from the dissociation of these states as "forward jets" is justified.

The validity of Eq. 7 at the S$\bar{p}$pS Collider and higher energies ($\sqrt{s} \geq 540$ GeV) will be discussed in section II.

The hypothesis of factorization of the diffractive vertex leads to scaling relations (factorization rules) among diffraction dissociation, elastic scattering and the total cross section. Figure 3 represents the reactions $h + a \rightarrow X + a$ and $h + a \rightarrow h + a$. Factorization implies that each amplitude is proportional to the product of the coupling constants $g(t)$ at each vertex. The ratio of diffraction dissociation to elastic scattering is then

$$\frac{d^2\sigma/dt\,dx}{d\sigma_{e\ell}/dt} = \left(\frac{g_{hX}(t) \cdot g_{aa}(t)}{g_{hh}(t) \cdot g_{aa}(t)} \right)^2 = C_h(s,x,t)$$
(8)

i.e., at given s,x and t, the ratio of the diffraction dissociation to the elastic scattering cross section of a hadron "h" interacting with a target particle "a" is a constant independent of the target particle.

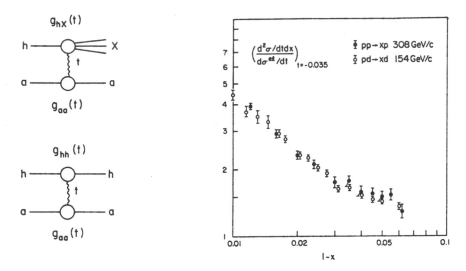

FIG.3: (Left) Elastic scattering and diffraction dissociation amplitudes.
(Right) Factorization test: The ratio of the diffractive to elastic
differential cross sections at fixed-t plotted against 1-x
for pp at 308 GeV/c and pd→Xd(pd) at 154 GeV/c (same s-value).

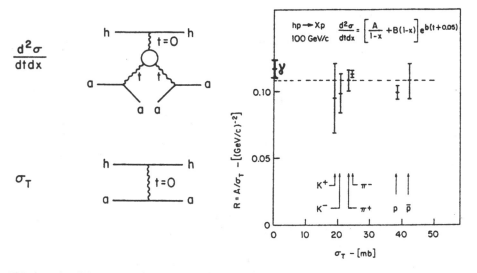

FIG.4: (Left) Pomeron exchange diagrams for single diffraction and for σ_T.
(Right) Factorization test: The ratio of the diffractive to the total
cross sections at 100 GeV/c for p^{+-}, K^{+-}, π^{+-} and γ incident
on protons.

When different hadrons dissociate on the same target, factorization leads to scaling relations only under specific models. For example, if the triple-pomeron amplitude is responsible for diffraction dissociation and simple pomeron exchange for the total cross section (see Fig. 4), it follows that

$$\frac{d^2\sigma/dt\ dx}{\sigma_T} \sim C_a(s,x,t) \qquad (9)$$

i.e., at given s, x and t, the ratio of the diffraction dissociation to the total cross section of different hadrons "h" dissociating on the same target particle "a" is independent of the dissociating hadron.

The finite mass sum rule (FMSR) provides a connection between the high mass and the low mass diffractive cross sections (including elastic scattering). It is an extension of the finite energy sum rule (FESR) which applies to total cross sections. Figure 5 represents the typical behavior of hadronic total cross sections as a function of s. Resonance-like structures, which are prominent at low energies, disappear as the energy increases and at high energies the cross section becomes a smooth function of s. The low energy region is described best by s-channel resonance production and decay while the high energy behavior is represented well by Reggeon t-channel exchange. The FESR states that the two descriptions are equivalent so that the extrapolation of the high energy behavior into low energies describes well the average behavior of the cross section in this region. This dual description of the total cross section in terms of s-channel resonance or t-channel Reggeon exchange is referred to in the literature as duality.

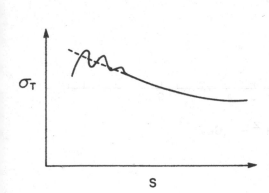

$a + b \longrightarrow$ anything

FIG. 5
Finite Energy Some Rule: The extrapolation of the smooth, high s cross section into the low s resonance region represents the average behavior of the cross section in this region.

$h + b \rightarrow X + b$

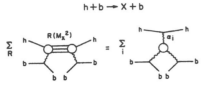

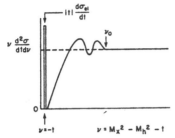

FIG. 6

The first moment finite mass sum rule: The extrapolation of $\nu \, d^2\sigma / dt \, d\nu$ from the high ν into the low ν region averages over the resonance-like structures, including elastic scattering.

Below: Experimental test of the finite mass sum rule. The fit to the high mass data for $\nu \, d^2\sigma / dt \, d\nu$ at fixed t, where $\nu = M_x^2 - M_p^2 - t$ and $d\nu \approx dM_x^2$ for small t, extrapolated into the low mass region averages over the values of $\nu \, d^2\sigma / dt \, d\nu$ in this region, including the value $|t| \cdot d\sigma/dt$ for elastic scattering which, for pictorial reasons, is represented here by the area of half a Gaussian function [1].

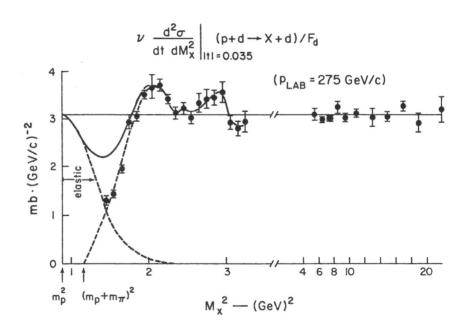

A similar description applies to the diffractive cross section. However, as shown in Fig. 6, the relevant variable is now M_x^2 rather than s and therefore one speaks of finite mass rather than finite energy sum rules. The first moment FMSR, which has been derived for triple-Regge amplitudes using analyticity and crossing symmetry [2], states that the extrapolation of the high ν behavior of the function ν (d$^2\sigma$/dt dν) into the low ν region, where $\nu = M_x^2 - M_h^2 - t$ is the cross-symmetric variable, represents the average behavior of the resonances (including elastic scattering) in this region (see Fig. 6). Quantitatively,

$$|t| \frac{d\sigma_{el}}{dt} + \int_0^{\nu_0} \nu \frac{d^2\sigma}{dt\,d\nu} = \int_0^{\nu_0} \nu \left(\frac{d^2\sigma}{dt\,d\nu} \right)_{fit\ at\ high\ \nu} d\nu \qquad (10)$$

where the value of ν_0 must lie beyond the "resonance region" but is otherwise arbitrary.

Experimental tests of factorization and of the FMSR are summarized in Ref. 1. A more recent test involving incident photons in the 100 GeV momentum range is described in Ref. 3. As mentioned earlier these tests give support to the hypothesis that the $1/M_x^2$ behavior of the differential cross section for single diffraction dissociation is due to a triple-Pomeron amplitude.

II. SINGLE DIFFRACTION CROSS SECTIONS

Fixed target experiments at Fermilab, $10 < \sqrt{s} < 28$ GeV, and collider experiments at the ISR, $20 < \sqrt{s} < 60$ GeV, have made accurate measurements of single diffraction siddociation cross sections using the "recoil technique". In this method, the t-value of the reaction and the value of M_x^2 are obtained from the angle and the momentum of the recoil particle, yielding differential cross sections, d$^2\sigma$/dt dM$_x^2$, in the region of t and M_x^2 within which the experiments are sensitive. The total cross section is obtained by extrapolation of the t and M_x^2 distributions to the unmeasured regions, followed by integration over t and M_x^2.

For pp $\rightarrow$ Xp, the differential cross section is given by Eq. 7. However, one should keep in mind that experiments usually quote as "diffractive" the entire cross section in the region of the diffractive peak. This cross section contains a non-diffractive part, which increases with increasing 1-x, as can be seen clearly in Fig. 2. A good parametrization of the entire cross section in the small 1-x region is given by [1]

$$\frac{d^2\sigma}{dtdx} = \frac{A(1+B/s)}{1-x} (b_{SD}\ e^{b_{SD}t}) + A' (1-x) (b'\ e^{b't}) \qquad (11)$$

The first term is due to single diffraction dissociation, while the second term is thought to be mostly due to non-diffractive contributions. This view is supported by the fact that the first term obeys factorization rules, but the second term does not. For pp → Xp, $A'/A \approx 100$, so that at $1-x = 0.1$ the diffractive and non-diffractive terms are approximately equal, while at $1 - x = 0.3$ the diffractive term is about 90% of the inclusive cross section.

The total diffractive cross section is obtained by integration of Eq. 7 or 11. Cross sections for pp → Xp are often quoted as $\sigma_D = 2 \sigma_{SD}$, taking into account the dissociation of both protons. These cross sections usually contain the non-diffractive contribution mentioned above, σ_{ND}, and the integration over M_x^2 is carried out from the inelastic threshold to a maximum value of $M_x^2 = M_p^2 + s (1-x_0)$. In comparing data from different experiments, care must be taken to refer to the same region of $1-x$. For clarity, we therefore write the inclusive integrated cross section for pp → Xp as

$$\sigma = \sigma_D + \sigma_{ND} \qquad (12a)$$

$$\sigma_D = 2\sigma_{SD} = 2A \left(1 + \frac{B}{s}\right) \ln \left[\frac{M_p^2 + s(1-x_0)}{1 \text{ GeV}^2}\right] \qquad (12b)$$

$$\sigma_{ND} \approx 100 \, A \, (1-x_0)^2 \qquad (12c)$$

We recall that the values of the parameters A and B in Eq. 7 were, respectively, 0.68 ± 0.05 mb and 36 ± 8 GeV^2. However, as it was pointed out, Eq. 7 as written is valid in the small t-region, where the data are described well by a simple exponential with a slope b_{SD}. The flattening of the differential cross section at larger t-values observed at the ISR and SppS Collider experiments [5,6] has the effect of increasing the integrated cross section by about 10 %. Therefore, the value of A to be used with Eq.12 is $A = 0.75 \pm 0.05$ mb.

The cross section contained within two given values of $1-x$ is, for large s, independent of the energy. Thus, the cross section for $0.05 < 1-x < 0.1$ is $\sigma = 1.6$ mb, out of which $\sigma_D = 1.0$ mb and $\sigma_{ND} = 0.6$ mb. The contribution of the diffractive component to the rise of the total proton-proton cross section is discussed in Ref. 4.

Figure 7 shows diffraction dissociation cross sections [5,7] σ_D as a function of $\sqrt{s}$ for $M_x^2/s < 0.05$. The values of the ISR experiments were corrected by subtracting the non-diffractive part from the data using Eq. 12c. This correction amounts to 0.2 mb. The band defined by the two solid curves represents the $\pm 1\sigma$ values derived from Eq. 12b using $A = 0.75 \pm 0.05$ mb, $B = 36 \pm 8$ GeV^2 and $1 - x_0 = 0.05$.

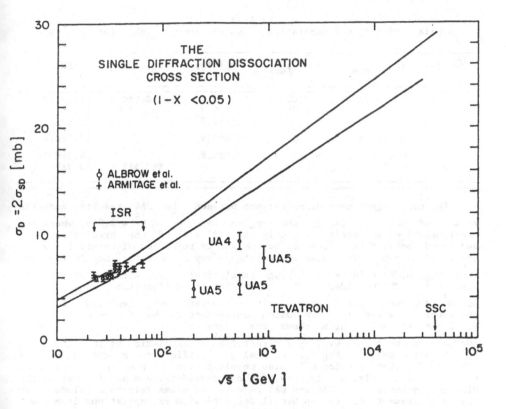

FIG. 7: The total pp(p̄p) single diffraction dissociation cross section as a function of √s for 1-x < 0.05. The two curves represent the ±1σ values derived from Eq. 12b, as discussed in the text.

The curves are in agreement with the ISR data. However, the SppS Collider data of UA4[6] and UA5[7] are not in good agreement with each other and they both fall considerably below the curves. The UA4 data were obtained using the "rapidity gap method", which is explained in detail in Ref. 6. The lower value for the cross section may be due to loss of high mass diffractive events resulting from closing of the rapidity gap caused by fluctuations in the rapidity width in a way not understood by the experiment. Loss of events may also occur at very small masses for which the acceptance is low and the corrections large. The UA5 data suffer from very large corrections, which have to be made by extensive Monte Carlo simulations. The table below presents the UA4 and UA5 results along with values obtained from Eq. 12b for comparison.

Table I
Single diffraction dissociation cross sections, σ_D(mb), for $\bar{p}p \rightarrow \bar{p}X$

$\sqrt{s}$(GeV)	$(M^2/s)_{min}$	$(M^2/s)_{max}$	σ_D(UA4)	σ_D(UA5)	σ_D(Eq. 12b)
200 UA5	threshold	0.05		4.8±0.5±0.8	11.4±0.9
540 UA4	"	0.05	9.4±0.7	5.2±1.0	14.4±1.1
	"	0.1	10.0±0.7		15.4±1.1
	16 GeV^2/s	0.05	6.4±0.4		10.3±0.8
	threshold	16 GeV^2/s	3.0±0.8		4.2±0.3
900 UA5	"			7.8±0.5±1.1	15.9±1.2

The most significant dissagreement between the UA4 results and the values of Eq. 12b is in the region 16 GeV^2 < M^2 < 0.05s, where the experimental cross section is only (62 ± 7)% of the prediction. As mentioned above, this deficiency may be due to rejection of events for which the rapidity gap closes in an unpredictable way. The deficiency in the low mass region, M^2 < 16 GeV^2, although statistically less significant, is also substantial. The Collider results imply that the differential cross section $d\sigma/dM^2$ does no longer go as $1/M^2$ with a constant coefficient independent of the energy. However, this conclusion contradicts another UA4 result [Ref. 6, M. Bozzo et al], which shows that data at $\sqrt{s}$ = 540 GeV and t = -0.55 (GeV/c)^2 can be fitted well by a direct $1/M^2$ extrapolation of the ISR data to higher masses (Fig.8). A smaller coefficient in the high mass differential cross section will also result in violation of the finite mass sum rule (see Fig. 6). In view of these considerations and in view of the discrepancy between the UA4 and UA5 results, it seems fair to conclude that the disagreement of the S$\bar{p}p$S collider data with extrapolations from lower energy experiments is not yet decisive.

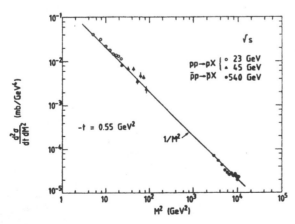

FIG.8: The differential cross section $d^2\sigma/dtdM^2$ for t=-0.55 (GeV/c)^2 and $\sqrt{s}$ = 540 GeV shown as a function of M^2 together with pp cross sections at lower energies [Ref.6, M. Bozzo et al].

III. DIFFRACTIVE MULTIPLICITY DISTRIBUTIONS

Charged multiplicities of high mass diffractive states have been measured at Fermilab [8], at the ISR [9] and at the SppS collider [10]. The data may be summarized as follows: The mean charged multiplicity as well as the multiplicity distribution at a given mass M are the same as those of non-diffractive inelastic collisions at $\sqrt{s}$ = M. This universality aspect between diffractive and non-diffractive charge multiplicities was first pointed out in 1982 [11]. In a more recent paper [12] entitled "A new statistical description of hadronic and e^+e^- multiplocity distributions", a model developed to fit inclusive non-diffractive data has been very successful in also fitting diffractive charged multiplicity distributions. Figure 9, taken from this paper, shows such a fit. In Fig. 10, taken from Ref. 10, the mean charged multiplicities of diffractive masses M and of non-diffractive inelastic collisions at $\sqrt{s}$ are plotted as a function of M or $\sqrt{s}$. These two figures summarize our knowledge of diffractive charged multiplicities. Clearly, more measurements of charged multiplicity distributions at high masses are in order.

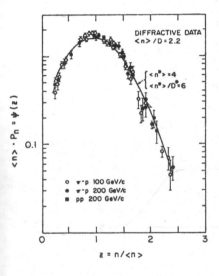

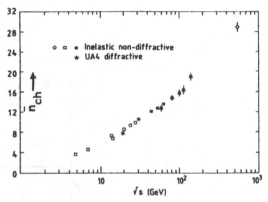

FIG. 10: The mean charged multiplicity of diffractive clusters as a function of diffractive mass M plotted at $\sqrt{s}$ = M together with data for non-diffractive inelastic events [10].

FIG. 9: The diffractive charged multiplicities of Ref.8 (R.L.Cool et al) fitted with the "modified gamma distribution" of Ref.12 [12].

IV. DIFFRACTIVE PSEUDORAPIDITY DISTRIBUTIONS

Pseudorapidity distributions of charged particles produced in high mass diffraction dissociation have been measured at the ISR[9] and at the $S\bar{p}pS$ collider [10].

Fig. 11 represents a schematic drawing of a diffractive cluster in $dn_{ch}/d\eta$ vs η space. The width of the cluster increases with M as $\ell n(M/M_p)^2$. Fig.s 12 and 13 summarize the available data. Diffractive clusters from the dissociation of masses M are centered at the expected average pseudorapidity $\bar{\eta} = \ell n(\sqrt{s}/M)$ and stretch from the maximum available value of rapidity, $\eta_{max} = \ell n(\sqrt{s}/M_p)$, to an inner edge given by $\eta_{min} = \ell n(M_p\sqrt{s}/M^2)$.

The best way to summarize the data is to say that the η - distribution of charged particles from the dissociation of a diffractive mass M is very similar to the distribution of particles from non-diffractive inelastic collisions at $\sqrt{s} = M$. Fig. 14, which shows the height of the pseudorapidity plateau for diffractive clusters along with that of non-diffractive inelastic collisions, supports this conclusion. These experimental facts find theoretical interpetation in the dual parton model [13].

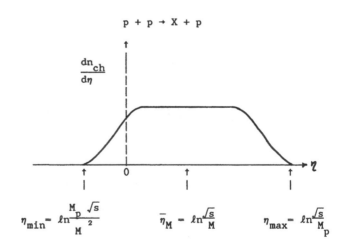

FIG. 11: Schematic drawing of the charged particle pseudorapidity distribution of a diffractive cluster of mass M created in a pp→Xp collision at $\sqrt{s}$.

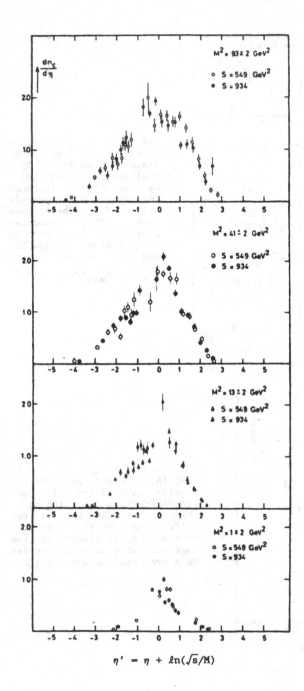

FIG. 12: The s-dependence of diffractive charged particle pseudorapidity distributions at the ISR. The center of the clusters has been shifted by $\ln(\sqrt{s}/M)$, so that they appear at $\eta=0$.

226

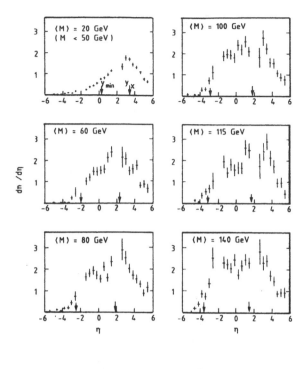

FIG. 13: Diffractive charged particle η-distributions at the $\bar{S}p\bar{p}S$ Collider($\sqrt{s}$=540GeV)[10]. Data are binned in mass intervals with average value < M >. The arrows show the expected position of the center of the cluster, $\eta=\ell n(\sqrt{s}/M)$, and that of the "inner edge", $\eta_{min}=\ell n(M_p\sqrt{s}/M^2)$.

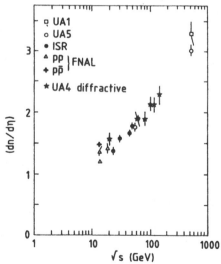

FIG. 14: Charged particle η-density for diffractive and non-diffractive inelastic events. The UA4 results for diffractive clusters evaluated around $\eta=\bar{\eta}$ are plotted at $\sqrt{s}$=M together with non-diffractive data evaluated around η=0 [10].

227

V. DOUBLE DIFFRACTION DISSOCIATION

In addition to single diffraction dissociation, diffractive clusters
may also be produced in double diffraction dissociation (DD), Eq. 2.

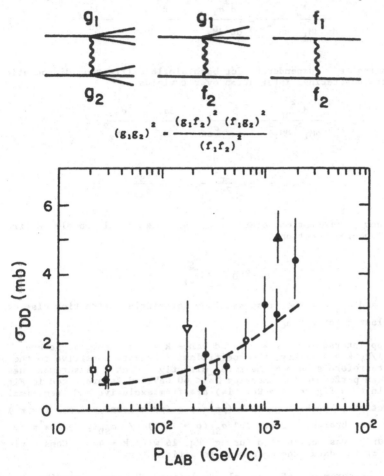

$$(g_1g_2)^2 - \frac{(g_1f_2)^2\,(f_1g_2)^2}{(f_1f_2)^2}$$

FIG.15:
Factorization and double diffraction dissociation
Above: Diagrams for DD, SD and elastic scattering. Factorization leads to
the formula indicated, from which one obtains Eq. 13 in the text.
Below: The inclusive proton-proton DD cross section as a function of P_{lab}.
The data are from exclusive and semi-inclusive cross sections scaled by the
"branching ratio" (see text). The curve represents $\sigma_{DD} = k\,\sigma_{SD}^2/\sigma_{el}$ with $k=\frac{4}{3}$.

The differential cross section for DD may be obtained from the cross section for single diffraction dissociation and for elastic scattering using factorization (see Fig. 15):

$$\frac{d^3\sigma}{dt \, dM_1^2 \, dM_2^2} = \frac{d^2\sigma_1}{dt \, dM_1^2} \cdot \frac{d^2\sigma_2}{dt \, dM_2^2} \cdot \frac{d\sigma_{el}}{dt} \qquad (13)$$

Assuming e^{bt} dependence for both single diffraction dissociation and for elastic scattering, integration over t yields

$$\frac{d^2\sigma_{DD}}{dM_1^2 \, dM_2^2} = k \frac{1}{\sigma_{el}} \frac{d\sigma_1}{dM_1^2} \frac{d\sigma_2}{dM_2^2} \qquad (14)$$

$$\text{where} \quad k = \frac{r^2}{2r-1} \; ; \quad r = \frac{b_{SD}}{b_{el}} \qquad (15)$$

Further integration over M^2 gives the total double diffraction dissociation cross section,

$$\sigma_{DD} = k \frac{\sigma_{SD}^2}{\sigma_{el}} \qquad (16)$$

where σ_{SD} and σ_{el} are the cross sections for single diffraction dissociation and for elastic scattering.

For pp interactions, $r \approx 2/3$ and hence $k \approx 4/3$. Notice, however, that for $r \approx 1/2$, $k \to \infty$. Thus, the coefficient k is very sensitive to the value of r and therefore should be measured directly. Such a measurement has, in effect, been performed for energies $\sqrt{s} < 60$ GeV and is presented in Fig. 15. The data in this figure (see Ref. 14) are from exclusive and semi-inclusive cross sections, such as $p + p \to (p\pi^+\pi^-) + (p\pi^+\pi^-)$ and $p + p \to (p\pi^+\pi^-) + X$, scaled by the "branching fraction" $\sigma_{SD}(p \to \text{all}) / \sigma_{SD}(p \to p\pi^+\pi^-)$. The dashed curve was calculated using Eq. 16 with $k = 4/3$. Considering the diversity of the data, the agreement is satisfactory.

The measurement of the all-inclusive DD cross section is very difficult. The DD events can only be identified as such by the observation of a rapidity gap between the two clusters. This method is not very effective in separating out DD from minimum bias non-diffractive events. This is illustrated in Fig. 16, in which we plot the expected charged particle η-distribution for two 10 GeV clusters produced in DD at $\sqrt{s} = 540$ GeV. The distribution for each cluster was taken from the data on single diffraction of Ref. 9 (Fig. 11). Clusters of higher mass would be centered at smaller $|\eta|$ and have wider distributions, closing the rapidity gap and rendering the overall distribution indistinguishable from that of a centrally produced single cluster. Taking $\sigma_{el} = 13.3$ mb for $\sqrt{s} = 540$ GeV and σ_{SD} from Eq. 12b evaluated for $1-x_0 = 0.05$, Eq. 16 yields $\sigma_{DD} = 5.1$ mb. The cross section for both masses being below 10 GeV is calculated to be 1.17 mb. Thus, the two clusters may be expected to be separated in only about 20% of the DD events. However, most of this cross section is due to low mass events, which do not trigger an apparatus such as that of UA5. In fact, in an attempt to measure DD using this method [7], UA5 selected a mass region $3 < M < 6$ GeV for each cluster, within which the expected DD cross section is ≈ 0.12 mb or 2.5% of σ_{DD}. A Monte Carlo simulation was then used to evaluate the total σ_{DD} from this measurement. It was found that for $1-x < 0.05$, $\sigma_{DD} = 3.5 \pm 2.2$ mb at 200 GeV and 4.0 ± 2.5 mb at 900 GeV. The calculated values, using Eq. 16 and the appropriate σ_{SD} and $\sigma_{e\ell}$, are respectively 4.4 mb and 5.4 mb. The agreement is satisfactory, however, the errors are large and clearly more work is needed in this area.

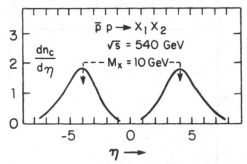

FIG. 16: Expected charged particle pseudorapidity distribution in double diffraction dissociation, $pp \to X_1 X_2$, for masses $M_1=M_2=10$GeV at $\sqrt{s}=540$ GeV.

A general method for obtaining the all-inclusive DD cross section is outlined below. For each event, one measures the maximum polar angle of the charged particles in each hemisphere and converts it to pseudorapidity using the formula

$$\eta_{min} = -\ln \tan \frac{\theta_{max}}{2} \tag{17}$$

The two values of η_{min}, one for each hemisphere, are then used to define a point in a scatter plot in the $\eta_{1,min}$ vs $\eta_{2,min}$ plane. For non-diffractive inelastic events, these points will form a peak centered at the origin. The distribution for DD events can be calculated from Eq. 14, using the relationship between η_{min} and the mass M of a diffractive cluster (Fig. 13):

$$\eta_{min} = \ln \frac{M_p \sqrt{s}}{\frac{M^2}{}} \tag{18}$$

Given that at high energies $d\sigma_{SD}/dM^2 = A/M^2$, the result is

$$\frac{d^2\sigma_{DD}}{d\eta_{1,min}\, d\eta_{2,min}} = k \frac{A^2}{\sigma_{e\ell}} \tag{19}$$

230

i.e. the distribution is flat. High mass events, for which η_{min} becomes negative, will have particles in the "other" hemisphere and therefore will yield points concentrated around the origin just as the non-diffractive events. A Monte Carlo simulation can be used to fit the peak at the origin. At high η_{min} the scatter plot will be dominated by DD events.

Figure 17 shows Monte Carlo generated scatter plots of $\eta_{1,min}$ vs $\eta_{2,min}$ for ND, SD and DD events at $\sqrt{s}$ = 2 Tev. The number of events in each plot is proportional to the corresponding cross sections, taken for this purpose to be σ_{ND}= 40 mb, σ_{SD}= 9 mb and σ_{DD}= 4 mb. The recoil proton in SD was ignored. It is evident that at high η_{min} the DD signal is dominant. Clearly, this is the most general way that can be and should be used to extract the inclusive DD signal.

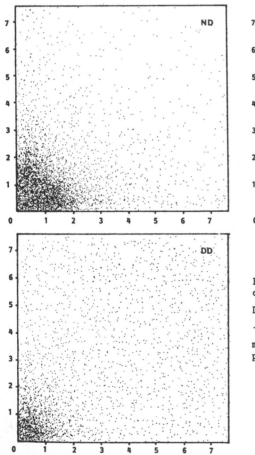

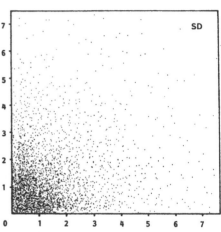

FIG. 17: Monte Carlo generated plots of $\eta_{1,min}$ vs $\eta_{2,min}$ for ND, SD and DD events at $\sqrt{s}$ = 2 TeV, where η_{min}= $-\ell n \, \tan(\theta_{max}/2)$ and $\theta_{i,max}$ is the maximum angle of the charged particles in hemisphere "i"(i=1,2).

VI. DOUBLE POMERON EXCHANGE

Double Pomeron exchange (DP), Eq. 3, has been studied at the ISR [14]. However, from the point of view of multiparticle production, the interest is at much higher energies, where the mass of the diffractive cluster X can be large. At present $\bar{p}p$ Colliders, where masses as large as $\sim$ 100 GeV can be reached, no DP studies have yet been conducted. Such studies require the use of leading particle spectrometers on both sides of the interaction region. Since the polar angles of the DP leading particles in $\bar{p}p$ Colliders are very small (less than 1 mrad), the spectrometers must use the bending magnets of the machine itself for momentum analysis and employ apparatus very close to the beam, located either within the beam vacuum pipe or in "Roman Pots" that can be moved close to the beam after injection. Such double-arm spectrometers have not been available at the $\bar{S}p\bar{p}S$ Collider and although they are in the plans [15] of the Collider Detector at Fermilab (CDF), they have not yet been fully commissioned. A design for spectrometers at the Superconducting Super Collider (SSC) was proposed [16] at the SSC Workshop in Berkeley, California, July 7-18, 1987 and will appear at the proceedings of the Workshop.

The DP process is interesting in that it probes the structure of the Pomeron. Comparison of the hadronization features of the mass X with those of single-diffraction dissociation and of non-diffractive inelastic collisions at $\sqrt{s} - M_x$ are essential in understanding the Pomeron in terms of QCD. We will not elaborate on this point here, but rather give a brief account of the kinematics and cross sections of DP exchange at high energies, which may be of some use to those contemplating experiments.

The differential cross section for double Pomeron exchange can be expressed in terms of the total and single diffraction dissociation cross sections using factorization (see Fig. 18):

$$DP - (SD)_1 \times (SD)_2 / \sigma_T$$

$$\frac{d^4\sigma}{dt_1 dt_2 dx_1 dx_2} - \frac{1}{\sigma_T} \frac{d^2\sigma_1}{dt_1 dx_1} \frac{d^2\sigma_2}{dt_2 dx_2} \tag{20}$$

At high energies, the pp single diffraction dissociation cross section at small t is given by (see Eq.11)

$$\frac{d^2\sigma}{dt\,dx} - \frac{A}{1-x} (b\,e^{bt}) \quad ; \quad A - 0.68 \pm 0.05 \text{ mb} \tag{21}$$

Assuming (from factorization) the same cross section for $\bar{p}p$ as for pp dissociation and substituting (21) into (20) yields

$$\frac{d^4\sigma}{dt_1 dt_2 dx_1 dx_2} - \frac{1}{\sigma_T} \frac{A^2}{(1-x_1)\,(1-x_2)} b^2 e^{b(t_1 + t_2)} \tag{22a}$$

$$\frac{d^2\sigma}{dx_1 dx_2} - \frac{1}{\sigma_T} \frac{A^2}{(1-x_1)(1-x_2)} \tag{22b}$$

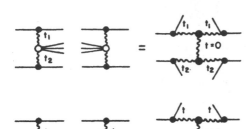

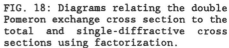

FIG. 18: Diagrams relating the double Pomeron exchange cross section to the total and single-diffractive cross sections using factorization.

The variables x_1 and x_2 may be replaced by the invariant mass M and the rapidity y of the cluster X using the equations

$$\frac{M^2}{s} = (1-x_1)\ (1-x_2) \tag{23}$$

$$y = \frac{1}{2}\ \ell n\ \frac{1-x_1}{1-x_2} \tag{24}$$

The result is

$$\frac{d^2\sigma}{d(M^2/s)dy} = \frac{1}{\sigma_T}\frac{A^2}{M^2} \tag{25}$$

By integrating over y, taking into account the limits in $(1-x)_i$ which are given by the coherence condition (Eq. 5b) and observing that $M^2_{max} = s(1-x)^2_{max}$, we obtain

$$\frac{d\sigma}{dM^2} = \frac{1}{\sigma_T}\frac{A^2}{M^2}\ \ell n\ \frac{M^2_{max}}{M^2} \tag{26}$$

The total double Pomeron exchange cross section is then given by

$$\sigma_{DP} = 2\ \frac{A^2}{\sigma_T}\ \ell n^2\ \frac{M_{max}}{M_{min}} \tag{27}$$

The differential cross section is expected to be suppressed for masses close to the two-pion threshold. Guided by the behavior of single diffraction dissociation cross sections near threshold, we believe that a value of M_{min} around 1 GeV, rather than $M_{min} = 2 M_\pi$, should provide a more realistic input into Eq. 27 for the calculation of σ_{DP}. The table below gives values for σ_{DP} vs $\sqrt{s}$, calculated from Eq. 27 with $M_{min} = 1$ GeV, $M_{max} = 0.15 \sqrt{s}$, A = 0.75 mb and σ_T from Ref. 4.

Table II

Double Pomeron exchange cross sections

Machine	$\sqrt{s}$(TeV)	σ_T(mb)	σ_{DP}(mb)
ISR	0.06	45	0.12
S$\bar{p}$pS	0.54	62	0.35
Tevatron	2	80	0.46
SSC	40	120	0.71

VII. CONCLUSION

The diffractive processes that have been discussed here are dominated by Pomeron exchange. The Pomeron plays an important role also in elastic scattering and in the total cross section, which is related to the forward elastic scattering amplitude through the optical theorem. Diffractive multiparticle production is the "minimum bias" physics of diffraction dissociation. The rest of diffractive physics is known as "hard diffraction". The QCD description of these processes will invariably involve a Pomeron-like theoretical structure. In order to formulate such a sructure, one needs to know the cross sections of the above processes, commonly referred to as "ℓns physics", and the nature and properties of the dissociation products, the physics of "hard diffraction". There is an abundance of theoretical models predicting large heavy flavor or gluon-jet content in hard diffraction, although the theoretical community is not unanimous in its opinion. In view of the small momentum transfer in diffractive processes, it has even been conjectured [17] that the "Pomeron energy" transferred in high mass SD may heat the proton receiving it to very high temperatures, leading to a QCD phase transition at relatively low diffractive masses (~100 GeV), a process that might explain the Centauro-type events observed in cosmic rays. Be as it may, the study of diffraction dissociation is certainly interesting and we may look forward to exciting results from the Tevatron and from the Superconducting Supercollider.

REFERENCES

1. K. Goulianos, Physics Reports Vol. 101, No. 3, December 1983.
2. A.I. Sanda, Phys. Rev. D6 (1972) 280.
    M.B. Einhorn, J. Ellis and J. Finkelstein, Phys. Rev.D5 (1972) 2063.
3. T.J. Chapin et al, Phys. Rev. D31 (1985) 17.
4. K. Goulianos, "Diffractive and rising cross sections", Proceedings of
    Workshop on Physics Simulations at High Energy, University of
    Wisconsin-Madison, 5-16 May 1986, World Scientific(Edited by V.
    Barger, T. Gottschalk and F. Halsen) pp. 127-140 ; Also to be
    published in "Comments on Nuclear and Particle Physics", Gordon and
    Breach, Science Publishers, Inc.
5. M.G. Albrow et al, Nuc. Phys. B108 (1976) 1.
    J.C.M. Armitage et al, Nuc. Phys. B194 (1982) 365.
6. UA4 Collaboration: M. Bozzo et al, Phys. Lett. 136B (1984) 217.
                        D. Bernard et al, Phys. Lett. 186B (1987) 227.
7. UA5 Collaboration: R.E. Ansorge et al, Z. Phys. C33 (1986) 175.
8. R.L. Cool et al, Phys. Rev. Lett. 48 (1982) 1451.
    F.T. Dao et al, Phys. Lett. 45B (1973) 399.
    S.J. Barish et al, Phys. Rev. Lett. 31 (1973) 1080.
    J.W. Chapman et al, Phys. Rev. Lett. 32 (1974) 257.
9. CHLM Collaboration: M.G. Albrow et al, Nuc. Phy. B102 (1976) 275.
10. UA4 Collaboration: V. Palladino, Proceedings of the First International
    Conference on Elastic and Diffractive Scattering, Chateau de Blois,
    France, 3-6 June 1985, Editions Frontiers (Edited by B. Nicolescu and
    J. Tran Thanh Van) pp.79-96.
    Also, D. Bernard et al, Phys. Lett. 166B (1986) 459.
11. K. Goulianos et al, Phys. Rev. Lett. 48 (1982) 1454.
12. K. Goulianos, Phys. Lett. 193B (1987) 151.
13. V. Innocente et al, Phys. Lett. 169B (1986) 285.
14. G. Alberi and G. Goggi, Physics Reports 74, No.1 (1981) 1-207.
15. K. Goulianos, "The CDF leading particle spectrometers", Proceedings of
    Antiproton 86, VIII European Symposium on Nucleon - Antinucleon
    Interactions, 1-5 September 1986, Thessaloniki, Greece, World
    Scientific (Edited by S. Charalambus, C. Papastefanou and R.
    Pavlopoulos) pp. 375-380.
16. Workshop on Experiments, Detectors and Experimental Areas for the
    Superconducting Supercollider, July 7-17, 1987, Berkeley, California.
17. K. Goulianos, "A Model for Centauro Production", Proceedings of
    Workshop on Physics Simulations at High Energy, University of
    Wisconsin-Madison, 5-16 May 1986, World Scientific(Edited by V.
    Barger, T. Gottschalk and F. Halsen) pp. 299-312 ; Also to be
    published in "Comments on Nuclear and Particle Physics", Gordon and
    Breach, Science Publishers, Inc.

BOSE-EINSTEIN CORRELATIONS:

FROM STATISTICS TO DYNAMICS

William A. Zajc
Physics Department
Columbia University
New York, NY

1   Introduction

Bose-Einstein correlations between identical particles are indisputably both multi-particle and quantum mechanical in nature. As such, they have evoked considerable interest from both theorists and experimentalists since the pioneering work of Goldhaber, Goldhaber, Lee, and Pais (GGLP) in 1960 [Gol60a]. However, it is only recently that these methods have been applied to test reaction *dynamics* rather than simply the consequences of Bose-Einstein (BE) statistics.

This review will concentrate on the experimental results themselves; drawing from the theoretical literature only those models needed to understand the presentation of the data. The perspective is a historical one, since that leads naturally from the early experiments, which were limited by the number of events, to the much more sophisticated present-day methods, where subtle systematic effects are the dominant source of error. The emphasis will be on collisions of elementary particles, with a few results from heavy ion collisions cited where appropriate. I will tend to use "pion" and "boson" interchangeably, since nearly all results obtained to date are for two-pion measurements, (although there is no difficulty in principle to extending BE methods to heavier mesons). It should be noted that at the outset that this review is not intended to be comprehensive. Instead, a few representative experiments for each topic will be discussed. The interested reader is referred to the reviews of Grishin [Gri77a], Giovannini, Mantovani and Ratti [Gio79a], Goldhaber [Gol85a], and Goldhaber and Juricic [Gol86a], for further information.

Two appendices are included for completeness. The first contains definitions of commonly used variables and notations. The second presents a "check list" of experimental requirements for both an *ideal* BE measurement and its presentation in the literature.

2   The Statistical Approach

This section describes analyses performed prior to the usage of correlation functions. Most such experiments relied on exact calculation of the phase space via the Fermi statistical model [Fer50a], hence the name of this section.

2.1   The Work of GGLP

The GGLP effect originated in the detailed study of the annihilation of 1.05 GeV/c anti-protons in a hydrogen bubble chamber. The reactions studied were of the form

$$\bar{p} + p \rightarrow n\pi^+ + n\pi^- + m\pi^0, \tag{1}$$

denoted as (n^+, n^-, m^0). The authors found that like-sign pions were more likely to be emitted with a small relative opening angle $\theta_{\pi\pi}$ than unlike-sign pions, in contradiction with a statistical model, which would predict only a small residual correlation from momentum conservation.

To explain this result, GGLP modified the usual formulation of the Fermi statistical model by explicitly introducing the symmetrization of the N-pion wavefunction, rather than simply dividing by $N!$ to count permutations. We present here a schematic version of their argument: Consider a N-pion state in a quantization volume V. Assume that some sub-volume Ω corresponds to the "reaction" volume. Then, the differential probability to find N pions in the reaction volume may be written as

$$dR_N \sim \frac{1}{N!} P_N(\Omega) \prod_{i=1}^{N} \left[\frac{V}{(2\pi)^3} \frac{d^3 p_i}{E_i} \right], \tag{2}$$

where

$$P_N(\Omega) = \int_\Omega d\vec{r}_1 \ldots d\vec{r}_N |\Phi_N|^2, \tag{3}$$

$$\Phi_N \equiv (\frac{1}{\sqrt{V}})^N \exp\{i \sum_{i=1}^{N} \vec{p}_i \cdot \vec{r}_i\}. \tag{4}$$

Clearly, in the case of an unsymmetrized wave-function, $P_N(\Omega)$ reduces to $(\Omega/V)^N$, and R_N therefore becomes simple phase-space with one parameter, the reaction volume Ω.

The *only* modification of Eqs. 2 and 4 required by GGLP to explain their data was the explicit symmetrization of Φ_N:

$$\Phi_N \to \hat{S}\Phi_N \equiv \frac{1}{\sqrt{N!}} (\frac{1}{\sqrt{V}})^N \sum_\sigma \exp\{i \sum_{i=1}^{N} \vec{p}_{\sigma(i)} \cdot \vec{r}_i\}, \tag{5}$$

where $\sigma(i)$ is the i-th element of the permutation σ on N objects, and the sum runs over all such permutations.

This simple *ansatz* permitted GGLP to explain both the like and the unlike pion opening angle distributions very nicely, after summing (incoherently) over the $(2^+, 2^-, 0^0)$, $(2^+, 2^-, 1^0)$, $(2^+, 2^-, 2^0)$ and $(2^+, 2^-, 3^0)$ reaction channels and integrating (laboriously) over all unobserved momenta. (See Fig. 1). The required value of the reaction volume corresponded to a radius of 0.82 fm, in accord with naive expectations for this process. (Due to the large number of different parametrizations of source densities, we follow the suggestion of Bartke and Kowalski [Bar84a] by quoting all radii as the equivalent root-mean-square radius. A conversion chart for various distributions may be found in Appendix A.)

However, as noted by GGLP, there are some problems with this formalism. First, the good agreement with data was obtained by using an invariant four-volume for Ω; the motivation being calculational ease rather than physical reasonableness. Nonetheless, indications were that the four-dimensional form fit the data slightly better than the more consistent three-dimensional

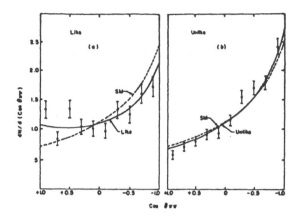

Figure 1: The distribution of pair opening angles observed by GGLP for like and unlike sign pion pairs, along with the predictions of the statistical model (SM) and a modified statistical model incorporating BE symmetrization with $R = 0.82$ fm.

form. Secondly, the projection from correctly symmetrized states of good isospin to the BE symmetrized states was handled only approximately. Thirdly, the *un*symmetrized statistical model required a reaction radius of 2.7 fm to account for the observed pion multiplicities; inclusion of the Bose-Einstein symmetrization did little to affect this result. Nonetheless, the relevant *two*-particle data were well described by the model with a radius of ~ 0.8 fm, a theme that we shall see repeated in later experiments: in many cases, Bose-Einstein analyses work better than they have any right to.

2.2 Other Work

Following GGLP, other authors noted the difference between like and unlike sign opening angles, although few were able extract a radius for the reaction volume. Perhaps this is due to increasingly more complicated final states (due to higher available energies), which prevented explicit calculation of Bose-symmetrized phase space. For example, Xuong and Lynch [Xuo62a], who studied the $(3^+, 3^-, m^0)$ channel in $p\bar{p}$ annihilation at 1.6 GeV/v, attributed large deviations from the phase space prediction of opening angles and invariant mass distributions to the influence of Bose-Einstein statistics for the three-like-pion systems, but were unable to perform the calculation of the appropriately symmetrized phase space. (Three and four like-pion correlations in πp interactions were also investigated by Bosebeck *et al.* [Bos73a], who concluded that the "spectacular" enhancements could be explained as a products of the second-order correlations.) A smaller like pair enhancement was seen by Ferbel *et al.* [Fer66a] for p$\bar{\text{p}}$ annihilation at 3.3 and 3.7 GeV/c, who noted that this is consistent with the trend predicted by GGLP for higher energy annihilations (due essentially to the increased phase space). This observation was extended to much higher energies by Firestone *et al.* [Fir75a], who examined opening angle distributions as a function of the (charged) invariant mass M of the system pp $\rightarrow$ pions $+X$ at 300 GeV/c. These authors found that the Forward/Backward ratio of the opening angle distribution was a strong function of M, but identical to the value obtained by GGLP at the appropriate invariant mass. The interaction region in K^+p interactions was found to be 0.8 fm by De Baere *et al.* [DeB70a], although these authors note little sensitivity of the data to the actual value assumed.

More detailed Bose-Einstein were performed by Bartke *et al.* [Bar67a] and by Donald *et al.* [Don69a]. Bartke *et al.*, foreshadowing future developments, plotted the opening angle enhancement as a function of $\Delta P \equiv |\,|\vec{p}_1| - |\vec{p}_2|\,|$, noting that if due to a true Bose enhancement, the enhancement should be maximal for small values of ΔP. Such a behavior was observed, again with a radius of ~ 0.8 fm (for $\pi^+ + p \rightarrow 6$ or 8 prongs at 8 GeV/c). Donald *et al.* found a radius of 0.52 fm for the $(2^+, 2^-, 1^0)$ channel in p$\bar{\text{p}}$ annihilation at 1.2 GeV/c, using a phase model that explicitly included vector resonance production along with Bose-Einstein statistics.

As the ISR and Fermilab provided significantly higher energies than previous machines, most of the correlation studies were performed in terms of rapidity and transverse momentum correlations [Bis75a,Bis76a,Oh75a]. These variables, while hardly ideal for a BE analysis, did indeed show enhancements for like-pairs as compared to unlike pairs. In fact, upon parametrizing their rapidity difference distributions in terms of $Q^2 \equiv -(p_1 - p_2)^2$, Biswas *et al.* [Bis76a] extracted a radius of 1.1 ± 0.3 fm. Although simple phase space arguments were no longer applicable at these energies, the simple inclusion of Bose-Einstein statistics in cluster models [Ran75a] or in Mueller-Regge Models [Biy74a,Miy75a] provided at least qualitative reproduction of the correlations by introducing correlations of length ~ 0.2 GeV ~ 1 fm^{-1}.

3   Introduction of the Correlation Function

It is clear from the discussion of the previous section that the more complicated dynamics of higher energy reactions created a severe limitation for direct application of the GGLP technique. The difficulty arose, not from the concept of Bose-Einstein symmetrization itself, but from the inability to calculate the production amplitudes that determined that relevant phase space densities. It is hard to imagine any fundamental information resulting from further studies using these methods, since the limiting factor came from the model-dependence of the underlying dynamics.

Fortunately, a way out of this situation was discovered, nearly simultaneously, by several independent researchers [Kop72a,Shu73a,Coc74a]. These authors realized that Bose-Einstein correlations in particle physics are analogous to second-order interference of photons in optics. In particular, the fundamental work of Hanbury-Brown, Jennison and Das Gupta in radio astronomy [Han52a,Han54a] and Hanbury-Brown and Twiss in optical astronomy [Han56a,Han56b,Han57a,Han57b], showed that the dimensions of a source of bosons (photons in their case) could be determined from the width of a second-order correlation function. (*Second-order* refers to the fact that the interference or correlation is observed in the intensity, rather than in the amplitude, as in the case of a Michelson interferometer. Hence the alternative name of intensity interferometry.)

3.1   Derivation of the Correlation Function

Rather than describing in detail the work of the above authors, I will present what has come to be the canonical "derivation" of BE correlations in particle physics. Assume two identical pions are produced with momentum $\vec{p_1}$ and $\vec{p_2}$ by a source $\rho(r)$ extended in space and time. Let a typical pair of production points be labeled r_A and r_B, and denote the positions of detecting the pion with p_i and r_i. If we further assume that the pion wave-functions may be described by plane-waves, the amplitude for this event may be written as

$$A_{12} = \frac{1}{\sqrt{2}}\left[e^{i p_1 \cdot (r_1 - r_A)}e^{i p_2 \cdot (r_2 - r_B)} + e^{i p_1 \cdot (r_1 - r_B)}e^{i p_2 \cdot (r_2 - r_A)}\right]. \qquad (6)$$

The second term is required due to the identity of the pions. Alternatively, we may think of it as resulting from the indistinguishability of the two processes

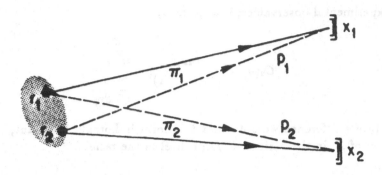

Figure 2: The two indistinguishable processes leading to the detection of identical pions with momentum p_1 at r_1 and p_2 at r_2.

leading to this final state, as shown in Fig. 2. The probability of such an event is then given by the square of A_{12}, integrated over all source points,

$$\mathcal{P}_{12} = \int d^4 r_A d^4 r_B |A_{12}|^2 \rho(r_A)\rho(r_B) = 1 + |\tilde{\rho}(q)|^2 \equiv C_2(q), \qquad (7)$$

assuming that the production amplitudes are independent of r and uncorrelated.

Equation 7 is the fundamental result of intensity interferometry. It states that the two-particle counting rate is proportional to the square of the Fourier transform of the source distribution. The Fourier transform is performed with respect to the relative (four)-momentum of the particles, with each component of relative momentum containing information of the source distribution in the conjugate spatial variable. The range of the enhancement for the i-th component q_i is of order $1/R_i$, as expected from uncertainty principle arguments. Note that the normalization of $\rho(r)$ enforces the condition $C_2(q = 0) = 2$, an appropriate result: The probability of finding two like-pions in the same state is 2! times greater than for unlike particles.

The function $C_2(q)$ is the correlation function for pair detection. In terms

242

of experimental observables, it is given by

$$C_2(q) = \frac{\langle n_\pi \rangle^2}{\langle n_\pi(n_\pi - 1) \rangle} \cdot \frac{\frac{dn}{d\vec{p}_1 d\vec{p}_2}}{\frac{dn}{d\vec{p}_1}\frac{dn}{d\vec{p}_2}} . \qquad (8)$$

For future reference we note that this form is Lorentz invariant, since the factors of E_i in the various $d\vec{p}_i/E_i$ cancel in the ratio.

3.2   Experimental Approaches to the Correlation Function

In practice, the correlation function is never calculated in its full four-dimensional form according to Eq. 8. Instead, the actual distribution A of pairs in relative momentum (more precisely, some projection of q into a spatial and temporal component: $q \to (q_s, q_0)$) is divided by the distribution of some "background" sample B:

$$C_2(q) \to C_2(q_s, q_0) \equiv \frac{A(q_s, q_0)}{B(q_s, q_0)} . \qquad (9)$$

The background is required to have all correlations induced by phase space, dynamics, experimental acceptance, etc. *except* those resulting from BE correlations.

In the absence of a prescription for calculating $B(q_s, q_0)$, this requirement is a tautology. However, there are several plausible schemes for generating B that allow one to proceed:

A. Use the relative momentum spectrum for unlike-sign pion pairs: $B(q_s, q_0) = B^{\pm\mp}$.

B. Use the relative momentum spectrum of like-sign pions taken from different events: $B(q_s, q_0) = B^{\text{diff}}$.

C. Calculate a relative momentum spectrum by explicitly including all dynamics except the BE symmetrization: $B(q_s, q_0) = B^{\text{MC}}$.

Clearly, some care must be employed when using any of these approaches. Method A requires particular attention to eliminating the effect of resonances in the $\pi^+\pi^-$ channel that do not exist in the $\pi^\pm\pi^\pm$ channels. In Method B, one must insure that the "fake" events so generated do not grossly violate applicable conservation laws, and that residual effects of the original BE correlation are removed from the background [Zaj82a,Zaj84a]. Method C, which really is the differential form of the original GGLP (integral) prescription, obviously requires detailed, usually unavailable, knowledge of the reaction dynamics. As such, it is seldom used, although a crude form may be found in the original GGLP paper (still in terms of the opening angle distribution, see Fig. 3). As an exercise, we have replotted the the opening angle distribution (Fig. 4) of a similar work [Xuo62a] in terms of a pair correlation function, using a very rough translation from $\cos\theta_{\pi\pi}$ to $|\vec{q}|$. This is shown in Fig. 5. A strong enhancement at low relative momenta is seen, corresponding to a radius of order 1 fm, demonstrating that even this very primitive approach provides a quick estimate of the source radius. It is also interesting to observe how the relatively uniform distribution in $\cos\theta_{\pi\pi}$ translates into the distribution of relative momentum, which illustrates the difficulty of obtaining good statistics at low relative momentum for untriggered experiments.

3.3 Fitting Procedures

Typical parametrizations for $\rho(r)$ and the corresponding $|\tilde{\rho}(q)|^2$ are found in Table A.4 of Appendix A. Assuming that some reasonable prescription may be found for the generation of the correlation function, it is then a straightforward matter to fit the resulting distribution to a a functional form and extract the source lifetime and/or radii. However, Deutschmann *et al.* [Deu78a] found it necessary to modify Eq. 7 by introducing the "coherence parameter" λ:

$$C_2(q) \rightarrow 1 + \lambda|\tilde{\rho}(q)|^2. \tag{10}$$

(An analogous parameter was used in phase space fitting by Biswas *et al.* [Bis76a].) This was motivated by the observation that typical $C_2(q)$'s generated by any of the above methods seldom reach the full value of 2 as q vanishes. Fitting to the form of Eq. 10 eliminates the systematic error associated with fitting the experimental correlation function to a function that

244

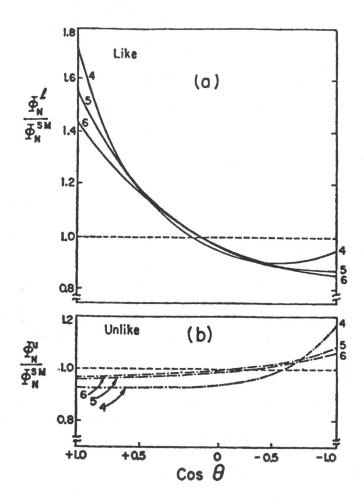

Figure 3: The ratio of the pair opening angle distribution predicted by Bose-symmetrized phase space divided by the corresponding unsymmetrized distribution for a.) like pairs and b.) unlike pairs as calculated by GGLP for 4,5 and 6 prong events.

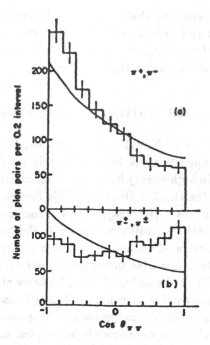

Figure 4: The pair opening angle distribution as measured by Xuong and Lynch [Xuo62a] for $p + \bar{p} \to 3\pi^+ + 3\pi^- + m\pi^0$ at 1.6 GeV/c.

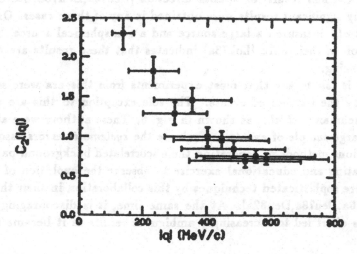

Figure 5: The pair correlation function estimated from the published data of [Xuo62a] .

can not describe the data. We shall have much to say about both the theoretical interpretations and experimental systematics that affect the value of λ in what follows. Before doing so, we examine in the next section some of the first results obtained by using these techniques.

4   Early Results using Correlation Functions

Following the suggestions of Kopylov, Podgorestkiï, Cocconi, and Shuryak, many experimenters applied some form of correlation function analysis to like-pion correlations in high energy hadron-hadron collisions [Ang77a,Ang77b], [Cal76a,DeW78a,Deu76a,Deu78a,Deu82a,Dri79a,Eze78a,Goo78a,Gra76a], hadron-nucleus   collisions   [Bec79a],   and   nucleus-nucleus   collisions [Ang79b,Ang80a,Fun78a]. We will concentrate in the hadron-hadron results here, deferring a discussion of nuclear results to a separate section.

It is quite simple to summarize the experimental data: Nearly all experimenters measured radii on the order of 1 fm, lifetimes of a similar scale, and found the coherence parameter λ to be substantially less than one (regardless of whether it was explicitly extracted). These results are summarized in Table 4. Several of these experiments attempted to determine the dependence of the extracted source parameters on energy or species [Deu76a], charged multiplicity [Goo78a, no dependence was observed], Lorentz frame [Gra76a, again a null result] or spatial direction [Deu76a,Deu78a,Eze78a]. No statistically significant results were obtained in any of these cases. Only Loktionov *et al.* measured a large source and a non-spherical source, but an examination of their data [Lok78a] indicates that these results are not terribly convincing.

In fact, it is fair to say that most experiments from this era were severly limited by the number of events. The sole exception to this are the data of Deutschmann *et al.*, as shown in Fig. 6. These authors were able to use their large sample of events to evaluate the *systematic* errors associated with various methods of generating the uncorrelated background pairs. It is a fascinating and educational exercise to observe the evolution of increasingly more sophisticated techniques by this collaboration in their three papers [Deu76a,Deu78a,Deu82a]. At the same time, it is discouraging to note that this effort led to increasingly ambiguous results as it became ap-

Reference	Reaction	Energy	R (fm)	$c\tau$ (fm)	λ
[Cal76a]	$\pi^- p$	11.2 GeV/c	0.9 ± 0.1	0.41 ± 0.15	—
[Deu76a]	πp	4-25 GeV/c	$0.87^{+.35}_{-.17}$	$0.70^{+.6}_{-.3}$	—
[Gra76a]	$K^+ p$	8.25 GeV/c	~ 1	0.9-1.2	—
[Ang77a]	$p\bar{p}$	0-.7 GeV/c	1.20 ± 0.09	1.4 ± 0.2	—
[Ang77b]	$\pi^- p$	40 GeV/c	1.5 ± 0.3	0.8 ± 0.5	—
[Deu78a]	$\pi^+ p$	4-25 GeV/c	1.60 ± 0.13	1.2 ± 0.3	0.4-0.5
	$K^- p$	"	"	"	"
	$p\bar{p}$	"	"	"	~ 1
[Goo78a]	$K^+ p$	32 GeV/c	0.8 ± 0.1	0.35 ± 0.15	0.3-0.6
[Eze78a]	pp	28.5 GeV/c	0.6 ± 0.1	0.6 ± 0.2	—
[Lok78a]	$p\bar{p}$	22.4 GeV/c	2.6 ± 0.4	3.1 ± 1.6	—
[Dri79a]	pp	53 GeV	1.10 ± 0.3	1.4 ± 0.6	0.4
[Deu82a]	$\pi^+ p$	4-25 GeV/c	1.2	1.0	0.8
	$K^- p$	"	"	"	"
	$p\bar{p}$	"	"	"	"

Table 1: Results from early interferometry experiments using correlation functions.

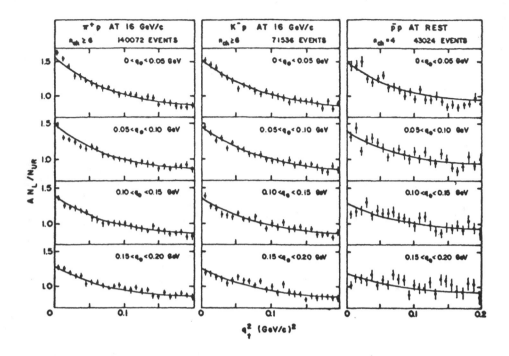

Figure 6: Ratio of like to unlike pairs versus q_T^2, as measured by Deutschmann *et al.* [Deu82a], along with fitted correlation functions. Note that their normalization does not require that the large q_T^2 value of the correlation function approach one.

parent that this measurement (and all other Bose-Einstein measurements) was dominated by systematic, not statistical errors.

5   Interlude Concerning Coherence vs. Correlation

It did not escape the attention of the theoretical community that the Bose-Einstein effect, while present, was not fully developed in all the measurements discussed in the previous section (i.e., $\lambda < 1$ in these experiments). This generated a flurry of speculation concerning the "coherence" of the pion field (notwithstanding the large uncertainties associated with the extraction of λ). In this section we provide a brief introduction to the role of coherence and its measurement in pion correlations.

5.1   Aspects of Partially Coherent Fields

Implicit in the derivation of Eq 7 for the correlation function is the assumption of *random phase* for the pion field. More precisely, the passage from Eq. 6 to Eq. 7 requires that the pion field at different points r_i in the source is completely *un*correlated in phase. If this is not the case, then the emission of a n-th pion depends on the "pre-existing" $n - 1$ pions, the appropriate language of description becomes that of (pure or mixed) coherent states, the simple argument leading to Eq. 7 no longer obtains, and we find that the predicted value of the coherence parameter λ is less than one, *even though the pion field contains appropriately Bose-symmetrized wave-functions.*

This point, although it has been made by many authors [Fow77a,Bar78a] [Fow78a,Fow79a,Fow85a,Gyu79a,Biy80a,Biy81a] is often a major stumbling block in understanding BE measurements. Exacerbating the situation is the close association in most physicists' minds between "coherence" and "correlation". If, after all, a coherent state is a highly correlated state of the field, how can it exhibit no Bose-Einstein correlations? Conversely, what is the origin of the pair-correlations observed from a random (i.e., chaotic) source?

We shall present some plausibility arguments to clarify this situation by considering the fluctuations of a classical (electromagnetic) field. The relevance of this results from the following:

250

- It is exactly the bosonic nature of photons that allows the creation of a (macroscopic) electromagnetic wave.

- A monochromatic classical electromagnetic wave is best described by a coherent state of the photon operator.

- The superposition of these waves leads to large fluctuations in intensity, which is the classical manifestation of the large fluctuations in cell occupancy expected for bosons.

Perhaps the most streamlined form of the argument is due to Bowler [Bow85a], although various versions of it have been presented by Gyulassy *et al.* [Gyu79a] and Hanbury-Brown and Twiss [Han56a,Han57a].

Consider two sources a and b of electromagnetic radiation. The amount of radiation emitted with wave-vector $\vec{k}_i$ is then given by

$$A_i = f_a e^{i\vec{k}_i \cdot \vec{r}_a} + f_b e^{i\vec{k}_i \cdot \vec{r}_b} \tag{11}$$

where f_a and f_b are phase factors for their respective sources. The intensity of radiation with wave-vector $\vec{k}_i$ is then of course

$$I_i = |f_a|^2 + |f_b|^2 + [f_a f_b^* e^{i\vec{k}_i \cdot (\vec{r}_a - \vec{r}_b)} + c.c] \ . \tag{12}$$

If a and b are mutually coherent sources, then the above expression is in its final form. On the other hand, if a and b are mutually incoherent, the cross-terms vanish if we average over a time long compared to the phase fluctuations:

$$\langle f_a f_b^* \rangle = 0 \ \Rightarrow \ \langle I_i \rangle = |f_a|^2 + |f_b|^2 \ . \tag{13}$$

Now consider the *average* of the *product* of the intensities, where the average is performed with respect to the phase factors of a and b. If they are coherent, then there is in effect no averaging to perform:

$$\langle f_a f_b^* \rangle \equiv f_a f_b^* \ \Rightarrow \ \langle I_1 I_2 \rangle = \langle I_1 \rangle \langle I_2 \rangle = I_1 I_2 \ , \tag{14}$$

so that

$$\frac{\langle I_1 I_2 \rangle}{\langle I_1 \rangle \langle I_2 \rangle} = \frac{I_1 I_2}{I_1 I_2} = 1 \ . \tag{15}$$

In the case of a and b being mutually incoherent, we find that

$$\langle I_1 I_2 \rangle = \{|f_a|^2 + |f_b|^2\}^2 + 2|f_a|^2|f_b|^2 \cos(\vec{k}_1 - \vec{k}_2) \cdot (\vec{r}_a - \vec{r}_b) \quad . \qquad (16)$$

Assuming $|f_a|^2 = |f_b|^2$, the uncorrelated source version of Eq. 15 becomes

$$\frac{\langle I_1 I_2 \rangle}{\langle I_1 \rangle \langle I_2 \rangle} = 1 + \frac{1}{2} \cos(\vec{k}_1 - \vec{k}_2) \cdot (\vec{r}_a - \vec{r}_b) \quad . \qquad (17)$$

The factor of $\frac{1}{2}$ before the interference term results from our considering only two sources; it is straightforward to demonstrate that for n sources the corresponding factor is $(n-1)/n$. Thus, in the limit of a large number of such uncorrelated sources (as in our integration in Eq. 7), the cosine term appears with coefficient unity, producing a correlation in the mutual intensities of an ensemble of *uncorrelated* classical sources, while for only one coherent source the cosine term vanishes.

This is the counter-intuitive aspect of intensity interferometry for many physicists. The preceeding derivation tells us that the ratio of intensities in Eq. 17 has a correlation term that *increases* as the number of *random* sources increases. Physically this results from the addition of N random phases to form a field amplitude $\vec{E}$ whose intensity can range from 0 to N^2 (even though its mean is of course $N/2$, where the 2 comes from the average of $\cos^2 \Phi$). An excellent discussion of this may be found in the book by Loudon [Lou73a]. Following his suggestion, I have explicitly modeled the fluctuations in intensity for 16 radiators, assuming that the phases are uncorrelated and each changing (to a random value) every t seconds, where t has an exponential distribution $e^{-t/\tau}$. Figure 7 shows the results of this exercise, which does in fact show large fluctuations about the mean value. In this case, the intensities at two points in time are correlated if the separation Δt satisfies $\Delta t \ll \tau$; this corresponds to the frequency-domain correlation implied by Eq. 17. To demonstrate this is intrinsically a *wave* phenomena, I have repeated the calculation for *intensities* randomly varying as $\cos^2 \Phi(t)$; the corresponding distribution (Fig. 8) shows only Poisson fluctuations about the mean.

The methods leading to Eq. 17 may be applied to the case of a mixture of a coherent source of average multiplicity $\langle N_{coh} \rangle$ and a incoherent source

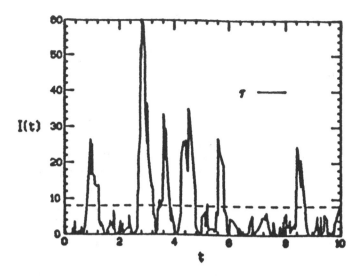

Figure 7: The intensity fluctuations from the sum of 16 radiators, each of strength $|\Re\{e^{i\Phi}\}|^2 = \frac{1}{2}$ with Φ randomly varying on the time scale labeled τ, assuming that the intensity is given by $|\Re\{\sum e^{i\Phi_i}\}|^2$.

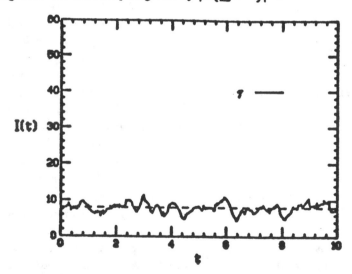

Figure 8: The intensity fluctuations from the sum of 16 random radiators, this time added according to $\sum |\Re\{e^{i\Phi_i}\}|^2$.

of average multiplicity $\langle N_{inc} \rangle$ by replacing Eq. 11 with

$$A_i = f_a e^{i\vec{k}_i \cdot \vec{r}_a} + \sum_{b=1}^{N} f_b e^{i\vec{k}_i \cdot \vec{r}_b} \quad , \tag{18}$$

where $|f_a|^2 = \langle N_{coh} \rangle$ and $\sum_{b=1}^{N} |f_b|^2 = \langle N_{inc} \rangle$. In the limit where $N \to \infty$, we recover the result of many authors, namely that the interference term is reduced by a factor

$$\lambda = 1 - \left[\frac{\langle N_{coh} \rangle}{\langle N_{coh} \rangle + \langle N_{inc} \rangle} \right]^2 \quad . \tag{19}$$

It is for this reason that λ is referred to as the coherence parameter, although incoherence parameter might be more apt. Note that this result has been derived from simple considerations of interfering classical waves– the introduction of (quantum mechanical) coherent states, while perhaps more appropriate, is not required. Similar, essentially combinatoric, extensions of these arguments have been made by Biyajima [Biy82a] and Lednicky *et al.* [Led82a] to the case of M independent coherent sources and N incoherent sources. In addition, these works carefully distinguish between the cases of *one* coherent source and *one* incoherent source with an arbitrary ratio of strength (as above), and several *independent* coherent sources of a given strength in the presence of a incoherent source.

5.2 Measuring Coherence

While the previous remarks may have done more to confuse than clarify, it should be clear that, ideally, measurement of λ is in principle sensitive to very interesting dynamical features of the pion field. In this section we explore those effects that obscure any simple such interpretation.

5.2.1 Other (Trivial) Source Dynamics

One obvious method whereby λ may be reduced occurs when different space-time regions emit different single-particle momentum distributions. Recall that in the derivation of Eq. 7 we assumed that the momentum distribution

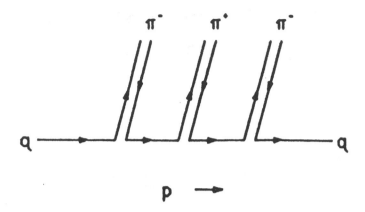

Figure 9: A production chain in momentum space which would produce no like-sign pions of equal momentum.

was independent of the point of emission. This is certainly not a true for a wide class of models, especially those incorporating longitudinal growth (see references in [Kol86a]), but even the most trivial sort of chain model has this feature (see Fig. 9). It also becomes a *non*-trivial matter to calculate the correlation function in these circumstances, since one is not allowed to simultaneously specify the momentum and position of each pion in the source. As pointed out by Pratt [Pra84a], the correct description requires the use of Wigner functions.

5.2.2 Final-State Interactions

An obvious final state interaction that reduces the value of λ is the two-pion Coulomb interaction. It is well known that the solution of Schrodinger's equation in a Coulomb potential modifies the two-pion relative momentum distribution according to

$$\frac{n_{\pm\pm}(q)}{n_{00}(q)} = \frac{2\pi\eta}{e^{2\pi\eta}-1} \quad , \quad \eta \equiv \frac{m_\pi e^2}{\hbar q_{\pi\pi}} \quad , \tag{20}$$

where $q_{\pi\pi}$ is the relative three-momentum of the two pions in their center of mass. Corrections based on this formalism were first made by Zajc [Zaj82a], who also noted that the invariant form of $q_{\pi\pi}^2$ is $Q^2 = |\vec{q}|^2 - q_0^2$. The implication for BE analyses is that these corrections can affect a large region of the relative momentum phase space, event though the value of some variable such as $|\vec{q}|$ is large. Such corrections are obviously acceptance dependent, and can produce substantial changes in the inferred value of λ.

A more subtle final-actions is the $\pi^\pm\pi^\pm$ strong interaction, which has recently been calculated by Suzuki [Suz87a]. Although the s-wave phase shift varies only from zero to -25 degrees in the range $0 < Q < 1$ GeV, nonetheless a substantial modification in the value of λ is found, suppressing the "true" value by $20_{-13}^{+5}\%$, where the errors result from uncertainties in the parametrization of the energy-dependence of the phase shifts. Furthermore, Suzuki notes that there is a fundamental ambiguity that obscures the separation of production dynamics from final-state interactions (for strong interactions) so that additional refinement of these estimates may require major advances in our understanding. He also notes that similar symmetry arguments apply in the production of $\pi^\pm\pi^0$ pairs, since the s-wave state is required to be $I = 2$. The study of "Bose-Einstein" correlations for such systems is both experimentally feasible and theoretically interesting.

Some of the ambiguity concerning the separation of production dynamics from final-state interactions may in fact be removed by the very nature of the spatially-extended pion source. In the case of the two-pion Coulomb interaction, the source size leads to a correction to Eq. 20 of magnitude $m_\pi R/137$ [Gyu81a,Zaj82a], which is negligible even for heavy ion collisions. On the other hand, recent work by Bowler [Bow87a] shows the effect of the final-state strong interaction is considerably modified, since the s-wave scattering length is ~ -0.2 fm, which is substantially smaller than the source-size for all reactions except e^+e^- collisions. Roughly speaking, the intercept λ is given by

$$\lambda = 1 + \frac{\sqrt{24}a}{R} + \mathcal{O}(a^2/R^2) \ , \tag{21}$$

where a is the s-wave $\pi\pi$ scattering length, and R is the size of the source (not the mean $\pi\pi$ separation). Bowler finds remarkable agreement with both the heavy ion and e^+e^- data using a simple formalism incorporating the known $\pi\pi$ phase shifts (although one must inquire what has become of the

256

string dynamics in the latter case).

5.2.3 Resonance Production

In our discussions of the pion source to this point, we have tacitly assumed that all pions are directly produced by the source. However, a large fraction of the final state pions are known to result from the decays of resonances such as the $\rho, \omega, \eta, ...$ and so on. The ρ, with a width $\Gamma_\rho = 153$ MeV, decays essentially in the production region. Such is not the case with ω $(\Gamma_\omega = 9.8$ MeV), the η $(\Gamma_\eta \sim 1$ keV), and the η' $(\Gamma_{\eta'} \sim 240$ keV). (The K*(892) is marginal in this regard.) As pointed out by Grassberger [Gra77a], a resonance with mass M and momentum $\vec{k}$ will travel a distance $\sim |\vec{k}|/\Gamma M$ before decaying, thereby creating an effective source for its decay products tens (for ω's) or thousands (for η's of fermis in extent.

Such a large source produces BE enhancements at only very small values of relative momentum, much smaller than the experimental resolution of any detector to date. If this narrow peak cannot be resolved, only the fraction f_d of directly produced pions appear to interfere, leading to a value of $\lambda \sim f_d^2$. Much more general arguments have been made by Lednicky and Podgoretskii [Led79a] for the general case of n different sources.

If one knows the primary abundances of the various resonances *and their primordial momentum distribution*, it is possible in principle to correct the observed value of λ for unobserved structure in the correlation function. Various groups have begun to estimate these corrections; see the section on recent e^+e^- data for a discussion.

5.2.4 Event Classification

The ratio of an average is not the average of a ratio. This innocuous observation motivated Gyulassy [Gyu82a] to study a class of models where the pair distribution depends on some exclusive parameter X such as impact parameter or jet production. If the correlation function is formed according to the prescription of Eq. 9, then clearly

$$\left\langle \frac{A(q;X)}{B(q;X)} \right\rangle \neq \frac{\langle A(q;X) \rangle}{\langle B(q;X) \rangle} , \qquad (22)$$

so that a correlation function formed as on the LHS will average over un-observed dynamical variables. Gyulassy was able to show that, in certain models, this can create large deviations of the correlation function from its "true" value, thereby distorting the value of λ.

Even if the ensemble-averaged correlation function retains the property of smoothly varying from 1 to 2, the averaging process may lead to a functional form so different from the unaveraged result that systematic errors may be introduced in the fitting process. A nice example of this is provided by Bowler [Bow85a], who assumes that the distribution of pion emitters in *one* event is Gaussian, $\rho_1(r; \sigma) \sim e^{-r^2/2\sigma^2}$, with some value of σ characteristic of that event. If the events have a distribution in their widths $f(\sigma) \sim e^{-\sigma^2/2\beta^2}$, so that the ensemble-averaged source distribution is given by

$$\rho(r) = \int f(\sigma)\rho_1(r\,;\sigma)d\sigma \quad , \tag{23}$$

it is straightforward to show that the averaged correlation function is given by $C_2(q) = 1/(1 + \beta^2 q^2)$. Attempting to fit this form to a Gaussian will tend to underestimate the value of λ (as well as distorting the extracted radius), since the Lorentzian form is more sharply peaked than a Gaussian. This is not just a curious mathematical result; this argument is qualitatively similar to the effect of impact parameter smearing in hadronic or nuclear collisions.

A similar problem can result from projecting the correlation function into a given set of relative momentum variables. Recall that $C_2(q)$ as defined in Eq. 8 is Lorentz invariant. Assume that, in the rest frame of the source, the source density is given by

$$\rho(r) = \frac{1}{\pi^2 R^3 \tau} e^{-|\vec{r}|^2/R^2 - t^2/\tau^2} \quad , \tag{24}$$

so that

$$C_2(q) = 1 + e^{-|\vec{q}|^2 R^2/2 - q_0^2 \tau^2/2} \quad . \tag{25}$$

This form is true only in the rest frame of the source. Yano and Koonin [Yan78a] showed that Eq. 25 can be made invariant by including an explicit dependence on the source four-velocity u:

$$C_2(q) \rightarrow 1 + e^{-[R^2 + \tau^2](q \cdot u)^2/2 + (q \cdot q)R^2/2} \quad . \tag{26}$$

258

If one chooses to treat $C_2(q)$ as a function of $Q^2 \equiv -(q \cdot q)^2$ only, then an implicit average is being performed over the distribution of q_0, $|\vec{q}|$ and $u = (\gamma_s, \vec{\beta}_s \gamma_s)$ accepted by the apparatus:

$$C_2(q) = 1 + \lambda e^{-\beta Q^2} \quad \Rightarrow \quad \lambda = \left\langle \exp\{-\gamma_s^2[R^2 + \tau^2](q_0 - \vec{\beta}_s \cdot \vec{q})^2\} \right\rangle \quad . \quad (27)$$

While some authors have examined the frame-dependence of fitting variables for $C_2(q)$, I am unaware of any investigations that have associated these searches with the decrease in λ predicted by the above argument.

5.2.5 Experimental Acceptance

The momenta of particles in an experiment are found either by visual scanning methods (for bubble or streamer chambers) or computer reconstruction algorithms (for electronic experiments). It is difficult enough to optimize these procedures for single-particle measurements.   Bose-Einstein analyses require detailed knowledge of the two-particle acceptance at small spatial separation. This is often a source of systematic error in determining the value of λ. While it is possible to calculate and correct for these effects (see, e.g., Althoff *et al.* [Alt85a]), a quite detailed model for both event generation and detector simulation is mandatory.

5.2.6 Contamination

This certainly qualifies as a trivial effect, although that does not prevent it from being bothersome. Typically some small fraction f of "pions" seen in an experiment are misidentified electrons, kaons or protons (depending on momentum). If one labels them as pions and proceeds with a BE analysis, the inferred value of λ will of course be $(1 - f)^2$ (assuming no dynamical correlations between the pions and the misidentified particles). At very low values of q one may also face a contamination from Dalitz pairs, as noted by Deutschmann *et al.* [Deu82a].

5.2.7 Background Calculation

As noted in Section 3.2, there are several prescriptions for generating the background relative momentum distribution $B(q)$. Ideally, one would like to try all legitimate contenders, and use the change in any fitted parameter to help estimate systematic errors in the fitting procedure. This has been done by various authors (see, e.g. [Deu82a,Bre85a,Arn86a]), with typical systematic errors of 20% on the extracted value of λ. One must also insure that the generated background is taken from the same class of events as the correlated pairs (cf. Section 5.2.4), that energy-momentum conservation is not grossly violated (if using a "mixed" background) [Kop75a], and that residual correlations from the mixing process are removed [Gra76a,Zaj82a,Aih85a]. It is probably these uncertainties that determine the ultimately attainable "true" value for λ, since they depend on the very definition of the correlation function.

6 Results from Lepton-Hadron Collisions

This section is devoted to a discussion of one experiment, that of the EMC collaboration [Arn86a], which studied Bose-Einstein correlations in deep inelastic μp interactions at 280 GeV/c. These results, while of intrinsic interest, are most notable for the careful and thorough treatment of backgrounds and biases present in a typical B-E analysis. As such, they can serve as a prototypical realization of many of the effects discussed in the previous section.

An exhaustive Monte Carlo simulation of the experiment (which combined visual and electronic tracking devices) allowed the authors to examine the two-particle tracking biases as a function of Q^2. Not surprisingly, they found that the required correction factor for like-sign pairs sharply increased with decreasing Q^2, due to misidentifications, decays, and reinteractions. They note that these problems are substantially worse for pair events than for single-particle distributions, as suggested by the consideration of Section 5.2.5. In addition, a significant (factor of ~ 2) overall correction to the value of λ was required due to both track and vertex misidentifications, as per the discussion of Section 5.2.6.

Three prescriptions for generating the two-particle background were considered by the EMC collaboration:

Ref(1): Real (π^+ , π^-) combinations generated from the same event.

Ref(2): Fake (π^+ , π^-) combinations generated by shuffling the compo-

260

nents of momentum transverse to the virtual photon axis. This scheme is designed to remove any dynamical correlations in the π^+ π^- channel while approximately conserving energy and momentum in the event.

Ref(3): Mixing of like-sign pions from different events, taking care to mix pairs from events of approximately the same effective mass (W).

The results of this exercise are shown in Fig. 10a (note that these authors use the quantity $\bar{M}^2 \equiv M_{\pi\pi}^2 - 4m_\pi^2 = Q^2$), which indicate that the three reference samples lead to vastly different correlation functions, showing that, as mentioned in Section 5.2.7, there are inherent uncertainties in determining the correlation function due to ambiguities in the definition of the background spectrum.

However, it is of some interest to form the same ratios for events generated by the Lund Monte Carlo (Fig. 10b), which suggests that these differences arise from different systematics in the background generation process *independent* of B-E effects. For example, the enhancement of Ref(1) at low Q^2 is attributable to the pion combinations from the decay of η's, η'''s and ω's. It is clear that the results from the Lund Monte Carlo should therefore be used to correct each background spectrum for the different biases associated with it. As demonstrated by Fig. 10c, this correction resolves most of the discrepancies between the various reference samples. The extent to which they still differ may therefore be interpreted as the systematic error associated with the entire Bose-Einstein analysis.

Finally, we turn to the extracted source parameters of this work. Fitting to the form

$$C_2(q) = 1 + \lambda e^{-Q^2 r^2} \quad , \tag{28}$$

they find (after conversion of the radii to the rms values as per the factors given in Appendix A) radii of 0.80 to 1.46 fm and λ of 0.60–1.08. The large range of these values reflects the care these authors have applied to determining the true errors induced by the calculation of the reference sample.

7   Results from Nuclear Collisions

Nearly all the data discussed to this point have agreed on one result– that the radius of the pion source is of order 1 fm. Optimists may regard this

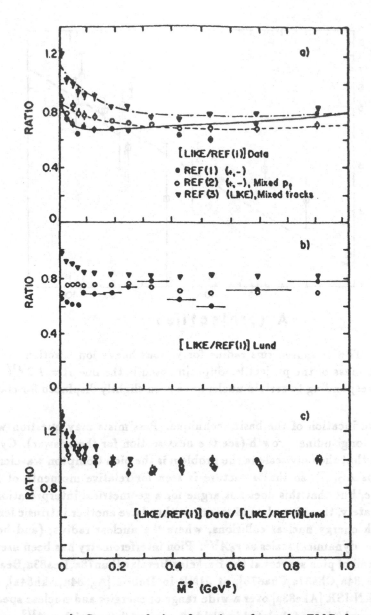

Figure 10: a.) Raw correlation functions from the EMC data using the three different reference samples discussed in the text. b.) The correlations predicted by the LUND Monte Carlo in the absence of B-E effects. c.) The data from a.) corrected by the LUND Monte Carlo.

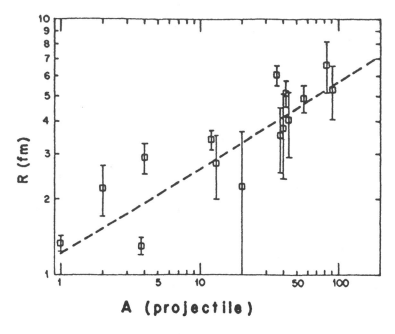

Figure 11: The measured rms radius for various heavy ion reactions versus the atomic mass of the projectile. Superimposed is the line $R = 1.2 A_P^{1/3}$ fm. Points corresponding to certain nuclei have been slightly displaced for clarity.

as a fine verification of the basic technique. Pessimists may question what became of longitudinal growth (see the next section for the answer). Cynics may argue that the only scale in the problem is the pion Compton wavelength (or perhaps Λ_{QCD}), so that structure is seen for relative momenta of 100-200 MeV/c, but that this does not argue for a geometrical interpretation.

Fortunately, there is a class of reactions that have another intrinsic length scale: High energy nuclear collisions, where the nuclear radius, (and hence the collision dynamics) scales as $r_0 A^{1/3}$. Pion interferometry has been used to measure radii of pion sources at the Berkeley Bevalac [Fun78a,Bea83a,Bea83b], [Bea84a,Zaj84a,Cha87a,Cha87b], at JINR in Dubna [Aga84a,Akh84a], and at the CERN ISR [Ake83a] over a wide range of energies and nuclear species. These data, which are tabulated in Table 7, are consistent with a $A_P^{1/3}$ scaling in the atomic mass of the projectile, as demonstrated in Fig. 11, thereby providing some indication the BE measurements do in fact have a geometrical interpretation.

Related to this are the results of Akesson *et al.* [Ake83a], who have

Reference	Reaction	Energy (per nucleon)	R (fm)	$c\tau$ (fm)	λ
[Fun78a]	Ar + BaI$_2$	2.57 GeV/c	3.74 ± 1.35	—	—
"	Ar + Pb$_3$O$_4$	"	4.04 ± 1.14	—	—
"	Ar + Pb$_3$O$_4$ (central)	"	4.87 ± 0.96	—	—
[Ake83a]	$\alpha + \alpha$	31 GeV	1.30 ± 0.10	1.1 ± 0.15	0.30 ± 0.07
[Bea83a]	Ar + KCl	2.25 GeV/c	6.04 ± 0.54	—	1.21 ± 0.22
[Bea83b]	Ar + KCl	1.92 GeV/c	4.65 ± 0.61	5.4 ± 1.8	0.74 ± 0.17
[Aga84a]	d + Ta	4.24 GeV/c	2.20 ± 0.50	—	—
"	α + Ta	4.24 GeV/c	2.90 ± 0.40	—	—
"	C + Ta	4.24 GeV/c	3.40 ± 0.30	—	—
[Akh84a]	C + C	4.24 GeV/c	2.75 ± 0.76	—	—
"	C + C(central)	"	3.76 ± 0.88	—	—
[Bea84a]	Kr + RbBr	2.45 GeV/c	6.6 ± 1.5	3^{+7}_{-3}	0.8 ± 0.3
[DeM84a]	p + Xe	200 GeV/c	1.33 ± 0.11	0.93 ± 0.16	1.27 ± 0.11
"	$\bar{p}$ + Xe	"	1.27 ± 0.10	0.95 ± 0.11	1.34 ± 0.08
[Zaj84a]	Ar + KCl → 2π^- + X	2.57 GeV/c	$3.52^{+0.6}_{-1.1}$	$3.29^{+1.4}_{-1.6}$	0.63 ± 0.04
"	Ar + KCl → 2π^+ + X	2.57 GeV/c	$5.14^{+0.5}_{-0.6}$	$1.54^{+2.4}_{-1.54}$	0.69 ± 0.09
"	Ne + NaF → 2π^- + X	2.57 GeV/c	$2.24^{+1.0}_{-2.0}$	$2.96^{+0.9}_{-1.0}$	0.59 ± 0.06
[Cha87a]	Fe + Fe → 2π^- + X	2.47 GeV/c	4.9 ± 0.6	1.7 ± 1.7	0.66 ± 0.05
[Cha87b]	Nb + Nb → 2π^- + X	2.4 GeV/c	5.3 ± 1.2	4.8 ± 1.8	0.78 ± 0.06

Table 2: Pion source radii measured in nuclear collisions.

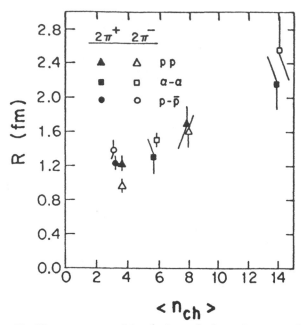

Figure 12: Radii as measured by [Ake83a] plotted versus the charged multiplicity.

measured radii for pp, pp̄ and $\alpha - \alpha$ collisions at the ISR. Their results (Fig. 12), when plotted versus the charged multiplicity in the collision, indicate that measured radius depends directly on this multiplicity, but not on the species in collision. (Recent results [Ake87a] suggest that the increase is confined to the longitudinal dimension of the source.) This effect has found a natural interpretation in the work of Barshay [Bar83a], who relates the measured radius to the impact parameter in the collision, and thereby to the multiplicity. The data for pp collisions at 63 GeV were verified by Breakstone *et al.* [Bre87a], who also showed that the correlation of radius with multiplicity weakened with decreasing $\sqrt{s}$. However, it is interesting to note that a similar dependence of radius on multiplicity was seen by Lu *et al.* [Lu81a] at vastly lower energies (and with much heavier ions) in Ar + Pb collisions at 1.8 A · GeV.

Also of interest in this regard is the work of De Marzo *et al.* [DeM84a], who used interferometry to study the pion source radius in pp, pp̄, p-Xe and p̄-Xe collisions at 200 GeV/c. They observed no target dependence for the pion emission region, i.e., the same radius was measured with either hydrogen or xenon as the target. These authors also see no significant change in the radius for a sub-sample of high multiplicity p-Xe events, but do observe that, where studied, the source is larger transverse to the collision axis than along it. Such data should play an important role in elucidating the interplay between geometry and dynamics in various longitudinal growth models [And87a].

8 Recent Results from e^+e^- Experiments

In this section I will discuss results from four experiments studying BE correlations over a range of energies in e^+e^- annihilation. Such experiments can provide rigorous tests of various (still open) questions in the field of intensity interferometry, since it is here that we have the best theoretical basis for describing the underlying dynamics.

8.1 Experimental Data

We begin with the work of Avery et al. [Ave85a], who studied Bose-Einstein correlations in the Υ region. These authors note that the traditional Kopylov-Podgoretskiĭ form of the correlation function (see Appendix A) is not Lorentz covariant, so that parameters extracted by fitting to it are frame-dependent. As an alternative, they choose to fit to

$$C_2(q) = 1 + \lambda e^{-q_T^2 r^2} , \qquad (29)$$

which, for their events, is the appropriate limit of the more general "modified Goldhaber" form

$$C_2(q) = 1 + \lambda e^{-\beta q_T^2 - \gamma(q_L^2 - q_0^2)} \qquad (30)$$

(Note that in the above equation β and γ are fit parameters, not the usual factors in a Lorentz transformation. Further note that the rms value of the radius is $\sqrt{3} \cdot r$ for the first parametrization.) Little change in the value of r (~ 1.5 fm) and λ ($\sim 04.$-0.5) is seen as $\sqrt{s}$ increases from below the $\Upsilon(1s)$ to above the $\Upsilon(4s)$. Based on (tuned Lund model) estimates of production rates for η's, ω's and charmed particles they find that λ is severely affected by the resonance decays, as per the considerations of Section 5.2.3. Their corrected value is $\lambda_{CORR} = 1.45 \pm 0.25$, thus indicating that with these assumptions the entire reduction of λ may be explained by resonance production.

Two further aspects of this paper are interesting. First, they have determined that the radius in the "weak half" of their events (defined as opposite the highest energy jet found in a triplicity analysis) is larger (1.9 fm) for $\Upsilon(1s)$ decays than for the continuum (1.3 fm). This is consistent with our understanding of $\Upsilon(1s) \rightarrow 3g$ versus the dominant two-jet structure in the continuum. The respective values of λ are also quite different, 0.69 ± 0.13 versus 0.43 ± 0.09. The second item of interest is their curious statement that the data show no dependence on q_0. This is contradicted by their Fig. 6, shown here as Fig. 13, which shows a decreasing BE effect with an increasing value of q_0 corresponding to a lifetime of order 0.8 fm.

We next turn to the work of Aihara *et al.*, who investigated like-pion correlations in e^+e^- annihilations at $\sqrt{s} = 29$ GeV using the TPC. These authors find $r = 1.13 \pm 0.07$ fm when fitting to the form of Eq. 29, and $1.10 \pm 0.06 \pm 0.07$ fm when fitting to the Kopylov-Podgoretskiĭ form. Both of these numbers have been converted using the procedures outlined in Appendix A; their near identity provides a gratifying confirmation of the conversion factors chosen. The value of λ extracted is $0.60 \pm 0.05 \pm 0.06$, where both statistical and systematic errors have been included. A very nice illustration of the points outlined in Section 5.2 are provided in their breakdown of the various contributions to the systematic errors in λ: 0.03 from iterative evaluation of the background (Section 5.2.7), 0.03 from Coulomb corrections (Section 5.2.2), and 0.05 from particle misidentification (Section 5.2.6). No attempt was made to explicitly correct for the behavior of known resonances, so that their value for λ is advertised as a lower limit. It is also worth noting that no variation of the radius with the charged multiplicity was seen, as opposed to the result of Akesson *et al.* [Ake83a]. However, this is consistent with the point-like nature of e^+e^- collisions, as opposed to the finite transverse extent of hadronic matter in the collisions studied by Akesson *et al.*

The Mark II collaboration has studied BE effects in e^+e^- for a number of years [Gol81a,Gol85a,Gol86a,Jur87a]. Their results extend from SPEAR to PEP energies, with a very large sample of events concentrated on the J/Ψ, and with smaller (yet still adequate) statistics in the SPEAR continuum (4-7 GeV), PEP energies (29 GeV), and 29 GeV $\gamma\gamma$ events at PEP. Essentially

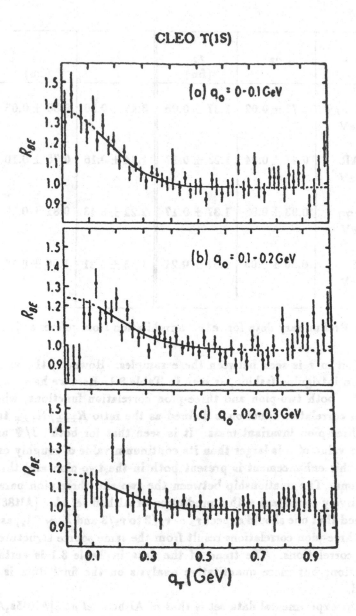

Figure 13: Pair correlation data of Avery *et al.* [Ave85a] for increasing cuts in the value of q_0.

DATA	λ_2	r_2 (fm)	λ_3	r_3 (fm)
SPEAR J/Ψ 3.1 GeV	0.77 ± 0.02	1.37 ± 0.05	3.45 ± 0.20	0.88 ± 0.05
SPEAR 4-7 GeV	0.47 ± 0.04	1.23 ± 0.17	1.44 ± 0.16	0.75 ± 0.10
PEP $\gamma\gamma$ 29 GeV	0.93 ± 0.08	1.37 ± 0.17	2.22 ± 0.43	0.81 ± 0.17
PEP 29 GeV	0.50 ± 0.05	1.51 ± 0.21	1.13 ± 0.21	1.06 ± 0.16

Table 3: Preliminary data for e^+e^- annihilation from reference [Jur87a].

no variation in r is seen between these samples. However, the value of λ varies in an intriguing fashion, as seen in Table 8.1. Here we have tabulated λ and r for both two-pion and three-pion correlation functions, where the three-pion correlation function is defined as the ratio $R_{\pm\pm\pm}/R_{\pm\pm\mp}$ taken at a given three-pion invariant mass. It is seen that for both J/Ψ and $\gamma\gamma$ events the value of λ is larger than its continuum value of roughly one-half, and that this enhancement is present both in the two-pion and three-pion distributions. The relationship between the two and three-pion parameters is qualitatively consistent with a prediction of Althoff et $al.$ [Alt86a], who have argued that one should expect $r_3 \sim r_2/2$ to $r_2/3$ and $\lambda_3 \sim 5\lambda_2$, assuming that the three-pion correlations result from the same source structure as the two-pion correlations. The trend of the data in Table 8.1 is certainly in this direction; but more quantitative analysis on the final data is clearly warranted.

Our last experimental data set is that of Althoff et $al.$ [Alt85a,Alt86a], who studied e^+e^- annihilation at PETRA energies (29 to 37 GeV). It is first important to note that, through a study of single-particle distributions and two-particle rapidity correlations [Alt85a], these authors achieved a superb "tuned" description of their events through the Lund Monte Carlo. This, along with appropriate detector simulation, allowed them to correct their like-pion distributions for two-particle acceptance variations and production kinematics, as shown in Fig. 14. In spite of the quality of the Monte Carlo event description used, the errors in the corrected value of λ are dominated by systematic uncertainties in the background: $\lambda_{CORR} = 0.70 \pm 0.06 \pm 0.09$, where the first error is statistical and the second reflects systematic errors resulting from uncertain proportions of kaons, Λ's and protons. No dependence of the measured radius was found on the pair Lorentz factor $\gamma_{\pi\pi}$ or the sphericity axis. An attempt was made to determine the "best" set of relative momentum variables describing the observed enhancements; it was found that Q^2 provided the best global fit, but that the data did not rule out more complicated dependences motivated by string models (see following).

8.2   String Models and Longitudinal Growth

The space-time dependence of particle production at ultra-relativistic energies follows naturally from simple considerations of relativity and quantum mechanics [Kol86a]. If we assume that the hadronization of any "cluster" into pions proceeds on a (proper) time scale of order $\tau_0 \sim 1/m_\pi$, then in a frame where the source is moving at rapidity y, the formation time occurs at a time $\tau \sim \tau_0 e^y/2$. Thus, position and momentum are highly correlated, with the highest momentum particles being emitted at the largest distances from the interaction "point". This is the basic idea of longitudinal growth.

One might expect interferometry to directly reveal the presence of the long, thin source predicted by such arguments. There is, of course, a practical limitation, in that no spectrometer built to date has the resolution required to resolve the BE enhancement at the very small values of relative momenta associated with a source many tens of fermis long. More important is a fundamental limitation: The very nature of longitudinal growth predicts that *pairs that are close together in momentum space were once close together in position space.* Conversely, pairs from opposite ends of the source never interfere, since they have large but opposite values of momentum, and thus a large relative momentum. Therefore, any pairs that have a small value of Q^2 typically were emitted within a space-time separation of order 1 fm.

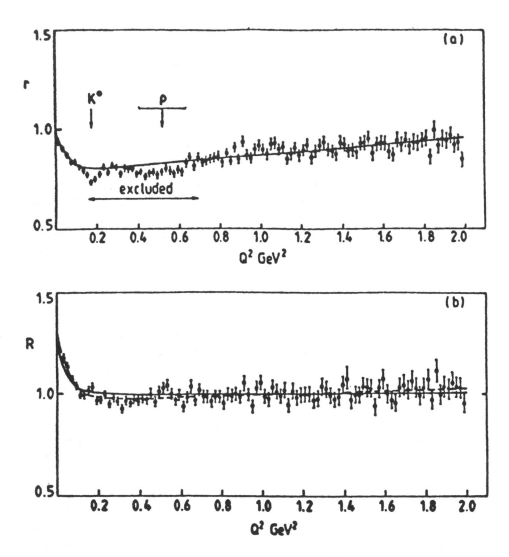

Figure 14: The correlation functions as measured by TASSO for a.) The uncorrected ratio of like to unlike-sign pairs (Here called r) and b.) The same ratio after MC correction for production kinematics and detector acceptance (Here called R).

These results have been incorporated into a model with an explicit solution for particle production with longitudinal growth due to Kolehmainen and Gyulassy [Kol86a].

In e^+e^- annihilation, string models predict production dynamics consistent with longitudinal growth. Normally, such models do not also contain the effects of like-particle symmetrization. These have been incorporated by Bowler [Bow85a] in the context of the Artru-Mennessier string model; the solid curve that appears on Fig. 14 shows the ability of this model to describe the data with quite reasonable string parameters. However, the dashed curve in the same figure represents a fit to the function first used by GGLP nearly 30 years ago: $C_2(q) = 1 + \lambda e^{-\beta Q^2}$. The near identity of these two curves indicates the level of accuracy that will be required to make quantitative distinctions between such vastly different models.

Andersson and Hoffman have also calculated Bose-Einstein correlations in a string model, by modifying the Lund Monte Carlo to include the "exchange" terms resulting from $\vec{p}_1 \leftrightarrow \vec{p}_2$. This action for the exchanged diagram is modified by a phase factor corresponding to the exchanged momenta multiplied by a real weight $e^{-b\Delta A/2}$, where ΔA is the area spanned by the loop containing the old and new production sites (see Fig. 15). These authors make the subtle and beautiful point that the coherent, point-like production of mesons in the Schwinger model is replaced by meson production that is extended in space-time in this model. Gauge invariance then requires the introduction of a loop integral, which leads to the term $e^{-b\Delta A/2}$.

Again, by using reasonable values for the string parameters, an excellent description of the data is obtained (Fig. 16), *provided the effect of resonances is ignored.* (Compare the dotted to the solid curves on Fig. 16). Recent work by Bowler [Bow86a] indicates that this result is not quite as paradoxical as it may seem, but is instead explainable by uncertainties in the production cross-section of η''s, since even a small η' production rate "contaminates" many of the potential pion pairs with at least one of its decay pions. This observation once more shows the sensitivity of these measurements to rather subtle aspects of the production dynamics.

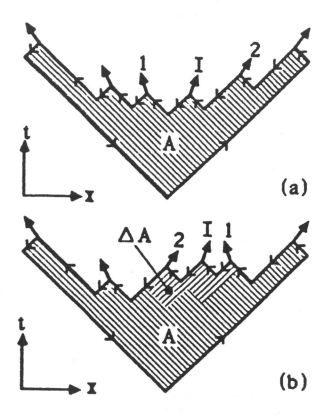

Figure 15: Space-time diagrams for particle production in the Lund string model: a.) The "normal" diagram, with weight $e^{-bA/2}$. b.) The "exchanged" diagram, which is suppressed relative to a.) by a factor of $e^{-b\Delta A/2}$.

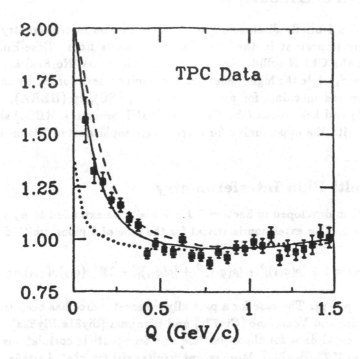

Figure 16: Model predictions of Andersson and Hoffman [And86a] for typical string parameters (solid and dashed curves), and after the inclusion of resonances (dotted curve), as compared to the TPC data for the correlation function in e^+e^- annihilation [Aih85a].

9 Future Directions

In this section I briefly discuss new directions in boson interferometry that may become important in the near future. Results from Bose-Einstein analyses at the CERN collider by the UA1 collaboration [Nor87a] are anxiously awaited, while the higher energies and luminosities afforded by current and/or proposed machines for pp (the Tevatron, SSC), ep (HERA), e^+e^- (SLC, LEP) and heavy ions (the CERN and BNL programs, RHIC) should provide us with the opportunity for increasingly sophisticated BE measurements.

9.1 Multi-Pion Interferometry

The formalism developed in Section 3.1 can easily be extended to n_π states, where $n_\pi > 2$. The extension is trivial for the case of 3 pions emitted incoherently:

$$C_3(p_1, p_2, p_3) = 1 + |\rho(q_{12})|^2 + |\rho(q_{23})|^2 + |\rho(q_{31})|^2 + 2\Re\{\rho(q_{12})\rho(q_{23})\rho(q_{31})\} \quad , \tag{31}$$

where $q_{ij} \equiv p_i - p_j$. The case for a partially coherent source has been treated by Giovannini and Veneziano [Gio77a] and Biyajima [Biy81a,Biy82a].

Experimental data for three-particle and four-particle correlations date back as far as 1973 [Bos73a]. More recent results exist for e^+e^- [Alt85a,Jur87a], pp [Ake87b] and heavy ion collisions [Liu86a]. It is fair to say that current situation is murky, especially in regard to the presence of true three-body correlations (i.e., those not resulting from the product of two-body factors). All results to date are consistent with no intrinsic three-body correlations. As an example, Fig. 17 shows the three-pion correlation (as a function of $Q^2 \equiv (p_1 + p_2 + p_3)^2 - 9m_\pi^2$) measured in jet events at the ISR [Ake87b]. If the background sample is weighted by Eq. 31, using the source parameters determined by a two-pion analysis on the same data, the dashed (flat) curve is obtained. This of course indicates that, for these data, no additional information is carried by the three-pion distribution. Nevertheless, it is important that such efforts be continued, since the three-particle distributions are sensitive to the *phase* of the source Fourier transform (in the incoherent limit), while in the case of substantial source coherence they explore higher-order moments of the pion density matrix [Gyu79a,Car83a,Car84a].

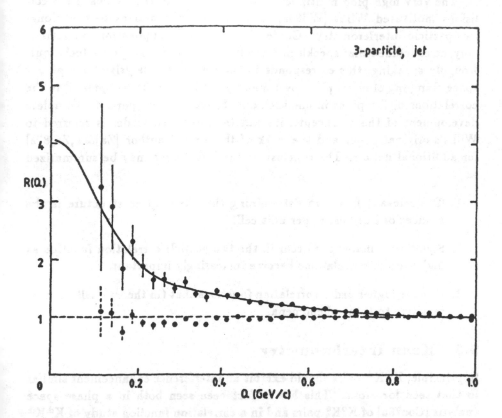

Figure 17: Solid curve: The three-pion correlations function $R(Q)$ versus the three-pion invariant relative momentum for jet events at the ISR [Ake87b]. Dashed curve: the three-pion correlation using a background sample weighted by the result expected from the known two-pion correlations.

The very high pion multiplicities expected in high energy heavy ion collisions motivated Willis [Wil84a] to investigate alternatives to traditional two-particle interferometry. Guided by analogy to applications in astronomy, he concluded that speckle interferometry was the appropriate technique. Roughly speaking, this corresponds to the statistical analysis of the phase-space clumping of many pions over many phase-space cells, as opposed to the correlations of few pions in one such cell. Space does not permit a complete development of these concepts, instead the interested reader is referred to Willis's original paper and the work of the present author [Zaj86a,Zaj87a] for additional details. The conclusions reached therein may be summarized as

1. The relevant factor in determining the phase space structure is the number of like-bosons per unit cell.

2. Significant distortions occur in the two-particle correlation function as higher-order correlations become increasingly important.

3. It is the higher-order correlation functions that (in theory) will be most sensitive to these phenomena.

9.2   Kaon Interferometry

In principle, $K^\pm K^\pm$ pairs should exhibit an interference enhancement similar to that seen for pions. This has in fact been seen both in a phase space analysis [Coo78a] of $K^0_S K^0_S$ pairs and in a correlation function study of $K^\pm K^\pm$ correlations [Ake85a]. These experiments, while severely limited by statistics, find source sizes in rough agreement with those from two-pion measurements.

Of course, the paucity of kaons relative to pions makes such work very difficult, particularly when one recalls that Bose-Einstein rates go as the square of the particle abundance. Nonetheless, kaon interferometry offers the chance to study bosons emitted from the same collisions as the pions, but with significantly different production dynamics (strangeness suppression, both quarks in the K are necessarily produced from the sea), reinteraction probabilities (especially for K^-'s) and resonance backgrounds (Φ's versus ρ's and ω's). Also, since the kaon mass is much larger than that of the pion, while the transverse mass distributions are roughly the same, $|\vec{q}|$ and q_0 for kaons are less coupled, which may be important in the determination of lifetimes.

9.3 Jet Interferometry

In principle, Bose-Einstein methods can elucidate the nature of jet production in high energy hadron reactions, just as they did for string dynamics in e^+e^- (See Section 8.2). A recent analysis by Akesson *et al.* [Ake87b] performed for events with jets ($E_T^{jet} > \sim 10$ GeV) formed in pp collisions at $\sqrt{s} = 63$ GeV indicates that B-E correlations within the jet region are best described in terms of Q^2, rather than the appropriate Lorentz invariants parallel and transverse to the jet axis (this result is quite similar to that of Althoff *et al.* [Alt85a]). Source sizes extracted in the regions orthogonal to the jets are larger than for minimum bias events, but are consistent with those of non-jet events of similar associated multiplicity [Ake83a]. However, the source size obtained from pions within the jet region is consistent with that for minimum bias events, especially when both data sets are analyzed in terms of the two-component model favored by the data. The authors find that a two-Gaussian model for the correlation function with $\lambda_1 \sim 0.3$, $R_1 \sim 1.1$ fm, and $\lambda_2 \sim 0.4$, $R_2 \sim 3.5$ fm describes both the jet and minimum bias data. Even more intriguing is their observation that these data are equally well parameterized as $1 + \lambda/(1 + \beta^2 Q^2$ (See Section 5.2.4) or $1 + \lambda e^{-\alpha Q}$. The last form suggests that the hadronization dynamics may already be apparent, since the results from string dynamics are nearly exponential in Q (See Fig. 16).

9.4 Final Thoughts

As stressed repeatedly in this review, present generation Bose-Einstein experiments are generally limited by systematic, not statistical, errors. This is certainly true for distributions in spatial momenta or Q^2. In the case of lifetimes, it *is* possible to simply run out of data at high values of q_0, particularly in e^+e^- and hadron-hadron collisions, where the small source size translates into a large relative energy difference. However, it is safe to say that the dominant uncertainty in extracting lifetimes results from confusion concerning the *Lorentz invariance* of the correlation function, as per the considerations of Section 5.2.4. Exacerbating the situation is the habit of fitting to source distributions whose theoretical justification is quite questionable (such as that for a disk with exponential lifetime, which for unfathomable reasons remains popular), or whose dynamics are not readily adaptable to a covariant formulation. Futhermore, fundamental issues will be obscured so long as radii are extracted by setting lifetimes to some arbitrary value, parameters are compared that differ by inessential factors of of $\sqrt{2}$ and discussions of systematic errors are ignored. Nonetheless, recent results suggest that increasingly large data samples, coupled with models based on the underlying source dynamics, will do much to ameliorate these problems, and move Bose-Einstein correlations further into the realm of quantitative science.

10 Acknowledgments

It is pleasure to acknowledge useful conversations and/or communications with D. Beavis, A. Caldwell, S. Nagamiya, M.I. Podgoretskiĭ, A. Shor and W. Lee.

A   Appendix A: Commonly Used Variables and Notation

A.1   Four-vectors

Four-vectors are written as simple variables, hence $q = (q_0, \vec{q})$. Four-products always appear with an explicit product sign: $a \cdot b \equiv a_0 b_0 - \vec{a} \cdot \vec{b}$.

A.2   Commonly used variables

- $q \equiv p_1 - p_2 \equiv (q_0, \vec{q})$.
  Note the sign of $q_0 = E_1 - E_2$ is arbitrary, so that $|q_0|$ appears in most expressions and is simply denoted as q_0.

- $\vec{P} \equiv \vec{p}_1 + \vec{p}_2$.

- $\vec{q}_L \equiv (\vec{q} \cdot \vec{P})\vec{P}/|\vec{P}|^2$.
  This is the component of $\vec{q}$ along the direction of the pion pair momentum vector.

- $\vec{q}_T \equiv \vec{q} - \vec{P}(|\vec{q}_L|/|\vec{P}|)$.
  This is the component of $\vec{q}$ transverse to the direction of pion pair momentum vector. Typically only $\vec{q}_T \equiv |\vec{q}_T|$ is retained.

- $Q^2 \equiv -(p_1 - p_2) \cdot (p_1 - p_2)$.
  This is the invariant relative momentum of the pair, which reduces to $|\vec{q}|^2$ in the pair rest frame.

- $M^2 \equiv (p_1 + p_2) \cdot (p_1 + p_2)$.
  Invariant pair mass. Note that $\bar{M}^2 \equiv M^2 - 4m_\pi^2 = Q^2$ is occasionally used.

- $\vec{\beta}_{\pi\pi} \equiv (\vec{p}_1 + \vec{p}_2)/(E_1 + E_2)$.
  Pion pair velocity in a given frame.

- $\gamma_{\pi\pi} \equiv (E_1 + E_2)/2m_\pi$.
  Lorentz gamma for pion pair in a given frame.

A.3   Useful Relations

- $q_0 = \vec{\beta}_{\pi\pi} \cdot \vec{q}_L$.

- $|\vec{q}|^2 = q_0^2 + |\vec{q}^*|^2$,
  where $\vec{q}^*$ is the three-momentum as seen in the rest frame of the two pion system. Note that this relation implies in any frame that $|\vec{q}| \geq q_0$.

- $Q^2 = \vec{q}_T^2 + q_0^2/(\gamma_{\pi\pi}^2 - 1)$,
  which demonstrates that for fast-moving pairs, Q^2 is essentially $\vec{q}_T^2$.

A.4   Source Parametrizations

Table A.4 contains the inter-relationships between the most commonly used source parametrizations. Note that the Fourier transforms are defined in terms of $x \equiv |\vec{q}|R$, $y \equiv QR$, and $z \equiv |\vec{q}_T|R$.

Reference	$\rho(\vec{r})$	$\langle r^2 \rangle^{1/2}$	$|\tilde{\rho}(\vec{q})|^2$
[Fun78a]	$\frac{1}{\sqrt{\pi}R^3}e^{-r^2/R^2}$	$\sqrt{\frac{3}{2}}R$	$e^{-x^2/2}$
—	Spherical shell of radius R	R	$\lvert \sin x/x \rvert^2$
[Gol60a]	Uniform sphere of radius R	$\sqrt{\frac{3}{5}}R$	$\lvert 3(\sin x - x\cos x)/x^3 \rvert^2$
[Gol60a]	$e^{-r^2/2R^2}$ in the two-pion rest frame	$\sqrt{3}R$	e^{-y^2}
[Kop72a] (aka K-P)	Uniformly illuminated disk of radius R	?	$\lvert 2J_1(z)/z \rvert^2$

Table 4: Common source density parametrizations.

No entry is provided for the rms equivalent to the uniformly illuminated disk, since this is a model-dependent interpretation of the spherical shell of radiators. However, we note that to a very good approximation we can write

$$\left|\frac{2J_1(z)}{z}\right|^2 \approx e^{-z^2/4},\tag{32}$$

so that, treated phenomenologically and without reference to the original source distribution, the parameter R in the Kopylov-Podgoretskiĭ form has an rms value given by $\sqrt{3}R_{KP}/2$. All radii fitted to the K-P form have been translated via this prescription when cited in this work. Similarly, those radii obtained fitting to the invariant form have been transformed using the factor of $\sqrt{3}$ listed in Table A.4, even though this implies a source of emitters expanding with the observed distribution of pair-velocities.

B Appendix B: The Perfect Bose-Einstein Paper

Here I provide a referee's check-list of criteria to be met by a perfect Bose-Einstein paper. Alternatively, this list represents the items necessary for the interested theorist to completely reconstruct the steps that went into the measurement of a given correlation function. This is an ideal case, independent of such mundane details as page-charges or individual diligence. Nevertheless, some of the best publications in the field approach this ideal; it is the norm that requires attention.

1. Motivation: What aspects of the reaction mechanism does this study hope to elucidate?

2. Experimental Details:

    (a) What region of phase space is measured?

    (b) How was the experiment triggered?

    (c) What is the single-particle momentum resolution?

    (d) What is the two-particle relative momentum resolution? How does it limit the range of source sizes detectable? Is it simply related to the single-particle resolution, or are there large ambiguities at small relative momentum?

(e) What is the acceptance in *relative* momentum? That is, what is the distribution in (q_0, q_L, q_t)?

3. Analysis Details:

   (a) What is the total number of events?

   (b) What cuts were applied to the data?

   (c) How was particle identification achieved, and what was the contamination?

   (d) How was the background generated? What steps were taken to apply the same cuts to the background pairs as to the real pairs? Was the effect of residual correlations in the background investigated?

4. Fitting Details:

   (a) What variables are used to define C_2, and what frame are they calculated in?

   (b) What function was fit? Does it have any theoretical origin (justified or unjustified)?

   (c) How was the fit performed? (I.e., one-dimensional or two-dimensional, least-squares or Principle of Maximum Likelihood, etc.)

   (d) How were errors in the background estimated?

   (e) How were errors from the ratio of small integers handled?

5. Corrections:

   (a) Was the final-state Coulomb repulsion important? If so, was it corrected for?

   (b) What is the sensitivity of the fit parameters to different prescriptions for generating the background?

   (c) Are there corrections for the presence of long-lived resonances? If so, how model-dependent are they?

   (d) Is the chi-squared of the fit reasonable? If not, are the discrepancies explained? How are errors on the extracted parameters assigned?

References

[Aga84a] G.N. Agakishiev *et al.*, Yad. Fiz. **39**, 543 (1983).

[Aih85a] H. Aihara *et al.*, Phys. Rev. **D31**, 996 (1985).

[Ake83a] T. Akesson *et al.*, Phys. Lett. **129B**, 269 (1983).

[Ake85a] T. Akesson *et al.*, Phys. Lett. **155B**, 128 (1985).

[Ake87a] T. Akesson *et al.*, Phys. Lett. **187B**, 420 (1987).

[Ake87b] T. Akesson *et al.*, CERN-EP/87-142, 1987, (submitted to Z. Phys. C).

[Akh84a] N. Akhababian *et al.*, Z. Phys. **C26**, 245 (1984).

[Alt85a] M. Althoff *et al.*, Z. Phys. **C29**, 347 (1985).

[Alt86a] M. Althoff *et al.*, Z. Phys. **C30**, 355 (1986).

[And86a] B. Andersson and W. Hoffman, Phys. Lett. **169B**, 364 (1986).

[And87a] B. Andersson, private communication.

[Ang77a] C. Angelini *et al.*, Lett. Nuovo Cim. **19**, 279 (1977).

[Ang77b] N. Angelov *et al.*, Sov. J. Nucl. Phys. **26**, 419 (1977).

[Ang79b] N. Angelov *et al.*, JINR-E1-12548 (1979).

[Ang80a] N. Angelov *et al.*, Sov. J. Nucl. Phys. **31**, 215 (1980).

[Arn86a] M. Arneodo *et al.*, Z. Phys. **C32**, 1 (1986).

[Ave85a] P. Avery *et al.*, Phys. Rev. **D32**, 2294 (1985).

[Bar67a] J. Bartke *et al.*, Phys. Lett. **24B**, 163 (1967).

[Bar78a] E.A. Bartnik and K. Rzazewski, Phys. Rev. **D18**, 4308 (1978).

[Bar83a] S. Barshay, Phys. Lett. **130B**, 220 (1983).

[Bar84a] J. Bartke and M. Kowalski, Phys. Rev. C30, 1341, (1984).

[Bea83a] D. Beavis et al., Phys. Rev. C27, 910 (1983).

[Bea83b] D. Beavis et al., Phys. Rev. C28, 2561 (1983).

[Bea84a] D. Beavis et al., in Proceedings of the 7th High Energy Heavy Ion Study, GSI Darmstadt, Oct. 8-12, (1984). (GSI-85-10).

[Bec79a] U. Becker et al., Nucl. Phys. B151, 357 (1979).

[Bis75a] N.N. Biswas et al., Phys. Rev. Lett. 35, 1059 (1975).

[Bis76a] N.N. Biswas et al., Phys. Rev. Lett. 37, 175 (1976).

[Biy74a] M. Biyajima and O. Miyamura, Phys. Lett. 53B, 181 (1974).

[Biy80a] M. Biyajima, Phys. Lett. 92B, 193, (1980).

[Biy81a] M. Biyajima, Prog. Theor. Phys. 66, 1378 (1981).

[Biy82a] M. Biyajima, Prog. Theor. Phys. 68, 1273 (1982).

[Bos73a] K. Bosebeck et al., Nucl. Phys. B52, 189 (1973).

[Bow85a] M.G. Bowler et al., Z. Phys. C29, 617 (1985)

[Bow86a] M.G. Bowler, Phys. Lett. 180B, 299 (1986).

[Bow87a] M.G. Bowler, Oxford preprint OUNP 65/87 (1987).

[Bre85a] A. Breakstone et al., Phys. Lett. 162B, 400 (1985).

[Bre87a] A. Breakstone et al., Z. Phys. C33, 333 (1987).

[Cal76a] A. Calligarich et al., Lett. Nuovo Cim. 16, 129 (1976).

[Car83a] P. Carruthers and C.C. Shih, Phys. Lett. 127B, 242 (1983).

[Car84a] P. Carruthers and C.C. Shih, Phys. Lett. 137B, 425 (1984).

[Cha87a] A.D. Chacon et al., LBL-18709 (1987). To be submitted to Phys. Rev. Lett. .

[Cha87b] A.D. Chacon, private communication.

[Coc74a] G. Cocconi, Phys. Lett. 49B, 459 (1974).

[Coo78a] A.M. Cooper *et al.*, Nucl. Phys. B139, 45 (1978).

[DeB70a] W. De Baere *et al.*, Nucl. Phys. B22, 131 (1970).

[DeM84a] C. De Marzo *et al.*, Phys. Rev. D29, 363 (1984).

[DeW78a] E. De Wolf *et al.*, Nucl. Phys. B132, 383 (1978).

[Deu76a] M. Deutschmann *et al.*, Nucl. Phys. B103, 198 (1976).

[Deu78a] M. Deutschmann *et al.*, CERN/EP/PHYS 78-1 (Jan. 1978).

[Deu82a] M. Deutschmann *et al.*, Nucl. Phys. B204, 333 (1982).

[Don69a] R.A. Donald *et al.*, Nucl. Phys. B11, 551 (1969).

[Dri79a] D. Drijard *et al.*, Nucl. Phys. B155, 269 (1979).

[Eze78a] C. Ezell *et al.*, Phys. Rev. Lett. 38, 873 (1978).

[Fer50a] E. Fermi, Prog. Theor. Phys. 5, 570 (1950).

[Fer66a] T. Ferbel *et al.*, Phys. Rev. 144, 1096 (1966).

[Fir75a] A. Firestone *et al.*, Nucl. Phys. B161, 19 (1975).

[Fow77a] G.N. Fowler and R. Weiner, Phys. Lett. 70B, 201 (1977).

[Fow78a] G.N. Fowler and R. Weiner, Phys. Rev. D17, 3118 (1978).

[Fow79a] G.N. Fowler, R. Stelte and R.M. Weiner, Nucl. Phys. A319, 349 (1979).

[Fow85a] G.N. Fowler and R.M. Weiner, Phys. Rev. Lett. 1373, 1985.

[Fun78a] S.Y. Fung *et al.*, Phys. Rev. Lett. 41, 1592 (1978).

[Gio77a] A. Giovannini and G. Veneziano, Nucl. Phys. B130, 61 (1977).

[Gio79a] A. Giovannini, G.C. Mantovani, and S.P. Ratti, Rivista del Nuovo Cimento 2, No. 10 (1979).

[Gol60a] G. Goldhaber et al., Phys. Rev. 120, 300 (1960).

[Gol81a] G. Goldhaber in Proceedings of the International Conference on High Energy Physics, Lisbon, Portugal, July 9-15, 1981.

[Gol85a] G. Goldhaber in Proceedings of the First International Workshop on Local Equilibrium in Strong Interaction Physics, D.K. Scott and R.M. Weiner, ed., World Scientific Publishing Co. (1985).

[Gol86a] G. Goldhaber and I. Juricic in Proceedings of the Second International Workshop on Local Equilibrium in Strong Interaction Physics, P.Carruthers and D. Strottman, ed., World Scientific Publishing Co. (1986).

[Goo78a] M. Goosens et al., Nuovo Cimento 48A, 469 (1978).

[Gra76a] F. Grard et al., Nucl. Phys. B102, 221 (1976).

[Gra77a] P. Grassberger, Nucl. Phys. B120, 231 (1977).

[Gri77a] V.G. Grishin, Sov. Phys. Usp. 22, 1 (1977).

[Gyu79a] M. Gyulassy, S.K. Kaufmann, and L.W. Wilson, Phys. Rev. C20, 2267 (1979).

[Gyu81a] M. Gyulassy and S.K. Kaufmann, Nucl. Phys. A362, 503 (1981).

[Gyu82a] M. Gyulassy, Phys. Rev. Lett. 48, 454 (1982).

[Han52a] R. Hanbury-Brown, R.C. Jennison and M.K. Das Gupta, Nature 170, 1061 (1952).

[Han54a] R. Hanbury-Brown and R.Q. Twiss, Phil. Mag. 45, 633 (1954).

[Han56a] R. Hanbury Brown and R.Q. Twiss, Nature 178, 1046 (1956).

[Han56b] R. Hanbury Brown and R.Q. Twiss, Nature 178, 1447 (1956).

[Han57a] R. Hanbury-Brown and R.Q. Twiss, Proc. Roy. Soc. A242, 300 (1957).

[Han57b] R. Hanbury-Brown and R.Q. Twiss, Proc. Roy. Soc. A243, 291 (1957).

[Jur87a] I. Juricic, Ph.d Thesis, University of California, Berkeley, in preparation.

[Kol86a] K. Kolehmainen and M. Gyulassy, Phys. Lett. 180B, 203 (1986).

[Kop72a] G.I. Kopylov and M.I. Podgoretskiĭ, Sov. J. Nucl. Phys. 15, 219 (1972).

[Kop73a] G.I. Kopylov and M.I. Podgoretskiĭ, Sov. J. Nucl. Phys. 18, 336 (1973).

[Kop75a] G.I. Kopylov and M.I. Podgoretskiĭ, JINR E2-9285 (1975).

[Led79a] R. Lednicky and M.I. Podgoretskiĭ, SJNP 30, 432 (1979).

[Led82a] R. Lednicky, V.L. Lyuboshitz, and M.I. Podgoretskiĭ, JINR E2-82-725, (1982).

[Liu86a] Y.M. Liu et al., U.C Riverside preprint FPC2 86-2 (1986). (Submitted to Phys. Rev. C).

[Lok78a] A.A. Loktionov et al., Sov. J. Nucl. Phys. 27, 819 (1978).

[Lou73a] The Quantum Theory of Light, R. Loudon, Clarendon Press, Oxford (1973).

[Lu81a] J.J. Lu et al., Phys. Rev. Lett. 46, 898 (1981).

[Miy75a] O. Miyamura and M. Biyajima, Phys. Lett. 57B, 376 (1975).

[Nor87a] A. Norton, private communication.

[Pra84a] S. Pratt, Phys. Rev. Lett. 53, 1219 (1984).

[Oh75a] B.Y. Oh et al., Phys. Lett. 56B, 400 (1975).

[Ran75a] J. Ranft and G. Ranft, Phys. Lett. **57B**, 373 (1975).

[Shu73a] E.V. Shuryak, Phys. Lett. **44B**, 387 (1973).

[Shu73b] E.V. Shuryak, Sov. J. Nucl. Phys. **18**, 667 (1974).

[Suz87a] M. Suzuki, Phys. Rev. **D35**, 3359 (1987).

[Xuo62a] N. Xuong and G.R. Lynch, Phys. Rev. **128**, 1849 (1962).

[Wil84a] W. Willis and C. Chasman, Nucl. Phys. **A418**, 413 (1984).

[Yan78a] F.B. Yano and S.E. Koonin, Phys. Lett. **78B**, 556 (1978).

[Zaj82a] "Two Pion Correlations in Heavy Ion Collisions", W.A. Zajc,
Ph. d Thesis, University of California, (1982).
Available as LBL-14864.

[Zaj84a] W.A. Zajc, Phys. Rev. **C29**, 2173 (1984).

[Zaj86a] W.A. Zajc, in Proceedings of the Second International Workshop
on Local Equilibrium in Strong Interaction Physics, P.Carruthers
and D. Strottman, ed., World Scientific Publishing Co. (1986).

[Zaj87a] W.A. Zajc, Phys. Rev. **D35**, 3396 (1987).

Multihadron events in Cosmic Rays

Todor Stanev
Bartol Research Foundation
University of Delaware
Newark, DE, 19716

Gaurang B. Yodh
Department of Physics
University of California, Irvine
Irvine, CA, 92717

1 Introduction

1.1 General Survey

Cosmic rays are nature's high energy particle beams that continuously impinge upon the earth's atmosphere providing a laboratory where high energy interactions occur from a few GeV to a few hundred TeV in the center of mass. The most dominant component of the cosmic ray beam consists of nuclei, from protons to iron. The energy spectrum of the total nuclear cosmic ray flux is shown Figure 1. The rapid decrease of flux is its main feature, although one can see two other features; a downward bend at about 10^6 GeV and an apparent flattening above 10^9 GeV[1].

The detailed composition of cosmic ray nuclei is only known well from direct measurements made above the atmosphere at energies below 10^5 GeV per nucleus. A compilation of experimental data between 1 TeV and 1000 TeV is shown in Figure 2 (Balasubrahmanyan, et al, 1986[2]). There is some indication that the cosmic ray beam is richer in nuclei heavier than protons and helium at 10^6 GeV.

Many important discoveries in particle physics have been made using the cosmic ray beam since 1930s. Experiments using cloud chambers and Geiger counters discovered the positron and the muon in the 1930s. After the war came the discovery of pions and strange particles using nuclear emulsion technique as well as cloud chamber methods. During this era, collisions of cosmic rays of tens to hundreds of GeV with nuclei were studied

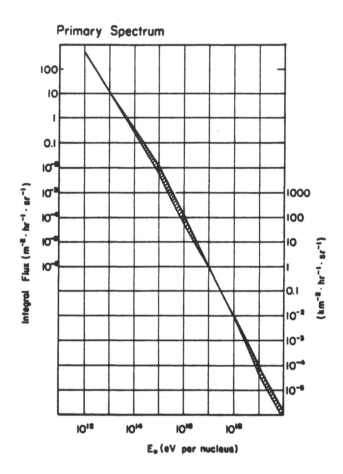

Figure 1: Composite integral energy spectrum of all nuclei of primary cosmic rays. The curve beyond 10^{15} eV is based upon air shower observations. The abscissa is energy per nucleus.

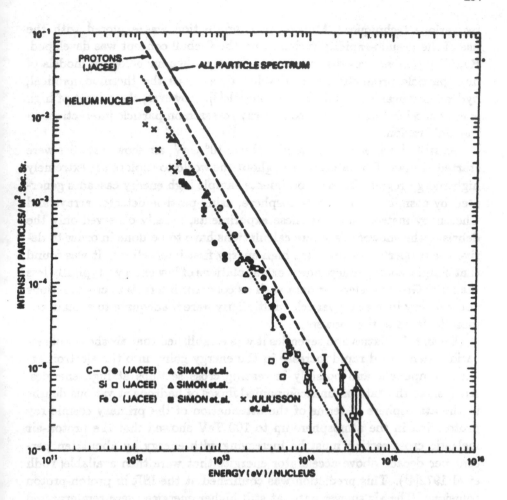

Figure 2: A compilation of the energy spectra of different nuclear groups in primary cosmic rays compared with the all particle spectrum. Note the tendency for the flux of heavier nuclei to approach the all particle spectrum above 10^{13} eV.

by various techniques. Multiparticle production was explored with the use of the pseudo-rapidity variable and the fireball concept was developed. Rapidity plateau was discovered and various phenomenological models of multiparticle production were developed, such as the thermodynamical, hydrodynamical and multi-fireball models[3]. However, the advent of high energy accelerators overtook cosmic ray research on particle interactions in the GeV region.

At still higher energies, energies above 10^6 GeV, air shower studies were started at many laboratories throughout the world to explore the extremely high energy region. These experiments sample high energy cascades generated by cosmic rays in the atmosphere, using particle detector arrays and Cherenkov instruments. As these experiments, usually observed only the debris of the shower, elaborate calculations have to be done in order to derive the properties of the ultra high energy first interactions. It was found that simple, energy independent extrapolations of low energy (typically less than 100 GeV) accelerator data such as constant interaction cross sections and a slowly increasing particle multiplicity were inadequate to explain the main features of the showers .

During the sixties and seventies it was established that air shower observations demanded rapid increase in the energy going into the electromagnetic component and possibly increasing cross sections at high energies. Analysis of the data on unaccompanied fluxes of hadrons at various depths in the atmosphere in terms of the attenuation of the primary cosmic ray proton flux in the atmosphere up to 100 TeV showed that the proton-air inelastic cross section must be increasing with energy by about ten percent per decade above accelerator energies that were then available(Yodh et al 1972[4]). This prediction was confirmed at the ISR in proton-proton collisions. The air shower data, at still higher energies, were reinterpreted including rising cross sections but one still needed further violation of scaling in order to explain various features of showers if the primaries were assumed to be mainly protons.

In parallel to the air shower studies, a new technique was developed to investigate individual interactions of cosmic rays at mountain and in balloon altitudes: the X-ray film emulsion chamber technique. With this method one could cover a large enough area with sensitive detectors to capture sufficient number of individual interactions in the 50 to 100 TeV energy range where target nucleus was defined and local atmospheric interactions in the 100 to 10000 TeV range.(Mt. Chacaltaya experiment[5], Mt. Fuji experiment[6], the Pamir experiment[7] , Mt. Kanbala experiment in China[8], the JACEE balloon experiment[9]). The study of jets in emulsion chambers from interactions produced locally in a target, during the 70s and 80s produced some very interesting results which were later confirmed by the collider experiments at CERN in 560 GeV proton-antiproton collisions. Three of these results were:

- the observation of clear correlation between multiplicity and transverse momentum,

- the violation of scaling in the central region i.e. the rise of the central plateau; and

- the increase of multiplicity with energy rising faster than logarithmically.

The most exciting result from these mountain level emulsion chamber experiments was the observation of exotic events such as the Centauro and binocular events. The Centauro events, first observed by the Mt. Chacaltaya collaboration, are the observation of an atmospheric jets from an interaction near the chamber which contain a small number gamma ray cascades and an extremely small ratio of electromagnetic to hadronic energy flow. The binocular events have been interpreted by the experimenters as events in which two or more jets are seen with very large transverse momenta. Another result, claimed by the Pamir collaboration from the analysis of the fractional energy carried by the observed *gamma* jets in the atmospheric interaction events is that scaling is violated strongly in the fragmentation region above a 1000 TeV. This interpretation is controversial, the Fuji group claims that the result indicates the presence of heavy nuclei in cosmic rays and does not require large violation of scaling in the fragmentation region. It is very important to measure inclusive particle production in the fragmentation region at colliders (such as was done recently by the UA7 experiment for gamma production[10]) in order to decide amongst the alternatives.

The JACEE experiment, in which emulsion chambers are flown at the top of the atmosphere, can examine individual collisions of primary nuclei with emulsion nuclei on an event by event basis up to almost 1000 Tev total energy. They have observed a reasonably large sample of nucleus-nucleus collisions at these extremely high energies. They interpret their observations to indicate the occurrence of a high p_t component and a central rapidity density which corresponds to an energy density greater than 2 GeV per cubic fermi, sufficient to cause a transition to the quark-gluon phase.

At still higher energies, only air showers experiments can be used to study high energy interactions. Because of the gross nature of these measurements only broad features of the energy dependence of interaction properties may be deduced from these observations. The Fly's Eye detector, which measures the longitudinal profiles of individual showers can be used to determine the energy dependence of the proton-air inelastic cross section above 10^8 GeV[11]. Conventional air shower experiments can also be used for determining the proton-air cross sections[12]. The results from these experiments show a continued rise of the cross section up to a $\sqrt{s}$ of about 30 TeV (Yodh 1986[13]).

With the turn on of every new high energy machine one of the first experiments that are carried out is to search for production of free quarks and other hitherto unseen particles. Cosmic ray beam also provides a laboratory where high energy hadron-nucleus and nucleus-nucleus collisions take place. Particle searches in the cosmic ray beam in the atmosphere, in air showers and in deep underground muon experiments have been done to search for fractionally charged quarks, heavy long lived particles (charged or neutral) and other exotic particles[14,15,16]. These cosmic ray experiments are sensitive to μb cross sections. No definitive evidence for either fractionally charged quarks or heavy long lived particles has been found.

Recently, ultra high energy astronomy has become a very active field [17,18]. Signals have been seen from astrophysical point objects, such as compact binaries, in the 1000 TeV energy range, which are correlated with the characteristic periods of the system observed in other frequency bands. Such correlations restrict the mass of the objects carrying these signals to be less than a few hundred MeV. If these carriers are photons then this cosmic radiation offers a laboratory where interactions of photons at $\sqrt{s} \sim 1$ TeV may be studied. Recent results for signals from Cygnus X-3 and Herculus

X-1 indicate that the showers of these signals are not muon poor, as would be expected for standard gamma rays, and may indicate occurrence of a threshold for new processes(results of the CYGNUS experiment [19,20]).

Cosmic ray experiments are difficult to interpret because of the mixed nature of the primary beam, the steepness of the energy spectrum and inherent fluctuations present in their interactions. Detailed simulations are needed to confront experimental data with theoretical model predictions.

1.2 What can be explored with cosmic rays ?

The conditions under which cosmic ray experiments are performed determine the type of results, which are accessible and define the relation between cosmic ray and accelerator results. Cosmic ray experiments are, of course, necessarily performed in the laboratory system. The experimental arrangement is usually positioned at large distance from the primary interaction and detects a mixture of the secondaries of that interaction and progeny from subsequent interactions, i.e. a fraction of the cascade that the primary particle has generated in the atmosphere.

For this cascade to penetrate to the depth at which the detector is placed and have sufficient energy to register in the detector above its threshold, only interactions with large enough energy transfer to secondaries can be studied. Detailed study show that secondaries created at $x_F < 0.02$ do not contribute significantly to the cascade development and thus cannot be sampled in a cosmic ray experiment. This situation is exactly opposite to an accelerator/collider center of mass experiment, where it is easy to study the small x region whereas the large x secondaries are difficult to study as they tend to go down the beam pipe. Cosmic ray results are, therefore, complimentary to collider results. Although the particles in the fragmentation region reflect mostly the behaviour of soft interactions and spectator partons, they do carry most of the interaction energy. Their study yields information on structure functions and reveals the behaviour of minimum bias events.

Within these limitations, the first physical result that comes from every cosmic ray experiment is the inelastic cross section. The cross section measurement is both an important result and a necessary tool for further analysis of the experiment. The measured quantity is an integral of the inclusive cross section for particle production for $x_F > x_{min}$, where x_{min} depends on the particular experimental arrangement and is also energy dependent. Many cosmic ray experiments, however, reach high enough energies so that they can in principle distinguish between model extrapolations, which represent equally well the behaviour of inelastic cross section at lower energy.

Cosmic ray experiments provide an opportunity to study unique beams, most notably heavy nuclei, muons and γ ray beams of very high energy. As mentioned earlier, a certain fraction of primary cosmic rays consists of heavy nuclei. Satellite and balloon experiments can intercept these nuclei outside or high enough in the atmosphere and study their interactions. The current statistics of interactions of energetic nuclei studied by cosmic ray experiments consists of several hundred events with energy greater than 10 TeV per nucleus with identified incident and target nuclei.

While the earth's atmosphere is a relatively thick target for strongly interacting particles it is not sufficient to stop or slow down energetic muons produced in high energy interactions in the upper atmosphere. At a depth of one interaction length, the atmosphere is so tenuous that a significant fraction of TeV pions and especially kaons decay before interacting. Muons from these and other (heavy flavor) decays provide a beam extending up to tens of TeV for ground level or underground studies of their properties and interactions.

1.3 Manifestations of particle physics in Cosmic Ray Experiments.

All cosmic ray experiments are sensitive to the value and the energy dependence of the inelastic cross section. While the exact sensitivity on the cross section depends on the triggering conditions of a particular experiment, the general idea in a cross section determination is to relate the intensity of the observed events to the intensity of the primary cosmic ray beam. A good example is the derivation of the cross section from the fraction of primary protons that reach the observation level without shower accompaniment, i.e. without suffering inelastic interactions. For a given energy and known atmospheric thickness this fraction will depend solely on the value of the cross section and will strongly decrease when the cross section rises[4].

For the majority of experiments, however, which detect a mixture of interaction products of several generations of the cascade, the cross section determination is mostly a measurement of energy dissipation in the early part of the atmospheric cascade. The bigger the cross section, the bigger the number of generations which contribute to the observed particle energy distribution. Such experiments, therefore, cannot reach unique conclusions for the behaviour of inelastic cross section without involving assumptions about the particle production spectra. A soft production spectrum of the secondaries will cause faster energy dissipation and imitate a larger cross section. This is one of the dualities in the interpretation of cosmic ray data. It decreases the sensitivity of a large class of experiments to either the inelastic cross section or the particle production spectrum. An additional complication is that the identity of the primary initiating the cascade is not known.

Typical cosmic ray experiments, and especially air shower experiments, are only sensitive to particle production down to a minimum x_F. The range over which integral over x_F has to be taken increases with primary energy. Thus a cosmic ray experiment is more likely to overestimate rather than underestimate the value of the inelastic cross section.

It is important, also, to remember that cosmic ray experiments almost never study interactions on proton targets. The average atomic number of the atmosphere is 14.5 and in the interpretation of cosmic ray results cross section and particle production spectra in hadron-nucleus interactions have to be used. Furthermore, in these interactions the relevant region for particle production is not the target hemisphere but rather the projectile region. The question that must be addressed is not how much the nuclear target alters the total multiplicity, but what is its influence on the secondaries with $x_F >$ the minimum for experiment under consideration.

About the only phenomenon, which is not masked, but instead is emphasized by the interaction-detection sequence, is the creation of particles at large transverse momenta. Because the distance between the interaction and the detector is sometimes measured in kilometers, the radial displacement of high p_t particles can be measured and not be easily obscured by the cascade process. While this general feature is valid in most experiments, detection of high p_t particles in underground muon experiments (with or without coincidence with an air shower detector) can be especially successful. The average height of production of a TeV muon is about 20km, so that a transverse kick of 1 GeV/c can cause a radial displacement as large as 20m.

2   Techniques of Experiments

Three different techniques used to investigate the properties of ultra high energy collisions with cosmic rays are discussed in this section:

- Emulsion chamber methods to study individual interactions,

- Underground experimental methods to study TeV muons and

- Air shower techniques to study the highest energy interactions.

2.1   Emulsion Chambers

Emulsion chambers are the most generally used detectors to study individual high energy cosmic ray interactions. They consist of multiple layers of absorber and photosensitive material, such as X-ray films, nuclear emulsions and plastic sheets(e.g. CR39). Several different types of chambers are in use in ground based and balloon experiments. By exposing emulsion chambers at high altitudes one can observe

- direct interactions of cosmic rays in the chamber and

- events that arise from a bundle of high energy showers coming from the same direction- so called families or A-jets.

These detectors have poor time resolution, but excellent spacial resolution. The chambers can measure the energy transferred into electromagnetic cascades by a high energy particle to about 20 percent. The energy threshold for detection of a jet can be made rather high, say greater than 200 GeV, so that the background obscuration due to low energy particles can be eliminated even for year long exposures needed to collect enough ultra high energy events.

2.1.1 Chambers for mountain experiment

Three main types of chambers are in use in ground based experiments: thin, thick and two chambers separated by an air gap[5,6,7]. Thin chambers, which contain about 10-30 radiations of lead absorber, register mostly electromagnetic component of A-jets. Thick chambers, which in addition to thin sheets of lead also have several interaction lengths of some other material such as carbon, are sensitive to electromagnetic and hadronic components of the jets. The ' air gap ' chambers allow for the spreading of particles produced in interactions in the upper chamber before they are detected in the lower chamber. It provides for better spacial resolution and better measurements of angles between the produced particles. The structure of typical emulsion chambers employed in the Pamir, Mt. Fuji and Mt. Chacaltaya experiments are shown in Figure 3.

The number of radiation lengths in a chamber vary from 30 to 70 and the number of interaction lengths from 0.6 to 2.4. Chambers are ' easy ' to handle and are stable for long exposures. Figure 4 demonstrates how an A-jet is detected by an emulsion chamber. Separate showers of the A-jet 'family ' are registered as dark spots in the X-ray film with a spread of several centimeters.

The energy of the shower of each ' jet ' is estimated by calculating the track length of the cascade through the emulsion chamber using measured optical density of spots, which gives a measure of number of particles in the cascade. The energy of each spot is called E_i and its position r_i. The energy estimate for the whole family is obtained by $E = \sum E_i$ where the sum is over the observed number of spots. In a thick chamber (or a chamber with an air gap such as used by the Mt. Chacaltaya group) hadrons and γ-rays are distinguished on the average using the depth distribution of starting points of cascades. Usually those cascades starting deeper than 4 radiation lengths are called hadronic cascades. One can thus divide the total visible energy of an A-jet into hadronic and electromagnetic parts: $E_{vis} = \sum E_\gamma^h + \sum E_\gamma^\gamma$. The threshold energy for detection of a dark spot in X-ray film is high, varying between 2 and 4 TeV for different experiments.

Emulsion chambers used in balloon experiments have a different configuration and are described next.

300

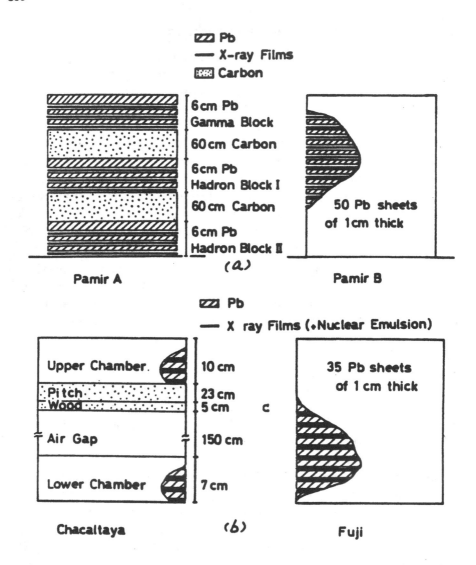

Figure 3: Schematic drawing of three types of emulsion chambers.

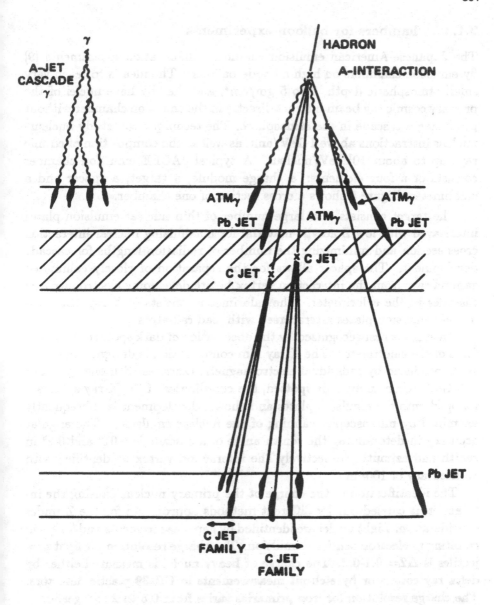

Figure 4: Illustration of A-jet families produced by photon and hadron interaction in the atmosphere.

2.1.2 Chambers for balloon experiments

The Japanese-American emulsion chamber collaboration experiments [9] fly emulsion chambers on high altitude balloons. The idea is to reach very small atmospheric depth, 3 to 5 gm/cm^2, and thereby have nuclei of the primary cosmic ray beam interact directly in the emulsion chamber without producing a cascade in the atmosphere. The technique can study nucleus-nucleus interactions above 1 TeV/amu as well as the composition of cosmic rays up to about 10^{15} eV/nucleus. A typical JACEE emulsion chamber consists of a four elements: a charge module, a target, a spacer and a calorimeter. Figure 5 shows a cross section of one chamber.

The target consists of a large number of thin nuclear emulsion plates interleaved with low Z acrylic material for maximizing nuclear interaction cross section and minimizing the number of radiation lengths for cascade development. The spacer allows for the separation of closely collimated gamma rays from the interaction vertex before developing electromagnetic cascades in the calorimeter. The calorimeter consists of X-ray films and nuclear emulsion plates interspersed with lead radiators.

An event is first recognized by the observation of dark spots in the X-ray films of the calorimeter. The X-ray film configuration is designed to detect spots produced by individual electromagnetic cascades with energy above 300 GeV. After an event is spotted, the coordinates of the X-ray spots are mapped onto the emulsion plates and shower development is subsequently examined by microscopic scanning of the nuclear emulsions. The angular accuracy in determining the zenith angle of a cascade is 0.01° and 0.2° in zenith and azimuth, respectively. The interaction vertex is identified with an accuracy of 100μm.

The identification of the charge of the primary nucleus causing the interaction is carried out by different methods appropriate for the Z under consideration. Light nuclei are identified by gap measurements and/or grain counting in electron sensitive emulsion. The charge resolution for light projectiles is $\Delta Z=$ 0.1-0.2. The charge of heavy nuclei is measured either by delta ray counts or by etch-pit measurements in CR-39 plastic detectors. The charge resolution for iron primaries varies from 0.5 to 2 charges.

The energy of each individual electromagnetic cascade in the calorimeter is determined by electron scanning, typically in a 50μm radius around the

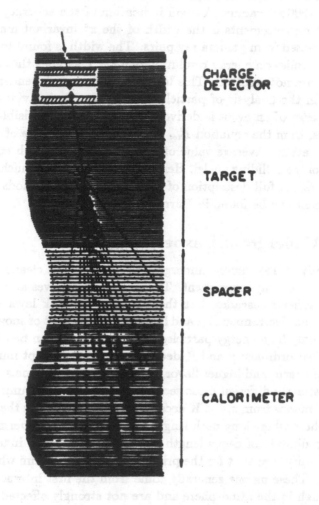

CHARGE DETECTOR

TARGET

SPACER

CALORIMETER

Figure 5: Schematic diagram of the JACEE emulsion chambers.

shower axis. For cascades that originate in the calorimeter the individual tracks are not separated and the event energy is measured by scanning in a larger, 200μm radius. A good indication of the accuracy in energy and position measurements is the width of the π° invariant mass distribution reconstructed from gamma ray pairs. The width is found to be 33% which in turn implies an energy resolution of 22%. Cascades that are closer than 12.5μm are not resolved. This leads to an energy dependent minimum p_t cut off in the analysis of particle distributions in individual events. The total energy of an event is derived from the observed visible energy in the cascades, from the equation $E_{tot} = \sum E_\gamma / K_\gamma$. The value of K_γ varies from event to event. Average value of K_γ may also vary with energy and with masses of the colliding nuclei. Hence E_{tot} is determined much less accurately than $\sum E_\gamma$. A full description of the experimental methods of the JACEE experiment can be found in Burnett et al[21].

2.2   Underground experiments

The study of TeV muons underground provides the cleanest possible high energy cosmic ray experiment. The muons themselves are of course result of atmospheric cascades, but the shielding of thick layer of rock totally absorbs the electromagnetic and hadronic component of showers and muons are the only high energy particles left. The muons can be divided into two categories: ordinary π and K decay muons and prompt muons due to the decay of charm and higher flavor particles. These channels produce muons with distinctly different cross section and energy spectrum.

TeV muons from π and K decay come primarily from the tenuous upper atmosphere where long path length before interactions permits decay. Relativistic dilation of decay length increases the spectral index of the muon flux by unity over that for the primary nucleon spectrum which produce πs and Ks. These muons generally come from the first interaction of the primary, high in the atmosphere and are not strongly affected by subsequent cascading. The interaction-decay competition also make the muonic yield dependent on the π to K production ratio in the initial interaction. In fact, about half of the TeV muons come from K decays for a $\frac{K_{ch}}{\pi_{ch}}$ ratio of 0.1. At high enough energies the muonic yield can be sensitive to charm and heavy flavor production.

The underground detectors consist of tracking chambers of relatively large size [22,23,24,25,26,27,28,29,30] with an angular resolution better than a few degrees, with multiple track capability and modest energy resolution. The energy is generally estimated from slant depth and single muon spectrum has been determined up to about 100 TeV. In principle, these experiments are sensitive to primary nucleon spectrum up to about 10,000 TeV and can not only explore such high energy collisions but also probe the composition of primary cosmic rays.

2.3 Air shower techniques

Above 10^{17} eV the flux of primary cosmic rays is small and one must use the air shower technique to utilize this beam to study ultra high energy interactions. The nuclear-electromagnetic cascade generated by cosmic rays spreads out to several hundred meters and can be detected by observing the Cherenkov light or atmospheric fluorescence generated by the shower particles or by sampling the shower at the earth's surface by detector arrays covering a large area. In either case, detailed properties of individual interactions are obscured by the cascading process. Typical quantities that are sampled by air shower detectors are[31]:

1. Longitudinal profiles of shower development through the atmosphere by time and pulse height measurement and analysis of Cherenkov light[32,33,34].

2. Time and pulse height profiles of muons in showers[35].

3. Muon content of air showers[36,37,38,39,40].

4. Time and energy structure of hadrons near shower cores [41,42].

5. Nitrogen fluorescence from air shower cascades[43].

6. Zenith angle variation of number of shower particles for fixed number of shower muons[44].

Interpretation of these experiments to derive unambiguous conclusions about particle interaction properties is difficult for several reasons. Firstly, because the primary spectrum is rapidly decreasing with energy, fluctuations in interactions play a very significant role. Lower energy primaries can mimic high energy showers through fluctuations. Secondly, air showers are fed primarily by the most energetic secondaries $x_F > 0.1$. In this fragmentation region we do not have sufficient information from collider experiments to be able to extrapolate production cross sections to air shower energies in a reliable manner. Thirdly, composition of primary cosmic rays is not known at these energies and air shower experiments do not directly measure the atomic mass of the primaries, therefore previously mentioned ambiguity between proton primaries with high inelasticity and heavy primary with small violation of scaling is difficult to resolve.

Inspite of these difficulties, it is possible to extract the interaction cross section of protons in air by the study of fluctuations in the location of the point of first interaction. This was first pointed out over twenty years ago by a Japanese air shower group[45,13].

3   Results on Multihadron production at Cosmic Ray energies

3.1   JACEE results

JACEE collaboration has had eight balloon flight exposures for a total exposure of 500 m²hr. Although the flux of nuclei with energy above 1 TeV per amu is extremely small, JACEE has observed a score of extremely interesting nucleus-nucleus interactions at very high energies. As an example we show in Figure 6 the pseudorapidity distribution of charged particles in a silicon nucleus interaction on AgBr target[46]. The energy of the incident nucleus was estimated to be 4 TeV/amu. The total number of secondary charged particles is 1010 ± 30 . It is difficult to measure the number of γ-ray tracks of the event since the interaction vertex is located in the nuclear emulsion layer of the calorimeter and the cascades have overlapping cores, however, they estimate that the total number of γ-rays is > 170. In Figure 6 the solid curve is for a multichain-model calculation; dashed curve for a wounded nucleon-model calculation; both for $< p_{t\pi} >=0.4$ GeV/c. The arrow indicates the unobserved region.

Although this event is particularly difficult to analyze, the three dimensional development of the overlapping cascades is consistent with a Monte

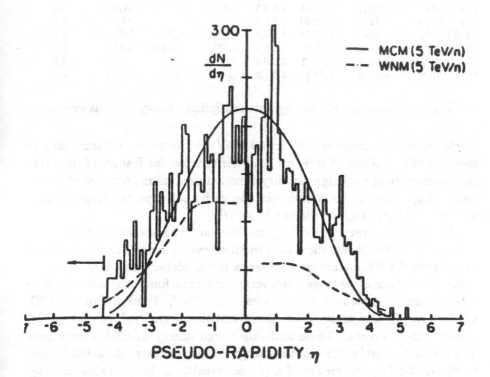

Figure 6: The CMS pseudorapidity distribution of charged particles in the Si+AgBr event. The solid curve, multichain-model calculation; dashed curve, wounded nucleon-model calculation; both for $< p_{tr} >$=0.4 GeV/c. The arrow indicates the unobserved region.

Event type (TeV/amu)	N_{ch}	$<p_{t\pi^\circ}>$ (GeV/c)	$[\eta_L]$	$<p_{t\pi}>$ (GeV/c)	$[\eta_L]$ ($|\eta|\leq 1$)	dN/dη	ϵ_{min} (GeV/fm^3)
V(1.5)+Pb	1050^{+300}_{-50}	0.95±.31	3.7-6.4	0.55±.05	≥6.4	258±12	2.4
Si(4.1)+AgBr	1010±30			$0.55^{+.12}_{-.05}$	≥6.1	183±12	2.7
Si(4.0)+Pb	790^{+40}_{-25}	1.3±.30	4.6-5.5	$1.0^{+.2}_{-.15}$	≥5.5	147±8	3.8
Ca(100)+CHO	760±30	0.53±.04	6.5-8.3			81±8	2.0
Ca(0.5)+Pb	670±40	1.03±.18	4.2-5.1	1.1±0.5	≥5.1	142±8	3.2
Ca(1.8)+Pb	> 452	$1.6^{+0.1}_{-0.3}$	3.5-6.0	$1.1^{+1.0}_{-.35}$	5.3-6.0	100±16	2.2
Ti(1.0)+Pb	> 416	1.0±0.2	3.8-6.6			134±8	2.9
C(>1.5)+Pb	400^{+15}_{-13}	$1.6^{+.4}_{-.6}$	4.0-7.0	≥ 1.2	≥6.0	81±7	4.5

Table 1: Summary of the highest-multiplicity heavy-nucleus events

Carlo simulation assuming 500 ± 100 MeV for the average transverse momentum of secondary π° mesons. Another spectacular feature of the event is the existence of very high rapidity density fluctuations, some of which are much bigger than random and coincide with that expected from possible generation of quark-gluon plasma bags[47]

The totality of very high energy nucleus-nucleus interactions detected by the JACEE experiment, show that in these events many more secondaries are observed with transverse momenta much higher than that expected from application of average transverse momenta found in proton-proton interactions to nucleus-nucleus interaction. Table 1 taken from JACEE publication[48] lists some of the characteristics of the highest multiplicity heavy nucleus events. These and other high energy JACEE events have also been analyzed in terms of the energy density at early interactions times to search for the occurrence of a phase transition. In the table, the $[\eta_L]$ following the $< p_t >$ values indicate the laboratory pseudorapidity region of the measured $< p_t >$ data.

The minimum energy density ϵ_{min} is determined from the Bjorken[49] formula. The formula relates ϵ to the pseudorapidity density for $\eta < 1$, the measured average transverse momentum $< p_t >$ and the minimum atomic mass of the projectile and target nucleus, A_{min} at the typical soft QCD interaction time τ_o by the equation:

$$\epsilon = \sqrt{< p_t >^2 + m_\pi^2} \times 1.5(dN/d\eta)/(2\pi\tau_o A_{min}^{2/3})$$

Figure 7 shows the correlation of p_t with ϵ_{min} value for heavy nucleus interactions and proton interactions. Full circles are individual nucleus events $(2 < Z < 26)$ and open squares are for proton-lucite events. The median $< p_t >$ values for events with $\epsilon_{min} < 1$ GeV/fm^3 and> 2 GeV/fm^3 are 490 and 850 MeV, respectively. All high multiplicity events from the table indicate high energy density and significantly higher $< p_t >$ values than highest energy nucleon collisions. A qualitative difference seems to appear at energy densities above 2 GeV/fm^3, which is almost an order of magnitude larger than naively expected deconfinement energy density of 1 GeV/$((4\pi/3)$fm$^3)$.

Some of the high multiplicity nucleus-nucleus interactions show unusually high fraction of photons [50]. The pseudorapidity distributions of neutral and charged particles often do not coincide, suggesting emission from uncorrelated centers. In general high multiplicity, high transverse momentum events are neutral rich, although the data shows big fluctuations. Figure 8 shows the scatter plot of ratio of neutral to charged pseudorapidity density, R, versus the charged pseudorapidity density for individual interactions. The plot suggests the existence of two separate populations, one of which with high R, is consistent with the trends of accelerator and JACEE p-carbon interactions, where R increases as $(dN_{ch}/d\eta)^{1/3-1/2}$. The other branch, however, might tend toward low content of neutrals, slightly resembling Centauro events[5].

3.2 Mountain emulsion chamber experiments

3.2.1 Method of Analysis

Mountain emulsion chamber(EC) experiments investigate individual interactions above 1000 TeV in the laboratory system. We first discuss problems associated with data acquisition and data analysis before presenting the results. Although the idea behind a mountain altitude EC experiment is simple, the data analysis is extremely complicated. The limited size of the chamber introduces quite a strong dependence of the detection efficiency on the altitude at which the observed family has been generated. During the propagation of the family in the atmosphere there is significant cascading, which masks the characteristics of the intial high energy interaction.

310

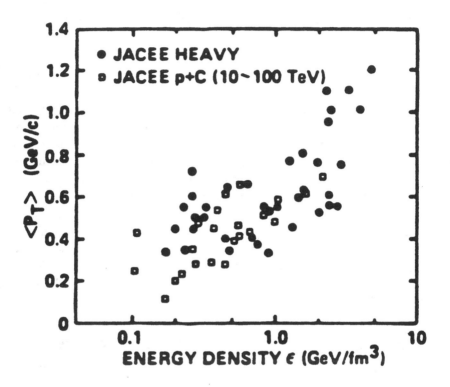

Figure 7: Correlation of $< p_t >$ with energy density ϵ. Full circles are individual nucleus events $(2 < Z < 26)$ and open squares are for proton-lucite events.

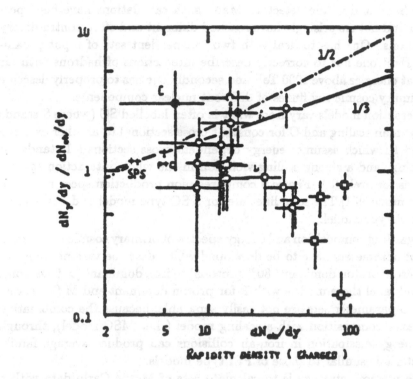

Figure 8: Scatter plot of ratio of neutral to charged psuedorapidity densities versus the charge psuedorapidity density.

Different experiments are performed at different altitudes and generally use X-ray films and nuclear emulsions of differing sensitivity, which makes direct comparison of results difficult.

In order to understand the properties of families one must carry out full Monte Carlo simulation of the hadronic-electromagnetic cascades in the atmosphere, and in the detector. Many such calculations have been performed and not surprisingly have reached different and often contradictory conclusions. One has to deal with two independent sets of input parameters: First one has to correctly describe interactions of hadrons with air nuclei at energies above 1000 TeV and secondly one has to properly describe the primary cosmic ray fluxes of different nuclear components.

Interaction models vary from what is often labelled SC (where S stands for Feynman scaling and C for constant cross sections) to another extreme labelled FI which assumes energy dependent cross sections (I stands for increasing) and a strong scaling violation in the whole interaction space (F stands for fireball). Figure 9 compares pion production spectra for two variant models[51]. The solid lines are for a SC type model and dashed line is for a F type model.

Regarding composition and energy spectra of primary cosmic ray species, a set of parameters have to be developed which describe variant compositions , from proton dominant(80% protons) to iron dominant (> 50% iron). We shall label these models with P for proton dominant and M for mixed. The two parameter sets are not easily separable, because the combination of a heavy composition with a scaling model (i.e. MSC model), through high energy dissipation in iron-air collisions can produce average family characteristic similar to those of PFI type models.

The research strategy is to calculate sets of Monte Carlo data with a variety of models, compare them with experimental data and draw conclusions on the combination of ingredients that best fits the experimental results. It is quite often that proper accounting for detector characteristics and biases outweighs the differences in the physics assumptions regarding interaction model and composition for some of the family properties.

The main experimental characteristics studied are:

- The visible energy spectrum of detected events, $E = \sum E_\gamma$: The slope of the E spectrum and its absolute intensity and attenuation of intensity

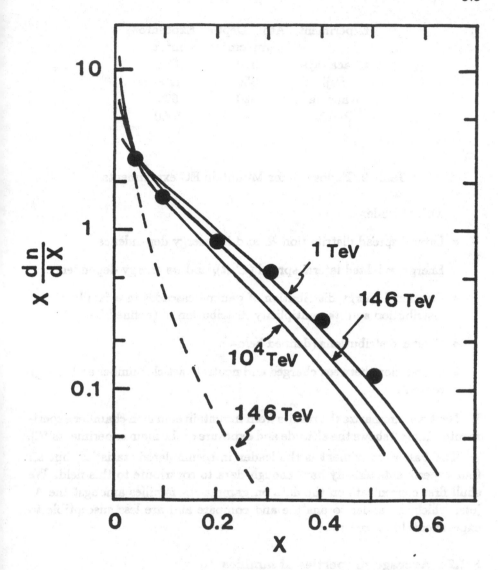

Figure 9: Comparision of pion production x distributions.

Experiment	Atm. Depth gm/cm^2	Exposure m^{2}yr
Chacaltaya	540	691
Fuji	650	1100
Kanbala	520	699
Pamir	596	6000

Table 2: Exposures for Mountain EC experiments

with altitude.

- Lateral spread distribution R_i and its energy dependence.

- Energy weighted lateral spread $(E.R)_i$ and its energy dependence.

- Fractional energy distribution of gamma cascades in a family, the f' distribution and its multiplicity distribution n' (defined below).

- Cluster distributions (defined below).

- Correlations between charged and neutral particle number and energy content.

Next we summarize the results from mountain emulsion chamber experiments. Table 2 shows the altitude and exposure of the main experiments[52].

The Pamir experiment is the leader in accumulated statistics, but all four experiments already have enough data to contribute to this field. We shall first concentrate on the data on gamma-ray families amongst the A-jets, which are easier to analyze and compare and are less susceptible to experimental systematics.

3.2.2 Average properties of families

The major experimental groups have succeeded in obtaining data which show no contradictions in their main average family characteristics.

The present consensus may be summarized as follows[53]:

1. The fluxes of gamma-ray families in the atmosphere attenuate faster than predicted by a pure proton primary cosmic ray flux and Feynman scaling. The detected intensity by all groups are almost a factor of 10 below the expectations in the above model.

2. The attenuation of the fluxes of individual gamma-ray showers with energy > 5 TeV is faster than would follow from a constant inelastic cross section. The observed Λ_{att} is about 100 gm/cm^2, which is fully compatible with a log$^2(s)$ rise of the proton-proton inelastic cross section.

3. The differential energy spectrum of the detected families can be described by a power law of $\sum E_\gamma$ with a spectral index of 2.4±0.1.

4. The lateral spread of family particles is significantly bigger than that which would be produced by a constant transverse momentum of 0.33 GeV/c and can be approximately described by a $< p_t > \sim 0.5$ GeV/c.

5. Production of QCD jets is needed to explain the high p_t tail of the lateral distribution of family particles.

3.2.3 Familiy Intensities

We shall return with more details to some of the above listed common properties of gamma-ray families, but let us now concentrate on item (1) - the observed family intensity, the analysis of which shows the duality in interpretation of the observations. Figure 10 shows the observed integral intensity of gamma-ray families by the Mt. Fuji group, as a function of $\sum E_\gamma$ together with that expected on the basis of different interaction and primary flux models. The energy, $\sum E_\gamma$ is a measure of the fraction of the total energy of the primary nucleus, which has survived in the form of high energy secondaries at observation level. The uppermost line shows the prediction of the scaling interaction model, combined with a pure proton primary flux. It is clearly above the experimental data points. In order to decrease the predicted intensity one has to dissipate energy faster. This can be achieved in several alternate ways: either make the inelastic interaction softer, or have a heavier primary nucleus, whose energy content is low in

316

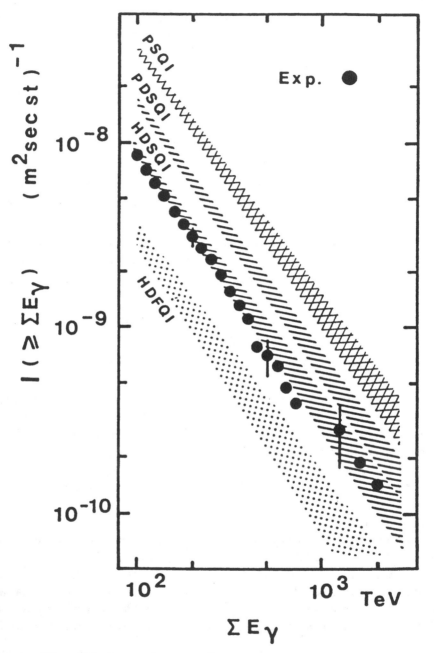

Figure 10: Integral energy flow spectrum of gamma families.

terms of its energy per nucleon. In Figure 10 the symbols P and PD refer to proton dominant compositions and HD is the same as M (or a composition with about 50% iron). S refers to scaling in the fragmentation region, F to scale violation, Q refers to inclusion of QCD jets and I to increasing inelastic cross sections. The data is below PSQI and PDSQI models, agrees well with HDSQI. A slight additional tuning of the primary composition and/or of the interaction model would make agreement with experiment perfect. A recent analysis of the China-Japan collaboration also shows that the data is consistent with an interaction model that uses scaling QCD violations, and a heavy dominant composition[51].

3.2.4 Energy distribution and multiplicity in families

The Pamir experimenters have developed a method to decrease the influence of systematic effects on the experimental data due to cascading called ' rejuvination '[54,55] . It consists in neglecting in the analysis all individual observed gamma-jets in a family that contain less than a certain fixed fraction of the total visible energy. This way most of the jets due to particles created during the cascading in the atmosphere are removed from the sample and the analysis is concentrated mostly on the products of the first interaction of the projectile. The energy distribution of the individual showers in the family is expressed in terms of the variable, $f'=E_i/\sum E_\gamma$, where f is required to be greater than f'_{min}, typically 0.04. For these remaining jets the average separation between the jet spots $< R >$ is calculated from measurements.

The idea of this analysis is contained in Figure 11 which shows the average value of n' of experimentally detected families with $\sum E_\gamma$ between 100 and 200 TeV, versus the average lateral spread of individual showers with respect to the energy weighted center of the family [56].Experimental data is given by the open circles for two different energy ranges. Crosses show values calculated in a quasi-scaling model and primary beams of pure composition. One can see that heavy compositions affect both n' and $< R >$, the latter due to the large nucleus-nucleus cross section, which allows longer path for family to spread out. In contrast, introducing strong scaling violation in the fragmentation region moves the simulated point vertically, i.e. increase n' without significantly affecting $< R >$. Large number of models, representing various degrees of scaling violation and different primary compositions have been inspected and conclusion is reached by the author[56] that only proton dominant composition with large degree of scaling violation in the fragmentation region is consistent with the experimental data. The degree of scaling violation is measured through the ratio of $\frac{1}{\sigma}(dN/dx)$ at x=0.3 and 1000 TeV with the same quantity at 100 TeV. The authors find a preferred value of this ratio of 0.25.

Recently, the Fuji-Kanbala[8] have performed a similar analysis. In contradiction with Pamir, they find a preferred model with small scaling violation in the fragmentation region, a value consistent with QCD models with increasing jet cross sections which explains well not only their own data also Pamir data for a composition dominated by heavy nuclei. Figure 12 shows the comparison of experimental results of the three experiments with calculated values in quasi-scaling (S) and fireball (F) models. Apart from the apparent contradiction between the calculated values of the Pamir and Fuji-Kanbala groups. the Fuji-Kanbala group finds that going from a scaling to a fireball model affects the value of n' very little.

3.2.5 Cluster Analysis and QCD jets

The Fuji-Kanbala group has chosen another method to decrease the influence of atmospheric cascading on the experimental data and draw conclusions on inelastic interactions above 1000 TeV. The method, called clusterization, and employed in different contexts by all groups is schematically explained in Figure 13, which shows an atmospheric cascade generating an A-jet[57].

In this example the number of particles created from the original interaction above the detection threshold is six. However, cascading increases that number to 19 gamma-ray spots of lower energy. A clustering procedure is developed to combine gamma-ray jets into clusters, each representing an original particle. One calculates $X_{i,j}=\sqrt{(E_i E_j)}R_{i,j}$ between the observed jets. If $X_{i,j}$ is less than a predetermined value, X_c (say 2 TeV.cm) then particles i, and j are combined into a cluster with energy $E_c=E_i+E_j$ and a new position vector R_c determined by their energy weighted centroid. One can see on the target diagram that the cluster algorithm enables one to

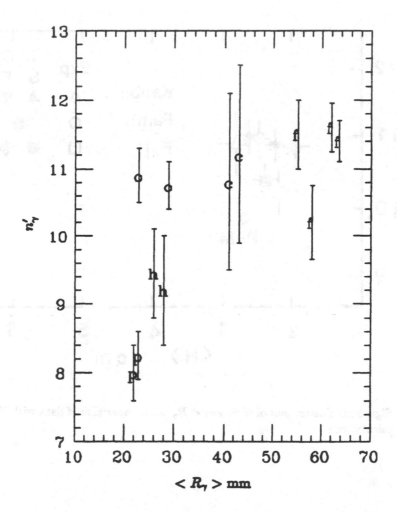

Figure 11: Scatter plot of n' versus $< R_\gamma >$. Pamir results and Monte Carlo.

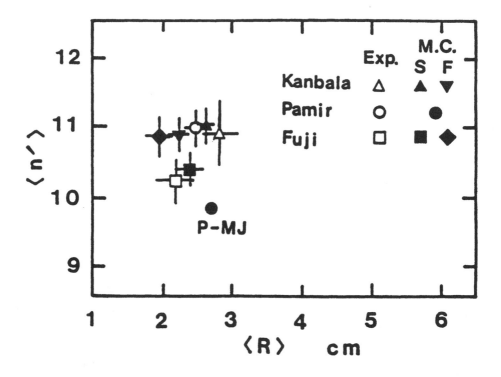

Figure 12: Scatter plot of n' versus $< R_\gamma >$. Comparision of data with Fuji calculations.

321

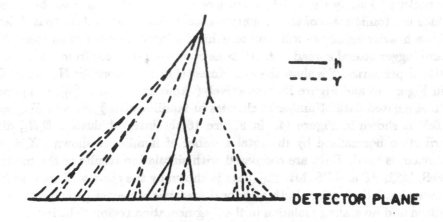

DETECTOR PLANE

CLUSTER ANALYSIS

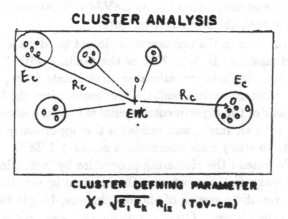

CLUSTER DEFINING PARAMETER

$$\chi = \sqrt{E_i E_k} \ R_{ik} \ \text{(TeV-cm)}$$

Figure 13: Schematic illustration of the clusterization method

make a better estimate of the number of jets from the original interaction.

Notice that by increasing X_c one can cluster more and more particles. The method was developed with the hope to pick up purely hadronic jet structure to study the QCD jet cross section and effects due to the difference in atomic mass of the primary particle. Simulations show that A-jets from heavier primaries will tend to exhibit a larger number of clusters, N_c, and bigger lateral spread $< E_c.R_c >$ as compared to those from proton initiated primaries. We show the experimental distributions for N_c and $E_c R_c$ in Figure 14 and Figure 15, respectively(Amenomori et. al.[6]); compared to simulated data. Number of clusters in families with $\sum E_\gamma$ and $E_{min} = 4$ TeV is shown in Figure 14. In Figure 15 the integral cluster, $E_c R_c$, distribution normalized by the total number of families is shown. $X_c = 40$ Tev.cm is used. Data are compared with simulation results of the models FeS, MSI, PF and PS. Mt. Fuji data is shown by the shadowed area, width shows statistical error. Both distributions favor a heavier primary composition and no scaling violation in the fragmentation region. The increase of average transverse momentum to 500 GeV/c will improve the agreement of the simulation with the experimental data.

Heavy primaries in the primary flux do not produce a large number of the observed families. In fact, 72% of the families with $\sum E_\gamma > 100$ TeV observed at Mt. Kanbala are estimated to originate from proton primaries and only 5% from iron primaries[51]. This emphasizes the fact that the average behaviour of the experimental events is indeed determined by proton interactions with air nuclei and validates the opportunity to use emulsion chamber data to study such interactions at $\sqrt{s} \geq 1$ TeV.

One could extend the clustering procedure by increasing X_c to several times the average distance between the clusters in the first clustering attempt. The resulting groups of shower clusters, bigger than the original clusters are called 'cores'. Other conditions imposed on this step, such as requiring that at least 80% of the family energy is contained in the cores, increase the probability that the core structure is representative of the initial interaction and is not artifact of the selection process. The percentage of multiple core formation in families with $\sum E_\gamma > 100$ TeV is found to be 12% and that for double cores is 7.5%. According to the Fuji-Kanbala analysis[51], the core distribution emphasizes the difference between in-

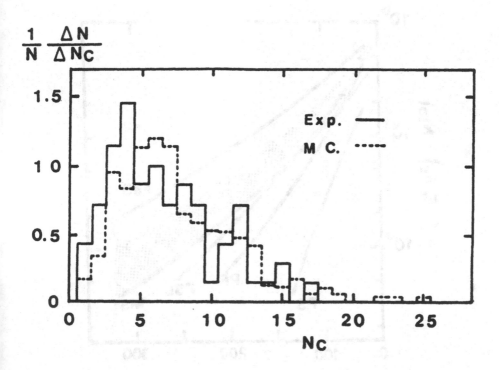

Figure 14: Distribution of number of clusters.

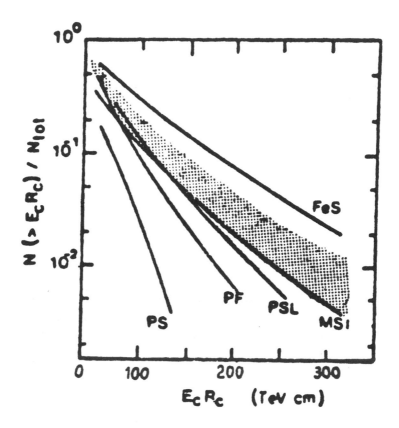

Figure 15: Normalized integral $E_c R_c$ distribution

teraction models. High multiplicity (scaling violation in the fragmentation region) models fail to produce multiple core events (5% of total statistics), while quasi-scaling models yield 11%, in good agreement with observations.

The X_{12} distribution for Fuji-Kanbala double core events is shown in Figure 16. High multiplicity models are ruled out by this distribution., because they fail to generate double cores with large values of X_{12}.

Quasi-scaling models perform much better, although the tail of the X_{12} distribution has to be explained by QCD jets, which contribute about 10% of the observed double core events. In contrast, the analysis of similar data by the Pamir group requires much larger QCD jet cross section to explain the tail of the distribution. Their analysis suggests that the probability for production of jets with $P_t >3$ GeV/c at $x_F=0.05$ is about $\frac{1}{2}$ per interaction above 1000 TeV[55].

3.2.6 Centauros and Mini-Centauros

Centauro events are A-jets which contain a small number of gamma-ray cascades and an extremely small ratio of electromagnetic to hadronic energy flow. The original Centauro event was observed by the Chacaltaya experiment. They have observed five more events of similar characteristics. The results are described in detail by Lattes, et. al.[5]. A family is qualified as Centauro if its total visible energy, $\sum E_\gamma$ is > 100 TeV and if the hadronic jets in the family carry at least half of the energy. Mini-Centauros are families with slightly smaller energy content ($E_{vis} > 50$ TeV) and contain unusually well developed hadronic component.

Such families can only be observed with thick chambers which can measure the energy carried by hadronic secondaries. Until recently six of a total of seven Centauro type events were observed by the Chacaltaya group, while the Pamir experiment only detected one. Now the two groups are jointly exposing thick emulsion chambers with 60 cm carbon absorber (or target, or spacer). One new Centauro event has been reported by the Pamir-Chacaltaya collaboration[58], out of a total of 41 families with $\sum E_{vis} > 100$ TeV, collected during an exposure of 164 m²yrs. Five more events show extremely well developed hadronic component[59].

Several independent Monte Carlo simulations [60,61] have been carried out to determine if Centauro type events can occur as fluctuations in the

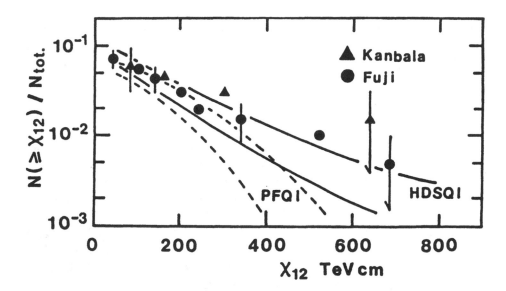

Figure 16: Integral distribution of double core events for the X_{12} variable.

family development. A comparison of data from Chacaltaya with simulations of Ellsworth, et. al.[60] is shown in Figure 17 and in Figure 18. Figure 17 shows the comparison of the calculated (line) and experimental distributions of the ratio of hadronic to electromagnetic energy. Note that the simulations can not account for events with $\sum E_h / \sum E_\gamma$ bigger than 10, where the Centauro events are concentrated. Figure 18 shows a scatter plot of $\sum E_\gamma$ versus $\sum E_h$. The dots are simulated families, circles represent Chacaltaya data.

The original Centauro events are clustered in the lower right region of the scatter plot, which is not populated by Monte Carlo points. Mini-Centauro, which are also included in this plot and in Figure 17, do not stand out on either plot and thus can be explained as being due to development fluctuations. The characteristics of the Centauro events are, however, too far from the average A-jet behaviour and simulations can not come close to reproducing them. On Figure 17 the new Pamir- Chacaltaya events are shown with crosses. One or may be two of them clearly show typical Centauro characteristics.

The visible energy of the Centauro events translates into a projectile energy of 2000 TeV or a $\sqrt{s}$ of 2 TeV. Extensive searches for Centauro type interaction have been performed without success at the Cern $\bar{p}$-p collider[62,63]. The Tevatron collider runs at the average Centauro energy and hadron rich Centauro type events should be observed at the collider. The cosmic ray event rate implies a rather large cross section for these events, several percent of the inelastic cross section. If no such events are found at the Tevatron collider we shall have to choose between the association of Centauro with nuclear target effects, or with development fluctuations and experimental biases that we do not take into account correctly or speculate on the existence of new particle physics processes of unusual nature.

3.3 TeV Muons

3.3.1 Single muon spectrum

We first discuss the depth-intensity curves measured by underground experiments. The decay channel, applied appropriately in cascade calculations,

328

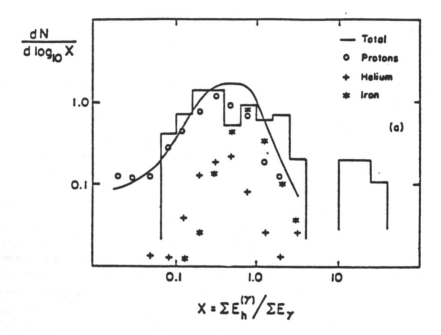

Figure 17: Comparision of simulation (solid curve) versus data (histogram) for the ratio of $\sum E_h$ to $\sum E_\gamma$.

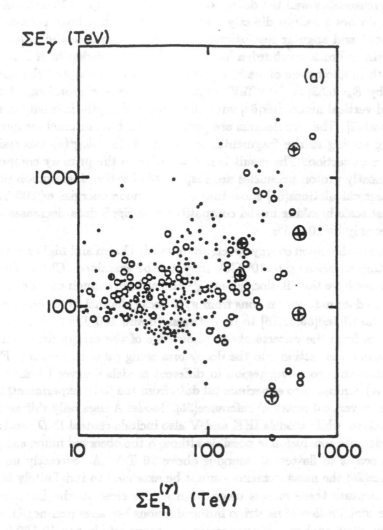

Figure 18: Scatter plot of $\sum E_\gamma$ versus $\sum E_h$ for simulated(dots) events and data (open circles).

describes reasonably well the detected fluxes of muons up to 10 TeV. Most detectors do not measure directly the muon energy but have reasonable good spacial and angular resolution to be able to correlate the muon direction with the rock overburden for the particular slant-depth. If T is the slant depth in kilometers of water equivalent then the energy of the muon is given by $E_\mu = 0.53(e^{0.4T}-1)$ TeV. Figure 19 shows a comparison of the calculated vertical muon flux[64] with the measured depth- intensity curve of reference[22]. The calculations are performed with an interaction model preserving scaling in the fragmentation region, with a $\log^2(s)$ increasing inelastic cross section. The result is not sensitive to the primary composition, as mostly proton primaries are responsible for the single muon flux. More recent calculations[65] show that even at muon energies of 100 TeV the biggest scale breaking model compatible with S$\bar{p}$pS data decreases the muon flux only by 10~15%.

The use of the muon energy spectrum to study charm and higher flavors at laboratory energies up to 10^6 GeV may now be examined. Charm decay muons do not have the dilation factor, $\frac{1}{E_\mu}$ and at high enough energies they will start to dominate over muons from π/K decay channel. This cross over energy is calculated[65,23,25] to be between 10 and 100 TeV. This uncertainty arises from the uncertainty in knowledge of the energy dependence of the charm cross section and the decay branching ratio into muons. Figure 20 shows the cross-over region in different models (curves labelled V, IKK and A) compared to experimental data from the KGF experiment[22], converted to vertical fluxes in reference[25]. Model A uses only diffractive Λ_C production, while models IKK and V also include central D-$\bar{D}$ production. No definite conclusion is possible although the observed muon energy spectrum seems to flatten at energies above 10 TeV. As correctly noted in reference[25] the measurements cannot be extended to indefinitely high energies, because these muons come from angles close to the horizontal where the contribution of neutrino induced muons becomes non-negligible, complicating the analysis. More precise measurements in the 10-100 TeV region are necessary to distinguish between models, which can only come from future large area detectors, such as the ones under construction in Gran Sasso[30].

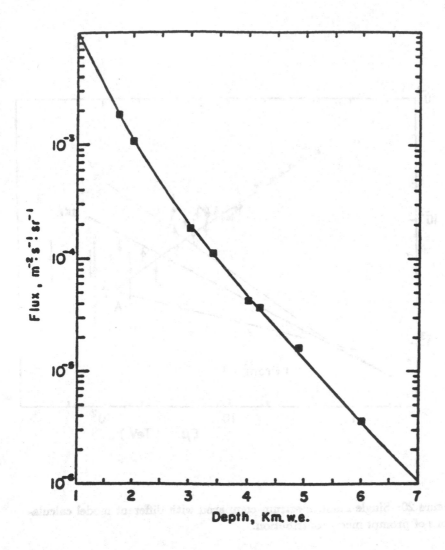

Figure 19: Comparison of calculated vertical muon fluxes and depth-intensity measurements.

332

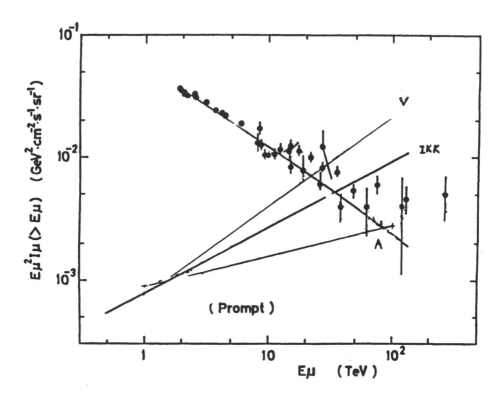

Figure 20: Single muon spectrum compared with different model calculations of prompt muon contribution.

3.3.2 Multiple muon groups.

Study of multiple muons in underground detectors has sensitivity to higher energies than that for single muons, above 1000 TeV. Muon groups provide one more parameter in addition to the muon intensity and energy spectrum - the lateral distribution of muons within the group. Multiple muon signal, however, is sensitive to primary composition, complicating the analysis.

Heavier nuclei, because of their large interaction cross section interact higher in the atmosphere and therefore are more prolific in generating muon groups once their energy per nucleon exceeds the muon energy. Thus the intensity of muon groups is a strong function of composition. Higher interaction altitude of heavy primaries also increases the lateral spread of a muon group by a geometrical factor related to the ratio of the mean free path of protons and nuclei. One has to be extremely careful in comparing calculations with experimental results on muon groups, because many existing detectors have dimensions only a few times the average muon separation, and the detected muon multiplicity is significantly smaller than incident muon multiplicity. One has to account for the dimension limitations even in big detectors, where edge effects can be important. Despite these problems one can separate to a certain extent through Monte Carlo simulations of atmospheric cascades the influence of the two factors.

An example of study of decoherence curve between pairs of muons in muon groups is shown in Figure 21. It compares the Monte Carlo calculations of reference[23] with data from the Homestake experiment[24]. The ordinate represents the lateral spread in muon groups calculated as the rate of R coincident events per unit time in two idealized small detectors of areas A_1 and A_2 separated by distance r; $D(r) = \frac{R}{A_1 A_2}$. Dots are experimental points and the boundaries of the shaded area are calculations for two distinctly different composition models. The two lines show the same calculated results when all muon distance are scaled with a factor of 50%, which is more than needed to account for the increase of average transverse momentum. Independently of the very high normalization, which this experiment requires (and which indicates a very heavy primary composition), the decoherence curve shows noticeable flattening at large distances.

It is difficult to relate this flattening to muons of prompt origin, because their contribution to the decoherence curve should be small. Let P_n

334

be the probability for a bundle of multiplicity n and P'_n be the probability for a bundle of same multiplicity containing one prompt muon. For Poisson distributions, $\frac{P'_1}{P_n} = \frac{P'_n}{<N_\mu>}$ and for small values of $< N_\mu >$, P'_1 is approximately equal to $< N_\mu >$. Then $P_n = P'_n$ at $n = \frac{<N_\mu>}{<N'_\mu>}$. Looking at the cross-over region in Figure 21 one could estimate the influence of prompt muon production to start affecting muon groups at energies above 10 TeV.

The flattening of the decoherence curve was recently confirmed by the Baksan underground telescope. The big statistics collected in this large detector has allowed the authors[26] to study the decoherence as function of muon threshold energy E_{th}, determined from the arrival direction of the event. The decoherence curves are fitted with lateral distribution of the form

$$f(r) = \int_{E'_{th}}^{\infty} E^{-0.7} \exp\left[-\left(\frac{rE\cos\theta}{r_o E'_{th}}\right)^{0.73}\right] dE$$

where E'_{th} is effective threshold energy, a parameter of the fit. For a constant cosmic ray composition, cross section and transverse momentum $E'_{th} \times r_o$ should be a constant, while the data show it rising approximately as $E^{0.2}$ which corresponds roughly to cross section increase of $E^{0.1}$ or transverse momentum must increase about 25% per decade of lab energy. Even taking into account a more reasonable cross section increase of $E^{0.05}$ (which corresponds roughly to a $\log^2(s)$ rise of the p-p inelastic cross section), the average transverse momentum is required to increase at a very rapid pace. An explanation must involve a change of the type of muon parents (e.g. increasing K/π ratio), or alternatively an increasing atomic mass of the primary composition.

3.4   Cross Section from Air Showers

Although the complex cascade of air showers obscures the study of individual interactions, it is possible to extract the interaction cross section of particles that initiate the showers by the study of fluctuations in the location of the point of first interaction. Two methods that have been used are:

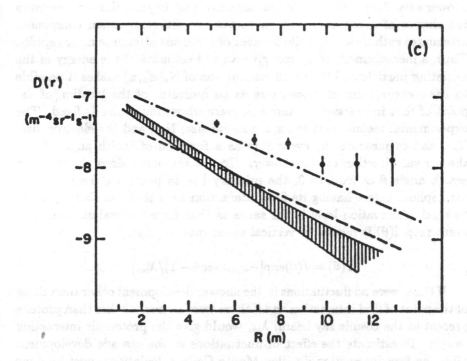

Figure 21: Decoherence curves; experiment and calculations, see text

1. Zenith angle distribution of air showers with fixed muon and electron sizes[12] and

2. Fluctuations in the depth of shower maximum [34,66,67,35,11].

In the first method one uses the fact that the longitudinal development of the muon component of a shower, $N_\mu(x)$, is essentially different from that of the electron component, $N_e(x)$, where x is the depth along the shower axis. Monte Carlo calculations show that beyond shower maximum the shapes of muon and electron curves are different, muon component attenuates rather slowly, while number of electrons attenuate more rapidly. Thus, a measurement of N_μ can give a good estimate of the energy of the initiating particle and the rapid attenuation of $N_e(E_0, x)$ makes it possible to use measurement of shower size as an indicator of the location of the point of first interaction for those showers whose muon size is fixed. The experimental technique is to fix the muon size, N_μ , and the electron size, N_e , and to measure the event rate as a function of zenith angle, θ , of the arrival direction of the shower. To get the same sized shower from zenith angle θ as from $\theta=0$, the primary has to penetrate deeper in the atmosphere before having its first interaction such that the slant distance to reach observation level is the same as that for a vertical shower. The event rate, $I(\theta)$ is related to vertical event rate $I(0)$ by:

$$I(\theta) = I(0)exp[-x_{obs}(sec\theta - 1)/\lambda_{obs}]$$

If there were no fluctuations in the shower development other than those of the point of first interaction and if there were no nuclei other than protons present in the cosmic ray beam, λ_{obs} would give the proton air interaction length. To estimate the effect of fluctuations in the cascade development after the first interaction detailed Monte Carlo calculations must be done to relate λ_{obs} to λ_{p-air}. In general, $\lambda_{obs} > \lambda_{p-air}$ since fluctuations tend to make for deeper penetrations of showers[13].

The Akeno group[12] has used this method to determine proton air interaction cross section above $\sqrt{s} > 5$ TeV. They have minimized the contamination due to nuclei heavier than protons by selecting showers with $N_e > N_{th}$ independent of zenith angle. This is so because for a fixed value of N_μ, air showers initiated by primaries heavier than protons give rise generally to a shower of a smaller size at deep observation depths than that due to protons, an effect emphasized by steep energy dependence of cosmic ray spectra. They have determined σ_{p-air}^{inel} from $\sqrt{s} \sim 5$ to 40 TeV. The values vary from 540 mb at low energy end to about 650 mb at high energies. This corresponds to about 100% increase in the cross section above the black disc value of 270 mb !

The second method uses the correlation between the depth of maximum and the point of first interaction. Several different techniques have been used to locate the depth of maximum of showers. These may be classified into two classes, A and B. Class A method is one in which the full longitudinal profile of showers is measured for each individual event and the location of shower maximum determined for each shower. This is the technique used by the Fly's Eye experiment[11] which measures the nitrogen fluorescence emitted along the track of the cascade through the atmosphere. The fluctuations in shower maximum location, X_m , can also be inferred from measurements of shower parameters made by observations of a slice of the shower at a fixed depth which are sensitive to longitudinal development, such as, pulse rise time[35,34], lateral distribution function[66] or fluctuations in Cherenkov light energy flow[67]. These indirect methods constitute the type, class B. We will only describe the results from class A experiment, as in this experiment it is possible to enrich the air shower sample in protons.

The Fly's Eye experiment reported a study of 1366 event sample of showers above about 10^{17}eV. For these well reconstructed showers longitudinal profile of each event was analyzed to determine the location of shower maximum to better than an accuracy of 100 gm/cm^2. The depth distribution of X_m is shown in Figure 22. The distribution beyond 800 gm/cm^2 is fitted to an exponential, exp(-X_m/Λ_m). This cut enhances the contribution from proton initiated showers as protons have the longest interaction length of all incident nuclei and therefore penetrate deepest. The value of the decrement Λ_m was found to be 73$\pm$9 gm/cm^2. This basic measurement must be related to proton-air interaction length and then to σ_{p-air}^{inel} by Monte Carlo simulations.

The simulations, again, need models to extrapolate the behaviour of proton-proton interactions to energies higher than currently available at colliders (beyond $\sqrt{s} \sim 900$ GeV). They also require models to relate p-p

338

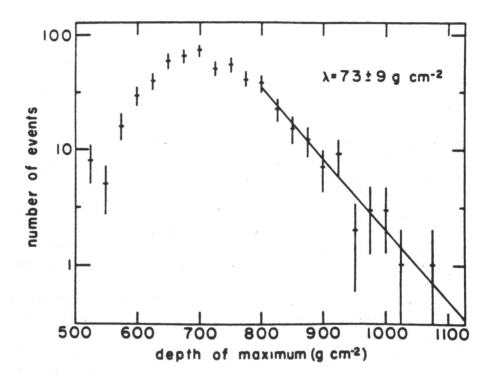

Figure 22: The distribution of depth of shower maxima for well reconstructed events.

interactions to p-air interactions. It is especially important to know how big is the violation of scaling in the projectile fragmentation region of the interaction. An elegant formula relating scaling violation in the central and fragmentation regions that has been suggested[68] uses a new scaling variable $x[s/s_o]^\alpha$ with the invariant cross section given by:

$$\frac{2E}{\sqrt{s}}\frac{1}{\sigma}\frac{d^2\sigma}{dxdp_t^2} = K(s,s_o)[s/s_o]^\alpha f(x[s/s_o]^\alpha,p_t)$$

This formula was used to fit the $S\bar{p}pS$ data (at $\sqrt{s}$ from 200 to 900 GeV) with $K(s,s_o)$ and α as parameters of the fit[69]. It works quite well with $\alpha = 0.25$ and $K(s,s_o) = (s/s_o)^{-.14}$. This fit has very strong energy dependence of the inelasticity coefficient K, which measures the fraction of the proton energy which goes into production of secondary particles. It would lead to an average proton elasticity of 0.85 at high energies. Other models for extrapolating to cosmic ray energies do not have this strong a violation of scaling in the fragmentation region. The results one can derive for the behaviour of proton-proton inelastic cross sections from cosmic ray data depend on these extrapolations. Stronger the violation of scaling that is assumed smaller is the value of cross section for proton-proton collisions at asymptotic energies that one derives. We discuss, briefly if such a model agrees with our understanding of mini-jet cross section in a QCD type model.

In our opinion, this strong increase of the proton elasticity serves to compensate for the overproduction of particles when such large amounts of energy are transferred from the fragmentation to the central region of the interaction. Indeed, one does not need to do that to understand the experimentally observed central rapidity density. In the mini-jet approach[70], for example, energy is transferred from the beam jets to the central region by QCD parton jets of $p_t \geq 1$ GeV/c. In a Monte Carlo program based on this approach[71] only 5% of the beam jet energy, transferred to the central region by mini-jets of $p_t > 1.4$ GeV/c, increase the central rapidity density from 2 charged particles per rapidity unit at $\sqrt{s} = 100$ GeV to 6 at $\sqrt{s} = 40$ TeV. Although the mini-jet cross section is very large, scaling violation in the fragmentation region is small because of the steepness of parton structure functions.

340

Nuclear target effects in interactions with air nuclei are not likely to seriously affect the production of fast secondaries. A recent perturbative QCD calculation[72] estimates that due to large formation time particles created at $y_A \approx 0.24 \ln A + 1.7$ in the laboratory system do not rescatter in targets with mass A.

The energy variation of σ^{inel}_{p-air} deduced from cosmic ray data[73] (above 10^6 GeV) is shown in Figure 23 which includes points based upon gamma family intensity at Mt. Fuji[6] as well as those from Akeno[12] and Fly's Eye [11] air shower experiments described above. The simulations used a model in which the violation of scaling in the fragmentation region was not extreme like in reference[68]. From these values of proton-air inelastic cross sections one derives σ^{tot}_{pp} by Glauber multiple-scattering technique[73] using a Chou-Yang relationship between the elastic scattering slope parameter and the p-p total cross section described in reference[73]. The value of σ^{inel}_{p-air} obtained by Fly's Eye experiment[11] corresponds to $\sigma^{tot}_{pp}=175^{+40}_{-27}$ mb at $\sqrt{s} \sim$ 30 TeV.

4   Summary

Cosmic ray experiments have provided a window on behaviour of particle collision properties above current collider energies. Analyses of these experiments are complicated and only gross features of multihadron interactions can be obtained. Steep cosmic ray energy spectrum, extremely high laboratory energies and limited capabilities of experimental techniques make most of these experiments sensitive to the projectile fragmentation region, a region which is hard to study at colliders. Certain special features make these studies important, such as: the study of individual interactions of heavy nuclei with energies up to several TeV/amu, study of proton-nucleus collisions up to $\sqrt{s} \sim$ 40 TeV, use of tagged beams from compact astrophysical sources to investigate photon interactions at $\sqrt{s} \sim$ 1 TeV. Some of the most interesting results so far obtained are summarized .

- Cross section shows no sign of saturating up to the very highest energies, $\sqrt{s} \sim$ 40 TeV.

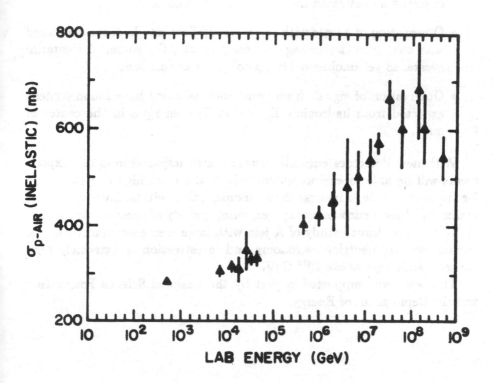

Figure 23: Energy variation of proton-air inelastic cross section.

342

- Average p_t continues to increase up to energies that can be studied by emulsion chambers and underground muon detectors. This is consistent with the continued increase of QCD jet cross sections.

- Large rapidity density fluctuations and energy densities have been observed in TeV/amu nucleus-nucleus collisions.

- Observation of events with large energy flow into hadrons compared with that into electromagnetic component , the so called Centauro events, as yet unobserved in $\bar{p}p$ collisions at colliders.

- Observation of signals from point sources which have muon content expected from hadronic collisions at TeV energies in the center of mass. 3

With new detectors currently under construction cosmic ray experiments will be able to explore quantitatively the ultra high energy regime for the next decade. Amongst the phenomena that will be investigated are search for objects such as monopoles, wimps; study of cores of air showers at 10^7 GeV, continued study of A-jets with large area emulsion chambers, gamma ray and neutrino astronomy and investigation of extremely high energy cosmic rays above 10^{10} GeV.

This work was supported in part by the National Science Foundation and the Department of Energy.

References

[1] T. K. Gaisser and G. B. Yodh, Annual Reviews of Particle and Nuclear Science, 30, 475-542, 1980.

[2] V. K. Balasubrahmanyan, B. V. Sreekantan and G. B. Yodh, Proc. of the Symposium on Cosmic Ray Super High energy Interactions, Beijing, PRC, 1986, edited by Ding Linkai, Huang Haohuai and Ren Jingru, pp 6-1 to 6-21, 1986.

[3] E. L Feinberg, Phys. Reports, 5, 237, 1972.

[4] G. B. Yodh, Y. Pal and J. S. Trefil, Phys. Rev. Letters, 28, 1005-1008, 1972; and Phys. Rev. D8, 3233-3236, 1973,

[5] C. M. G. Lattes et. al.,Physics Reports, vol 65, no. 3, 151-229, 1980.

[6] M. Akashi, et. al., Phys. Rev. D24, 2353, 1981(this paper discusses composition); and M. Amenomori, et. al., Phys. Rev., D25, 2807, 1982 (this paper discusses interactions).

[7] Pamir Collaboration; Proc. of the Symposium on Cosmic Ray Super High energy Interactions, Beijing, PRC, 1986, edited by Ding Linkai, Huang Haohuai and Ren Jingru, pp 3-1 to 3-23, 1986.

[8] China-Japan collaboration; J. R. Ren, et. al., Proc. of the Symposium on Cosmic Ray Super High energy Interactions, Beijing, PRC, 1986, edited by Ding Linkai, Huang Haohuai and Ren Jingru, pp 3-24 to 3-36, 1986.

[9] T. H. Burnett, et. al., Phys. Rev. Letters, 51, 1010, 1983; and T. H. Burnett, et. al., 19th International Cosmic Ray Conference, La Jolla, CA, 1985, edited by F. C. Jones, J. Adams and G. M. Mason, (NASA conference publication 2376), 2, 48, 1985.

[10] UA7 collaboration;M. Haguenauer, et. al., 20th Int. Cosmic Ray Conference, Moscow, USSR, 1987, edited by V. L. Kozyarivsky, et. al.,(NAUKA publication), vol. 5, 23-26, 1987.

[11] R. M. Baltrusaitis, et. al., Phys. Rev. Letters, 52, 1380-1383, 1984.

[12] T. Hara, et. al.,Phys. Rev. Letters, **50**, 2058,1983.

[13] G. B. Yodh, Annals of the New York Academy of Sciences, **461**, 239-259, 1986

[14] L. W. Jones, Rev. Mod. Phys., **49**, 717, 1979.

[15] S. C. Tonwar, et. al., J. of Phys,**A5**, 569, 1972; see also S. C. Tonwar, et. al.,Pramana, **8**, 50, 1977.

[16] A.I. Mincer, et. al., Phys. Rev., **D32**, 541, 1985.

[17] R. J. Protheroe, Rapportuer talk, 20th Int. Cosmic Ray Conference, Moscow, 1987.

[18] D. E. Nagle, T. K. Gaisser and R. J. Protheroe, Rev article on UHE gamma ray astronomy, to appear in Annual Reviews of Particle and Nuclear Science, 1988

[19] B. L. Dingus, et. al; Phys. Rev. Letters, **60**, 1928, 1988.

[20] G. B. Yodh, Proc. of the Int. Conference on Quark Gluon Plasma, Bombay, India, Feb 1988 (to be published).

[21] T. H. Burnett, et. al.,Nuclear Inst. and Methods, **A251**, 583, 1986.

[22] M. R. Krishnaswamy, et. al.,15th Int. Cosmic Ray conference, Plovdiv, Bulgaria, **6**, 85, 1977.

[23] J. W. Elbert, et. al., Phys. Rev., **D27**, 1448,1983.

[24] M. L. Cherry, et. al., Phys. Rev., **D27**, 1444,1983.

[25] H. Inagawa, et. al., J. Phys. G: Nuclear Phys., **12**, 59, 1986.

[26] E. V. Budko, et. al., 20th Int. Cosmic Ray Conference, Moscow, USSR, 1987, edited by V. L. Kozyarivsky, et. al.,(NAUKA publication), vol. 6, 221, 1987.

[27] B. Degrange, and S. Tisserant, 10th European Cosmic Ray Symposium, Bordeaux, France, edited by J. N. Capdeville, 1986, to be published. See also, Frejus Collaboration, 20th Int. Cosmic Ray Conference, Moscow, USSR, 1987, edited by V. L. Kozyarivsky, et. al., (NAUKA publication), vol. 6, 216, 1987.

[28] G. Battistoni, et. al.,Proc. 18th. Int. Cosmic Ray Conference, Bangalore, India, edited by N. Durgaprasad, et. al., Indian National Document Center publication, 11, 466,1983.

[29] J. Bartelt, et.al., Phys. Rev. Letters, 54, 19,1985; M. L. Marshak, et. al., Phys. Rev. Letters, 54, 1965, 1985; K. Ruddick, et. al., Phys. Rev. Letters, 57, 531,1986.

[30] B. Barish, Proc. of Japan-US seminar on Cosmic Ray Muon and Neutrino physics/astrophysics, ed. by Y. Ohashi and V, Z. Peterson, Inst. of Cosmic Ray Research Publication, Tokyo, 176, 1986.

[31] G. B. Yodh, Lectures at the ICFA school on Instrumentation for High Energy Physics, edited by C. Fabjans, to be published by World Scientific, 1987.

[32] G. Thornton and R. W. Clay, Phys. Rev. Letters, 43,1622,1979; and Phys. Rev. D23, 2090, 1981.

[33] A. A. Andam, et. al., 17th Int. Cosmic Ray Conference, Paris, France, 11, 281, 1981.

[34] K. J. Orford and K. E. Turver, Nature, 264, 727, 1976.

[35] R. Walker and A. A. Watson, J. of Phys. G7, 1297, 1981.

[36] B. S. Acharya, et. al., Proc. 18th. Int. Cosmic Ray Conference, Bangalore, India, edited by N. Durgaprasad, et. al., Indian National Document Center publication, 9, 162,1983.

[37] G. B. Khristiansen, et. al.,16th Int. Cosmic Ray conference, Kyoto, Japan, 14, 360, 1979.

[38] J. N. Stamenov, et. al., 16th Int. Cosmic Ray Conference, Kyoto, Japan, **8**, 330, 1979.

[39] Yodh G. B., et. al., Proc. 18th. Int. Cosmic Ray Conference, Bangalore, India, edited by N. Durgaprasad, et. al., Indian National Document Center publication, **6**, 70,1983.

[40] B. Dingus, et. al., Proc. of 2nd Int. Conf. on the Intersection between particle and Nuclear Physics, Lake Louise, Canada, AIP Proc. No. 150, Donald Geesman, editor, page 1078, 1986.

[41] B. V. Sreekantan, et. al.,Phys. Rev. **D26**, 1050,1983.

[42] J. A. Goodman, et. al., Phys. Rev., **D26**, 1026, 1982.

[43] R. M. Baltrusaitis, et. al., Nucl. Inst. and Methods, **A240**, 410, 1985; Phys. Rev. **D31**, 2192 ,1985.

[44] T. Hara et. al.,Phys. Rev. Letters, **50**, 2058, 1983.

[45] S. Fukui, et. al.,Prog. of Theor. Phys. Suppl., **16**, 1, 1960.

[46] T. H. Burnett, et. al., Phys. Rev. Letters, **50**, 2063,1983.

[47] T. Stanev, Phys. Rev.,**D31**, 101, 1985.

[48] T. H. Burnett, et. al., Phys. Rev. Letters, **57**, 3250, 1986.

[49] J. D. Bjorken, Phys. Rev., **D27**, 140, 1983

[50] T. H. Burnett, et. al., Proc. of the Symposium on Cosmic Ray Super High energy Interactions, Beijing, PRC, 1986, edited by Ding Linkai, Huang Haohuai and Ren Jingru, p 7-17, 1986.

[51] J. R. Ren, et. al., Institute of Cosmic Ray Physics of the University of Tokyo internal report; ICR Report No:153-87-7, 1987.

[52] E. H. Shibuya, Rapportuer talk, 20th Int. Cosmic Ray Conference, Moscow, 1987.

[53] S. G. Bayburina, et. al., Nuclear Phys. , **B191**, 1-25,1981.

[54] S. G. Bayburina, et. al., Lebedev Physical Instittue report, Trudy, FIAN, v154, 1984.

[55] A. M. Dunaevsky, et. al., Proc of Int. Symposium on Cosmic Rays and Particle Physics, Tokyo, 1984, (Institute of Cosmic Ray Physics at the University of Tokyo), p184, 1984.

[56] S. Dunaevsky, Proc. of the Symposium on Cosmic Ray Super High energy Interactions, Beijing, PRC, 1986, edited by Ding Linkai, Huang Haohuai and Ren Jingru, p 7-17, 1986.

[57] G. B. Yodh, Proc of the Int. Conf. on the Intersection of Particle and Nuclear Physics, Steamboat Springs, Colorado, 1984.

[58] Pamir-Chacaltaya joint Collaboration, Phys. Letters, B190, 226-233, 1987.

[59] Pamir-Chacaltaya joint Collaboration, 20th Int. Cosmic Ray Conference, Moscow, USSR, 1987, edited by V. L. Kozyarivsky, et. al.,(NAUKA publication), vol. 5, 334, 1987.

[60] R. W. Ellsworth, et. al., Phys. Rev., D23, 771,1981.

[61] B. S. Acharya, et. al., Phys. Rev., D24, 2437, 1981.

[62] G. Arinson, et. al., Phys. Letters, 122B, 189, 1983.

[63] J. G. Rushbrooke, Proc. of Int. Europhysics , edited by L. Nitti and G. Preparata, (Laterza, Bari,1985),p 839, 1985.

[64] T. K. Gaisser and T. Stanev, Nucl. Inst. and Methods, A235, 183, 1985.

[65] L. V. Volkova, et. al., Il Nuovo Cim., 10C, 465, 1987.

[66] R. N. Coy, et al.,Proc of Workshop on Very High Energy Interactions of Cosmic Rays, edited by M. Cherry, K. Lande and R. L. Steinberg, (Univ. of Pennsylvania Press), 120, 1982.

[67] M. N. Dyakonov, et. al.,17th. Int. Cosmic Ray Conference, Paris,(Centre d'Etudes Nucleares de Saclay, Gif-Sur-Yvette, France), 6, 43,1981; 18th Int. Cosmic Ray Conference, Bangalore, India, ed. by N. Durgaprasad, et.al.,(Indian National Scientific Documentation Center, New Delhi, India),6,110,1983.

[68] J. Wdowczyk and A. W. Wolfendale, Nature, 306, 347,1983; J. Phys. G10, 257, 1983; Nuovo Cim., 54A, 433,1977.

[69] P. Carlson, Proc. of Euoropean Cosmic Ray Symposium, Bordeaux, France, edited by J. N. Capdevielle, 149, 1986.

[70] T. K. Gaisser and F. Halzen, Phys. Rev. Letters, 54, 1754, 1987.

[71] T. K. Gaisser and T. Stanev, Proc. of Workshop on Radiation in SSC experimental Halls, Lawrence Berkeley Lab, Ca, USA, 1988 (unpublished).

[72] E. M. Levin and M. G. Ryskin, LINP preprint No:1246, Leningrad, 1987.

[73] T. K. Gaisser, U. P. Sukhatme and G. B. Yodh, Phys. Rev.,D36, 1350, 1987.

Theory

THE POMERON IN QCD

Alan R. White

High Energy Physics Division

Argonne National Laboratory

Argonne, IL 60439

Abstract

An analysis of the Pomeron in QCD is presented which begins with the reggeon diagrams of the spontaneously broken theory. Infra-red divergences which occur when SU(2) gauge symmetry is restored produce a winding-number condensate which is identified with the Pomeron condensate of the Super-Critical Pomeron. The Critical Pomeron is produced as the gauge symmetry is restored to SU(3) if the asymptotic freedom constraint on the number of fermions is saturated.

Section 1: Introduction

The "Pomeron" came to the forefront[1] of strong-interaction physics during the 60's and early 70's as part of the phenomenology developed to codify and correlate the wide-class of multiparticle phenomena observed at the Fermilab and CERN accelerators.[2] It is therefore particularly appropriate to the purposes of this volume to discuss the current understanding of the Pomeron in terms of the now established "theory" of the strong interactions—that is QCD. There is a widely held opinion that the Pomeron, and diffraction phenomena in general, is simply a phenomenological problem that will be understood with sufficient effort in QCD but which is unlikely to provide significant new insight into the theory. As I shall elaborate in this article my own view is essentially the polar opposite of this negative viewpoint.

It was always thought that because the Pomeron, as a Regge pole, carries the quantum numbers of the vacuum it should be intimately related to the vacuum properties of a theory. Since the vacuum properties of QCD are believed to be quite complex it should therefore be expected that tracking down the Pomeron is comparably complicated. I hope that my analysis demonstrates that the nature of the Pomeron is a very important, if not the most important, consistency constraint on a theory whose study can actually provide a direct route to the underlying "vacuum properties" of the theory. (Note that the critical dimension of string theories was first discovered[3] by studying the unitarity properties of the "Pomeron".) The Pomeron describes interactions in a mixed infra-red and ultra-violet region of phase-space. It is also subject to very precise unitarity constraints—the cross-channel constraint of reggeon unitarity[4,5] in particular. Within QCD we can expect therefore that the Pomeron will make contact with the infra-red properties of confinement and chiral symmetry breaking, with the ultra-violet parton model and asymptotic freedom properties, and will confront the combination of these properties directly with unitarity.

One of the most obvious questions that we would expect a study of the Pomeron in QCD to answer is—does QCD explain rising cross-sections? Over the years there have been many attempts to explain the physical origin of rising cross-sections, for example high-mass diffraction[6] baryon-pair production,[7] the eikonal expanding disc,[8] and hard-parton scattering[9]—the latest version being[10] "QCD minijets". The presence of vector gluon exchange in QCD certainly seems to provide a potential explanation in that the cross-section given by single vector gluon exchange is energy-independent and higher-order exchanges easily provide a logarithmic increase (which is essentially the minijet increase). Unfortunately this argument, like all such "explanations" of rising cross-sections, suffers from the criticism that large "unitarity corrections" to the basic process proposed as an explanation, cannot be controlled.

The Critical Pomeron was discovered[11] well over a decade ago as a strong-coupling solution of the Reggeon Field Theory. It remains the only known description of *asymptotically* rising cross-sections which completely satisfies all cross-channel (and direct-channel) unitarity constraints. In deriving the conditions under which the Critical Pomeron occurs in QCD we effectively argue that the vector gluon is indeed the explanation of (asymptotically) rising cross-sections but only under very special circumstances. That is the Critical Pomeron occurs under the special condition that a smooth limit can be taken—giving the high-energy behavior of QCD—in which a "physical" gluon state becomes massless and decouples from the true physical states. In this sense the rising cross-section is due to the fact that vector gluons are "almost" in the physical spectrum. Or alternatively at asymptotic energies gluons are "almost deconfined". However, part of our conclusion is that the true asymptotic scale for Critical Pomeron behavior is so large that it is not a relevant description of currently observed rising cross-sections.

The program which this article describes is an extensive one which I have been pursuing (with varying intensity) over many years. The nature of this volume determines that only an outline of the variety of ideas and techniques needed for the program can be presented. The absence of any exposition of multiparticle complex angular momentum theory will be particularly frustrating to the reader since it underlies the major features of the analysis we describe. A much more detailed review is long overdue, and indeed I have been promising one for some time. I do have the basic material assembled and it is my intention to publish a review as soon as possible. I am acutely aware of the inadequacy of the kind of overview presented here as a justification for my claim to an understanding of the major theoretical problem posed by the "Pomeron in QCD".

There are many ways of formulating the Reggeon Field Theory (RFT) of the Pomeron. To set the subject in the context of this volume I shall begin in Section 2 by briefly reviewing the phenomenological formulation of RFT as an effective field theory of multiplicity fluctuations. The Critical Pomeron then appears as a "phase-transition" phenomenon producing KNO scaling and many other diffractive scaling laws. We describe the nature of the *Super-Critical phase* in Section 3. A *Pomeron condensate* develops and a *vector* Regge trajectory degenerate with that of the Pomeron *enters the theory*.

We briefly describe the derivation of gluon RFT for spontaneously broken Yang-Mills theories in Section 4. The singular nature of the reggeon interactions prevents any direct application of the renormalization group. Infra-red analysis of reggeon diagrams as the gauge symmetry of spontaneously broken QCD is restored to SU(2) is described in Section 5. Confinement is shown to arise from a gluon condensate which in turn is identified with the Pomeron condensate of the Super-Critical Pomeron theory. The full structure of the super-critical phase is also shown to emerge—the vector trajectory degenerate with the Pomeron is that of an SU(2) singlet gluon. The gluon condensate is identified as a winding-number condensate in Section 6 and its physical significance is elaborated on by comparing with the massless Schwinger model.

The occurrence of the Critical Pomeron as the full SU(3) gauge symmetry of QCD is restored is discussed in Section 7. To avoid dependence on the transverse momentum cut-off it is necessary to have asymptotic freedom of the theory when the Higgs sector breaking the gauge symmetry to SU(2) is still present. This implies the Critical Pomeron occurs only when asymptotically free QCD is "saturated with fermions". As we briefly describe QCD saturated with fermions is a very special theory in many respects.

Section 8 is devoted to the possibility that color sextet quarks are responsible for dynamical electroweak symmetry breaking. When added to three families of color triplet quarks this would produce precisely the fermion saturation of QCD that we argue is required for the Critical Pomeron. the implications for strong and electroweak unification are also described very briefly.

Section 2: Multiplicity Fluctuations and Reggeon Field Theory

The starting-point here is[12] the idealization that "most" particles are produced in events where the particle production is spread uniformly across the rapidity axis (apart from short-range correlations) with a cut-off in transverse momentum. In impact parameter this production process can be described as a random walk and when the unitarity integral is computed for the elastic amplitude the output is automatically a Regge-pole— the Pomeron. This is pictured in Fig. 2.1. The representation of the production process can be thought of as a plot of the rapidity distribution or (essentially equivalently) as a multiperipheral amplitude. From this starting point therefore the Pomeron describes the bulk of the events generating the average multiplicity.

If we consider events which contain twice the average multiplicity of particles, also spread uniformly across the rapidity axis, then we expect these to generate two Pomeron exchange as illustrated in Fig. 2.2. Events with the average density of particles on half the rapidity axis and twice this density on the other half generate the Pomeron graph of Fig. 2.3. Similarly events with arbitrarily high fluctuations of the particle density on sections of the rapidity-axis generate comparably complicated Pomeron graphs as illustrated in Fig. 2.4. The simplest way to make this correspondence specific would be to use a simple field-theoretic picture of the Pomeron (strong-coupling $\lambda\phi^3$, for example). However, it is believed that the correspondence should hold under quite general circumstances.

The propagator for the Pomeron is that of a simple Regge pole

$$\Gamma(y,\underline{b}) = e^{-\Delta_0 y - \underline{b}^2/4\alpha' + \cdots},\tag{2.1}$$

or in angular-momentum $(j = 1 - E)$ and transverse momentum $(\underline{k})$

$$\Gamma(E, \underline{k}^2) = \left[E - \Delta_0 - \alpha'\underline{k}^2\right]^{-1}. \tag{2.2}$$

The interactions are, in first approximation, taken to be constants whose phase is determined by general arguments involving the role of absorption. For example, Pomeron absorption of the basic production process can be represented as in Fig. 2.5. Interference of this process with the basic production process also generates the two Pomeron graph, as does the *diffractive* process involving a rapidity-gap. Both processes are illustrated in Fig. 2.6. We shall not elaborate on this point but all field theory and string theory models, and general arguments, agree that the relative weights of these contribution are determined uniquely (the AGK cutting rules[12]). The full RFT includes all such contributions in addition to those with the simple multiplicity fluctuation interpretation discussed above.

The full RFT lagrangian is

$$\mathcal{L}(\bar\psi, \psi) = \bar\psi \overleftrightarrow{\tfrac{\partial}{\partial y}} \psi + \alpha' \nabla \bar\psi \nabla \psi + \Delta \bar\psi \psi + ir\left(\bar\psi^2 \psi + \bar\psi \psi^2\right) + \lambda\left(\bar\psi^3 \psi + \bar\psi \psi^3\right) + \cdots, \tag{2.3}$$

where $\bar\psi$ and ψ are respectively creation and destruction operators for the Pomeron and $\alpha_{\mathbf{P}}(t) = 1 - \Delta + \alpha' t + \cdots$. The Critical Pomeron[11] is a "strong-coupling", infra-red fixed-point, solution of the theory defined by (2.3) in which $\Delta = 0$, that is $\alpha_{\mathbf{P}}(0) = 1$, and in the infra-red region $(E, \underline{k} \to 0)$ all couplings beside α' and r are driven to zero. The scaling behavior of Pomeron Greens functions at this fixed-point then determines the high-energy Regge behavior of the total cross-section. The full set of scaling laws implied by the Critical Pomeron is described in several reviews.[13] Some examples are

$$\sigma_T \sim (\ln s)^\eta + \cdots \tag{2.4}$$

$$\frac{d\sigma}{dt} \sim (\ln s)^{2\eta} \left[f\left(t \ln^\nu s\right)\right]^2 \tag{2.5}$$

$$\sigma_y \sim y^\eta (Y - y)^\eta \tag{2.6}$$

$$\langle n^p \rangle \sim C_p \langle n \rangle^p + \cdots. \tag{2.7}$$

The last result (2.7) is KNO scaling and so a particular view of the Critical Pomeron which the above description provides is that it is a critical phenomenon description of the long-range multiplicity correlations implied by KNO scaling. That is RFT is an effective field theory (in impact parameter and rapidity) description of short-range multiplicity fluctuations in which KNO scaling represents a long-range "phase-transition effect".

A much more formal derivation of reggeon field theory is obtained by starting with the reggeon unitarity equations which follow directly from the application of multiparticle complex angular momentum theory.[5] That is it follows from the full set of multiparticle t-channel unitarity equations that if a theory contains a Regge pole with trajectory $j = \alpha_{\mathbb{P}}(t)$ then it must also contain an infinite set of Regge cuts with trajectories

$$j = \alpha_{N\mathbb{P}}(t) = N\alpha_{\mathbb{P}}\left(\frac{t}{N^2}\right) - N + 1. \tag{2.8}$$

Furthermore discontinuity formulae for the Regge cuts, closely analogous to momentum-space unitarity equations, must be satisfied.[4,5] RFT can be regarded as a (very general) Feynman diagram-like solution of these "reggeon unitarity equations".

From this last viewpoint the phenomenological significance of the Pomeron reggeon field theory is that it represents the full requirements of t-channel unitarity imposed on the phenomenological success[2,13] of a single Regge pole description of diffraction phenomena. The attraction of the Critical Pomeron is then that it both satisfies t-channel unitarity and describes asymptotically rising total cross-sections. It is worth emphasizing that the Critical Pomeron is the *only* known description of rising total cross-sections which does satisfy t-channel unitarity. The expanding-disc solution[14] obtained for $\alpha_{\mathbb{P}}(0) > 1$ has never been shown to satisfy reggeon unitarity and arguments have been given[15] which suggest very strongly that it does not.

Section 3: The Super-Critical Pomeron

An obvious question raised by the last Section is—what is the phase that the Critical Pomeron represents a phase-transition to? This appeared to be a very unphysical question when it was first raised. However, with the advent of QCD and the study of its phase-transitions it is clear that there are certainly candidate "physical" phase-transitions which the Critical Pomeron might describe. We shall return briefly to this topic in later Sections (it is discussed in much more detail in Ref. 16), for the moment we proceed to look for the new phase in the traditional manner by looking for a "new-vacuum" of RFT.

Near the critical point the Pomeron potential is

$$V(\bar{\psi}, \psi) = \Delta\bar{\psi}\psi + ir\left(\bar{\psi}^2\psi + \bar{\psi}\psi^2\right), \tag{3.1}$$

which has two stationary points at

$$\left(\bar{\psi}, \psi\right) = \left(0, \frac{-\Delta}{2ir}\right), \left(\frac{-\Delta}{2ir}, 0\right). \tag{3.2}$$

Shifting to the first stationary point, say, produces the new lagrangian

$$\mathcal{L}_s(\bar{\psi}, \psi) = \bar{\psi}\frac{\overleftrightarrow{\partial}}{\partial_y}\psi + \alpha'\nabla\bar{\psi}\nabla\psi - \Delta\bar{\psi}\psi - \Delta\bar{\psi}^2, \tag{3.3}$$

and so if the original "mass term" $\Delta\bar{\psi}\psi$ is negative (giving $\alpha_{\mathbf{P}}(0) > 1$) then this shift reverses the sign as we would require for a new vacuum. However, there is also a Pomeron "source" term $-\Delta\bar{\psi}^2$ and no corresponding "sink". The resulting asymmetry of Pomeron Greens functions would actually violate Lorentz invariance and so this field shift has to be rejected.[15] The appropriate vacuum shift is, in fact, to modify ψ on the negative rapidity-axis and $\bar{\psi}$ on the positive rapidity axis.[17] A complete derivation of the corresponding perturbation expansion involves some subtleties which we will not dwell on. We can briefly summarize the relevant issues as follows.

The reggeon field theory vacuum is not really a well-defined concept. For example, Pomeron Greens functions could all acquire constant components without changing any hadron amplitudes. Only the singularities of Greens functions are physically significant. Consequently it is necessary to determine the modification of the Pomeron perturbation expansion given by a "change of vacuum" as a (multiparticle) modification of the external hadron states in order to precisely determine the high-order perturbation theory diagrams. In practice this means defining "vacuum-shifted" Pomeron amplitudes as sums over the original multi-Pomeron amplitudes with zero energy and momentum for the additional external Pomerons, as illustrated in Fig. 3.1. The important point is that the multi-Pomeron amplitudes are defined from higher multiparticle amplitudes. To discuss this in detail would require the full formalism of multiparticle complex angular momentum theory which is in fact also required to give full details of the Yang-Mills analysis described in the following Sections. The vital feature is, however, that the relevant multiparticle pomeron amplitudes do not contain certain classes of diagrams which naively enter the vacuum-shifted perturbation expansion—for example those shown in Fig. 3.2. Such graphs would destabilize the new perturbation expansion if they were present.

The low-order diagrams are straightforwardly evaluated. They involve a Pomeron with $\alpha_{\mathbf{P}}(0) = 1 + \Delta$, and a two Pomeron *source and sink* (due to the Pomeron vacuum condensate) as illustrated in Fig. 3.3. The effect of the source and sink is to introduce transverse momentum singularities into the diagrams. For example, the one loop diagram of Fig. 3.3 has the form

$$\int d^2k \frac{1}{\left[E + 2\Delta - \alpha'\underline{k}^2 - \alpha'(\underline{k} - \underline{q})^2\right]\left[\alpha'\underline{k}^2 + \Delta\right]\left[\alpha'(\underline{k} - \underline{q})^2 + \Delta\right]} \tag{3.4}$$

which differs from the familiar two-Pomeron diagram by the presence of the additional "propagators" $\left[\underline{k}^2 + \Delta\right]^{-1}$ and $\left[(\underline{k} - \underline{q})^2 + \Delta\right]^{-1}$. In fact these propagators are just the particle poles, or "signature factors", to be expected from an odd-signature Regge pole and indeed they clearly generate a "two-particle" threshold at

$$q^2 = (2M_v)^2 = \frac{-4\Delta}{\alpha'}. \tag{3.5}$$

A complete analysis of the unitarity properties of the source (and sink) diagrams shows that we can characterize the Super-Critical Pomeron by

i) the presence of a "Pomeron condensate" (whose physical significance we have yet to determine),

ii) a Pomeron Regge pole with trajectory $\alpha_P(t) = 1 + \Delta + \alpha' \underline{k}^2 + \cdots$,

iii) an exchange degenerate odd-signature Regge pole with trajectory $\alpha_v(t) = \alpha_P(t)$ which produces a vector particle with mass

$$M_v^2 = -\frac{\Delta}{\alpha'}. \tag{3.6}$$

This last property suggests that to find a theory containing the Super-Critical Pomeron we should look for a theory containing a massive, reggeized, vector particle. The natural candidate is, therefore, a spontaneously broken gauge theory. This is, in fact, how we shall make contact with QCD. Indeed the appearance of a massive vector particle as the Pomeron becomes Super-Critical suggests that the corresponding phase-transition may involve *deconfinement* within QCD.

Section 4: Massive Yang-Mills Reggeon Field Theory

Using the Higgs mechanism to give all gluons a mass it is straightforward, in any Yang-Mills theory, to obtain the "leading logs" ($g^2 \ln s$) in the Regge limit.[18] It is by now well-known that such logarithms sum up to "reggeize" the elementary gluons and quarks of the theory. For example higher-order corrections to one gluon exchange give

$$\frac{s}{t - M^2} + g^2 K(t) \ln s + g^4 (t - M^2) K(t) \ln^2 s + \cdots \tag{4.1}$$

$$\equiv \frac{s^{\alpha(t)}}{t - M^2} \quad \text{with} \quad \alpha(t) = 1 + g^2 (t - M^2) K(t), \tag{4.2}$$

where

$$K(q^2) = \frac{1}{16\pi^2} \int \frac{d^2 k}{[(k-q)^2 + M^2][k^2 + M^2]} \tag{4.3}$$

From the work of Sen[19] we know that *all logarithms* generated in the Regge limit have as coefficients (in general very complicated) transverse momentum diagrams. From reggeon unitarity we expect to be able to organize all of these logarithms into a reggeon field theory of massive gluons (and quarks). Indeed *all* calculations of non-leading logarithms have so far borne out this expectation.[20,21] This RFT differs from that of the Pomeron in that singularities due to gluon particle states are also present. For example, the propagator is

$$\Gamma(E, \underline{k}^2) = \left[E - 1 + \alpha(\underline{k}^2) \right]^{-1} \left[k^2 - M^2 \right]^{-1}. \tag{4.4}$$

which differs from the Pomeron propagator by just the pole factor which we identified in (3.4) as due to a vector particle.

The lowest-order (gluon) reggeon interaction vertices[20,21] are known and also have a singular transverse momentum dependence (in contrast to the Pomeron vertices of Section 2). For example, the four-reggeon vertex has the form

$$\underline{k} - \underline{k}_2 \overset{\underline{k}_1}{\underset{}{\bigcirc}} \underline{k}_2 = \frac{(k_2^2 - M^2)(k_1^2 - M^2)}{(k^2 - M^2)} + \cdots. \tag{4.5}$$

As a result of such singular transverse momentum dependence all gluon reggeon interaction vertices have the same dimension. Consequently although it is formally possible to organize the high-energy behavior of spontaneously broken Yang-Mills theories into a reggeon field theory of reggeized gluons the renormalization group cannot be used to find a fixed-point involving only a small number of interactions (as was the case for the Pomeron theory). Equivalently we must find the Pomeron degrees of freedom before we can utilize the renormalization group in the high-energy Regge region. As we shall see in subsequent Sections this means we must first analyze the infra-red divergences, in transverse momentum, of reggeon diagrams. In effect the emergence of a Pomeron with the desired properties can then be viewed as a consequence of the restoration of gauge invariance.

The full infra-red analysis we describe in the next Section actually requires much more than the simple development of a (massive) gluon RFT for elastic scattering amplitudes. The formalism must be extended to quite complicated multiparticle amplitudes. The necessary multiparticle dispersion theory and complex angular momentum theory is developed in Refs. 2,22. It has been applied to the Pomeron RFT in some simple cases[23] and the necessary Yang-Mills formalism was developed in Ref. 24. (It will, of course, be described in my forthcoming review.) It is technically quite complicated and the details are, unfortunately, important dynamically. The appearance of the "anomalous" fermion vertices described in the next Section is clearly a vital feature. However, any attempt to describe the formalism would again take us far beyond the purposes of this article!

Section 5: Infra-Red Analysis of Reggeon Diagrams

To discuss the removal of the gluon mass in reggeon diagrams we must first consider whether we can expect to reach the unbroken gauge theory smoothly in the process. The lattice gauge theory principle of *complimentarity*[25] states that we can indeed go smoothly from the Higgs region to the confinement region of a gauge theory provided we use *fundamental representation* Higgs scalars. We assume that we can take this principle over to our reggeon diagram analysis if we use a transverse momentum cut-off in all the diagrams (the analogue of a lattice). Note that with the transverse momentum cut-off the reggeon diagrams should be (Borel) summable and effectively define a reggeon theory "non-perturbatively".

Using fundamental (triplet) Higgs scalars we can break the SU(3) gauge symmetry of QCD to SU(2) with one triplet vacuum expectation value and break it completely with a second such expectation value. We can therefore consider restoring the SU(3) symmetry of QCD in two stages:

i) SU(2) gauge symmetry is restored as a triplet of gluons having mass $M_1 \to 0$—the remaining five massive gluons form two SU(2) doublets and one singlet.

ii) SU(3) gauge symmetry is restored as, in particular, the SU(2) singlet mass $M_2 \to 0$.

We consider first the limit i) characterized by $M_1 \to 0$.

The starting point in analyzing the infra-red divergences of reggeon diagrams as $M_1^2 \to 0$ is to note that the exponentiation of reggeization is also an exponentiation of infra-red divergences, that is

$$S^{\Delta(k^2)} \underset{M^2 \to 0}{\sim} \exp\left[\frac{-g^2}{16\pi^2} \ln s \ln\left(\frac{k^2}{M^2}\right) + O(g^4)\right]. \tag{5.1}$$

At first sight this exponentiation sends *all* reggeon diagrams to zero! However, the reggeon interactions also produce exponentiating divergences which can potentially cancel those due to reggeization. The full divergence structure can be analyzed[18,21] by defining interaction "kernels" for fixed numbers of (elementary) gluons which have definite SU(2) color and which combine the reggeization and interactions e.g.

$$I = 0,1,2 \quad \boxed{K_I} \quad = \quad \cdots\Box\cdots + \cdots\Box\cdots + \cdots\bigcirc\cdots \qquad \cdots\Box\cdots = \frac{\Delta(k^2)}{k^2 - M^2} \tag{5.2}$$

It is found that reggeization dominates *all* color non-zero kernels, that is

$$K_I \underset{M^2 \to 0}{\longrightarrow} -\infty \qquad \text{all } I \neq 0. \tag{5.3}$$

Also any color-zero kernel

$$I = 0 \left\{ \begin{array}{c} k_1 \text{---} \\ k_2 \text{---} \\ \vdots \\ k_N \text{---} \end{array} \boxed{K_0} \begin{array}{c} \text{---}\, k_{N+1} \\ \text{---} \\ \vdots \\ \text{---}\, k_{2N} \end{array} \right. \tag{5.4}$$

diverges if any subset of all the transverse momenta go to zero. K_0 is finite (and scale-invariant) *only* if *all momenta* are taken uniformly to *zero*.

Further exponentiation can result from interaction with the remaining massive reggeized gluons (and quarks) of the theory. For example, if ∿ denotes the SU(2) singlet gluon which remains massive, exponentiation is produced by

$$\text{[diagram]} + \text{[diagram]} + \text{[diagram]} + \cdots \tag{5.5}$$

This removes all zero transverse momentum gluon configurations except for those without the appropriate local reggeon interaction. Since such interactions can be computed from a (multiparticle) dispersion relation it can be shown that there will be an interaction provided only that the gluon configuration involved can couple to on-shell physical states—that is

$$\text{on-shell states} \; \text{[diagram]} \; \left.\right\} \text{zero transverse momentum gluons} \neq 0. \tag{5.6}$$

This is the case *except* for the "unnatural" color charge parity configuration

$$\tilde{K}^0 \equiv \cdots\cdots \equiv \text{[diagram]} \; \left.\right\} \begin{array}{l} \text{color zero, } axial \text{ vector configuration} \\ \text{with } all \text{ transverse momenta zero} \end{array} \tag{5.7}$$

We claim that the infra-red divergence associated with the $\tilde{K}^\circ$ configuration of SU(2) gluons dominates the $M_1^2 \to 0$ limit and in effect produces *confinement*. To analyze its effect systematically it is necessary to go to multiquark amplitudes in which (potentially) bound-state quark reggeon states can be formed which can themselves scatter by exchanging multigluon configurations which will ultimately produce the Pomeron. For example, amplitudes of the form illustrated in Fig. 5.1 would be expected to give, in the appropriate region of phase-space, reggeized pions scattering by exchange of a Pomeron. Our argument is that the $\tilde{K}^\circ$ divergence selects the surviving amplitudes in the limit $M_1^2 \to 0$. That is reggeized "hadron" scattering amplitudes are selected out from the full multi-quark S-Matrix by this infra-red divergence. The resulting "hadrons" have as lowest approximation (in terms of quarks q and antiquark $\bar{q}$)

$$\begin{array}{ll} q & \overline{} \\ \bar{q} & \underline{} \\ & \cdots\cdots\cdots \end{array} \leftrightarrow \text{``meson''}, \qquad \begin{array}{l} q & \overline{\phantom{xxxxxxx}} \\ q & \underline{\phantom{xxxxxxx}} \\ q & \\ & \cdots\cdots\cdots \end{array} \leftrightarrow \text{``baryon''} \qquad (5.8)$$

and they scatter by exchange of the "Pomeron"

$$\begin{array}{c} \sim\!\sim\!\sim\!\sim \\ \cdots\cdots \end{array} \leftrightarrow \mathbb{P} \qquad (5.9)$$

A vital feature of the infra-red analysis is that the reggeon vertices coupling the "pion" and the "Pomeron" in Fig. 5.1 contain fermion loops in which some of the fermion lines are effectively placed on the mass-shell by the high-energy limits involved. As a result fermion loop anomalies occur in the vertices and it is this which enables the "anomalous" gluon configuration $\tilde{K}^\circ$ to couple. To elaborate on this feature (which is clearly the essential dynamics of our analysis) would unfortunately require the full apparatus of multiparticle Regge theory which we are avoiding presenting in this article. Another essential idea which we will not attempt to prove is that the $\tilde{K}^\circ$ divergences can be factorized off from the Pomeron and associated with the definition of the hadronic states. The zero transverse momentum of all gluons involved means that effectively we have a "gluon condensate" involved in our definition of hadrons as reggeons.

The simplest set of divergences actually generating pion scattering by the exchange of a Pomeron are shown in Fig. 5.2 with the fermion loop couplings also shown. More complicated diagrams of the form of Fig. 5.3 will also contain the $\tilde{K}^\circ$ divergences and these clearly generate the higher-order diagrams of the Super-Critical Pomeron. Indeed since the Pomeron is defined as the SU(2) singlet reggeized gluon together with the $\tilde{K}^\circ$ configuration it will have a *Regge trajectory degenerate with that of the gluon.*

A complete analysis of higher-order diagrams would be required to fully identify the Super-Critical Pomeron with spontaneously broken QCD. However, since all the features of the Super-Critical Pomeron are present we feel confident in assuming that

spontaneously broken QCD with the gauge symmetry restored to SU(2)—in the presence of a transverse momentum cut-off—gives the Super-Critical Pomeron

The transverse momentum cut-off may seem redundant in the above discussion. However, without it there would be an ambiguity in the infra-red limit of infinite sums of gluons. The most intriguing feature of the above is clearly that the (mysterious) Pomeron condensate of the Super-Critical Pomeron is apparently to be identified with a gluon condensate. In fact the gluon condensate has a rather straightforward physical significance which we now elaborate on.

Section 6: The Physics of the Gluon Condensate

In the Schwinger model (two-dimensional QED) physical states are meson bound-states and can be described as (using a suggestive notation)

$$\text{⎯⎯⎯⎯⎯⎯} \Big\} \text{electron-positron pair plus "photon" condensate} \qquad (6.1)$$

From Manton's analysis[26] it is clear that the photon condensate $\langle 0|A_x|0\rangle = \frac{1}{2}$ should be regarded as a winding-number condensate (in a particular gauge $\epsilon^{\mu\nu}A_\nu \equiv A_x$). Furthermore this condensate can be viewed as resulting from regularizing the *massless* fermion sea in the presence of the (two-dimensional) fermion loop anomaly. the same physical results can be obtained[27] simply by summing perturbative infra-red divergences coupling through fermion loops (after appropriate regularization of the anomaly).

From his analysis of the Schwinger model, Manton suggested that an analogous winding-number condensate could arise in QCD when the massless quark sea is regularized in the presence of instanton interactions and the fermion anomaly. This would imply

$$\langle 0|W|0\rangle \neq 0 \qquad (6.2)$$

where

$$W = \exp \int d^3 \underline{x} K^\circ(x) \qquad (6.3)$$

and

$$K^\mu(x) = \frac{g^2}{8\pi^2}\epsilon_{\mu\alpha\beta\gamma}\mathrm{Tr}\left[A^i_\alpha \partial_\beta A^i_\gamma - \frac{2ig}{3}\epsilon^{ijk}A^i_\alpha A^j_\beta A^k_\gamma\right]. \qquad (6.4)$$

The lowest-order term appearing in the infra-red gluon configuration $\tilde{K}^\circ$ of the last Section is in fact three gluons in the combination produced by the second term in (6.4). This second term is the infra-red dominant part of (6.4)—since it contains no derivative—and is also the only part that gives a non-zero contribution to W for pure gauge field configurations. Assuming that the sum over higher-order numbers of gluons produces the exponentiation of (6.3) (as gauge invariance requires) we see that the gluon condensate to be identified with the Super-Critical Pomeron condensate is a "winding-number" condensate that would be expected to arise from the instanton interactions of massless fermions.

At first sight it might seem bizarre that the ("non-perturbative") effects of instanton interactions can be seen in perturbative reggeon diagrams. However, we are actually discussing the infra-red limit of infinite sums of reggeon diagrams and this can indeed generate "non-perturbative" effects. In fact the multi-quark S-Matrix calculation that underlies Fig. 5.3, for example, proceeds by first assuming that the reduction to transverse momentum diagrams in each Regge channel places all but the fermion lines shown on mass-shell. Calculating the t-channel imaginary part then places these lines on-shell also. The corresponding real part which is what really corresponds to an instanton interaction is then effectively induced by a dispersion relation for the reggeon vertex. Consequently instanton interactions are inferred only very indirectly.

That we see a winding-number condensate which would be expected to be associated with regularizing the massless quark sea is presumably a consequence of studying high-energy (or "infinite momentum") scattering. The hadron bound states scattering via Pomeron exchange clearly contain "valence quarks" that are effectively massless at infinite momentum. The consistent definition of such states not surprisingly encounters effectively the same problem as the regularization of the massless fermion sea. Our reggeon diagram analysis of this problem is "semiperturbative" and is analogous in many respects to the diagrammatic solution[27] of the Schwinger model. It is interesting that while the winding-number condensate produces confinement in the Schwinger model when the electron is massless, the condensate disappears[28] as the electron becomes massive. In the massive theory confinement is understood as due to the growth of the charge with distance analogous to the conventional understanding of confinement in QCD. The spectrum of the massless and massive theories are nevertheless smoothly related. We suggest that similarly our winding number condensate is an infinite momentum understanding of confinement which can smoothly relate with the conventional description at finite momentum.

Section 7: SU(3) Gauge Symmetry, Fermion Saturation and the Critical Pomeron

Since the Pomeron trajectory of Section 5 is exchange degenerate with the vector trajectory $\alpha_v(t)$ on which the SU(2) singlet gluon with mass M_2 lies, it follows that

$$\alpha_{\mathbf{P}}\left(M_2^2\right) = \alpha_v\left(M_2^2\right) = 1 \tag{7.1}$$

and hence

$$\alpha_{\mathbf{P}}(0) \underset{M_2^2 \to 0}{\longrightarrow} 1 \tag{7.2}$$

Apparently then it is straightforward to take the limit $M_2^2 \to 0$, which restores SU(3) gauge symmetry, and obtain the Critical Pomeron from the Super-Critical Pomeron of Section 5. However, we stressed that it was important that the infra-red limit generating the Super-Critical Pomeron be carried out with a transverse momentum cut-off $\Lambda_\perp$. This cut-off is a relevant variable at the Critical Pomeron critical surface and in particular renormalization effects[29] move the critical point $\Delta = 0$ to

$$\Delta_c^0 = \frac{r^2}{\alpha'} \ln \left[\frac{r^2}{\alpha' \Lambda_\perp} \right] + O(r^2) \qquad (7.3)$$

If we can indeed take the limit $M_2^2 \to 0$ at fixed $\Lambda_\perp$, where M_2 is the *renormalized* gluon mass, then we would expect to obtain the Critical Pomeron. Variation of $\Lambda_\perp$ would nevertheless immediately move the theory away from the critical point. In particular if we assume that Δ, r and α' are dominated by infra-red transverse momenta and are relatively insensitive to $\Lambda_\perp$, then it follows from (7.3) that increasing $\Lambda_\perp$ increases $|\Delta_c^0|$ and hence the theory becomes sub-critical. This argument suggests that QCD with the SU(3) gauge symmetry restored will always be sub-critical provided the transverse momentum cut-off is taken sufficiently large.

This last conclusion can be avoided only if we take the ultra-violet limit $\Lambda_\perp \to \infty$ before taking the infra-red limit $M_2^2 \to 0$. In this case the mass M_2 becomes an unambiguous physical parameter and subsequent variation of $\Lambda_\perp$ will not be able to destroy the critical limit $M_2^2 \to 0$. To take $\Lambda_\perp \to \infty$ in the presence of the Higgs sector (breaking the gauge symmetry to SU(2)) we must have asymptotic freedom. Without this, ultraviolet contributions to the sums of reggeon diagrams will introduce ambiguities in the $\Lambda_\perp \to \infty$ limit. However, asymptotic freedom is possible for the color triplet Higgs sector only if the usual asymptotic freedom constraint on the number of fermions is saturated.[30] That is if, for example,

$$N_f = 16 \qquad (7.4)$$

where N_F is the number of flavors of color triplet quarks. We conclude therefore that the *CriticalPomeron* only *occurs* in *QCD* when the theory is *"saturated with fermions"*.

There are additional properties of QCD saturated with fermions which correlate with the occurrence of the Critical Pomeron. Most important is the presence[31] of an infra-red fixed-point in the β-function when all quarks are massless—this is illustrated in Fig. 7.1. Technically this fixed-point is important in allowing the infra-red analysis of Section 5 to go through in the presence of quarks. (The analysis should be first carried out with massless quarks with the massive quark theory constructed by mass (chiral) perturbation theory.) The fixed-point is also important in minimizing the divergence of perturbation theory (there are no infra-red renormalons). This implies that our "semi-perturbative" construction of the high-energy behavior via reggeon diagrams can give the correct answer. The only non-perturbative ambiguity of the theory will be due to instantons and as we have discussed we believe that we are accounting for them via the anomalous fermion loop interactions. The infra-red fixed-point is probably also physically significant in the sense that the many of the scaling properties of the Critical Pomeron (KNO scaling etc.) are indirectly related to this fixed-point.[32]

Although we have "tracked-down" the Pomeron in QCD using the Super-Critical phase all of our analysis is consistent with the conventional expectation that the Pomeron will correspond to a closed-string configuration of color-flux when a small number of quark flavors are present. The Critical Pomeron we obtain by restoring SU(3) gauge symmetry is, of course, a complicated sum over all Pomeron interaction diagrams, the lowest approximation to which is four-gluon exchange with anomalous color charge parity. This is actually the lowest number of gluons approximating the imaginary part of a Wilson loop. Consequently we anticipate that in the sub-critical phase the Pomeron we have identified does have the conventional perturbation expansion, as described in Section 2, and the bare Pomeron has all the multiperipheral properties of closed-string (of color flux) exchange. Note that from the multiperipheral viewpoint the bare Pomeron intercept is a function of the number of hadrons in the theory and increases with each new quark flavor.[7,33] Therefore it is natural that critical behavior is obtained just as the maximum number of flavors is added to the theory.

Section 8: Electroweak and Collider Physics

It seems that although our analysis does identify a sub-critical Pomeron of the conventional kind in QCD the theoretically attractive Critical Pomeron occurs only in the unrealistic situation that a very large number of quark flavors are present in the theory. However, we can also consider adding higher-color quarks to the theory since they have a much greater effect on the β-function. In particular color sextet quarks contribute like five triplet quark flavors so that

$$N_F = 16 \leftrightarrow 6 \text{ flavors of color triplet quarks}$$
$$+2 \text{ flavors of color sextet quarks.} \tag{8.1}$$

This case is particularly interesting because a chiral doublet of color sextet quarks has been much discussed[34] as a candidate for producing dynamical electroweak symmetry breaking.

Two color sextet quarks (U,D) will produce a triplet of "sextet pions" π_6^+, π_6^-, π_6° which can become the longitudinal components of the W^+, W^- and Z° respectively, generating a mass

$$M_W \sim g F_{\pi_6}, \tag{8.2}$$

where F_{π_6} is the sextet quark chiral constant. If F_{π_6} satisfies Casimir scaling as expected,[35] then

$$C_6 \alpha_s \left(F_{\pi_6}^2 \right) \sim C_3 \alpha_s \left(F_{\pi_3}^2 \right) \tag{8.3}$$

where C_6 and C_3 are respectively sextet and triplet Casimirs. Since $C_6 = \frac{5}{2}C_3$ it is clear that a reasonable evolution of α_s will give the physical value of M_W via (8.2).

The idea that color sextet quarks are responsible for electroweak symmetry breaking has many attractions not the least of which is the explanation of the electroweak scale as a QCD scale. An immediate consequence of giving longitudinal W's and Z's the QCD interactions of sextet pions is that we can expect a much larger pair production cross-section[36] than is predicted in the Standard Model. Recent events seen[37] at CERN suggest that this might indeed be the case. We would like to emphasize that this possibility was predicted by us[38] before the CERN events were announced.

From the point of view of the present article the attractive aspect of sextet quarks producing the electroweak Higgs sector is clearly that the full compliment of quarks required for Critical Pomeron high-energy behavior is introduced at a currently accessible energy scale. Unfortunately this does not mean that the Critical Pomeron becomes observable at the electroweak scale. Instead it implies that it becomes observable at a scale that is *asymptotic* with respect to the electroweak scale. A very pessimistic conclusion for any hope of directly applying Critical Pomeron scaling behavior at any foreseeable accelerator. There are, however, a number of compensating comments that we should make.

Firstly a more complete understanding of the parameters of the Pomeron in terms of QCD parameters may imply that the theory is not too far from the critical point without the sextet quarks. In this case Critical Pomeron scaling *together with approach to scaling terms* could still provide a valuable phenomenology of diffraction. Secondly we believe that the true significance of the Critical Pomeron will ultimately be its theoretical importance in relating short distance and large distance properties of QCD. We have not elaborated the arguments in this article but in Ref. 39 we have argued that the Critical Pomeron is actually necessary for the parton model to apply to hadron scattering in QCD. If this is the case the entrance of the sextet quarks into the theory may be a necessary condition for true compatability of perturbative QCD with the confinement low-energy hadron S-Matrix. This would imply that it is not mere coincidence that short-distance jet physics emerges in hadron colliders at a scale comparable with the electroweak scale!

We have concentrated on QCD with SU(3) color gauge symmetry in this article. However, the analysis can be extended to higher gauge groups by decoupling a succession of fundamental Higgs representations. The result will be a more complicated spectrum of "Pomeron" trajectories. A particularly interesting feature is that the phase-transition which occurs for the maximum number of fermions in QCD will occur for less than the maximum for higher gauge groups. Consequently the dynamical breaking of the gauge symmetry which "almost occurs" in QCD will occur when enough fermions are added in a higher gauge group. Especially interesting in this respect is the SU(5) theory we have proposed in Ref. 40. This theory is uniquely selected as a unified theory containing the necessary sextet quark doublet to produce electroweak symmetry breaking. This theory is also "saturated with fermions" and the analysis we have presented here extended to this case would imply the gauge symmetry is indeed dynamically broken, perhaps to SU(3) × U(1). The photon would then be part of the "Pomeron spectrum" in this theory!

References

1. Early work referred to the "Pomeranchukon" or "Pomeranchon", by the early seventies "Pomeron" was well-established, see for example "The Pomeron"—Proceedings of the VIIIth Rencontre de Moriond published by CNRS, France (1973).

2. For a recent review of the vast amount of data obtained see K. Goulianos, Physics Reports, **101**, 169 (1983).

3. C. Lovelace, Physics Letts. **34B**, 500 (1971).

4. V. N. Gribov, I. Ya Pomeranchuk and K. A. Ter Martirosyan, Phys. Rev. **B139**, 184 (1965).

5. A. R. White in Structural Analysis of Collision Amplitudes, edited by R. Balain and D. Iagolnitzer (North Holland, Amsterdam, 1976).

6. A. Capella, M. S. Chen, M. Kugler and R. D. Peccei, Phys. Rev. Lett. **31**, 497 (1973), W. R. Frazer, D. R. Snider and C. I. Tan, Phys. Rev. **D8**, 3180 (1973).

7. T. K. Gaisser and C. I. Tan, Phys. Rev. **D8**, 3881 (1973); M. Suzuki, Nucl. Phys. **B64**, 486 (1973); D. Sivers and F. von Hippel, Phys. Rev. **D9**, 830 (1974).

8. H. Cheng, J. K. Walker and T. T. Wu, Phys. Lett. **44B**, 97 (1973).

9. D. Cline, F. Halzen and J. Luthe, Phys. Rev. Lett. **31**, 491 (1973).

10. T. K. Gaisser and F. Halzen, Phys. Rev. Lett. **54**, 1754 (1985), L. Durand and H. Pi, Phys. Rev. Lett. **58**, 303 (1987).

11. A. A. Migdal, A. M. Polyakov and K. A. Ter Martirosyan, Zh. Eksp. Teor. Fiz. **67**, 84 (1974); H. D. I. Abarbanel and J. B. Bronzan, Phys. Rev. **D9**, 2397 (1974).

12. V. A. Abramovskii, V.N. Gribov and O. V. Kancheli, Sov. J. Nucl. Phys. **18**, 308 (1974); A. H. Mueller, Proceedings of the European Physical Society Conference, Ain-en-Provence (1973).

13. H. D. I. Abarbanel, J. B. Bronzan, R. L. Sugar and A. R. White, Phys. Repts. **21C**, 119 (1975); M. Moshe, Phys. Repts. **37C**, 255 (1978); A. R. White, Proceedings of the Proton-antiproton Collider Physics, 1981 Madison, Wisconsin.

14. D. Amati, M. Ciafaloni, M. LeBellac and G. Marchesini, Nucl. Phys. **B112**, 107 (1976); D. Amati, M. Ciafaloni, G. Marchesini and G. Parisi, Nucl. Phys. **B114**, 483 (1976).

15. J. B. Bronzan and R. L. Sugar, Phys. Rev. **D16**, 466 (1977).

16. A. R. White, Proceedings of the Tenth Workshop on High Energy Physics and Field Theory, Protvino, Moscow Region USSR (1987), Argonne preprint ANL-HEP-CP-87-68.

17. A. R. White, Proceedings of the Marseilles Conference on Hadron Physics at High Energies (1978), Proceedings of the 2nd International Symposium on Hadron Structure and Multiparticle Production, Kazimierz, Poland (1979).

18. E. A. Kuraev, L. N. Lipatov and V. S. Fadin, Zh. Eksp. Teor. Fiz. **71**, 840 (1976); H. Cheng and C. Y. Lo, Phys. Rev. **D15**, 2959 (1977).

19. A. Sen, Phys. Rev. **D27**, 2997 (1983) and extensive unpublished work on non-abelian gauge theories.

20. J. B. Bronzan and R. L. Sugar, Phys. Rev. **D17**, 2813 (1978).

21. J. Bartels, Nucl. Phys. **B151**, 293 (1979), **B175**, 365 (1980).

22. H. P. Stapp, in Structural Analysis of Collision Amplitudes, edited by R. Balain and D. Iagolnitzer (North Holland, Amsterdam 1976), H. P. Stapp and A. R. White, Phys. Rev. **D26**, 2145 (1982).

23. J. L. Cardy, R. L. Sugar and A. R. White, Phys. Lett. **55B**, 384 (1975).

24. A. R. White, CERN preprint TH 2976 (1980) unpublished.

25. E. Fradkin and S. H. Shenker, Phys. Rev. **D19**, 3682 (1979); T. Banks and E. Rabinovici, Nucl. Phys. **B160**, 349 (1979).

26. N. S. Manton, Ann. Phys. **159**, 220 (1985).

27. Y. Frishman, Springer Lecture Notes in Physics, Vol. 32 (1975).

28. S. Coleman, R. Jackiw and L. Susskind, Ann. Phys. **93**, 267 (1975).

370

29. R. L. Sugar and A. R. White, Phys. Rev. **D10**, 4074 (1974).

30. D. J. Gross and F. Wilczek, Phys. Rev. **D8**, 3633 (1973); T. P. Cheng, E. Eichten and L. F. Li, Phys. Rev. **D9**, 2259 (1974).

31. T. Banks and A. Zaks, Nucl. Phys. **B196**, 189 (1982).

32. A. R. White, Proceedings of the XVIIIth Rencontre de Moriond (1983).

33. J. W. Dash, E. Manesis and S. T. Jones, Phys. Rev. **D18**, 303 (1978).

34. W. J. Marciano, Phys. Rev. **D21**, 2425 (1980), E. Braaten, A. R. White and C. R. Willcox, Journal of Modern Physics **A1**, 693 (1986).

35. J. Kogut *et al.*Phys. Rev. Lett. **48**, 1140 (1982).

36. A. R. White, Mod. Phys. Letts. **A2**, 397 (1987).

37. A. Norton, Proceedings of Physics in Collision, Chicago, Illinois (1987).

38. A. R. White, Proceedings of the Workshop on Physics Simulations at High Energy, Madison, Wisconsin (1985).

39. A. R. White, Phys. Rev. **D29**, 1435 (1985), Proceedings of the Workshop on Elastic and Diffractive Scattering at the Collider and Beyond, Blois, France (1985).

40. K. Kang and A. R. White, Journal of Modern Physics **A2**, 409 (1987).

Figure Captions

Fig. 2.1 The Pomeron and the correspondence with average multiplicity events.

Fig. 2.2 Two Pomeron exchange corresponds to events with twice the average multiplicity.

Fig. 2.3 A Pomeron interaction graph.

Fig. 2.4 Many Pomeron interactions.

Fig. 2.5 Pomeron absorption.

Fig. 2.6 Other processes contributing to two Pomeron exchange.

Fig. 3.1 The shifted Pomeron propagator.

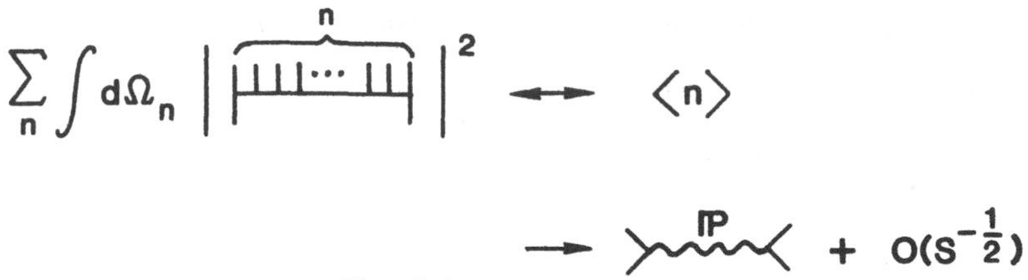

$$\sum_n \int d\Omega_n \left| \overbrace{\underbrace{\text{||||} \cdots \text{||||}}}^{n} \right|^2 \longleftrightarrow \langle n \rangle$$

$$\longrightarrow \quad \gtrsim\!\!\!\wedge\!\!\!\wedge\!\!\!\!\stackrel{I\!P}{}\!\!\!\wedge\!\!\!\wedge\!\!\!\lessdot \quad + \quad O(S^{-\frac{1}{2}})$$

Fig. 2.1

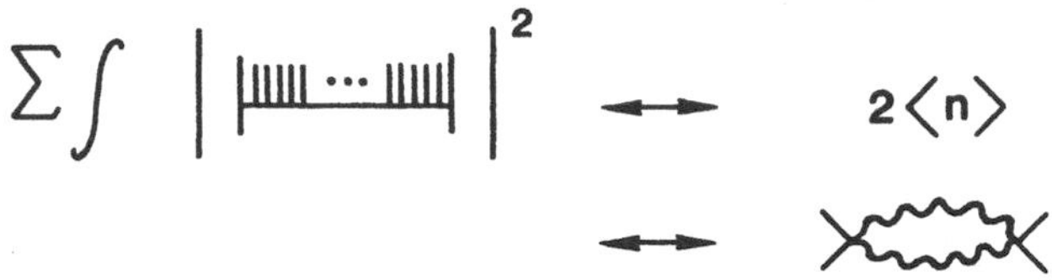

$$\sum \int \left| \text{||||||} \cdots \text{||||||} \right|^2 \longleftrightarrow 2\langle n \rangle$$

$$\longleftrightarrow \quad \gtrless\!\!\!\bigcirc\!\!\!\lessgtr$$

Fig. 2.2

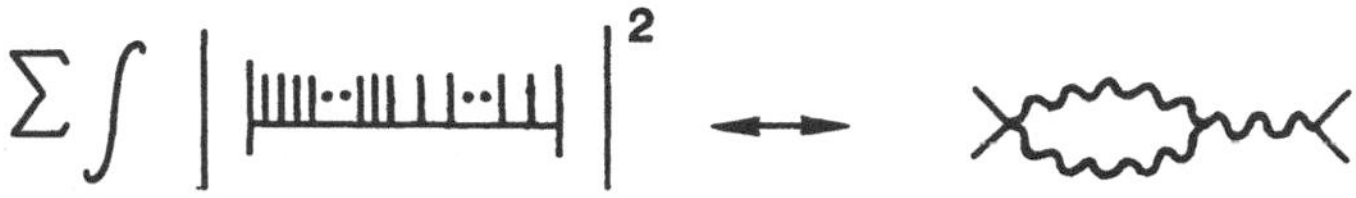

$$\sum \int \left| \text{|||||}\cdot\text{|||||}\cdot\text{||||} \right|^2 \longleftrightarrow \quad \gtrless\!\!\!\bigcirc\!\!\!\wedge\!\!\!\wedge\!\!\!\lessdot$$

Fig. 2.3

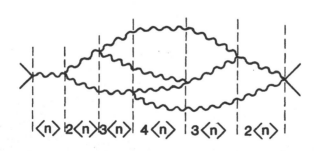

$$\langle n \rangle \quad 2\langle n \rangle \quad 3\langle n \rangle \quad 4\langle n \rangle \quad 3\langle n \rangle \quad 2\langle n \rangle$$

Fig. 2.4

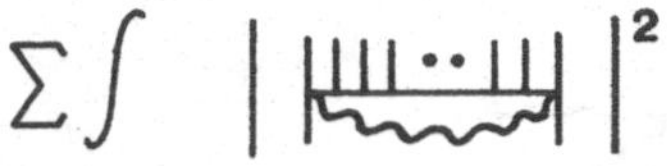

$$\Sigma \int \left| \text{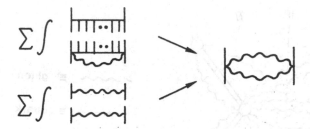} \right|^2$$

Fig. 2.5

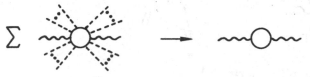

Fig. 2.6

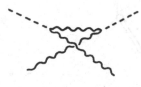

- - - - - Pomeron with $E = \underline{k} = 0$

Fig. 3.1

Fig. 3.2

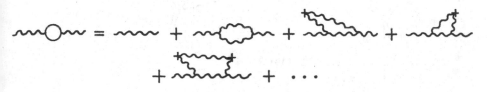

+ $\quad$ + $\cdots$

Fig. 3.3

374

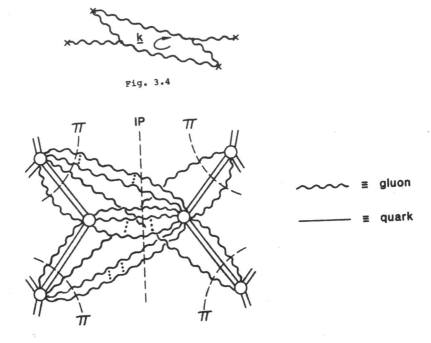

Fig. 3.4

≡ gluon

≡ quark

Fig. 5.1

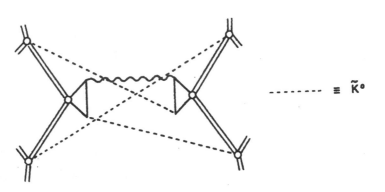

------- ≡ K̃°

Fig. 5.2

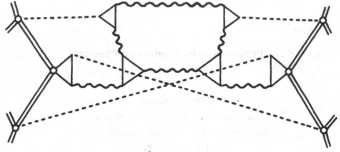

Fig. 5.3

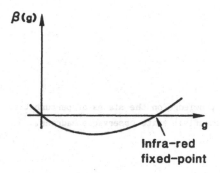

Fig. 7.1

March 1988
LU TP 88-2

Problems in Hadronization Physics

Bo Andersson
Department of Theoretical Physics,
University of Lund, Sölvegatan 14A,
S-223 62 Lund, Sweden

Abstract:

I briefly review our present knowledge on the states of perturbative QCD and
how these states decay into the final state observable hadrons.

Introduction

This review will deal with our present understanding (or lack of understanding, very often) of hadronization in multiparticle processes. I feel that it is necessary to start with a brief review of how the subject has developed (we are nowadays far away from the situation ten years ago) and also to introduce a set of notions to be used in the following. In particular in order to obtain the final state observable hadrons, we must have some basic structure, providing a framework for the energy-momentum and quantum number flows which are observed. That is the parton model.

A long time ago we were happily engaged in the predictions of multiparticle phenomena based upon the parton model. We used the nonabelian gauge theory QCD as a tentative motivation for its amazing qualitative success.

We knew that at its roots was the Weizsäcker-Williams approximation (WWA), a timehonoured tool of the abelian QED [1]. This relationship was further elaborated in the beautiful paper by Altarelli and Parisi [2]. They showed that the scaling deviations (first exhibited in the somewhat prohibiting framework of operator product expansions and the renormalization group) could be understood as iterations of the WWA, applied to the colour field quanta. The asymptotic freedom of QCD together with a set of "factorization theorems" served us as a guide for the drill exercise with the structure functions and the fragmentation functions, i.e. the initial and final state "flux factors".

We were dreaming of pencillike partonic jets and we sent our experimental friends out to measure these flux factors "once and for all" (i.e. "do it well in a given kinematical region"). We promised them rewarding predictions from the WWA and the "QCD evolution".

We are nowadays sadder and wiser men and women. We now know that there is a much longer and more strenuous way left to go before we can give *quantitative* predictions in high energy hadronic physics. There are no free luncheons and there are no pencillike jets giving simple interpretations of a partonic activity. One of the basic reasons is the complex relationship between the partonic and the final state hadronic stage, i.e. the hadronization phenomena. There are recent review papers both on the experimental [3] and the theoretical [4] details of the subject and I will in general be satisfied to exhibit some illustrative properties and to discuss some open problems.

The complexity of multihadron physics is both of a principal and a more trivial kinematical character. Most high energy physicists tend to agree that there are three principal timescales in the interactions and the subsequent development of the final (observable) state. I will call them the partonic scale (below fractions of a fm), the fragmentation scale (from 1~5 fm) and the resonance decay scale (which may vary from the fragmentation scale all the way to the detector observations - and beyond).

The physics is not well understood on anyone of these timescales but equipped with the resources of the Particle Data Group (masses, branching ratios, Clebsch-Gordan coefficients etc) the third stage can be rather easily implemented in Monte Carlo simulation processes given the two earlier ones. Therefore I will concentrate on the different approaches to the physics of these two.

I would however like to mention to all those of my colleagues which frown at Monte Carlos that the amazing difficulty to filter out useful dynamic information would be even more frustrating if there were no useful implementations of this almost purely kinematical stage. Actually those who are interested solely in "all charged particles", "all pions" - inclusive cross sections can go happy through life using this complexity to their gain by preparing nonstructured "longitudinal phase space " statistical models to describe almost 90 % of the cross section at least up to the energies presently available.

This review will, however, not deal with the techniques of the Monte Carlo implementations but with the possible physics frameworks.

In order to understand the remaining few percent of the cross section it is necessary to filter out the dynamical information from the first two stages of the interaction. We believe that most phenomena on the partonic time scale should be possible to compute from perturbation theory. As I will discuss below (section 2), it is, however, not necessarily a straightforward task (the fact that presentday QCD perturbation theory is hideously complicated and only attainable in the lowest few orders even after months of hard work is still "straightforward" in this sense!).

For the second stage, the fragmentation process, we have to take recourse to phenomenological models and I will discuss some properties of the most popular ones below (section 3).

Let me end this maybe somewhat sombre Introduction with a few general words on the subject. It is true that we have had to sober down from the euphoria of the early parton model days and it is true that there is a tedious journey in front before we will be satisfied with our knowledge on QCD and on hadronic physics. But whoever expected that our first encounter with a theory without a classical macroscopic background (which e.g. QED has), i.e. a totally quantum mechanical confining theory, should be easy? This is probably the largest intellectual adventure our generation of particle physicists is up for. There should be lots of exciting physics to learn about out there, especially if we remember the many-splendoured beauties encountered in the "simpler" QED.

2. The Partonic Timescale

In this section I will briefly discuss the properties of the partonic states, such as we know them from present day calculations and considerations in QCD. This is a necessary prerequisite for the discussion on fragmentation models in the next sections.

I will start with a problem, the question of whether the colour force field contains a direction, "the colour flow direction" also as a quantum field theory. My next subject will be the treatment of the severe infrared divergencies in the theory and how to obtain finite cross sections. I will then say a few words on the calculation of exact matrix elements before bringing up via a brief discussion of some coherence problems the multiparton emission models of parton cascade type. After a brief discussion of both the problems of gauge-choice and recoil in the emission, I will end with a few words on the use of multiparton states in hadron-hadron collisions.

2A. The Problem of Colour Flow

There are lots of intricate problems in the perturbative treatment of a field theory. QCD is even more complicated by the fact that we have no real knowledge about the properties of the basic states and also by the strong infrared singularities of the theory.

Let me start with one principal problem for the states. Usually the states of QCD are characterized by a set of partons (quarks (q, colour 3), antiquarks (q̄, colour 3̄) and gluons (g, colour 8)) together with their energy momentum vectors and the properties of the underlying forcefields are neglected.

It is sufficient for an abelian gauge field theory to specify the field quanta in order to obtain all possible information on the fields - in this case there is no particular ordering necessary. For a nonabelian gauge field theory this is no longer so obvious. In particular for a confining theory where the fields have a thin tubelike extension one expects that a colour ordering should be necessary - the fields should stretch from one charge to another etc.

Actually, this is also noticeable in the transition matrix elements of perturbative QCD (PQCD). They can all be arranged in a gauge invariant way into groups such that there is a particular *colour flow ordering*.

The simplest case is a four parton state like an e^+e^--annihilation event into a $(q\bar{q}g_1g_2)$-state (which then is in an overall colour singlet configuration). In this case the matrix-elements (to any order in PQCD) can be grouped as belonging either to the $(qg_1g_2\bar{q})$- or the $(qg_2g_1\bar{q})$-colour flow configuration.

In the cross sections, which are quadratic in the matrix elements, there will, however, occur interference terms between the colour flows. They are generally nonvanishing, although of the order $1/N_c^2$ (N_c = number of colours) compared to the rest. Therefore the occurrence of such interference terms is more a qualitative than a quantitative problem (in some cases the size is however almost 30 % of the rest).

In the spirit of perturbation theory, e.g. in leadinglog approximations (LLA) etc, these terms can be neglected or eventually redistributed among the terms containing a proper colour ordering. But anybody doing fragmentation physics should be aware that these terms, without classical correspondence, have no simple interpretation for covariant fragmentation models.

The eventual occurrence of a *colour flow direction quantum number* should be remembered in this connection. This possibility may be related to the structure of the QCD vacuum state and its response to an outside colour force field, as I have discussed in another context [5].

I will nevertheless from now on for the fragmentation models assume that there is a given colour ordering.

2B. Infrared Problems

For all calculations in PQCD one must be aware of the sincere infrared problems, i.e. there are all kinds of divergencies in the long wavelength limit. The physical reason for these problems is that we are actually calculating inclusive cross sections which contain the ordinary total cross section together with the multiplicity of the included particles. Then the divergencies signalize that an infinity of soft and collinear partons may be involved. It is in general not known whether one obtains e.g. 50 soft gluons in each event or maybe 1 in most and an enormous amount in the few remaining ones.

The ordinary assumption is that the computed cross section (normalized with e.g. the Born cross section) denotes the mean multiplicity per unit phase space. Further one assumes that the emission is independent, i.e. Poissonian in nature so that one may renormalize with the probability of emitting nothing "before" a certain value (Sudhakov formfactor). To be precise, if the cross section is $d\sigma$ then one introduces the mean number

$$d\bar{n} = \frac{d\sigma}{\sigma_0} \tag{1}$$

Then with the independence assumption we conclude that the probability to emit no parton at all inside the phase space region Ω is

$$\exp - \bar{n}(\Omega) \tag{2}$$

$$\bar{n}(\Omega) = \int_\Omega d\bar{n}$$

Similarly exactly one emission inside Ω will occur with the probability

$$\bar{n}(\Omega) \exp - \bar{n}(\Omega) \tag{3}$$

etc.

In some few cases when two particle correlations may be neglected the introduction of such form factors may solve the divergence problems. Care should, however, in general be used because of the typical quantum interference phenomena which may occur (cf below).

In practice one must, however, introduce different cutoff procedures (in particular because of the intrinsic scale in the running coupling constant, cf below). In that case it is necessary to require *infrared stability* (besides obvious Lorentz covariance and gauge invariance constraints).

Infrared stability means that no physical observable should be influenced by the cutoff procedure. In other words, the occurrence of one or many partons, which are soft on the scale of the detector resolution, should have no influence on the predictions.

It is useful to note that in QCD, in contradiction to QED, one is in general not free to choose this resolution power. For QED it is always possible to imagine the use of a sufficiently fine grained and distant detector so that the difference between an electron state and a state with an electron and an almost collinear photon may be resolved. For QCD on the other hand there is an intrinsic scale, the hadronic mass scale below which no resolution is possible. Two colour connected partons with a common invariant mass below this scale obviously end up in the same hadron.

In Monte Carlo simulation programs the commonly used cutoff is just the invariant mass of two adjacent partons m_{ab} or its scaled equivalent $y_{ab}=m_{ab}^2/W^2$, with W the total cms energy of the state. "Preconfinement" [6] cluster masses, which are used in some fragmentation models are of this type and similarly the dipole masses in one of the latest developments of the Lund Model.

Other devices e.g. of Sterman-Weinberg type may be useful but in general the partonic state looses its significance in these cases.

2C. Matrix Elements

As of now almost all the next to leading order QCD-corrections to the cross section have been computed. The amount of work that has gone into these projects means that it is very doubtful that we will obtain even higher order corrections with present day techniques, except in a few simple situations. It is an unfortunate fact, however, that these precise calculations hardly are sufficient to describe in a consistent way e.g. the present (rather sophisticated) PETRA/PEP data in e^+e^--annihilation events.

To give an example, in the so-called energy-energy correlations, introduced by the Washington group [7] it is necessary to choose [3,8] a very small mass cutoff m_{ab} in order to describe the asymmetry by means of the first and second order QCD corrections (3- and 4-jet events). This means that there is only a small amount of 2-jets, i.e. back-to-back qq-jets (~5 %) and the α_s and α_s^2 correction take ~95 % of the cross section! An outsider will under those circumstances ask about the size of the next order correction.

It is further in general difficult to describe the transverse momentum fluctuations out of the main event plane by means of only 2-, 3- and 4-jet events.

2D. Coherence Problems

It is therefore, nowadays, generally believed that one needs at least approximate methods to compute the rate of multiparton emission. This is provided by the QCD branching processes which I will consider next.

I would like to start by considering the corresponding situation in QED bremsstrahlung. If an e^+e^- pair is produced in a point and goes apart ("charge separated") with rapidities y_+ and y_- ($>y_+$), then it is wellknown that bremssttrahlung is emitted inside the rapidity (or angular) range

$$y_+ < y < y_- \hspace{5cm} (4)$$

If we consider the Feynman diagrams contributing to the process (cf fig. 1) we note that only the sum of the two contributions is gauge invariant. It is perfectly feasible to choose a gauge such that one of them dominates, e.g. the one corresponding to emission from the "e-leg". Or to choose a gauge such that there is no interference between the two diagrammatic contributions. *It is*, however, *wrong* physically to talk about this gauge choice as "more physical" than any other choice. We all know that none of the particles emits "by itself". It is because of the fast charge separation i.e. the quick changes in the field configuration that there is emission and as I have already said the

*charges emit together. The rapidity range in eq. (4) is characteristic of a
dipole emission.*

Exactly the same coherence phenomena occur in QCD but the situation is further
complicated by the fact that the emitted radiation (in general gluons or qq-
pairs) is charged in itself. These coherence phenomena were first pointed out
for QCD by Mueller, Marchesini and collaborators [9] and by the Leningrad
group [10] and they are basic to any gauge field theory.

2E. Parton Cascade Models

Early attempts to obtain models for multiparton emission ("jet calculus" [11])
did neglect the coherence phenomenon. But both these and later efforts have in
general been based upon the Altarelli-Parisi (AP) (or if you like the QCD WWA)
evolution equation:

$$df(x,Q^2) = dt \int \frac{\alpha_s}{2\pi} \, dz \, P(z) \, f \, (y,Q^2) \, dy \, \delta(x-yz) \qquad (5)$$

i.e. the change in any number distribution of partons f during a small change
in the evolution parameter t is given by the way the splitting function P
change f integrated over all allowed values of the splitting variable z. P
contains conventionally the colour Casimir factors and it is useful to note
for the running coupling constant α_s that for large values of the number of
colours N_c

$$N_c \frac{\alpha_s}{2\pi} \to \frac{1}{2\ln(Q^2/\Lambda^2)} \qquad (6)$$

The parameter Λ depends upon the choice for the variable Q^2 but is usually of
the order of a few hundred MeV. Different groups use different interpretations
of the evolution parameter t and the off-shellness Q^2, as well as of the
splitting variable. (All definitions are, of course, equivalent up to the next
order in α_s, or more correctly up to subleading contributions in the LLA of
the relevant Feynman graph functions.)

The reason to put the α_s-contribution inside the integral in eq. (5) is that often the off-shellness Q^2-variable is identified with the transverse momentum squared in the splitting $k_T^2 \sim z(1-z)m^2$ with m^2 occuring in $t=\ln(m^2/\Lambda^2)$.

Generally in applications one defines a Sudhakov formfactor corresponding to the probability that no splitting occurs inside a certain t-range and using Monte Carlo programs evolve the equation (5) either towards smaller masses m ("final state bremsstrahlung") [12] or (with due caution) "backwards" ("initial state bremsstrahlung") [13] with spacelike masses back to the original wavefunction.

The different conventions are described in detail in e.g. ref. [4] but they all lead to a branching process in which different partons occur either as early "trunks" decaying into finer (and less off-shell) "branches" and "twigs".

Implementation of the coherence conditions discussed above was first done in the Marchesini-Webber algorithm but is nowadays available as options in e.g. the Lund Monte Carlo (Bengtsson-Sjöstrand).

An instructive, different approach is the dipole approximation [14] which at the moment is worked out in the Lund group (based upon ideas first suggested by the Leningrad group [10]). When an original colour $(3\bar{3})$ goes apart (creating a colour dipole) a first gluonic emission corresponds to the production of two new "colour antennae" between the $(3\bar{8})$ and the $(8\bar{3})$.

Then new gluonic emission from these dipoles produce new antennae and new emission etc until each dipole mass is down below some cutoff mass.

One funny thing about the method is that the coherence condition in this language means: "do not double-count!", i.e. every gluon emission is allowed from one and only one "earlier" dipole (in accordance with eq. (4)).

There is further a non-trivial ("local") solution to one of the basic problems in the AP-approach, i.e. the recoil- and the subsequent gauge-choice problem. The AP-approach is actually only valid in the collinear limit and every emission evidently produces both a transverse momentum of the emitted parton and a corresponding recoil for the emitter.

When repeated emissions occur it is evident that "collinear" may no longer be a well-defined notion and the use of a particular gauge related to the original direction may create problems. In e.g. a Marchesini-Webber cascade a certain partonic system with a given mass may decay differently in different parts of the system. In particular the decay properties depend upon the velocity of the subsystem.

This is not so in the dipole approach where the decay properties are fixed by the dipole mass. The orientation of a decaying subsystem has been choosen so that the external colour flow, i.e. the colour flow into and out of the subsystem is changed as little as possible by the recoils. This choice creates non-trivial azimuthal correlations and the whole procedure has a nice self-similarity built in.

2F. Parton Cascades and Scattering in Hadronic Reactions

Several models are now available for hadron-hadron reactions in which total event generators are available.

The Pythia model [15] contains one scenario which is extreme in the sense that everything is worked out from (a common-sense interpretation of) perturbative QCD. The wellknown (inclusive) large p_T cross-section (colour-Rutherford scatterings) are (after division with a normalizing cross-section) interpreted as the mean number of collisions of a particular kind. They are arranged by Sudhakov formfactors into descending "hardness" order, each equipped with relevant "initial" and "final state" bremsstrahlung radiation. When no scattering with transverse momentum above a certain p_{Tcut} occurs, the event is treated as soft or diffractive with a reasonable parametrization. On the other hand by means of an impact parameter method [15] there is a bias such that when one hard scattering occurs then there are in general several more. Fragmentation is done by string methods and although there are several options to draw the strings and a consequent lack of uniqueness the authors have worked hard to minimize these problems.

The Pythia model is a rather complete model in the sense that one may obtain everything from Higgs production, W and Z formation to Drell-Yan events and the above mentioned Rutherford scatterings. A corresponding general and versatile all-purpose model is Isa-Jet [16]. The differences to Pythia is from a principal point of view that much more of the cross-section is treated by a soft parametrization (based upon Gribov calculus like the Dual Parton Model [17]) and that the fragmentation of the partonic state is done by independent fragmentation.

Hadronic collisions in the Fritiof model [18], which is based upon a string dynamics scenario, is extreme in the opposite direction to Pythia. The basic dynamics is nonperturbative and the Rutherford scatterings (only the hardest one and only if it is above a bremsstrahlung "noise level" is treated) occur as a perturbation on this longitudinal momentum transfer process. Just as in the Dual Parton Model (DPM) string fragmentation is used. But the fluctuation is for large energies in the Fritiof model correlated to transverse momentum emission while the DPM contains many small central and longitudinally stretched strings. Multigluon emission and Rutherford scatterings are at the moment introduced into DPM and it will be very interesting to see how the basically longitudinal dynamics scenario will survive with "minijets" etc.

There is one final problem in the partonic interaction scale, brought up by the Leningrad group [19], which I have not treated yet. The phase-space density of partons in cascade calculations may eventually be so large that the interactions between the emitted partons can no longer be neglected.

In e.g. ref. [19] there is an instructive picture of the resolution power of an external probe, a "partonometer" used on a hadronic state. A small wavelength, large Q^2 partonometer will resolve a fraction of the state $x \sim 1/Q^2 R^2$ with R a typical length parameter of the state. The corresponding structure functions will contain partons on this Q^2-scale which have larger energies (larger Bjorcken x_B-fractions) than x, due to fluctuations in the energy-density of the state. It may also contain x_B-partons with $x_B < x$, due to evolution à la the AP-equation above. If the AP-evolution continues too far, however, the authors of ref. [19] expect interactions of the emitted partons. In practice the AP-equation, which for small x-values is basically a classical probabilistic "gain-equation", will then also obtain "loss terms".

Considerations to the same effect, although with a different dynamical motivation occur in the Fritiof model. It is expected that, while colour separation occur in all hadronic interactions, the energy density in general is small in the state. Emission of radiation of a given wavelength λ can only occur from an "antenna region" of the same or a smaller size. Short wavelength radiation, like large transverse momentum gluons, will then need a small antenna region. If it is too small, the available energy is not sufficient.

This "soft radiation" will consequently be limited in size but it would take us too far to consider the detailed consequences.

3. The Fragmentation Scale

In this section I would like to briefly cover a set of problems in the presently available phenomenological models. We have little precise knowledge of the structure of the hadronic bound states, although some models like the MIT-bag and its successors "the little Brown bag", "the Cheshire Cat bag" etc indicates with growing sophistication some general properties. Our lack of knowledge is unfortunately reflected in the occurrence of several phenomenological parameters.

Most of my considerations will be on three different frameworks, the iterative cascade jet algorithm with the Lund string model as a "collective" model and Independent Jet Fragmentation as an opposite "individual" approach and then finally the Cluster Fragmentation models with the prime example in the Marchesini-Webber model.

I will try to specify the dynamical assumptions behind these models leaving many of the detailed descriptions for the interested reader to find out from the original literature or from the excellent review in ref. [4]. I will further leave out the comparisons to experimental data due to space limitations referring to refs. [3] for detailed comparisons.

3A. Longitudinal Phase-Space

In the beginning was longitudinal phase-space. Thus already the first cosmic ray event studies noted that energy and momentum flows as well as the multiplicity densities of the final state hadrons had a preferred axis, "the beam direction". Any deviations in the transverse directions were small and limited.

The longitudinal phase space production models, very often termed "the pion gas in rapidity space", contains basically two parameters, one related to the multiplicity per unit rapidity and the other to the width of the transverse momentum distribution. Such models are simple and easily implemented in a Monte Carlo framework and as I have said before, these models may take you a long way as long as you have little interest in large fluctuations, precise quantum number flows and the basic dynamics behind the final state charged hadron distributions. The decay of many resonances with the corresponding redistribution of energy and momentum and the many kinematical complications means that a "statistical" scenario is not far away from different mean or inclusive observations.

3B. Iterative Cascade Models and Their Problems

The first models containing a bit more sophistication were the jet cascade fragmentation models [20]. The basic dynamical idea is again a longitudinal distribution but of an iterative character.

A jet containing a given energy (-momentum) (e.g. $W_+=E+P$) decays into a single particle (with mass m_1 and fractional energy momentum $z_{+1}=e_1+p_1/W_+$) and a new jet carrying $(1-z_+)W_+$. Remembering that

$$e_1^2 - p_1^2 = m_1^2 \tag{7}$$

(for a single longitudinal space direction) the particle energy and momentum is fixed by m and z_{+1} and we may repeat the process (cf fig. 2) and produce

$$z_{+2} = \eta_2(1-z_{1+}) \ , \ m_2 \tag{8}$$
$$z_{+3} = \eta_3(1-\eta_2)(1-z_{1+}) \ , \ m_3 \ \text{etc}$$

with the values of η_j ($\eta_1=z_{1+}$) eventually distributed in accordance with a common "inclusive distribution" $f(\eta)d\eta$ for $0<\eta<1$.

While the first jet cascade models were used for full fledged final state hadrons, it was suggested in ref. [21] that they may be successfully used in a framework where a new qq-pair appears at each "vertex", with the quantum numbers, such as flavour and transverse momentum etc, "locally conserved" in the pair production. Several such models were suggested [22] although the Feynman-Field scenario tended to be the most popular one.

One particular property of all these models is the appearance of a central rapidity "plateau", i.e. for large energies the multiplicity distribution fractional energy-momentum behaved as

$$C \frac{dz}{z} = C \, dy \tag{9}$$

The constant C can very often be related to the behaviour of f(z) for large z-values e.g. as

$$f(z) \sim (1-z)^{C-1} \tag{10}$$

It should be noted, however, that in order to saturate eq. (9) i.e. to obtain a real central plateau it is necessary to have very large energies. Monte Carlo simulation studies show that for an ordinary particle composition it is necessary to have cms energies above 100 GeV before it is developed. At these energies, lots of other processes, like multiparton bremsstrahlung emission and "hard Rutherford scatterings" are around and tend to obscure such a saturation.

Even before the FF model appeared there was however some work done on a "collective" decay scenario [23]. It is evident that "no jet is alone", i.e. even a $q_0 \bar{q}_0$-pair jet produced in e^+e^--annihilation contains more than one "end", which is influenced by the quantum numbers occurring originally. In general at least for large energies it is perfectly feasible to produce one jet from the q_0-side and another from the $\bar{q}_0$-side and afterwards try to "join them" in the centre by means of some procedure. This turns out to be a difficult problem to do in a consistent way, however. It is easy to show that for almost all distribution functions f(η) the central distribution depends upon in which order the jets are formed, i.e. the procedure is not left-right symmetric.

There is further the somewhat complicated question of "order". FF introduced the notion of "rank ordering", counting e.g. from the q_0-end "the 1st rank" hadron as the one containing the q_0 together with - for a meson - another $q=q_1$ from "the first" vertex, and similarly for the 2nd rank meson $(q_1 q_2)$ etc.

It was quickly recognized that rank ordering could not easily be understood as an ordinary time ordering in the process. From the Landau -Pomeranchuk notion of formation time in bremsstrahlung processes it was known that solely due to relativistic covariance a "fast particle" (large fractional energy-momentum z) takes a much longer time than a corresponding low z-hadron to get disentangled.

Therefore the iterative process, which seemed to produce an "outside-in" cascade with the fastest mesons generally produced "first" was claimed to be wrong dynamically. Inside a string framework, however, this paradox can be solved surprisingly at the same time as the question of left-right symmetry also obtains a solution. In order to see how this comes about we will consider some properties of the Lund string model [24].

3C. A Solution to the Problem

A piece of string with a (massless) q- and a q̄-particle at each endpoint may as in fig. 3a stretch out with the end-points moving apart along the lightcones. The q- and q̄-particles loose energy momentum at a constant rate $\kappa \sim 1$ GeV/fm.

After some time (in the cms after $m/2\kappa$ with m the mass of the stringstate) both endpoints are stopped and turned around so that they afterwards are accelerated towards each other by the string tension κ. This procedure is repeated over and over again and these states, termed "the yo-yo modes" are taken as models for stable meson states in string models.

In order to see how the relativistic covariance works in the model we consider fig. 3b, in which the state described above is viewed from a moving frame. Then the q-particle energy momentum is increased by the factor exp y and the q̄-particle (note that the q̄-particle has a "negative" component ω-k) by exp -y. Consequently for large y the q̄-particle is stopped long before the q and afterwards dragged along by the field with the q loosing and the q̄ gaining energy-momentum at the rate κ. After the q is stopped and the q̄ has "passed" it the situation reverses and the yo-yo state as such is seen to move away with rapidity y. The only Lorentz invariant is the surface-area spanned by the string during a period and this can be proved to be proportional to the squared mass m^2.

We next consider a qq̄-state with a large mass $W^2 = W_+ W_-$ (fig. 3c). When the q- and the q̄-particles move apart more and more energy is stored into the forcefield. Nature never liked such a situation and the stored potential energy may be decreased by the breaking of the string-field i.e. by the production of new qq-pairs (new endpoints) along the string. The newly produced endpoints move apart each dragged in the opposite direction by the forcefield.

We note that in this simple but truly confined situation the string forcefield ends on the q- and q-charges. This is different from electrodynamics where a produced e^+e^--pair may well go apart but continue in principle to interact (producing common bremsstrahlung e.g.).

As time passes more and more qq-pairs are produced. A q- and a q-particle from adjacent "vertices" will together with the connecting string piece form a yo-yo meson and in the final state the original large mass string will break up into many small mass yo-yo's, moving apart with different rapidities.

We firstly note that in this simple one-space-dimensional model only one variable, like the fractional energy-momentum z_+ in eqs. (7)-(8) (or for that matter $z_-=e-p/W_-$) together with the meson mass m is sufficient to describe the state of each yo-yo.

Secondly, adjacent vertices must in order that $m^2>0$ be on a spacelike distance with respect to each other. This implies that although in the frame used to describe the breakup in fig. 3c, the vertex 1 is in time before the vertex 2 etc in another frame such time-ordering will be different. What can be said is only that in each frame the slowest mesons are the first to be produced while the fast ones obey the rules found by Landau and Pomeranchuk, i.e. they have formation times $\sim e/m^2$.

Rank ordering is in this way understood as an ordering in descending light-cone coordinates for the vertices, the first rank vertex corresponding to the first vertex e.g. along the positive (or for that matter the negative) lightcone. It is therefore perfectly feasible to arrange the breakup process as an iterative jet cascade in this kind of model and still keep the notion that a jet is formed "inside-out".

A jet cascade with rank ordering means consequently to prescribe z_{+1}, z_{+2}, z_{+3} etc for the fractional energy-momentum from the original q-side or z_{-1}, z_{-2}, etc from the q-side.

3D. A Symmetric Fragmentation Model

A third and more farreaching conclusion is that all vertices must have an equal significance in the production process. This means that the production process in a Lorentz covariant and causal framework must be such that one should be able to describe it starting at any vertex and proceeding "up or down" along the lightcones. In ref. [25] it is shown that there is only a single statistical process of this kind available. Together with the space-time picture above this result is the main one for the one space-dimensional Lund string model.

The process can be described most easily in exclusive terms. The probability to obtain n final state mesons with (p_1, m_1), (p_2, m_2) etc is then proportional to

$$dP_n(p_1, \ldots, p_n) \sim \prod_{j=1}^{n} Ndp_j \delta(p_j^2 - m_j^2) \delta(\Sigma p_j - p) \times exp-bA_n \tag{11}$$

Thus dP_n is proportional to the n-particle phase-space factor with a scale parameter N for each meson together with the negative exponent of the connected space-time area A_n (hatched in fig. 3c) of the string before the breakup (the parameter b is a scale parameter for this area).

The result in eq. (11) may be interpreted as phase-space times a squared matrix element M ("the golden rule"). While the Lund model in the beginning often was thought of as a kind of Schwinger model [26] there is actually a major dynamical difference. For 1+1 dimensional QED, Schwinger showed that the theory was (excluding some mathematical problems) equivalent to a local free field theory - a dipole density - with massive quanta. An external current as in ref. [27] will then produce a coherent state of such quanta but the locality of the field means that these mesons are actually produced as total entities.

In the Lund string model a q_1- and a q_2-particle from adjacent vertices will form a final state meson and such a production process will require for gauge invariance a gauge connector

$$exp \; ig \int_1^2 A^\mu dx_\mu \tag{12}$$

with A^μ the gaugefield. Therefore the total matrixelement must contain a factor exp $ig\int A^\mu dx_\mu$ with the integral around the closed "breakup curve". Such a Wilson operator is in a confined theory expected to behave as

$$\exp i\ \xi A \qquad\qquad (13)$$

with A the enclosed (spacetime) area and with the parameter ξ such that $\mathrm{Re}\xi=\kappa$. The imaginary part of ξ may as in other production processes with a complex dielectricity constant be related to the production rate. A tentative answer would then (using another Schwinger calculation on the tunnelling in an external force field [28]) relate the parameter b in eq. (12) to the transverse width of the force field.

The phase relation above has been used to explain the so-called Bose-Einstein correlations observed in jet production [29].

From eq. (12) it is possible to calculate the inclusive distribution for the first rank meson:

$$f = \frac{1}{z}\ N(1-z)^a\ \exp\ -b\ \frac{m^2}{z} \qquad\qquad (14)$$

The parameter a is a function of N and bm^2 such that the distribution f is normalized. It plays a role in the theory similar to a Regge parameter because it is rather easy to show that for the quantities dP_n in eq. (12) one has [25]

$$\sum_n \int dP_n \sim (W^2)^a \qquad\qquad (15)$$

Further the distribution in squared momentum transfer Γ between a string with mass W_1 and a string with mass W_2 obtained by breaking the original string with mass W in a vertex:

$$\Gamma \approx \frac{W_1^2 W_2^2}{W^2} \qquad\qquad (16)$$

can be shown to be

$$\Gamma^a \, \exp \, -b\Gamma \qquad\qquad (17)$$

As $\Delta y = -\ell n(\Gamma)$ has the obvious meaning of the rapidity gap produced between the string jets 1 and 2 this distribution can be interpreted in Mueller-Regge terms as (this was first pointed out by Artru)

$$\exp \, -(a+1)\Delta y \qquad\qquad (18)$$

While the Lund string scenario has a nice uniqueness in a 1+1 dimensional world, it needs for a confrontation with the 1+3 dimensional experiments to be endowed with a lot of non-unique features, just as the other iterative cascade models.

3E. Notes on the Parameters in a Realistic $q\bar{q}$ Jet Situation

While string motion and yo-yo states are easily generalized to the case of massive qq-pairs (which move along hyperbolas with the massless orbits-lightcones as asymptotes) the production of such qq-pairs as well as the introduction of transverse momentum in the breakup process needs a true quantum-mechanical treatment.

(The interested reader may evidently ask whether the 1+1 dimensional model presented above has anything to do with a quantum mechanical reality. The answer is that most of the considerations are bona fide for longitudinal motions at least as a semiclassical framework. In a paper by Gottfried and Low [30] it is shown that the Heisenberg undeterminedness relations allow a specification of both the rapidity and the space coordinates for a fast particle.)

In the Lund model the production of massive (mass μ) qq-pair with transverse momentum k_T is treated as a tunnelling process [28,31] with a production probability

$$\exp \, - \frac{\pi}{\kappa} \, (\mu^2 + \hat{k}_T^2) \qquad\qquad (19)$$

Using "ordinary" values for the quark masses one expects flavour rates of d:u:s:c ≈ 1:1:0.3:10⁻¹¹. Thus while strangeness shall occur on the level of 10-15 % (which seems confirmed by the present e^+e^--annihilation data), charm and heavier quarks are strongly suppressed in a scenario governed by the scale of κ.

The transverse momentum spectrum for the final state mesons, obtained provisionally by adding the results of the gaussian spectrum from the q- and the q-sources is then in the model also of a gaussian character. From studies in e^+e^--annihilation reactions and (likewise in the applications to inelastic leptoproduction and hadron-production) it is found that this is a good approximation albeit with κ/π ≈ (250 MeV/c)² exchanged to a σ²=(350 MeV/c)². Monte Carlo studies indicate that unresolved (soft gluon) radiation may eventually account for this difference. One should also note the appearance of $m_T^2=(m^2+p_T^2)$ in the fragmentation function in eq. (15).

For meson production it is necessary not only to prescribe an s/u ration but also a Vector/pseudoscalar (V/P) ratio. It may be expected from statistical considerations that V/P ≈ 3 but arguments have been given (also from the tunnelling process [32]) that there should be a suppression due to the larger difficulty of tunnelling into a vector than a pseudoscalar bag (the ratio may be different for ρ/π and κ*/κ e.g.). Phenomenologically V/P is around 1. With the assumption that tensor meson production may be neglected (it is known to be less than a few percent anyway and in a string framework it should be low due to angular momentum considerations) we now have a complete model for the decay of a qq-state into mesons.

It is known however that the baryon-antibaryon (BB̄) production rate is non-negligible inside partonic jets. The simplest scheme [33] for BB̄-production is to assume that diquarks (di) and antidiquarks (d̄ī) (colour 3̄- and 3-charges) may be produced in the field similarly to qq-pairs.

This does not necessarily mean that di-d̄ī pairs should be taken as elementary field excitations but only that di-d̄ī pairs may be dragged out in the tunnelling process in one way or another.

Another scheme "the popcorn models" [34] is obtained by assuming the existence of elementary colour-excitations "of the wrong kind". Thus an original (rr̄) field is transformed by a gḡ-excitation inside a certain region into a [(rg)=b̄,(r̄ḡ)=b] field which may break again producing a (rgb) B-state and a (r̄ḡb̄) B̄-state (eventually with one or more meson states in between).

The general rate of baryon production is in the simplest model regulated by a di to q-ration qq:q ≈ 0.09:1. For the popcorn model there is a somewhat larger rate (due to the generally assumed production of one meson in between the B$\bar{\text{B}}$ in half of the cases the total state needs a somewhat larger phase space).

In any detailed jet fragmentation model for baryon production there is a fair amount of parameters. If on the other hand only protons, neutrons and Λ-particles (as well as the corresponding antibaryons) are observed (this is usually the case), then the results are rather insensitive to all these parameters. This is so at least for models with the SU(6)-symmetry constraint that a baryon is a *symmetric* state of the three quarks (neglecting the colour degree of freedom). Then it is necessary to weight the different flav our and spin states with the probability for a given di and a given q to form a symmetric (qqq)-system.

For a detailed description of all baryons and hyperons with their resonances it is, however, necessary to specify the ratio of spin 1 to spin 0 diquarks the ratio of strange to nonstrange diquarks and the ratio of decuplet to octet baryons. For the popcorn model there is further freedom for the relative probabilities of different BM$\bar{\text{B}}$ combinations (i.e. when a meson M occurs in between the B$\bar{\text{B}}$).

In leptoproduction and hadroproduction events with a baryon target there is a necessity to provide fragmentation also for a forcefield with a more complex colour charge at the endpoint than a "simple" qq-particle.

In the Lund string model there are dynamical reasons to expect a final state "one-dimensional" baryon. Thus whatever original configuration the colour charges have, the final state situation when one of them is affected by a momentum transfer is that they all tend to stretch together along the momentum transfer direction [35].

We note that in a force-field with charges like fig. 4 (the three valence constituents denoted by the colour charges r, b, g are distributed longitudinally) there is one colour flow direction from the r to the (bg)=$\bar{\text{r}}$ and the opposite direction from the g to the (rb)=$\bar{\text{g}}$. Consequently if pairs are produced in these regions, the q-particles are always dragged "inwards", i.e. towards the third valence constituents and there will always be a baryon produced around this original q (baryon number conservation!).

The actual distribution of the three valence constituents (and the even more complex question of how to treat events with "sea quark" interactions) have in the Lund model been treated by simple (rather ad hoc) phenomenological parametrizations.

It is probably necessary to come back to these questions in the future (for presently available data there has been little need of sophistication) and to try to provide a more complete and consistent framework for baryon breakup and $B\bar{B}$-production.

I have in this section treated the Lund model as an example of the iterative cascade jets. Any such model will need many parameters and as long as we have so little knowledge of the actual binding properties of the hadrons (e.g. masses of di- and q-particles if any of these notions have an operative meaning!?) the parameters cannot be calculated. All of the parameters σ^2, s/u, V/P, di/q etc tend to have "reasonable" values and as I have said before most observable features are (if state symmetrization is duly accounted for) well described in terms of these four parameters mentioned.

3F. On Multijet Fragmentation in String Models

While q- and q-particles (with their flavours) are associated with the endpoints of a string , in the Lund model the gluons are treated as massless pointlike excitations in the interior of the string field. Just as for an ordinary classical string there are modes corresponding to a "pluck" or if you like a pointlike "hammer blow". The interesting property of the massless relativistic string is that such excitations are fairly "stable" in the sense that such an excitation will continue to drag along the string as long as some of the original energy-momentum is left.

A gluonic excitation is acted upon by the force 2κ, which means that it looses its energy momentum at double the rate as compared to the endpoints (there are two adjoining string pieces to deliver it to). This means that for a $qg\bar{q}$-state (fig. 5) the two string pieces (qg) and ($g\bar{q}$) move outwards in different directions. One useful way to think about it is to say that half the gluon energy momentum combines with the q-energy momentum respectively the $\bar{q}$-energy momentum to form these pieces.

When the gluon is stopped (having used up its original energy momentum) its delivered energy momentum continue to flow outward and a new segment, dragged out solely by the remaining qq energy momentum, is formed (fig. 5). For a wee central gluon the situation will evidently be very similar to a one-dimensional qq-string, i.e. there are two tiny segments delivered by the gluon moving outwards in the two directions of the q and q together with an ordinary "flat" piece stretched longitudinally. For a collinear gluon there will be a collective dragging out from the g and the q (or q). It is easily seen that while the velocity of a string segment spanned by e.g. the gq is $c \cos(\theta/2)$ the mass of the system is $4E_qE_g \sin^2(\theta/2)$ (with θ the opening angle). Consequently a sufficiently collinear pair contains together so little mass that they will in general end up in the same hadron. In other words in a model of this kind there are often hadrons from the fragmentation process, which have larger energies than the individual partons. This is *infrared stability* at work.

Let me end with two remarks. The first one is that all parameters of the string fragmentation in a multiparton system are fixed by the ones of a single qq-system, and although the Monte Carlo version [36] of the Lund model for multiparton events is not unique with respect to symmetric fragmentation etc, the possible deviations in different approaches are essentially not observable, i.e. the method is very stable.

The second comment is that there are strong experimental indications that something similar to "the string effect" is needed in any model. The string effect was predicted [37] and means that there should be an observable asymmetry in the phase space multiplicity distribution in the angular segments of a qgq-event. There are particles produced in the (qg) and the (gq) segments where there are string fields but much fewer observed in the qq segment where there is no field. Both the qualitative and quantitative properties (more multiplicity and more string field push for larger meson masses and for particles with larger momentum components out of the event plane - such particles take larger pieces of the field!) have been experimentally confirmed [38].

3G. Independent Fragmentation

The asymptotic freedom of QCD, which implies the possibility to treat the partons as noninteracting entities during a scattering process, may in some situations be extended also to the hadronization phase.

There are several difficulties in such a scheme, however. Basically one can divide these difficulties into problems related to Lorentz covariance and problems on infrared stability. In a truly Lorentz covariant model one should be able to treat any features of the model independent of the frame of reference. In connection with most independent fragmentation (IF) models it is, however, necessary to define a particular frame in order to conserve energy and momentum. Usually some common cms frame is chosen.

After that one or another of the iterative cascade jet algorithms is chosen for the fragmentation of each parton.

One is then in such a simple scheme faced with two basic difficulties. The first is that the cascades must be stopped somewhere and it is not evident at what point this should happen. Thus one must introduce some almost arbitrary cutoff procedure.

The second difficulty is that the original QCD parton state is usually formulated in terms of massless partons, i.e. the energy momentum vectors fulfil $e = |p|$. When masses and transverse momentum is given to the hadrons of the cascade, this relation is no longer possible to fulfil. Therefore some "reshuffling scheme" must be used e.g. a rescaling of the energy and changes of the momenta or eventually some twisting of the angles between the parton momenta.

There is of course also the question of how to conserve flavour etc for the low energy particles at the end of the cascade.

The problems of infrared stability are severe for most IF models. If taken to its extreme an IF model would treat two collinear partons with energies e_1 and e_2 and produce a number of particles of the order $\bar{n}_j = C \ln(e_j/m)$ from each of them. For large values of the energies e_j it is evident that this will lead to a large overproduction if we compare to the result of producing particles from the two partons together, i.e. $\bar{n}_{12} = C \ln(e_1 + e_2/m) < \bar{n}_1 + \bar{n}_2$.

For the details of how different groups have chosen their ad hoc procedures I refer to the original papers [39] or to an excellent review [4].

In my opinion it is hardly viable to consider these models as more than possible parametrizations of data and I do not think that the authors of most IF models intended anything more than that.

There is, however, one scheme suggested by Montvay [40], which would solve all the problems mentioned above. Nobody has, however, yet picked on that possibility (although there is in the Lund Monte Carlo a simplified option of the Montvay model).

The idea is to use separate gluon-type (colour 8) and quark-type (colour 3) strings with different values of the string tension at rest κ_8 and $\kappa_3=\kappa$. Strings of the 8-type and the 3-type may be joined in junctions and the motion of such a junction is given by the composant of the string tensions acting upon it. For the case when $\kappa_8 > 2\kappa$ the 8-type of strings are never dragged out and one recovers the ordinary Lund model.

For the case when $\kappa_8 < 2\kappa$, however, one obtains a set of new models with properties very similar to those intended in IF models. To see that we may boost ourselves into the restframe of one junction where the three string segments from the joining partons extend in directions determined so that the sum of the tensions vanishes.

There will obviously be a large amount of different junctions and junction places available for a multiparton state. The parton state could be visualized like a polype with arms extending in all directions from a skeleton of 8-strings. Two collinear partons would (if they are colour connected) go away and pull with them a junction and then only be connected to the rest of the system by a single string. Therefore the Montvay scheme is both Lorentz covariant and infrared stable. Although all quantum numbers in principle can be locally and globally preserved an implementation of the Montvay model would probably be exceedingly difficult.

3H. Cluster Models of Fragmentation and Comparisons to Strings

Cluster models for particle production have been with us since the first analysis of high energy multiparticle interactions started. The notion of "cluster" is not a precise one, however. Sometimes clusters have been associated with the known resonances and sometimes with purely mathematical entities with a continuous mass spectrum and even a rather broad decay multiplicity distribution.

In PQCD parton cascade models there is, due to the expected "preconfinement" a natural mass distribution (with a strong peak for low masses) for colour adjacent partons. It is then natural to use a statistical phase-space decay and I will as an example consider the Webber model [12] in some detail.

For the Marchesini-Webber cascade model [12], the coherence properties of the parton emission is taken into account in a somewhat complicated way. The evolution algorithm is not explicitly Lorentz covariant, being defined in a particular frame and the kinematics is consequently difficult to see through.

The evolution is carried out to rather small (virtual) masses with a cutoff m_{min} of the order 1.4 GeV ($= 2m_g$ with m_g a fictitious gluon mass). There is also similarly fictitious quark masses (for u and d and 300 MeV and for s around 500 MeV) so that at the end of the cascade most gluons decay directly to qq-pairs (those who "refuse" are finally "forced" to do it). It should be noted that in the middle of any cascade the probability for gluons to decay into gluons is almost a factor of 10 larger than for the decay into qq-pairs.

Thus, at the end the cascade contains a large amount of colour connected qq-pairs stemming in general from different decay chains. These pairs are then combined into colourless clusters. The clusters are finally allowed to decay into final state hadron pairs according to a statistical isotropic phase-space (relative probabilities $(2s_1+1)(2s_2+1)2p^*/m$ with s_j j=1,2 the spins and p^* the common momentum of the hadrons in the restframe of the cluster).

Therefore, at first sight, both the p_T-properties and the flavour properties of the final state hadrons seem to come out of the parton cascade together with a simple statistical ansatz. There is, however, as always a large amount of extras needed to understand the detailed properties of the experimental data.

Firstly, although most of the clusters have masses of the order of 1-3 GeV, there is a tail out to essentially larger masses (I have been told that this tail is basically independent of the cms energy). Therefore a method to split large masses has to be introduced.

It is interesting to note that due to the coherence properties of the cascade and this splitting operation the distribution of the final clsuters is very similar to a "chain" stretched along the same momentum space curves as a string would be stretched.

Secondly, in order to reproduce the baryon spectrum (remember the many parameters needed to fit detailed properties of baryon fragmentation in jet cascade algorithms!) it is necessary to allow the decay of gluons also into di-di-pairs with a rate of about 5 % as compared to qq-pairs. In this way a $B\bar{B}$-pair may stem from two different clusters (there will consequently be a "drag-effect" in a sense similar to the $B\bar{B}$ production out of a stretched out string).

Thirdly, if all clusters should decay into 2-particle states, then there is a severe underestimate of very fast (large fractional energy momentum) single hadrons and particular care must be taken to include such effects.

Finally it has been pointed out that there is a necessity to be careful with respect to infrared stability in any cluster algorithm based upon a parton cascade. If a single wee gluon is emitted or not emitted in between two branches of the cascade then the final state cluster mass may be very different. To see that we note that if the gluon is not emitted then the qq-pair forming the cluster may in their common cms go apart with $e+p=\bar{e}-\bar{p}=M$. If a gluon is emitted (with energy ω in this frame) then a new q'q'-pair is formed and we obtain two clusters each with the approximate energy $M\omega/2$. This example exaggerates the problem to some extent, but some means to ensure that there is no sudden large breakdown in the mass spectrum is nevertheless needed.

With these notes in mind it should be said that cluster decay has a simple beauty and that the Webber model is a very useful and versatile model, which seems to explain almost all the data from precise hadronization studies (just as well as string fragmentation algorithms do it).

It is interesting to note the large similarities of the two models now - a string or a chain of clusters stretched in a similar way - despite the fact that they were conceptually very different when conceived. It is my own feeling that (except for hyperon physics, for Bose-Einstein correlations and eventually for polarization properties - there is a "natural" polarization mechanism in string theories [41]) there is for the immediate future more to be done on the basic properties of the QCD parton model than on the fragmentation physics per se. When the energies increase there is a larger and larger phase-space available for radiation. It will be most interesting to see whether the present day parton cascade models are able to understand the possible activities in e.g. the transverse momentum space.

References

1. C.F. v.Weizsäcker, Z. Phys. $\underline{88}$ (1934) 612; E.J. Williams, Phys. Rev. $\underline{45}$ (1934) 729; M.S. Chen, P. Zerwas, Phys. Rev. $\underline{D12}$ (1975) 187

2. G. Altarelli, G. Parisi, Nucl. Phys. $\underline{B26}$ (1977) 298

3. W. Hofmann, Invited Talk at the 1987 International Symposium on Lepton and Photon Interactions, Hamburg 1987
    H. Yamamoto, Proceedings of the 1985 International Symposium on Lepton and Photon Interactions, eds. M. Konuma, K. Takahashi (Kyoto 1986)
    S. Bethke, Proceedings of the XXIII International Conference on High Energy Physics, ed. S.C. Lohen (World Scientific, Singapore 1987)
    S.L. Wu, Phys. Rep. $\underline{107}$ (1984) 59
    Mark II Collaboration, A. Petersen et al., SLAC-Pub-4290(LBL 2343 (1987)

4. T. Sjöstrand, LU TP 87-18 (1987), to be published in International Journal of Modern Physics A

5. B. Andersson, LU TP 87-20 (1987), to be published in Proceedings of the Multiparticle Symposium 1987 (World Scientific, Singapore)

6. D. Amati, G. Veneziano, Phys. Lett. $\underline{83B}$ (1979) 87
    G. Marchesini, L. Trentadue, G. Veneziano, Nucl. Phys. $\underline{B181}$ (1981) 335

7. C. Basham, L. Brown, S. Ellis, S. Love, Phys. Rev. Lett. $\underline{41}$ (1978) 1585

8. W. Bartel et al. (JADE Collaboration), Phys. Lett. $\underline{101B}$ (1981) 129, $\underline{134B}$ (1984) 275, Z. Physik $\underline{C21}$ (1983) 37
    A. Petersen, in Elementary Constituents and Hadronic Structure (ed. J. Tran Thanh Van, Editions Frontière Dreux 1980)
    H. Aihara et al. (TPC/2γ Collaboration), Z. Physik $\underline{C28}$ (1985) 31
    M. Althoff et al. (TASSO Collaboration), Z. Physik $\underline{C29}$ (1985) 29

9. A.H. Mueller, Phys. Lett. $\underline{104B}$ (1981) 161
    B.I. Ermalaev, V.S. Fadin, JETP Lett. $\underline{33}$ (1981) 269
    A. Basetto, M. Ciafaloni, G. Marchesini, Phys. Rep. $\underline{100}$ (1983) 201

10. L.A. Gribov, E.M. Levin, M.G. Ryskin, Phys. Rep. $\underline{100}$ (1983) 1
    Ya.I. Azimov, Yu.L. Dokshitzer, V.A. Khoze, S.I. Troyan, Phys.Lett. $\underline{165B}$ (1985) 147, Leningrad preprint 1051 (1985)

11. K. Konishi, A. Ukawa, G. Veneziano, Nucl. Phys. B157 (1979) 45

12. G. Marchesini, B.R. Webber, Nucl. Phys. B238 (1984) 1
    B.R. Webber, Nucl. Phys. B238 (1984) 492
    M. Bengtsson, T. Sjöstrand, Phys. Lett. 185B (1987) 435, Nucl. Phys.
    B289 (1987) 810
    T.D. Gottschalk, D.A. Morris, Nucl. Phys. B288 (1987) 729
    D.A. Morris, Ph.D. thesis Caltech, CALT-68-1440 (1987)

13. T. Sjöstrand, Phys. Lett. 157B (1985) 321
    M. Bengtsson, T. Sjöstrand, M. van Zijl, Z. Physik C32 (1986) 67
    B.R. Webber, Am. Rev. Nucl. Part. Sci. 36 (1986) 253

14. G. Gustafson, Phys. Lett. 175B (1985) 453
    G. Gustafson, U. Pettersson, LU TP 87-9 (1987)

15. H.-U. Bengtsson, Comp. Phys. Comm. 31 (1984) 323
    T. Sjöstrand, Fermilab Pub 85/119 T (1985)
    T. Sjöstrand, M. van Zijl, Lund preprint LU TP 86-25, LU TP 87-5
    H.-U. Bengtsson, T. Sjöstrand, Comp. Phys. Comm. 46 (1987) 43

16. F.E. Paige, S.D. Protopopescu, in Physics of the Superconducting
    Supercollider 1986, eds. R. Donaldson, J. Marx (1987) p.320

17. A. Capella, J. Tran Thanh Van, Phys. Lett. 114B (1982) 450, Z. Physik
    C18 (1983) 85, C23 (1984) 212
    P. Aurenche, F.W. Bopp, J. Ranft, Z. Physik C23 (1984) 67, C26 (1984)
    279, Phys. Lett. 147B (1984) 212

18. B. Andersson, G. Gustafson, B. Nilsson-Almqvist, Nucl. Phys. B281 (1987)
    289, LU TP 87-6 (1987)
    B. Nilsson-Almqvist, E. Stenlund, Comp. Phys. Comm. 43 (1987) 387
    B. Andersson, Talk at the Shandong Workshop, China (Proceedings to be
    published by World Scientific, Singapore)

19. L.A. Gribov, E.M. Levin, M.G. Ryskin, Phys. Rep. 100 (1983) 1

20. A. Krzyvichi, B. Petersson, Phys. Rev. D6 (1972) 924
    J. Finkelstein, R.D. Peccei, Phys. Rev. D6 (1972) 2606

21. F. Niedermayer, Nucl. Phys. B79 (1974) 355

22. R.D. Field, R.P. Feynman, Nucl. Phys. B136 (1978) 1
    B. Andersson, G. Gustafson, C. Peterson, Z. Physik C1 (1979) 105

24. B. Andersson, G. Gustafson, G. Ingelman, T. Sjöstrand, Phys. Rep. 97 (1983) 31

25. B. Andersson, G. Gustafson, B. Söderberg, Z. Physik C20 (1983) 317

26. J. Schwinger, Phys. Rev. 128 (1962) 2425

27. A. Casher, J. Kogut, L. Susskind, Phys. Rev. D10 (1974) 732

28. W. Heisenberg, H. Euler, Z. Physik 98 (1936) 714
    J. Schwinger, Phys. Rev. 82 (1951) 664
    E. Brezin, C. Itzykson, Phys. Rev. D20 (1979) 179
    B. Andersson, G. Gustafson, T. Sjöstrand, Z. Physik C6 (1980) 235

29. B. Andersson, W. Hofmann, Phys. Lett. 169B (1986) 364

30. K. Gottfried, F.E. Low, Phys. Rev. D17 (1978) 2487

31. B. Andersson, G. Gustafson, T. Sjöstrand, Z. Physik C6 (1980) 235

32. B. Andersson, G. Gustafson, LU TP 82-5 (1982)

33. B. Andersson, G. Gustafson, T. Sjöstrand, Nucl. Phys. B197 (1982) 45
    T. Meyer, Z. Physik C12 (1982) 77

34. A. Casher, H. Neubeyer, S. Nussimov, Phys. Rev. D20 (1979) 179
    B. Andersson, G. Gustafson, T. Sjöstrand, Physica Scripta 32 (1985) 574

35. B. Andersson, G. Gustafson, O. Månsson, LU TP 82-13 (1982)

36. T. Sjöstrand, Phys. Lett. 142B (1984) 420, Nucl. Phys. B248 (1984) 469

37. B. Andersson, G. Gustafson, T. Sjöstrand, Phys. Lett. 94B (1980) 211

38. A. Petersen, Elementary Constituents and Hadronic Structure (ed. J. Tran Thanh Van, Editions Frontière, Dreux 1980)

39. P. Hoyer, P. Osland, H.G.Sander, T.F. Walsh, P.M. Zerwas, Nucl. Phys. B161 (1979) 349

A. Ali, J.G. Körner, G. Kramer, J. Willrodt, Nucl. Phys. B168 (1980) 409

A. Ali, E. Pietarinen, G. Kramer, J. Willrodt, Phys. Lett. 93B (1980) 155

40. I. Montvay, Phys. Lett. 84B (1979) 331

41. B. Andersson, G. Gustafson, G. Ingelman, Phys. Lett. 85B (1979) 417

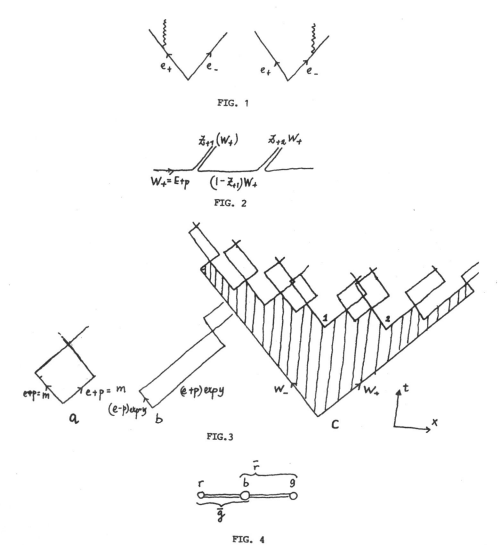

FIG. 1

FIG. 2

FIG. 3

FIG. 4

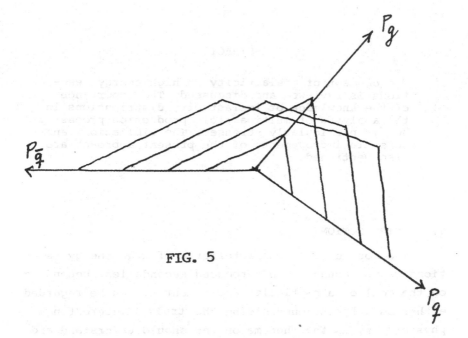

FIG. 5

THE CONCEPT OF INELASTICITY IN HIGH ENERGY REACTIONS

G. Wilk

Institute for Nuclear Studies
ul. Hoża 69; 00-681 Warsaw
Poland

ABSTRACT

The concept of inelasticity in high energy reactions is reviewed and discussed. The importance of the knowledge of inelasticity distributions in the analysis of multiparticle production processes is particularly stressed. The limitations and possible improvements of the present approach are also mentioned.

1. INTRODUCTION

The most obvious characteristic of high energy reactions is the abundance of produced secondaries. Depending on the goal of a particular experiment it can be regarded either as a "background hiding the truly interesting physics" or, as the phenomenon one should understand and keep under the control. It is this latter attitude we are committed to in what follows. There is a number of experimental observations, which persists through the whole history of multiparticle production[1] (concerning the behaviour of mean multiplicity $\bar{n}$, transverse momentum p_T of secondaries, multiparticle distributions $P(n)$, Bose-Einstein correlations etc.) which point strongly towards the common statistical approach to multiparticle reactions.

Here the concepts of local equilibrium, hydrodynamics,
quantum statistics, chaoticity and coherence etc. form
the backbone of our thinking. In every approach of this
type the central variable is the energy available in a given event W, not the total energy $\sqrt{s}$ of the reaction.
This leads us directly to the concept of inelasticity,
which is the central issue of this review.

2. THE CONCEPT OF INELASTICITY

Long ago, in cosmic ray experiments, an interesting
phenomenon was observed, namely that not the whole available c.m. energy $\sqrt{s}$ can be found among the observed secondaries, but only a fraction K of it:

$$K = \frac{W}{\sqrt{s}} \; ; \qquad K \cong 0.3 \div 0.6, \qquad\qquad (1)$$

which then defines the energy $W = K\sqrt{s}$ actually "used up"
to produce the observed particles.[2] For a long time K
was regarded as a constant parameter. The subsequent analysis of cosmic ray data suggested, however, that K should
decrease with energy $\sqrt{s}$ in order to explain the observed
significant scaling violations in the beam fragmentation
region.[3] It is claimed to be true also in UA-5 data.[4]
An independent analysis of inelasticity K based on accelerator data also points towards the decrease of K from
K = 0.5 (ISR) to K = 0.25-0.3 (SPS).[5]

If the centraly produced particles consume only a fraction K of the available c.m. energy $\sqrt{s}$, the rest of it,
$(1-K)\cdot\sqrt{s}$ has to be found outside the central region of
particle production. This has led to the concept of leading particle (LP) (or particles). Their spectrum

$$f(x_L = \frac{2p_L}{\sqrt{s}}) = \frac{d\sigma}{\sigma dx_L}$$

has been measured up to ISR energies[6,7] and found to be
reasonably flat. Apparently two uncorrelated LP are pro-

duced with $<x_L> = 0.5$ at ISR energies. This means that we must have a whole spectrum of K, $\chi(K)$ with maximum at $K \cong <K> = \int_0^1 dK K \chi(K)$, which decreases with $\sqrt{s}$ ($\chi(K)$ is normalized to unity). The only experimental information so far, which allows for direct extraction of $\chi(K)$ is provided by the bubble chamber measurement performed by Brick et al. [8] at $\sqrt{s} = 16.5$ GeV (the quoted experimental $\chi(K)$ was obtained from it in Ref. 9). So far we have just introduced and defined inelasticity K as one more possible observable. Where and why is it so important as to deserve a special study? These are dynamical questions and can be answered only in a dynamical context. Already in Sec. 1 we outlined our indications by stressing the apparent necessity of employing statistical methods to multiparticle production. Here, all observables depend crucially on the energy in a given event, W, rather then the total energy of reaction $\sqrt{s}$. This energy W is released into a sort of strongly interacting central lump of hadronic matter, whose properties we are interested in. It is then crucial to know the distribution of W from event to event in order to learn something reliable about the properties of hadronic matter.

The statistical models were, in fact, first models of high energy collisions, proposed just after cosmic rays experiments revealed their multiparticle character [10] (understood as an excitation of many degrees of freedom). Later on, we shall say a bit more on the most successful of them, Landau Hydrodynamical Model (LHM), the only model which - after incorporating some modern, QCD based, ideas of hadronic structure - has successfully coped with data for more than 3 decades. [11]

Despite all their successes, the statistical models were neither popular nor fashionable until the possibility of the accurance of phase transition between the hadronic matter and the deconfined quark-gluon plasma (QCD plasma) was discovered. Then their usefulness was acknowledged and enormous progress in their understanding and development was achieved[12]. Indeed, only these models can accomodate such phase transition[13]. The currently popular phenomenological models able to describe all possible sets of experimental data, such as DTU[14] or LUND[15] models, simply do not admit of such a notion at all, moreover, in order to adequatelly address or describe it, they would have to be profoundly modified by introducing a kind of equilibriation. But it also means that they would then become just another kind of statistical models, with inelasticity emerging again as a most important variable.

Our interest in the inelasticity was triggered by the problem of initial conditions encountered in LHM[16]. To solve the equations of hydrodynamics one has to specify the initial energy density $\varepsilon = \frac{W}{V}$. With the $\chi(K)$ we know that $W = K\sqrt{s}$. But what concerns the initial volume V the situation is far from clear. It is supposed to be the volume in which the initial stopping and equilibration of the energy W takes place. (It is precisely because of the unsatisfactory situation in this field and because of the belief that finally we should observe a clear plateau in the rapidity distribution, that, so called scale-invariant hydrodynamical model has been formulated[17]. In fact, this is the most popular version of hydrodynamical model in use now). In the original LHM the initial volume $V = V_o/\gamma$ depends on the volume of the hadron at rest, V_o, and on the Lorentz contraction factor $\gamma = \sqrt{s}/2m$, where $\sqrt{s}$ is the total energy and m is the mass of the hadron. The mean multiplicity in an event where $W = K\sqrt{s}$ is then proportional to the total entropy S of the system and reads:

$$\bar{n} \sim S \sim V \cdot T^3 = \text{const } K^{3/4} \cdot s^{1/4} \qquad (2)$$

(where T is the temperature and for the equation of state

used $\varepsilon \sim T^4$).

Unfortunately, precisely this relation was found[18] to be incompatible with data at fixed inelasticity K[6,19], i. e. the $\sqrt{s}$ dependence (for fixed K) was confirmed whereas the K dependence (for fixed $\sqrt{s}$) not. The experimental data show rather that

$$\bar{n} = \text{const } W^{1/2} = \text{const } (K\sqrt{s})^{1/2} \qquad (3)$$

strongly suggested[18] that things happen as if the Lorentz contraction is not determined by the total energy $\sqrt{s}$ but rather by the affective energy W. Indeed, with this changes eq. (2) emerges. The analysis in ref. 18 was performed in the picture of the throughgoing valence quarks and stopping glue of the incoming hadron, adapted to the LHM from ref. 20. This would mean, that only the gluonic (interacting) part of the hadron suffers Lorentz contraction which takes place after the leading particles (emerging from valence quarks) have left the system. The apparent contradiction with the standard parton picture, in which the valence partons, not the wee ones (identified with gluons), are Lorentz contracted is then obvious. What is, however, not so obvious, is whether the Lorentz factor for a quark-gluon system has just its classical value. The analysis performed by Mueller[21] has shown that, in fact, γ depends on the dynamics of the system and can, in principle, take any value.

2. INELASTICITY DISTRIBUTION

The first attempts to provide inelasticity distribution $\chi(K)$ with the desired properties discussed above were merely simple parametrizations.[22-24] The QCD-motivated derivation of $\chi(K)$ was presented in ref. 25 and its generalization, taking into account also the distribution in the rapidity Δ of the system of secondaries, is described in

ref. 26. It is based on the gluon-gluon interaction picture for hadronic reactions,[20] which received special attention[18,27] when the intrinsically chaotic properties of
the gluon fields were discovered[28] (they allow for a much
faster dissipation of the initial energy into gluonic component than it was claimed before on the basis of perturbative QCD only[12]).

Let us assume that in every event at least two uncorrelated leading particles (LP) are produced and that they
arise from the throughgoing valence quarks of the incoming
hadrons. The glue components of those hadrons are expected
to interact strongly and finally produce the secondaries,
populating the central rapidity region. The dominance of
the gluon interactions over constituent quark interactions,
implicit here, is just a consequence of QCD (and the apparent intrinsic chaoticity of the gluonic field mentioned
before). We have then effectively the picture presented in
Fig. 1. The strongly interacting gluons produce what we
call "minifireballs" (MF), which, in turn, are supposed to
merge and equilibrate[27,28] and form a central fireball
(CF) of energy W and momentum P (or invariant mass
$M = \sqrt{W^2 + P^2}$ and rapidity $\Delta = \frac{1}{2} \ln\{(W+p)/(W-p)\}$ in the

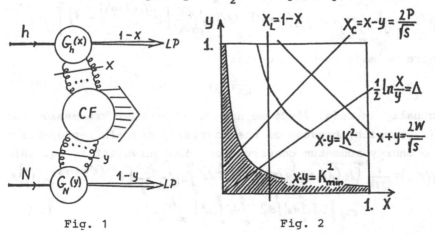

Fig. 1   Fig. 2

overall c.m.s.). Neglecting masses we have

$$W = \frac{\sqrt{s}}{2} (x+y), \quad P = \frac{\sqrt{s}}{2} (x-y) \tag{4}$$

or

$$M = \sqrt{xys}, \quad \Delta = \frac{1}{2} \ln \frac{x}{y}. \tag{5}$$

Here, and in what follows, (x,y) are the fractions of the energies taken off the corresponding incoming hadrons and allocated to the CF (via an indefinite member of MF's). We define then the probability function to form such a CF as

$$\chi(x,y) = \sum_{\{n_i\}} \delta(x - \sum_i n_i x_i) \delta(y - \sum_i n_i y_i) \prod_{\{n_i\}} \tilde{P}(n_i) \tag{6}$$

where the summation is over the number of all possible MF's distributed according to $\tilde{P}(n_i)$. For simplicity we shall assume that MF's are produced independently[29], i.e. that

$$\tilde{P}(n_i) = \bar{n}_i^{n_i} \exp(-\bar{n}_i)/n_i. \tag{7}$$

(In this respect our approach resembles hadron[30], gluon[20] or photon bremsstrahlung[31] models.) Expressing δ functions via Fourier integrals one can perform all summations and arrive at the general formula:

$$\chi(x,y) = \frac{1}{(2\pi)^2} \int_{-\infty}^{+\infty} dt \int_{-\infty}^{+\infty} ds \cdot \exp\left\{ i(xt+ys) + \int_0^1 dx' \int_0^1 dy' \frac{d\bar{n}}{dx'dy'} \cdot \left[e^{-i(x't+y's)} - 1 \right] \right\} \tag{8}$$

where we made a substitution:

$$\sum_i \bar{n}_i (\ldots) \to \int_0^1 dx' \int_0^1 dy' \frac{d\bar{n}}{dx'dy'} (\ldots). \tag{9}$$

Actually in refs. 25–26 we have used another expression for $\chi(x,y)$ which, because of the Fourier integrals representing the energy momentum conservation, is equivalent to eq. (8):

$$\chi(x,y) = \frac{1}{(2\pi)^2} \int_{-\infty}^{+\infty} dt \int_{-\infty}^{+\infty} ds \cdot \exp\left\{ i(xt+ys) + \int_0^x dx' \int_0^y dy' \frac{d\bar{n}}{dx'dy'} \left[e^{-i(x't+y's)} - 1 \right] \right\} \cdot$$
$$\cdot \exp\left\{ -\left[\int_0^1 dx' \int_0^1 dy' - \int_0^x dx' \int_0^y dy' \right] \frac{d\bar{n}}{dx'dy'} \right\} \cdot \tag{10}$$

It arises when in the substitution (9) we remember that for
x'>x and y'>y the actual number of MF produced has to be
zero, i.e. in this region of phase space $P(n_i=0) = e^{-\bar{n}_i}$,
which immediately leads to eq. (10). Of course, both equations (8) and (10) are equivalent and would lead to identical numerical results, if no further approximations were performed. But as we have to resort to some approximation scheme in order to proceed with numerical calculations, they will correspond to slightly different parametrizations.

In both cases the central ingredient is the spectral
function for the production of MF's: $\frac{d\bar{n}}{dxdy}$. As they are formed in the process of the interaction of gluons from h and
N, it is natural to assume that $\frac{d\bar{n}}{dxdy}$ is proportional to
the number of gluons in h and N: $G_h(x) \cdot G_N(y)$ and that the
interaction is represented by the gluon-gluon cross sec-
tion $\sigma_{gg} = \sigma_{gg}(W=K\sqrt{s})$ and normalized to the inelastic
cross section for hN scattering $\sigma_{hN} = \sigma_{hN}(\sqrt{s})$:

$$\frac{d\bar{n}}{dxdy} = \frac{\sigma_{gg}(K\sqrt{s})}{\sigma_{hN}(\sqrt{s})} G_n(x)G_N(y)\, \theta(xy-K^2_{min}) = \omega(x,y)\, \theta(xy-K^2_{min});$$
$$K_{min} = \frac{2\mu}{\sqrt{s}}. \qquad (11)$$

The θ-function, which fixes our phase space and cuts-off
the potentially dangerous region of $x,y \to 0$, originates from
the requirement, that CF has to contain at least one MF,
which in turn has to decay to, at least, two particles
(clusters?) of effective mass μ each; μ is then a free parameter. The simplest possible form of $\sigma_{gg}(w)$ which incorporates the most general trends of the cross sections is:

$$\sigma_{gg} = \frac{\alpha}{w^2} + \delta \ln \frac{w^2}{\bar{\mu}^2}. \qquad (12)$$

In this way we have three more parameters: α, δ and a scale $\bar{\mu}$. As for $G_{h,N}$ we have assumed that they are equal to

the standard gluon distribution functions; we have used G(x) of ref. 32 with scaling violations included and governed by Q^2 = xys.

Because $\frac{dN}{dxdy}$ is strongly peaked at small x,y it is reasonable to make an approximation:

$$\exp\{-i(x't+y's)\}-1 \cong -i(x't+y's)-\frac{1}{2}(x't+y's) \qquad (13)$$

This leads to a Gaussian integral for $\chi(x,y)$ which can easily be evaluated and gives (χ_0 is the normalization factor ensuring that $\int dx \int dy \, \chi(x,y)\theta(xy-K_{min}^2)= 1$):

$$\chi(x,y)=\chi_0 \frac{\exp(-I)}{2\pi \sqrt{h^2}} \cdot \exp\left\{-\frac{1}{2h^2}\left[I_{02}(x-I_{10})^2+I_{20}(y-I_{01})-\right.\right.$$
$$\left.\left. - 2 I_{11}(x-I_{10})(y-I_{01})\right]\right\} \qquad (14)$$

where

$$I_{nm}(x,y) = \int_0^x dx'x'^n \int_0^y dy'y'^m \, \theta(x'y'-K_{min}^2)\omega(x'y')$$

$$I(x,y) = \{ \int_0^1 dx' \int_0^1 dy' - \int_0^x dx' \int_0^y dy'\} \, \theta(x'y'-K_{min}^2)\omega(x'y')$$

$$h = (I_{20} \cdot I_{02} - I_{11}^2)^{1/2}. \qquad (15)$$

From $\chi(x,y)$ one can immediately get distributions in other variables and calculate single-variable distributions. They probe the allowed phase space, $x\in(0.1)$, $y\in(0.1)$, $xy>K_{min}^2$, in different ways, c.f. fig. 2.

$$\chi(K) = \int_0^1 dx \int_0^1 dy \, \theta(xy-K_{min}^2) \, \delta(\sqrt{xy}-K) \, \chi(x,y); \qquad (16)$$

$$\chi(\Delta) = \int_0^1 dx \int_0^1 dy \, \theta(xy-K_{min}^2) \, \delta(\frac{1}{2}\ln\frac{x}{y} - \Delta) \, \chi(x,y); \qquad (17)$$

$$\chi(x_c) = \int_0^1 dx \int_0^1 dy \, \theta(xy-K_{min}^2) \, \delta(x-y-x_c) \, \chi(x,y); \qquad (18)$$

$$f(x_L) = \int_0^1 dx \int_0^1 dy \, \theta(xy-K_{min}^2) \, \delta(1-x-x_L) \, \chi(x,y); \qquad (19)$$

$$\chi(K,\Delta) = 2K\,\chi(x=K\exp(\Delta),y=K\exp(-\Delta)),\qquad(20)$$

$$\chi(W,P) = \frac{2}{s}\,\chi(x=(W+P)/\sqrt{s},y=(W-P)/\sqrt{s}).\qquad(21)$$

All the relevant moments can now be easily evaluated by using corresponding distributions, eqs.(16-21).

The comparison with available data for pp collisions and predictions for higher energies are presented in fig.3.

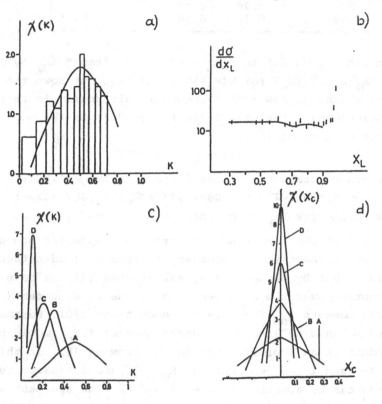

Fig.3. a) (K) at ISR. Comparison with data from ref.8. b) (d /x) for the leading particle at ISR energies. Comparison with data from ref.6. c) χ(K) at s=16.5 – 62.5 GeV (A), 540 GeV (B), 2 TeV (C) and 40 TeV (D). d) The same for $\chi(x_c)$.

Table I contains the results for selected moments for

various energies (the values with asterix served as an input)(c.f. ref.26).

Table I $(D(...) = [\langle(..)^2\rangle - \langle(..)\rangle^2]^{1/2}$

s(GeV)	K	D(K)	$D(x_c)$	$D(\Delta)$
16.5-62.5	0.49*	0.19	0.20	0.20
540	0.30*	0.12	0.11	0.18
2000	0.20	0.10	0.07	0.17
40000	0.10	0.06	0.04	0.15

To get a good fit to $f(x_L)$ one has to "tame" σ_{gg} by let-
ting $\sigma_{gg}(W) = \sigma_{gg}(W_o)$ for $W < W_o = 5GeV$ and also cut down the number of gluons at low x (c.f.,however, discussion in ref.33)
by introducing a factor $\beta < 1$ (in fact $\beta = 0.1$) in the most divergent part of G(x), i.e.

$$G(x) \sim \frac{\beta}{x} \quad \text{for } x \ll 1. \tag{22}$$

As to other parameters: $\bar{\mu}$ was fixed at value $\bar{\mu} = 1.0GeV$, $\mu =$
0.3GeV, $\alpha = 9.4GeV^2 \cdot \sigma$, $\delta = 0.08 \cdot \sigma$, ($\sigma = \sigma_{NN}(16.6GeV) = 32mb$).With
these parameters σ_{gg} at low energies turns out to be 25 mb.

A word of caution is needed here. This apparent abundance of the parameters is necessary to take full advantage of
the fact,that by using eq.(10) rather then (8), we have better control over the x_L dependence of the LP spectrum (as
now all moments I_{nm} are (x,y)-dependent and $I \neq 0$). Parameters
W_o and β "shaping" the (x,y)-dependence of I,I_{nm} would be
redundant in an approach originating from eq.(8). In this
case I=0 and I_{nm} depend only on $K_{min} = \frac{2\mu}{\sqrt{s}}$, α and δ (parameters
W_o and β can be absorbed into μ, α and δ). The approach sim-
lifies enormously (in fact, for simpler choices of G(x),
everything can be estimated analytically),the price is the
deterioration of the fit to $f(x_L)$. But, apart from this, the
$\chi(K)$ and energy dependence of $\langle K \rangle$ can be fitted quite satis-
factorily with results almost identical to those in fig.3.
In this approach one can show that, approximately (for h=N):

$$\chi(x,y) \simeq \exp\left\{-\frac{(x-I_{10})^2}{2I_{20}} - \frac{(y-I_{01})^2}{2I_{02}}\right\} . \tag{23}$$

The energy dependence of $\chi(x,y)$ can now be easily traced down to the relative importance of the different terms in $\sigma_{\eta\eta}$. For example, keeping only leading terms in x^{-1} and $\sqrt{s}$ one gets

$$I_{10} = I_{01} \simeq \frac{\alpha}{\sigma_{pp}(s)} + \frac{1}{2}\delta\frac{\ln^2\frac{s}{\mu^2}}{\sigma_{pp}(s)} ,$$

$$I_{20} = I_{02} \simeq \frac{1}{2} I_{10} . \tag{24}$$

These asymptotic relations are also true for I_{nm} from eq. (10). It then means that asymptotically

$$\chi(x,y) \xrightarrow[\sqrt{s}\to\infty]{} \delta(x-I_{10}) \, \delta(y-I_{01}) \tag{25}$$

with I_{10}, I_{01} given by eq.(25). If $\delta > 0$, $<K>$ will eventually stop to decrease and converge to $<K> \sim \ln^2(s/\mu^2)/\sigma_{pp}(s)$.[34]

3.   COMMENTS AND SUMMARY

The approach presented here is an attempt to isolate the part of the dynamics responsible for the production in the central region in rapidity. Such separation can never be perfect as the definition of the central region is not really fixed, it rather depends on the conditions of a given experiment. This is a common weakness of all such approaches and should be kept in mind. By now we have a fairly well established picture of the hadron, its composition in terms of valence quarks and gluons. It should then form a basis of every attempt to describe the multiparticle processes. Our $\chi(x,y)$ or $\chi(K,\Delta)$ are supposed to do just this, to provide for every event in which (invariant) energy $K\sqrt{s}$ is released at (mean) rapidity Δ, the probability of its occurence in terms of what is known about hadronic structure and basic interactions. (Notice,that we, sort of, "convert" all eventual sea of $q\bar{q}$ pairs back into an equivalent glue component; an opposite attitude was advocated in some recombination models[36]).

As we have already discussed, there is an implicit statistical picture behind $\chi(K,\Delta)$ and not everyone is expected to agree with it. So far, however, our considerations were mostly model independent since nowhere did we use the assumption that we have one, equilibrated CF. The nature of CF comes into play once we want to calculate any other measured observable besides the ones presented above. Then we would prefer to have precisely one CF, whose detailed properties, like multiplicity distribution, p_T dependence of secondaries, single particle rapidity distributions etc. are already given by some other model for this CF in terms of its (invariant) mass $M=K\sqrt{s}$ and rapidity Δ. Then our $\chi(K,\Delta)$ smears it out over the available phase space in a highly controlled way. We can then concentrate on deducing the (unknown) dynamics of the lump of matter forming CE[24,37].

There are also other approaches using the concept of inelasticity, although none of them provides (or even utilizes) its distribution. It is, however, reassuring to find that the apparent decrease of $\langle K \rangle$ with energy up to $\langle K \rangle =0.3$ at SPS energies, was also found in the independent analysis using geometrical model and partition temperature concept[38]. This raises the question of an interrelation between the impact parameter description (geometrical models[39]) and that with inelasticity. They seem to be complementary, stressing different aspects of reactions and there is no one-to-one correspondence between them. The reason is quite obvious: the inelasticity K is (in principle) a measurable quantity whereas the impact parameter b is not (being a conjugate variable to the transverse momentum transfer). We can then have $\chi(K,\Delta;\vec{b})$ but not $\chi(b)$ instead of $\chi(K,\Delta)$ (or vice versa). In fact, this is the only way to connect inelasticity with the cross-sections[40].

The concept of inelasticity distributions, as presented here, will most surely have to develop and reach higher le-

vel of sophistication once new applications are discovered. For one thing, the leading particle which we so sharply separated from a central region, can be (and most probably is) a kind of a leading jet (composed of valence quarks and the remnants of glue). It can then produce its own particles through bremssthralung or fragmentation. This leads naturally to a two-component model. The question is: what kind of observables can tell us the difference between those two components? It turns out that there is such an observable, namely the chaoticity of the system produced[41]. If it is really the case then one should rather have:
$$\chi(x_c, x_{ch}; y_c, y_{ch}) \cdot \delta(x-x_c-x_{ch}) \cdot \delta(y-y_c-y_{ch})$$ describing the allocation of the given total fraction of the colliding hadron momenta x to the coherent (x_c, y_c) and chaotic (y_c, y_{ch}) components. Such a picture, gradually emerging as the most obvious and natural from many analysis of data and models, is actually under consideration. (For other recent two-component approaches see ref.42).

The concept of inelasticity ought to be implemented in hadron-nucleus and nucleus-nucleus reactions too.This seems to be quite straightforward once the parameters of the model are fixed from the pp scattering data. But this apparent simplicity is deceptive if one wants to get a better understanding of some details of nuclear collision (for example, where is the energy taken off the leading particles actually stored, how many CFŝ we have, does nucleus act collectivelly or is it just a system of mostly independently interacting nucleos etc.)[43]. In ref.44 we have presented first, obviously oversimplified, attempt to fit the leading proton spectra in pA reactions (as measured in ref.7). The only new parameter was the factor $\bar{\nu}$ multiplying the moments I_{nm}, I. It represents an effective number of nucleons participating in the scattering (so we are assuming here that nucleus acts collectivelly) and is given by

$$\bar{\nu} = \frac{A\sigma_{pr}}{\sigma_{\rho A}} + \nu_0 \qquad\qquad (26)$$

where ν_0 is a parameter and it was found to be $\nu_0 = 1.8$; the $(\bar{\gamma} - \gamma_0)$ is taken from nuclear scattering data.(The fact that $\nu_0 \neq 0$ is a measure of importance of all other nuclear effects which we have ignored here).The average inelasticities $\langle K \rangle$ for C and Pb nuclei were found to be 0.57 and 0.69, indicating the increase of the overall energy loss.

In order to get a better understanding of the stopping power of the nuclei (or the degradation of energy of the LP) one has, however, to resort to a much more detail picture of the nuclear collision than that presented above. The one we have adopted is, in fact, simplified version of the so called coherent tube[45] or effective target[46] models (as the distribution over the number of participating nucleons P_ν was assumed to be given by simple δ function: $P_\nu = \delta(\nu - \bar{\nu})$). These are not the only ones possible and surely not the ones mostly advocated[47]. Investigation along these lines is currently in progress[48].

I would like to express my gratitude to G.N.Fowler, F.Navarra,M.Plumer,A.Vourdas and R.M.Weiner on collaboration with whom this work is based.(It is a part of the project supported by the German Federal Minister for Research and Technology (BMFT)). My sincerely thanks go also to S.Raha for valuable discussions and reading of the manuscript. The hospitality of the Department of Physics of the Marburg University where this work was finally finished is also gratefully acknowledged.

REFERENCES

1. C.f. Weiner,R.M. in "Hadronic Matter in Collision", ref.12;

2. Cocconi,G., Phys.Rev. 111, 1699 (1958);

3. Wdowczyk,J. and Wolfendale,A.W., Nature 306, 347 (1983), Nuovo Cim. 54A, 433 (1979) and J.Phys.G10, 257 (1983);

4. UA5 Coll., Alner,G.J. at al., Z.Phys. C33, 1 (1986);

5. Friedlander,E.M. and Weiner,R.M., Phys.Rev. D28, 2903 (1983);

6. Basile,M. et al., Nuovo Cim. A73, 329 (1983), Lett.Nuovo Cim. 38, 359 (1983) and 41, 298 (1984) and references therein;

7. Barton,D.S.et al., Phys.Rev. D27, 2580 (1983);

8. Brick,D. at al., Phys.Lett. 103B, 242 (1981) and Phys. Rev. D25, 2794 (1982);

9. Fowler,G.N., Friedlander,E.M., Plumer,M. and Weiner,R. M., Phys.Lett. 145B, 407 (1984);

10. Heisenberg,W., Z.Phys. 126, 569 (1949); Fermi,E., Prog. Theor.Phys. 5, 570 (1950); Pomeranchuk,I., Dokl.Akad. Nauk,SSSR, 78, 889 (1951);

11. Collected Papers of L.D.Landau, ed. Ter Haar, Gordon and Breach, New York 1965; see also: Carruthers,P. Ann. N.Y.Acad.Sci. 229, 91 (1974); Daibog,E.I., Rosental,I.L. and Tarasov,Yu., Fortsch.Phys. 27, 313 (1979); Shuryak, E.V., Phys.Rep. 61, 71 (1980); Clare,R.B. and Strottman, D., Phys.Rep. 141, 177 (1986);

12. C.f. for example: "Quark Matter ´86", eds. L.S.Schroeder and M.Gulassy, Nucl.Phys. A461 (1987); "Hadronic Matter in Collision" (LESIP II), eds. P.Carruthers and D.Strottman, World Scientific,1986; "Hadronic Matter under Extreme Conditions", eds. G.M.Zinovjew and W.P. Shelest, Kiev, Naukova Dumka, 1986; "Local Equilibrium in Strong Interaction Physics" (LESIP I), eds. D.K.Scott and R.M.Weiner, World Scientific, 1985; "Quark Matter´84" ed. K.Kajantie, Lecture Notes in Physics 221, Springer Verlag 1985; Cleymans,J., Gavai,R. and Suhonen,E., Phys. Rep. 130, 217 (1986); Shuryak,E.V., Phys.Rep. 115, 151 (1984);

13. Mann,A. and Weiner,R.M, Lett.Nuovo Cim. 2, 248 (1971); Hagedorn,R. in "Quark Matter´84" and "Hadronic Matter under Extreme Conditions" (ref.12); Gorenstein,M.I., Zinoview,G.M. and Shelest,V.P. in "Hadronic Matter under Extreme Conditions" (ref.12); - and references therein;

14. Capella,A. and Tran Thanh Van,J., Nucl.Phys. A461, 501c (1987), Kaidalov,A.B. and Martirosyan,K.A., Sov.J.Nucl. Phys. 29, 979 (1984) and 40, 135 (1984); Bopp,F.W., Aurenche,P. and Ranft,J., Phys.Rev. D33, 1867 (1986);

15. Andersson,B. Nucl.Phys. A461, 513c (1987) and references therein;

16. Weiner,R.M. in "Hadronic Matter under Extreme Conditions" (ref.12); Feinberg,E.L., ibidem;

17. Bjorken,J.D. Phys.Rev. D27, 140 (1983); von Gersdorff, H., Mc Lerran,L.D., Kataya,M. and Ruuskanen,P.V., Phys.Rev. D34, 794, 2755 (1986) and references therein;

18. Carruthers,P. and Ming,D.V., Phys.Rev. D28, 130 (1983);

19. Basile,M. et al., Nuovo Cim. A67, 244 (1982);

20. Pokorski,S. and Van Hove,L., Acta Phys.Polon. B5, 229 (1974) and Nucl.Phys. B86, 243 (1975);

21. Mueller,A., Phys.Rev. D2, 224 (1970);

22. Takagi,F., Z.Phys. C13, 301 (1982) and C19, 213 (1983);

23. Basile,M. et al., Lett.Nuovo Cim. 41, 298 (1984);

24. C.f. ref.9 and Plumer,M. in LESIP I, ref.12;

25. Fowler,G.N., Weiner,R.M. and Wilk,G., Phys.Rev.Lett. 55, 173 (1985);

26. Wilk,G. in LESIP II, ref.12; Fowler,G.N., Vourdas,A., Weiner,R.M. and Wilk,G., Phys.Rev. D35, 870 (1987);

27. Shuryak,E.V., Phys.Rep. 61, 71 (1980); Carruthers,P. in "Quark Matter 84" and LESIP I, ref.12, and references therein;

28. Matinyan,S.G., Sov.J.Part. and Nucl., 16, 226 (1985);

29. One can eventually use also Bose-Einstein distribution but its meaning is by no means clear; c.f. Navarra,F. and Weiner,R.M., "Inelasticity Distribution from Chaotic Sources",Marburg University preprint, 1987;

30. Stodolsky,L., Phys.Rev.Lett. 28, 60 (1972);

31. Etim,E., Pancheri,G. and Touchek,B., Nuovo Cim. B51, 276 (1976);

32. Gluck,M., Hoffmann,E. and Reya,E., Z.Phys.,C13, 119 (1982); This parametrization is good even for small values of x where perturbative QCD calculations are not applicable. For very high energies ($\sqrt{s}$=40 TeV) the parametrization of Duke and Owens was found more suitable: Duke,D.W. and Owens,J.F., Phys.Rev. D30, 49 (1984);

33. Eichten,E., Hinchliffe,I.,Lane,K. and Quigg,C., Rev. Mod.Phys. 56, 579 (1984); Frautschi,S. and Krzywicki, A., Z.Phys. C1, 43 (1979);

34. It can be shown that a recent calculation (ref.35) which leads to opposite conclusions concerning the s dependence of $\langle K \rangle$ is not only contradicted by the experimental data on $\langle K \rangle$ obtained from rapidity distributions[4,5], but leads also to a flat K distribution,

which is again in conflict with experiment[8,9];

35. Barshay,S. and Chiba,Y., Phys.Lett. 167B, 449 (1986);

36. Hwa,R.C., Phys.Rev. D22, 759, 1593 (1980); c.f. also Proc. XIV Int.Symp.Mult.Dynamics, Granlibakken,Lake Tahoe,Cal.,June 22-27,1983, eds. P.Yager and J.F.Gunion, World Scientific, and references therein;

37. Fowler,G.N., Friedlander,E.M., Weiner,R.M. and Wilk,G., Phys.Rev.Lett. 56, 14 (1986); c.f. also Fowler,G.N. and Weiner,R.M., this volume;

38. Chou,T.T., Yang,C.N. and Yen,E., Phys.Rev.Lett. 54, 510 (1985); Chou,T.T. and Yang,C.N., Phys.Rev. D32,1692 (1985);

39. See, for example, in: "Geometrical Pictures in Hadronic Collisions", a reprint volume, ed. S.Y.Lo, World Scientific, 1987;

40. First attempts in this direction are are already made by Plumer,M. and Hama,Y., in preparation;

41. Fowler,G.N.,Friedlander,E.M.,Weiner,R.M. and Wilk,G., Phys.Rev.Lett. 57, 2119 (1986); Carrithers,P.,Friedlander,E.M. and Weiner,R.M., Physica 23D, 138 (1986); c.f, also Fowler,G.N. and Weiner,R.M., this volume;

42. Cai Xu, Chao Wei-qin,Meng Ta-chung and Huang Chao-shang, Phys.Rev. D33, 1287 (1986); Pancheri,G. and Srivastava, Y., Phys.Lett. 155B, 69 (1986); Banerjee,H., De,T. and Syam,D., Nuvo Cim. 89A, 353 (1985);

43. The most recent review on this subject is: Nikitin,Yu. P. and Rosental,I.L., "High Energy Physics with Nuclei", Studies in High Energy Physics Series, Vol.5, ed. N.M. Queen, Harwood Acad.Press., 1986;

44. Fowler,G.N., Vourdas,A.,Weiner,M. and Wilk,G., Proc. XV Int. Workshop on Gross Properties of Nuclei and Nuclear Excitations, Hirschegg, Kleinwalsertal, Austria, Jan. 12-17, 1987, ed. H.Foldmeier, p.225;

45. Berlad,G., Dar, A. and Eilam,G., Phys.Rev. D13, 161 (1976); Fredriksson,S., Nucl.Phys. B111, 167 (1976);

46. Cai Xu, Chao Wein-qin and Meng Ta-chung, Phys.Rev. D36, 2009 (1987);

47. Pajares,C. and Ramallo,A.V., Phys.Rev. D31, 2800 (1985); Date,S., Gyulassy,M. and Sumiyoshi,H., Phys.Rev. D32, 619 (1985), and references therein;

48. Fowler,G.N., Navarra,F.,Plumer,M. Vourdas,A., Weiner, R.M. and Wilk,G., in preparation.

DUAL PARTON MODEL*

A. Capella

Laboratoire de Physique Théorique et Hautes Energies,
Université de Paris XI, Orsay, FRANCE

U. Sukhatme

Physics Department, University of Illinois, Chicago, IL, USA

C-I Tan

Physics Department, Brown University, Providence, RI, USA

J. Tran Thanh Van

Laboratoire de Physique Théorique et Hautes Energies,
Université de Paris XI, Orsay, FRANCE

* Supported in part by the U.S. Department of Energy under
contracts DE-AC02-76ERO3130.A9013-Task A and DE-FG02-
84ER40173.

I.   INTRODUCTION

When two high energy hadrons collide, the bulk of the cross section consists of events in which a large number of particles are produced with small transverse momenta with respect to the collision axis. Since quantum chromodynamics (QCD) is generally accepted as the theory of strong interactions, one would ideally like to be able to compute these "soft" multiparticle production processes directly from the QCD Lagrangian. However, the running coupling constant is clearly much too large for ordinary perturbation theory to be sensible, and consequently a non-perturbative procedure must be adopted.

Much qualitative understanding on the structure of QCD in the non-perturbative regime has been gained in recent years through the large-N studies of non-abelian gauge theories.[1-3] It has become increasingly clear to many that, because of confinement, the effective degrees of freedom for QCD at the soft hadronic scale can best be expressed in terms of gauge invariant composite variables such as those describing strings. An interacting string theory naturally leads to a topological expansion. At high energies, such an expansion can be represented by a Reggeon-Pomeron calculus, i.e., a perturbative Reggeon Field Theory (RFT).[4] More specifically, we shall work in the framework of the dual topological unitarization (DTU) scheme, which, in addition to being a topological expansion, also explicitly implements the constraints of duality and unitarity.[5,6] This scheme provides the basic ideas and motivation underlying the dual parton model (DPM),[7] which is the main subject of this review.

DPM was first developed at Orsay in 1979, by incorporating partonic ideas into DTU. Gradually, it has evolved as a result of the work of many people and now provides a rather complete description of all the dominant features of soft multiparticle production in high energy collisions. The model was also successful in anticipating several new features of the data at CERN $\bar{p}p$ collider energies, which were subsequently confirmed by experiment.

In DPM, a meson is a low-lying excitation of an open string with valence quark and anti-quark residing at its ends. When a string is "stretched", it decays into mesons by breaking into short strings. The dominant contribution to the high energy scattering of two mesons comes from a closed string (a Pomeron) exchange, having a cylinder topology. A baryon, in a topological expansion, can also be treated as an open string with a quark and a diquark at its ends. A unitarity cut of the cylindrical Pomeron shows that the sources of multiparticle production are two hadronic chains. (Fig. 1) At higher energies, cut multi-Pomeron diagrams (which correspond to more complicated topological structures and are of higher orders in a $1/N$ expansion), consisting of several chains with relative weights constrained by unitarity, become increasingly important (Fig. 2). The dual parton model provides a systematic way of evaluating these multi-chain diagrams. Therefore, <u>a sequence of energy scales is naturally present.</u> As the energy is raised, more complicated topologies gradually come into play and become important.

Within each diagram, the x-distributions for the ends of a chain are determined from momentum distribution · functions and the production of particles in any chain is given by fragmentation functions. Because of this factorization property, the model is rather constrained. DPM has proved capable of quantitatively describing inclusive single-particle rapidity distributions, rising rapidity plateaus, charged-particle multiplicity moments and violations of KNO scaling, charge distributions in rapidity, correlations between $<p_T>$ and multiplicity, long-range rapidity correlations, antiproton-proton annihilation, diffraction dissociation processes, two-particle inclusive reactions, production of heavy flavors, minijet production data from the UA1 collaboration, etc. Furthermore, the model can be used for both hadronic as well as nuclear beams and targets.

It is important to note that the multi-chain diagrams of DPM usually give a simple intuitive understanding of many experimental features. For example, rising rapidity plateaus in inclusive processes come from increasing overlap of the two chains in the single Pomeron exchange diagram and additional centrally located contributions from higher order corrections in the topological expansion. Another example is the delayed threshold phenomenon in heavy flavor production. This arises because the initial energy in a hadronic collision is shared by two chains both of which have center of mass motion, thus leaving much reduced energy available for making heavy flavors. Whenever possible, we will attempt to provide a simple qualitative understanding of most experimental features before proceeding to show successful quantitative phenomenology.

Finally, it should pointed out that DPM has proved to be of interest to a wide variety of people. Physicists who study soft processes have an obvious interest, whereas those who are looking for rarer events often find the dual parton model to be a reliable and useful way of estimating background.

In this article, we describe DPM and its numerous successes. The plan of the article is as follows. Sec. 2 contains a brief description of the theoretical basis of DPM and, in particular, its relation to 1/N expansions of QCD. The model is fully described in Sec. 3. Two key features of the model are emphasized in Sec. 4 and their phenomenological consequences are spelled out in Sec. 5 and Sec. 6. The effects of incorporating transverse momentum dependence so as to include the so-called "semi-hard" processes in the model are given in Sec. 7. Diffractive dissociation processes can also be described in

DPM framework. This is discussed in Sec.8. Finally, Sec. 9 contains some concluding remarks.

II. THEORETICAL FRAMEWORK

QCD is classically a scale invariant theory characterized by a single dimensionless gauge coupling constant, g. However, scale invariance is broken due to renormalization; as a consequence, the gauge coupling becomes scale dependent. For QCD, due to "anti-screening", g increases with distance; hadron dynamics involves QCD at a scale where the gauge coupling is of order unity. Therefore, soft hadronic physics requires a non-perturbative treatment of QCD.

DPM is a phenomenological realization of a topological expansion of QCD where one incorporates both partonic ideas and Regge concepts. A topological expansion of QCD is a consequence of a large-N expansion, and is therefore intrinsically non-perturbative. A large-N expansion was first introduced by 't Hooft in the particle physics context when he generalized QCD from SU(3) to SU(N_c), N_c being the number of colors.[1] Because of the absence of a small coupling parameter, it has been articulated succintly by E. Witten[3] who wrote in 1978: "Most likely, we can expect to solve exactly a four-dimensional theory only in some limit in which the physical degrees of freedom decouple from each other,........ In QCD the large-N limit is such a limit, and it is probably the only such limit."

Consider a SU(N_c) gauge theory coupled to N_f flavors of quarks in the fundamental representation. In the limit where N_c is large, and N_f and $g^2 N_c$ are fixed, one finds that each Feynman graph can be characterized by a two-dimensional orientable surface; the Feyman expansion can be regrouped as a sum over the surface topologies, each weighted by an additional factor of N^{-2} as the genus increases by one. As a consequence, scattering amplitudes are dominated by planar graphs. Recent realization that this limit formally corresponds to a semi-classical limit, with the factor N^{-2} playing the role of an effective Planck constant, has raised the hope that this approach can eventually lead to practical calculational schemes for QCD. For instance, the meson spectrum can be obtained by analyzing small fluctuations of the gauge invariant string variables about their large-N vacuum expectation values.[8]

In large-N QCD, because of confinement, the zeroth order leads to noninteracting mesons, which, together with their high-spin recurrences, can be conveniently labelled by Regge trajectories. It is generally believed that these Regge trajectories can be associated with the excitations of a yet-to-be-specified smooth string model. Indeed, the most likely scenario for making further progress towards "solving" QCD at hadronic scale in the foreseeable future is to treat QCD in terms of effective string degrees of freedom. Because the propagation of a string is characterized by a two-dimensional surface, an interacting string theory naturally gives rise to a topological expansion. Therefore, instead of literally having N large, the expansion can be interpreted as a topological expansion for the S-matrix elements.

Aside from its theoretical attractiveness, the major reason for believing in the large-N expansion is because it leads to qualitatively correct phenomenology, e.g., the OZI rule for resonance decays. Equally important is the fact that, to the leading order, meson scattering amplitudes are given by sums of tree diagrams with exchanges of infinite towers of physical mesons, which naturally leads to Regge phenomenology.

It was pointed out by Veneziano that, phenomenologically, it is more appropriate to consider a generalized large N-limit, where both N_c and N_f become large but with their ratio N_c/N_f and $g^2 N_c$ fixed.[2] In this limit, planar fermion loop insertions are allowed at each order without altering the global topology of a diagram. As a consequence, desirable features of the conventional large-N expansion (such as the OZI-rule) are unaffected. However, by allowing planar fermion loop insertions in the leading order, the expansion satisfies "planar unitarity".[2,5,6] For instance, the discontinuity of a forward four-point function having the topology of a disk is precisely the sum of production cross sections given by planar 2-to-n amplitudes having the same disk topology, (provided that the momentum lines in the unitarity sum are arranged in a planar fashion). Since phenomenological amplitudes must be unitary, this topological expansion of Veneziano is much more suitable for phenomenology. In particular, it allows a consistent simultaneous description of elastic scattering and production processes. It is in the context of this expansion that we shall develop the DPM, where unitarity play a crucial role. By emphasizing the topological expansion, we shall keep $N_c = 3$ in what follows, and, treat a baryonic string as having a quark and a diquark attached to its ends.[9]

Comprehensive reviews of topological expansions can be found in Refs. 2, 3, and 6. Here we only mention certain key features. The leading order is planar, with external particles attached to a single external quark-loop boundary (valence quarks residing at the ends of strings). Open string excitations are ordinary mesons which are ideally mixed and obey an OZI-rule. For 2-to-2 amplitudes, the next order involves the cylinder topology--a two-dimensional planar surface with two quark-loop boundaries. For instance, for an elastic process $a + b \rightarrow a + b$, with type-a particles attached to one boundary and the b's to the other, the forward scattering limit is described by the exchange of leading "closed string" excitations, or, the Pomeron.

Prior to the general acceptance of QCD as the fundamental theory of strong interactions, a successful phenomenology of hadron collisions has been the Reggeon-Pomeron calculus, i.e., perturbative RFT, which is a multiple scattering model based on Regge theory and on general S-matrix principles of unitarity and analyticity.[4] Analyses during the seventies have revealed that the essential qualitative features of high energy production processes can be understood in terms of a multiperipheral picture which automatically leads to limited transverse momenta, short-range-ordering in rapidity, etc.[10] These are the desired features which follow if forward amplitudes are dominated by the planar and the cylinder topologies, with higher order configurations treated perturbatively. RFT precisely provides such a perturbative treatment.

Interestingly, Veneziano[2] has pointed out that there is an one-to-one correspondence between terms in his topological expansion of QCD and those in RFT. Therefore, RFT provides a phenomenological parametrization of the topological expansion in high energy collisions as Regge behavior sets in. Of course, RFT contains free parameters, e.g., Regge trajectories and their couplings. On the other hand, once these parameters are fixed from a fit to total, elastic and diffractive cross sections, a quantitative description for particle production is obtained. Unfortunately, multiparticle production often takes place in regions where Regge behavior can at best be approximate, e.g., regions involving low subenergies. A successful model must therefore allow a reliable extrapolation from the Regge limit to the regions involving resonance production. This is accomplished by incorporating duality.

The duality concept was first recognized in two-body reactions where one finds that the extrapolation of Regge behavior works equally well in an average sense in the low energy region. The introduction of the dual model provides an explicit mathematical realization of this phenomenon. Guided by n-point dual amplitudes, a generalization of the duality concept to multiparticle production can also be carried out. More generally, duality has now been recognized as a key ingredient of a string theory. Since an interacting string theory always leads to a topological expansion, it is therefore not surprising that duality naturally enters into such an expansion.

In several earlier attempts to unitarize the dual resonance model, it was recognized that the duality features can best be maintained in a topological expansion,[5,6,11] and a topological expansion naturally emerges in the large N_f limit. This program has been referred to as dual topological unitarization (DTU); it was first carried out systematically along phenomenological lines by Chan H. M. and his collaborators[5] and subsequently by Chew and his collaborators[6] purely within an S-matrix framework. The work of Veneziano showed how this approach could be considered as a concrete phenomenological realization of the topological expansion of QCD. Since such a unitary scheme does incorporate Regge behavior, it therefore is also a RFT. Although one does not explicitly make use of the color degrees of freedom in DTU, their presence is implicit through the assumption of confinement. Consistent with the large-N limit, hadrons are treated as excitations of strings with confined quarks located at their ends. Indeed, our dual parton model, in which we introduce explicit color partonic ideas, has its roots partly in the DTU framework.

III. DUAL PARTON MODEL

The discussion of the previous section indicates that the scattering amplitude for high energy hadronic collisions can be expanded in a topological expansion. The dual parton model provides a concrete recipe for converting each diagram in the expansion into an experimentally observable cross section.

Due to the unitary structure, associated with each topological configuration for the forward elastic amplitude is a well-defined mechanism of multiparticle production through the notion of "cut Pomeron". For instance, consider the leading cylindrical Pomeron contribution to the forward absorptive amplitude. It's s-channel discontinuity contains 2 sheets of cut-cylinder surface. This single inelastic collision (cut-Pomeron) in RFT corresponds to having two chains of hadrons produced along the cut sheets. (Fig. 1) These chains are stretched between valence quarks of the colliding hadrons.[7,12-15] In general, a k-cut-Pomeron diagram of RFT corresponds to a production diagram involving 2k chains of hadrons.[2] (Fig. 2) In RFT, because of unitarity and analyticity, once the basic Regge parameters are known, the contribution to the total cross section for having k cut-Pomerons, σ_k , can be found. A detailed description of the method is given in Ref. 16, (see also Refs. 29 and 34).

To proceed, we next make use of the partonic structure of hadrons, and treat each contributing diagram as a two step process (i) separation of color in the hadronic collision, and (ii) fragmentation of colored objects which results in the production of hadronic chains. Consequently, the two ingredients one needs are momentum distribution and fragmentation functions.[7]

For definiteness, let us first consider the leading contribution to multiparticle production in high energy proton-proton collisions. (As discussed before, the leading contribution is a two chain diagram. This comes from a unitarity cut of a single dual Pomeron having the topology of a cylinder -- the simplest topology with vacuum quantum number exchange in the t-channel.) The collision separates the valence quarks of each incident proton into two colored systems - a quark and a diquark. The fragmentation occurs in the form of two quark-diquark chains. Note that each chain is an overall color singlet which contains a quark fragmentation region and a diquark fragmentation region.

In order to make the dual parton model quantitative, it is necessary to specify the probability $\rho_1(x_1)\rho_1(x_2)$ that the interaction separates the protons into two quarks with momentum fractions x_1 and x_2 and two diquarks with the remaining momentum fractions $(1-x_1)$ and $(1-x_2)$. We shall shortly see that the main feature of the momentum distribution function, $\rho_1(x)$, is its $x^{-1/2}$ behavior at $x=0$, which follows by a Regge consideration. This leads to a peaked behavior near $x=0$, resulting in two "held-back" quarks near x_1, $x_2=0$.

Typically in a pp collision,[7] the "effective" values of x_1 and x_2 are around 0.05. The total energy $\sqrt{s}$ in the overall pp center of mass (CM) frame is shared between the two chains shown in Fig. 1. Denote $\sqrt{s_1}$ and $\sqrt{s_2}$ as the energies of chain-1 and chain-2 in their respective CM frames; we have $s_1 = s\, x_2\, (1 - x_1)$, $s_2 = s\, x_1\, (1 - x_2)$. The rapidity shifts Δ_1 and Δ_2 necessary to go from the overall pp CM frame to the CM of chain-1 and chain-2 are given by $\Delta_1 = (1/2)\ln[(1-x_1)/x_2]$ and $\Delta_2 = -(1/2)\ln[(1-x_2)/x_1]$ respectively.

The other ingredient which one needs are fragmentation functions of quarks and diquarks. They cannot be calculated, at the present stage, from first principles. One of the following alternatives can be adopted: (1) Use parton model fragmentation functions obtained from hard processes at moderate Q values (universality ansatz.). (2) Use iterative cascade model (Lund model, independent fragmentation model, etc.)[17] The second approach is useful mainly for a Monte Carlo treatment of DPM; this is briefly discussed in Appendix 1. In what follows, we mostly follow the first approach. In some situations, we use dimensional counting rules[18] in order to determine the $x \to 1$ behavior of fragmentation functions. Regge arguments can also help in determining their behavior near $x=1$ as well as at $x=0$.[19] The precise form used in our calculation can be found in the original publications.

At this stage, one can combine the above ingredients to obtain physically observable quantities. For example, assuming one-cut Pomeron dominance, the single- particle inclusive cross section for $pp \to h + X$ is given by the superposition of two chains (see Fig. 1)

$$dN^{pp \to h}(s,y)/dy = N_1{}^{qq-q}(s,y) + N_1{}^{q-qq}(s,y) \tag{1}$$

The contribution from each chain can be obtained by folding the momentum distribution function with the fragmentation functions, e.g.,

$$N_1{}^{qq-q}(s,y) = \int_0^1 \int_0^1 dx_1 dx_2 \, \rho(x_1,x_2) \, dN^{qq-q}(y-\Delta_1;s_1)/dy \tag{2}$$

where dN^{qq-q}/dy is the rapidity distribution for the particle-h in the CM of the qq-q chain obtained from q and qq fragmentation functions.

Corrections to the leading two chain diagram come from multiple cut-Pomeron exchanges. For example, a two-cut-Pomeron exchange diagram gives rise to four chains. The two new chains are assumed to terminate on sea quarks and antiquarks, whose momentum distribution functions will be shown later to have typically a $1/x$ dependence at small x. In a general diagram involving k-cut-Pomeron exchanges, one needs the probability for the incident protons to break into 2k colored constituents. The momentum distribution for such a breakup can be expressed in terms of 2k variables, $\rho_k(x_1,x_{2k};x_2,x_3,\dots\dots,x_{2k-1})$, where x_1,x_{2k} denote the momentum fractions taken by the valence quark and diquark, and x_i (i=2,3,...,2k-1) are the momentum fractions of the sea quarks and antiquarks, subject to the constraint that the sum of these 2k x's is unity.

The relative weights of various diagrams involving k-cut-Pomerons is the cross section σ_k, which, as we mentioned before, can be determined by unitarity requirements; a detailed analysis can be found in Ref. 16. The general formula for the inclusive rapidity distribution can now be written down. For instance, the inclusive spectra of charged particles in pp collisions is

$$dN^{pp}/dy = (1/\sigma_{in})d\sigma^{pp}/dy$$

$$=(1/\Sigma\sigma_k)\Sigma\sigma_k[N_k^{qq-q}(s,y)+N_k^{q-qq}(s,y)+(2k-2)N_k^{q_s-\bar{q}_s}(s,y)] \qquad (3)$$

where $N_k^{qq-q}(s,y)$, etc., are inclusive spectra of charged hadrons produced in each chain. This formula extends in a natural manner to other hadronic collisions. For more details, see Ref. 7.

IV. BASIC FEATURES OF DPM

The dominant features of the momentum distribution functions, $\rho_k(x_1,x_{2k};x_2,x_3,.......,x_{2k-1})$, can be determined to a large extent by Regge consideration. More precisely, when the rapidity gaps between various constituents in a given graph are all large, one can apply Regge arguments to determine the cross sections for this configuration.[13] This allows one to obtain the limiting behavior of ρ_k when any one of its 2k arguments goes to zero.[7,16,20] The behavior turns out to be singular in all cases except when the x of the diquark goes to zero. After extracting these singular behaviors by a product, we shall in practical applications approximate the remaining smooth factor by a constant, which can then be fixed by requiring normalization.

Let us illustrate in some detail for the case of a single cut-Pomeron, i.e., k=1 with two chains. For a meson projectile, one finds that ρ_1 behaves as $x^{-1/2}$ as $x \to 0$ for both the valence quark and the antiquark.[7,20] This is due to the fact that the rapidity gap created between the quark and the antiquark is controlled by an ordinary Reggeon, with a trajectory intercept 1/2. (Fig. 3a). One therefore has

$$\rho_1{}^m(x) = c_1{}^m \, x^{-1/2}(1-x)^{-1/2} \qquad (4)$$

In the case of an incoming proton, the same x behavior of course applies when the x value of the quark goes to zero. (Fig. 3b) However, when the x of the diquark goes to zero one gets instead a non-singular behavior x due to the fact that the corresponding rapidity gap is governed by an exotic trajectory ($qq\bar{q}\bar{q}$) with typically a negative intercept, $\alpha_{exot.} = -1.5$, (Fig. 3c). Therefore, we have[7,20]

$$\rho_1{}^P(x) = c_1{}^P \, x^{-1/2}(1-x)^{1.5} \qquad (5)$$

where x is the momentum fraction of the valence quark. (Note that the small x behavior is Eqs. 2 and 3 is the same as the corresponding one in the parton model structure functions, whereas the power of 1-x is smaller. However, in hard processes, this power depends on Q^2 and, according to the Altarelli-Parisi equation, decreases when Q^2 decreases. Eqs. (4) and (5) can thus be regarded as the low Q^2 limit of the deep inelastic structure functions.)

Proceeding in the same fashion, we obtain[16] for the general case involving sea-quarks

$$\rho_k(x_1, x_{2k}; x_2, x_3, \ldots, x_{2k-1})$$

$$= c_k{}^p \; x_1{}^{-0.5} \; x_2{}^{-1} \; \ldots \; x_{2k}{}^{1.5} \; \delta \; (1-x_1-x_2-\ldots-x_{2k}). \tag{6}$$

Note that all x-dependences are singular except for that of the diquark, x_{2k}. (For a more precise expression for the case involving sea-quarks, see Appendix 1.)

Our parametrization for ρ_k as a product of its singularities, (6), is of course an approximation, reflecting our current ignorance. For k=1, we have checked that the numerical results change very little when Eqs. (4) and (5) are multiplied by smooth functions of x. They are also insensitive to the power of (1-x) in Eq. (5) when the latter is changed slightly, (between 1 and 3.) For larger k values, this factorization   assumption   is of course more questionable. One hopes that it captures the main properties of the true momemtum distribution functions. [Note however that the weights σ_k decrease rapidly with increasing k, so that the errors committed for larger values of k are less and less important.]

Although the mechanism of particle production, as discussed earlier, depends on both the momentum distribution functions and the fragmentation functions, much of the qualitative features of particle production data can be understood in terms of two basic properties, which are consequences of the momentum distribution functions only. These are:

(1) Held-back valence quark:

Due to the x behavior of the valence quark momentum distribution function, the valence quarks of each colliding hadron is slowed down by the interaction. In the case of a meson, the held-back quark can be either the valence quark (with the antiquark being fast) or the antiquark (with the quark being fast). From Eq. (4) one finds that both configurations have equal probabilities, while the configuration with both quark and antiquark sharing equal momenta is suppressed. In the case of a proton, Eq. (5) leads to the preferred configuration where only a valence quark is slowed down with the diquark remaining fast. Consequently, in nondiffractive events, the valence quark and its companion antiquark from an incoming meson prefer not to recombine into the same final hadron. Similarly, a quark tends to break away from an incoming proton and not to recombine with its original diquark. These important consequences of the DPM have received recent spectacular confirmation.(21)

(2) Short, central q-q̄ multi-chains:

Corrections to the leading two chain diagram come from multiple Pomeron exchanges. For example, cutting a two Pomeron exchange diagram gives rise to four chains. The two new chains are assumed to terminate on sea quarks and antiquarks, which have momentum distribution functions which have a 1/x dependence at small x. Hence these chains span a smaller range of rapidity than those involving valence quarks and diquarks. They are forward-backward symmetric; on average, they are centered around $y^*=0$ and span the central rapidity region. It is clear that each chain must at least have a minimum threshold CM energy in order for physical hadrons to materialize from it. Therefore, on average, up until the highest CERN - ISR energies ($\sqrt{s} < 63$ GeV), the two chain contribution dominates. However, at higher energies (like those at the CERN $\bar{p}p$ collider or the Fermilab Tevatron) there is sufficient energy available to share between four (or more) chains, and multiple Pomeron exchanges give a important contribution. They are responsible for the continuous rise of the central plateau height with energy and violations of KNO scaling.

V.   CONSEQUENCES OF THE HELD-BACK EFFECT

The $x^{-1/2}$ behavior of the valence quark momentum distribution function corresponds to a distribution in the rapidity gap Δy between the valence quark rapidity y and the maximal available rapidity y_{max}, of the form exp $(-1/2\ \Delta y)$. Thus the average length of this gap is independent of the energy. The calculation shows that this length is about 2.5 units. Therefore, up to ISR energies, the average rapidity of the held-back valence quark is very close to $y^* = 0$, and will influence the central rapidity region. In this section, we discuss some experimental consequences of such a situation :

a) <u>Particle ratios in pp and $\bar{p}p$ collisions</u>

The Axial Field Spectrometer Collaboration has measured the ratio of inclusive distributions in pp and $\bar{p}p$ collisions for a variety of detected charged particles (pions, kaons, protons and antiprotons) The measurements were made at the CERN-ISR at $\sqrt{s} = 53$ GeV in the rapidity range $|y|<0.8$ and transverse momentum $p_T<1.5$ GeV/c. The experimentally obtained values of the ratios

$$R^h \equiv [dN^{\bar{p}p \to h}\ (|y|<0.8)/dy]\ /\ [dN^{pp \to h}\ (|y|<0.8)/dy] \qquad (7)$$

for various detected particles h are shown in Table 1.

The results obtained[22] in DPM with standard fragmentation functions are also given in Table 1. At a qualitative level, these results are a straight forward consequence of the valence quark-diquark separation with the valence quark being held-back near x=0. Let us consider for definitness the case in which the produced hadron is either a π^+ or a π^- . In a pp collision, we have 4u and 2d valence quarks. The fragmentation of the u-quark (d-quark) into π^+ (π^-) is favored whereas d-quark (u-quark) fragmentation into $\pi^+(\pi^-)$ is unfavored. In a $\bar{p}p$ collision we have three quarks (u,u, and d) with favored fragmentation into π^+ and three antiquarks ($\bar{d}$, $\bar{u}$, and $\bar{u}$) with unfavored fragmentation. Since the ratio of favored to unfavored fragmentation functions is smaller than 1, one can see immediately that $R^{\pi^+} < 1$, $R^{\pi^-} > 1$, and $R^{\pi^{\pm}} = 1$. The other values of R in Table 1 can be understood qualitatively in a similar way.

With increasing energies, the held-back valence quark will move away from the central region and all the above ratios will go to one.

b) Charge Distributions

A remarkable consequence of valence quark held-back effect is provided by some striking results on the charge distributions. (See Figs. 4 and 5). Let us consider a proton-pion collision. The results obtained[23] in DPM, with standard fragmentation functions, are also shown in these figures. The experimental results are very well reproduced by the model calculations. A striking feature of these results in the forward-backward asymmetry of the charge distribution in fig. 4, and the fact that the charge in the π^- and K$^-$ hemisphere is much smaller than one (in absolute value) at moderate energies. (Fig. 5) It is most interesting to understand the origin of these results.

Let us consider pion-proton interaction. On the proton side, as a consequence of the momentum distribution function in Eq. (5), there are three dominant configurations, which are on the average equally probable, $(uu)_F$-d_S, $(ud)^0_F$-u_S and $(ud)^1_F$-u_S, where F stands for fast and S for slow. The average charge of the slow valence quark is then $Q^P_S = (Q_d + 2Q_u)/3 = 1/3$ and that of the fast diquark $Q^P_F = (2Q_d + 4Q_u)/3 = 2/3$. Applying the same argument to the pion side, one finds that here the average charge of the slow and fast constituents are equal : $Q^{\pi^+}_F = Q^{\pi^+}_S = 1/2$, and $Q^{\pi^-}_F = Q^{\pi^-}_S = -1/2$.

The forward-backward asymmetry observed in Fig. 4 is just a consequence of the fact that $Q^P_F > Q^{\pi^+}_F$ (the measured ratio between the maxima of dQ/dy in the proton and $\pi+$ hemispheres is equal to the ratio $Q^P_F/Q^{\pi^+}_F$).

The fact that the integrated charge in the π^- hemisphere is, in absolute value, smaller than one, is a consequence of the average charges of the slow systems (valence quarks and antiquarks) which can cross from its own hemisphere to the opposite one, (with a probability given by Eq. (4)), and deposit there its charge. Depending on whether none, one or both of the slow constituents cross the $y^* = 0$ line, one has four different values for the integrated charge in the π^- hemisphere : $Q^{\pi^-}_F + Q^{\pi^-}_S = -1$, $Q^{\pi^-}_F = -1/2$, $Q^{\pi^-}_F + Q^{\pi^-}_S + Q^P_S = -2/3$ and $Q^{\pi^-}_F + Q^P_S = -1/6$. If all these configurations were equally probably the average charge in the π^- hemisphere would be $(-1-1/2-2/3-1/6)/4 = -7/12$.

In fact, as a consequence of Eqs. (4) and (5), the probability that a valence quark crosses from one hemisphere to the other is smaller than 0.5 and goes to zero as $s \to \infty$. Therefore the observed charge will be larger than 7/12 (in absolute value) and increase towards 1 as s increases, in agreement with the data. (See Fig. 5).

Note that the above considerations are entirely based on the momentum distribution functions (Eqs. (4) and (5)) and therefore qualitatively, independent of the fragmentation functions. These results demonstrate in a nice way the influence of the valence quark in the central rapidity region at moderate energies, and provide striking confirmation of the particle production mechanism in DPM.

c) One and Two-Particle Inclusive Distributions

The Lund group has observed [24] that the hadronic spectra in the fragmentation region ($x_F > 0.2$) can be described (at a 10-20% level of accuracy) by the following combinations of quark -and diquark fragmentation function :

$$(1/\sigma)d\sigma^{\pi^+ p \to \pi^+}/dx = (1/2)[D^{u \to \pi^+}(x) + D^{\bar{d} \to \pi^+}(x)] = D^{u \to \pi^+}(x) \quad (8)$$

$$(1/\sigma)d\sigma^{\pi^+ p \to \pi^-}/dx = (1/2)[D^{u \to \pi^-}(x) + D^{\bar{d} \to \pi^-}(x)] = D^{u \to \pi^-}(x) \quad (9)$$

$$(1/\sigma)d\sigma^{pp \to \pi^+}/dx = (1/3)D^{u \to \pi^+}(x) + (2/3)D^{ud \to \pi^+}(x) \quad (10)$$

$$(1/\sigma)d\sigma^{pp \to \pi^-}/dx = (1/3)D^{u \to \pi^-}(x) + (2/3)D^{ud \to \pi^-}(x) \quad (11)$$

These relations are just the ones expected from the held-back valence quark effect. Thus, it is not surprising that the hadronic spectra in the fragmentation regions can be described in DPM (7,19,20,25) using standard fragmentation functions. Of course, Eqs. (8-11) refer to the fragmentation regions only. However, DPM, in the formulation given in Ref. 7, also allows to describe the central rapidity region (see section 6a). Moreover, using the cascade model[17] to obtain the quark and diquark fragmentation functions into a pair of hadrons and assuming that particles produced in different chains are independent from each other, one can also describe various two-particle inclusive spectra[25]. Some of these spectra are presented in Fig. 6 and compared with available data.

d) Heavy Particle Production in pp Collisions

An intriguing feature of high energy particle production is the so-called "delayed threshold" for heavy particle production. For exemple, the production of antiproton in proton-proton collision is almost negligible for $\sqrt{s}$ <10-12 GeV even though it can occur at $\sqrt{s} = 4 \, m_N \approx 4$ GeV. Indeed, this delayed threshold could be qualitatively understood in term of t_{min} suppression for massive particles in a peripheral production. However, accounting for the delayed threshold quantitatively in a systematic fashion has been difficult to achieve and it poses a test for any viable scheme of particle production from FNAL to collider energies.

In the dual parton model, such a delayed threshold is a simple consequence[26] of the two chain structure and the asymmetrical energy distribution between the valence quark and diquark. A part of the initial energy goes into the motion of the center of mass of the two chains and furthermore the remainder is shared by the two chains. Let us make an estimate for $\bar{p}$ production in pp collisions.

In pp collisions, the energy of each chain is given by $s_{chain} = x_q x_{qq} s$ and only this amount of energy is available for particle production. It turns out that the average value of the c.m. energy required to produce a q-qq chain of energy $3 \, m_p$ (p, $\bar{p}$ threshold) is $\sqrt{s} \approx 13$ GeV.

Such a high threshold, which is natural with the two chain structure, is hard to obtain with a single chain.

Some results obtained in DPM for the energy dependence of heavy particle production, are shown in Figs. 7 and 8 and compared with experimental data.

Due to a different structure of the chains in pp and $\bar{p}p$ (existence of a diquark-antidiquark chain which takes almost all the energy in the case of pp collision), we predict a larger nucleon-antinucleon pair production in $\bar{p}p$ than in pp collision especially at low energy. More precisely, we expect that the effect of delayed threshold is less pronounced in $\bar{p}p$ collision.

VI. CONSEQUENCES OF THE MULTI-CHAIN CONTRIBUTIONS

The above results have been obtained using two chain contribution (only the first term in Eq. (3) is taken into account with $\sigma_1 = \sigma_{in}$). Indeed, for this type of data and at the considered energies, the contribution of the multi-chain diagrams is very small. This is due to the smallness of the average squared mass of the q_s-$\bar{q}_s$ chains up to ISR energies which is a consequence of the $1/x$ behavior of the sea quark momentum distribution functions. This implies that these chains have a very small length in rapidity space. Moreover, as a result of the convolution between momentum distribution and fragmentation functions (and of the existence of physical thresholds in the chains), it turns out that the short chains have also a small value of $(dN/dy)_{y*=0}$. However, with increasing energies the invariant mass of the q_s-$\bar{q}_s$ chains will increase and their relative contribution will become increasingly important. In the following we shall review the experimental consequences of the onset of these multichain configurations. Since the q_s-$\bar{q}_s$ chains are rather short in rapidity space and symmetric with respect to $y^* = 0$, their effects will be mainly found in the central rapidity region.

a) Rapidity Distributions

An obvious consequence of the increasing contribution of the q_s-$\bar{q}_s$ multichains with increasing energy is that the central "plateau" will rise, whereas the fragmentation region will not be affected by the (centrally located) q_s-$\bar{q}_s$ chains. This will, of course, prevent the multiplicity distribution from becoming more and more "plateau-like" with increasing energy. Such an unexpected behavior was one of the first surprises from the CERN pp collider. Numerically, it was predicted in Ref. 15 that the value of dN/dy at $y^* = 0$ at $\sqrt{s} = 540$ GeV would be about 3.5, in good agreement with the subsequent measurements by the UA1 and UA5 collaborations. Here the contribution of the multichains is about 30%. In the fragmentation regions, we have found Feymann scaling from ISR to collider energies within 5%. This is supported by the data[27]. The shape of the rapidity distribution is also reproduced[28] in DPM, using standard fragmentation functions.

b) Multiplicity Distributions

In order to compute the multiplicity distribution in DPM, we make the following assumptions[29] :

(1) In each chain, particles and resonances are randomly produced (Poisson distribution);

(2) Particles produced in different chains are uncorrelated.

The latter is a simplifying assumption, which has already been made when writing Eq. (3) and is in fact used in all works on DPM. The first assumption is a very natural one in low-p_T physics, since all fragmentation models lead to a short-range order picture of the hadronic spectra with Poisson-like distributions.

With these two assumptions, we obtain the following expression for the multiplicity distribution

$$P(N) = \sum_{k=1}^{\infty} \sigma_k \, P_k(N) \Big/ \sum_{k=1}^{\infty} \sigma_k \qquad (12)$$

where $P_k(N)$ is the (Compound Poisson Distribution) obtained by convoluting the Poisson distributions for each of the 2k chains.

In this way, the shape of the multiplicity distributions, as well as the rise with energy of the high multiplicity tail (KNO scaling violation) are well described.[29,30,31,32] (See Fig. 9.) The origin of the KNO scaling violation is quite obvious. The contribution of the multichain graphs becomes increasingly important when s increases and since they contribute mostly to high multiplicities, they will push upwards the high multiplicity tail. Note that the increase with s of the multichain contribution is mostly due to the increase with s of the invariant mass of the q_s-$\bar{q}_s$ chains (rather than to the s-dependence of the weights σ_k).

Another result of DPM is that the multiplicity distributions are broader when selecting central rapidity intervals of limited length, in agreement with experiment. (See Fig. 10).

Another interesting consequence of DPM is that the normalized moments $C_i = (N^i)/(N)^i$, increase with energy faster in a central rapidity interval $|y|<y_0$ than in the complete rapidity region. This striking prediction[29] of DPM has been confirmed recently by comparing the UA5 collaboration non-diffractive data, with data from the NA22 collaboration at $\sqrt{s}$ = 22 GeV - for the same type of events and taken under the same experimental conditions as those of the UA5 collaboration. The results of taking the ratios of the measured moments, in various rapidity intervals is shown in Fig. 11 together with the predictions[29] of DPM.

The physical origin of these results is the following : By restricting to central rapidity intervals of limited length, one increases the relative effect of the multichains and thereby the energy dependence of the moments C^i is also increased. This is no longer true when the length of the considered rapidity interval is very small. In this case, the average multiplicity is very small and the short-range correlation becomes dominant.

c) <u>Long-Range Rapidity Correlations</u>

The UA5 collaboration has measured an important long-range correlation between the forward and backward hemispheres. The value of this correlation is about 40% in pp collision at $\sqrt{s}$ = 540 GeV. At ISR, such a correlation was also observed but it only of 15% at $\sqrt{s}$ = 63 GeV and consistent with zero at lower ISR energies.

In order to measure this correlation, one plots the value of the average multiplicity in the backward hemisphere $<N_B>$ as a function of the event multiplicity in the forward one, N_F. The data[33] can be described by a linear expression

$$<N_B> = a + b \ N_F \qquad (13)$$

The slope b, (0<b<1), measures the strength of the correlation. In order to suppress the short range correlation, one cuts out a central rapidity region $|y|<1$. The slope $b_{L,R}$ measures, then, a genuine long-range correlation. It is easy to show that the definition of b in Eq. (13) is equivalent to the following one

$$b = \frac{<N_F N_B> - <N_F> <N_B>}{<N_F^2> - <N_F^2>} \qquad (14)$$

The value of b has been computed in DPM, from the expression for the average multiplicity using assumptions (1) and (2) in Sect. 6b. The result[34,35] is in good agreement with experiment.

As in any multiple-scattering model, the origin of the long-range correlation in DPM (where one assumes short-range order correlations within the individual chains), is the superposition of a fluctuations number of component[36]. In order to present the argument in a simple way, let us neglect the fluctuations in the position of the chain end points. One can then see that the contribution from individual chains (uncrossed terms) to the numerator of Eq. (14) is identically zero (since $<N_F^{chain-i} N_B^{chain-i}>$ $= <N_F^{chain-i}> <N_B^{chain-i}>$ as a consequence of asssumption (1) in Sect. 6b). Furthermore the crossed terms linear in the number of chains, 2k, also vanish (since $<N_{chain}^i N_{chain}^j> = <N_{chain}^i> <N_{chain}^j>$ for different chains, $i \neq j$, as a consequence of assumption (2) in Sect. 6b). There is, however, a non-vanishing contribution coming from the quadratic terms in k :

$$<N_F N_B> - <N_F><N_B> = (1/4)(<k^2> - <k>^2) <N_{qs-\bar{q}s}>^2 \qquad (15)$$

Here $<k^i>$ $= \Sigma \ k^i \sigma_k / \Sigma \ \sigma_k$. We see from (15) that the long-range correlation is due to fluctuation in the number of chains and it is proportional to the square of the average multiplicity of the rescattering chains q_s-$\bar{q}_s$. This has two important consequences :

1) Since the q_s-$\bar{q}_s$ chains contribute only to the central rapidity region, DPM predicts[34] that the value of $b_{L,R}$ will decrease very fast when one decreases the length $2y_0$ of the suppressed rapidity interval $|y|<y_0$. This prediction is confirmed by UA5 data. (See Fig. 12).

2) Fixing the event multiplicity amounts essentially to fixing the number of chains, 2k, (large multiplicity fluctuations within each chain are strongly suppressed and therefore large multiplicity events are mainly reached by increasing the number of chains). Thus, the long-range correlation vanishes at fixed multiplicity. This is indeed one of the main features of the data[33] and a crucial test for models.

d) Short-Range Correlations and Cluster Size

Since the long-range correlation vanishes in event samples of fixed multiplicities, this is obvious the best place to study the short-range correlation. This is usually done by measuring[33] the semi-inclusives 2 particle correlations - and also the multiplicity distributions in N_F at fixed values of $N = N_F + N_B$. Both sets of data are well described by DPM.[37] (See Figs. 13 and 14.) The results in these figures are obtained assuming that the "clusters" (that are randomly produced within each chain), are just a realistic mixture of directly produced particles and resonances and we have taken a charged average multiplicity per cluster $<K>_{eff} = 1.4$[38,39]. On the other hand, the analysis of the short-range correlation performed by the UA5 collaboration shows that the clusters are substantially bigger and have a broad decay with $<K>_{eff} \cong <K^2>/<K> \approx 2.6$[33,38]. Consequently, they conclude that real clusters (i.e. a clusters of known resonances) are needed in order to reproduce the short-range correlation. The difference between the two results is due to the chain fluctuations - which are present in DPM and not in the UA5 analysis. In DPM, the q_s-$\bar{q}_s$ chains already act like clusters and give a substantial contribution to the semi-inclusive two-particle correlation, $C_N(y_1, y_2)$. (See Fig. 13).

More precisely, the expression for C_N in cluster models depends on the value of $<K>_{eff}$. By comparing this expression with the experimental data one gets the value of $<K>_{eff}$. In DPM, the cluster model expression for C_N has to be modified to take into account the fluctuations in both the number of chain and their rapidity positions. Comparing this modified expression with the data, one obtains a much smaller value of $<K>_{eff}$, (around 1.4). We conclude in this way that no clustering of resonances is required by the present data. A similar conclusion has also been reached[39] from a study of e^+e^- and l p data, which involves only a single non-fluctuating chain.

VII. SEMI-HARD PROCESSES : MINIJETS

a) Cross-Section

The minijet analyses due to the UA1 collaboration[41] has indicated that there is a substantial part of the total cross-section, increasing rapidly with energy, which is due to semi-hard processes, i.e. processes in which the transverse momenta are relatively large (p_T >3-4 GeV). The corresponding cross-section is about 18 mb at $\sqrt{s}$ = 900 GeV. In these processes, partons carrying a very small longitudinal momentum fraction $x \approx p_T/\sqrt{s}$ are involved. Heuristically, gluons are expected to play the dominant role in these regions, and it has been suggested that the corresponding cross-sections can be computed in perturbative QCD[42,43]. Although the validity of this speculation is questionable to some of us, it does serve as a useful guide for estimating the magnitude of the minijet cross-section.

Since the increase with energy of the minijet cross-section is comparable to the increase of the total cross-section, it has been speculated[43] that the rise with energy of the latter is due to the onset of the semi-hard processes. This conjecture can be tested by including the semi-hard cross-section, computed in QCD, together with a soft component, with free parameters, into an unitarization (eikonal) scheme[44,45]. Performing a fit to the experimental data, one can show[45] that, although the energy dependence of the soft component is considerably reduced by the presence of the semi-hard component, still most of the increase of the total cross-section between ISR and collider energies is due to the soft component. The results of the fit[45], showing the net contribution to the total cross-section of the minijet events, are shown in Fig. 15.

b) <u>Multiplicity Distributions</u>

The UA1 collaboration has separated the observed events into two sets : the jet sample, containing at least one minijet, and the non jet one. They have observed that the jet sample has a larger average charged multiplicity $<N>_{jet} \approx 2 <N>_{no\ jet}$, and a narrower multiplicity distribution[41]. In a KNO plot, ($<N> P_N$ as a function of $z=N/<N>$), the multiplicity distributions of both the jet and no jet samples seem to obey KNO scaling between $\sqrt{s} = 200$ and 900 GeV. However, the latter observation is not conclusive, since, in the same energy range and in the same window of rapidity, they do not observe a KNO scaling violation for the complete event sample either.

In spite of that, it has been assumed in several papers[43,46] that the separate jet and no jet samples obey exact KNO scaling. In these papers, the authors "explain" the KNO scaling violation as being due to the increase with s of the minijet contribution which, with its large average multiplicity pushes upwards the high multiplicity tail of the distribution. If such a mechanism were to be added to the one at work in DPM, it would spoil the nice description of the multiplicity distributions given in Sect. 6b.

Before discussing this point in more detail it should be noted that this mechanism for the KNO scaling violation is rather similar to the one at work in DPM. As discussed in Sect. 6b, the contribution of the (high multiplicity) multichain configurations in DPM increases with increasing energy and pushes upwards the high multiplicity tail of the distribution. In fact it has been argued in ref.[45] that the minijets are just a part of the multi-chain contribution, and consequently, are implicitly included in DPM.

Indeed, as discussed in Sect. 7a, the dominant contribution to the minijet cross-section comes from a semi-hard gluon-gluon interaction. The dominant particle production diagram corresponding to such an interaction has four chains : the two chains initiated by the sea quark and antiquark resulting from the decay of the hard scattered gluon, plus the two valence chains resulting from the hadronization of the (octet color) proton. This four chain diagram is quite similar to a standard 4 chain diagram of DPM (see Fig. 2), except for the fact the sea quarks and antiquarks resulting from the decay of the hard scattered gluons, will have a relatively large p_T.

Since the total contribution of the two-chain plus multi-chain diagrams is equal to the non-diffractive inelastic cross-section, it is clear that the minijets events can not be introduced in DPM without reducing the contribution of the no jet multi-chain configurations. In fact, minijet events with the measured cross-section have been introduced[45] in DPM in a consistent way following the approach described in Sect. 7a. One can, in this way, split the weights σ_k of the multichain configurations into σ_k^{jet} and $\sigma_k^{no\ jet}$ (with $\sigma_k^{jet} + \sigma_k^{no\ jet} = \sigma_k$) and perform all the calculation of Sect. 6 for the jet and no jet event samples, separately. We get[45] :

$$<N>_{jet} = 1.6\ <N>_{no\ jet}$$

The multiplicity distributions for the complete (jet plus no jet) sample are practically identical to those obtained in Sect. 6 without the minijet events. The multiplicity distributions of both the jet and no jet sample violate KNO scaling. However, in the collider energy range this KNO scaling violation is rather small. In a KNO plot, the multiplicity distribution of the jet sample turns out to be narrower than that of the no jet sample, in qualitative agreement with UA1 data.

c) $\underline{<p_T>\ Behavior}$

It has been found that the average p_T, which is constant up to top ISR energies, increases from ISR to colliders energies. Moreover, the value of $<p_T>$ for an event sample of fixed multiplicity N, increases with N[47]. Both phenomena could be naturaly explained as a consequence of the increase with s of the minijet samples - which has relatively large values of both p_T and N. However, it appears that the $<p_T>$ versus N correlation is also present in the no jet event sample[41].

In DPM, the $<p_T>$ versus N correlation, has been explained[48,49] by the intrinsic p_T of the constituents at the chains end points, which produces a rotation of the chain axis relative to the C.M. If these intrinsic p_T have a gaussian distribution and are independent from each other (except for a trivial δ-function expressing the p_T conservation), one gets an average intrinsic p_T per constituent in the presence of 2k constituents (k chains) given by

$$< p_T >_{2k} = [(2k-1)/(2k)] <p_T^2>_{k \to \infty} \tag{16}$$

By choosing samples of events with increasing multiplicities, one is chosing configurations with increasing number of chains, and we see from Eq. (16) that the value of $<p_T^2>$ will increases with N until it reaches a saturation. This is precisely what is observed experimentally. In order to explain the measured value of the correlation one needs an intrinsic $<p_T^2>_{k \to \infty}$ for the sea quarks and antiquarks between 1 and 2 GeV^2. At the time of publication of[48,49], this value could seem unreasonably large. At present it can be understood as a consequence of the abundance of minijet events at collider energies. Although a thorough analysis of this problem is not yet available, it has been shown in Ref. 45 that if one incorporates the minijet events in DPM framework in the way described in Sect. 6b, one gets a reasonable description of the $<p_T>$ versus N correlation, including the observed differences between the jet and no jet event samples. The results are shown in Fig. 16.

Before closing this discussion of the $<p_T>$ versus N correlation, we would like to say that the above behavior is not valid in the fragmentation region. Here only the chains initiated by valence quark and diquark do contribute. Therefore, by increasing the multiplicity of the event sample, the $<p_T>$ will not increase but will in fact decrease by kinematical reasons (increasing the multiplicity without increasing the number of chains will reduce the value of $<p_T>$ per particle).

IIX. DIFFRACTION DISSOCIATION IN THE DUAL PARTON MODEL

In DPM, high mass single diffractive dissociation is described by the two-chain diagram shown in Fig. 17. These two chains or strings are stretched between valence constituents of the excited proton and sea constituents of the non-excited antiproton (which have to form a color singlet). Due to the x^{-1} behavior of the momemtum distribution function of a q_s-$\bar{q}_s$ pair (see Appendix 1), the antiproton inclusive spectrum will have[50] the characteristic $1/(1-x_p)$ behavior, typical of a triple-Pomeron approach, well verified by the data. Moreover, at fixed $x = 1-x_p = M^2/s$, the two chains in the diagram of Fig. 17 are identical to those appearing in the dominant (two-chain) component of inelastic $\pi^0 p$ scattering at center-of-mass energy $\sqrt{s} = M$ shown in Fig. 1. Thus, at fixed M, the hadronic spectra resulting from the hadronization of the two chains in Fig. 17 is identical to the hadronic spectra in a $\pi^0 p$ collision at $\sqrt{s} = M$.

The computation of the rapidity distribution, $dN/dy(M^2)$, of the diffractively produced state of mass M can be performed using the standard convolution between momentum distribution functions and fragmentation functions for each of the two chains of Fig. 17 and adding them together. The results[51] are compared with the experimental data in Fig. 18.

IX. CONCLUSION

QCD calculations based on perturbative expansions in the coupling constant are not applicable on a hadronic scale of energies. However, the QCD Lagrangian also contains the parameters N_c and N_f, which are the number of colors and flavors respectively. A suitable large-N expansion of QCD gives rise to a topological classification of the contributing gaphs. Such a topological expansion, when coupled with the concepts of duality, unitarity, Regge behaviour and the parton structure of hadrons, provides the motivation underlying the dual parton model. In this paper, we have reviewed extensively evidence that DPM can quantitatively describe essentially all the important features seen in multiparticle production in hadronic collisions. Special attention was paid to two key features of the model-- the held back effect and multi-chain diagrams. DPM has also been successfully extended so as to handle nuclear beams and targets, but these developments are not described in this paper.

DPM can also account for other qualitative features such as ratios of total cross sections. For concreteness, let us focus on the ratio r of πp to pp total cross sections. As discussed previously, it is not possible at present to compute in QCD the cross sections of various graphs in the $1/N$ expansion. Therefore, one cannot compute the ratio r rigorously. (An interesting recent attempt can be found in Ref. 52). However, a qualitative argument leading to the experimental value $r \approx 2/3$ is as follows: If one considers the differential cross section $d\sigma/dx_1 dx_2$ corresponding to the dominant two-chain diagram (where x_1 and x_2 refer to the momentum fractions of the held-back valence quarks), then we can see that there are two dominant contributions for a pion (q_F-$\bar{q}_S$ and $\bar{q}_F$-q_S) and three for a proton (($uu)_F$-d_S and 2 $(ud)_F$-u_S), where subscripts S and F stand for slow and fast respectively. (see Sec. 5b). If one makes the simplest assumption that all these possibilities give equal contributions to the total cross section, then the value $r = 2/3$ is obtained. This argument is clearly very crude and at the same level as the additive quark model.

The model also naturally incorporates the phenomenon of "Pomeron renormalization", or "flavoring", which leads to rising cross sections at collider energies.[53,54] At the present time, the dual parton model certainly provides the most complete overall phenomenological framework for quantitatively describing soft hadronic collisions.

FOOTNOTES AND REFERENCES

(1)   G. t'Hooft, Nucl. Phys. B72, 461 (1974).

(2)   G. Veneziano, Nucl. Phys. B74, 365 (1974), B117, 519 (1976), and "Color Symmetry and Quark Confinement", p. 113, ed. J. Tran Thanh Van, (Editions Frontieres, 1977).

(3)   E. Witten, Nuc. Phys. B160, 57 (1979); S. Coleman, in "Pointlike Struture Inside and Outside Hadrons", Erice lectures, p.11, ed. A. Zichichi (Plenum, New York, 1982).

(4)   See, e.g., V. N. Gribov, Sov. Phys. JETP 26, 414 (1968); V.A. Abramowski, V.N. Gribov and O.V. Kancheli, Yad. Fiz. 18, 595 (1973); Sov. J. Nucl. Phys. 18, 308 (1974).

(5)   Chan Hong-Mo et al., Nucl. Phys. B86, 470 (1975) and B92, 13 (1975).

(6)   G. F. Chew and C. Rosenzweig, Nucl. Phys. B104, 290 (1976), and Phys. Rep. 41C, 263 (1978); G. F. Chew et al., "GeV Partons and TeV Hexons from a Topological Viewpoint", p. 143, ed. J. Tran Thanh Van, (Editions Frontieres, 1984).

(7)   A. Capella, U. Sukhatme, C-I Tan and J. Tran Thanh Van, Phys. Lett. 81B, 68 (1979); A. Capella, U. Sukhatme and J. Tran Thanh Van, Z. Phys. C3, 68 (1980).

(8)   A.A. Migdal, Phys. Rep. C102, 200 (1983). For further reviews, see for instance, A. Jevicki, "Large N Loop Space Methods in QCD", Proceedings of the XXII International Conference on High Energy Physics, Leipzig, 1984; S. R. Das, Rev. Mod. Phys. 59, 235 (1987), and C-I Tan, "Meson Dynamics in the Large-N Limit", in

Proceedings of the First Asia-Pacific Physics Conference, p. 1231, (World Scientific Publishing Co., Singapore, 1983).

(9)   In the large-N limit, one must treat baryons differently from mesons. While a meson can always be made out of a quark-antiquark pair, irrespective of N, a baryon must be made out of N quarks. If we regard QCD at large N as a weakly coupled field theory of mesons, it has been suggested in Ref. 3 that baryons are in some sense solitons in such a theory. This has in turn led to the successful "skyrmion" picture. In a strictly large-N framework, the baryon structure is determined by solving a set of non-linear Hartree equations, with each quark interacting with a collective quark sea having the color of an antiquark. For a topological expansion, we represent this "quark sea" by a diquark.

(10)   See, for instance, "Hadron Physics at Very High Energies" by D. Horn and F. Zachariasen, Frontiers in Physics Lecture Note Series, W. A. Benjamin, Inc. (1973).

(11)   H. Lee, Phys. Rev. Lett. 30, 719 (1973) and G. Veneziano, Phys. Lett. 43B, 413 (1973).

(12)   X. Artru and G. Menessier, Nucl. Phys. B70, 93 (1974).

(13)   P. Aurenche and L. Gonzales Mestres, Phys. Rev. D18, 2995 (1978); C. Chiu and S. Matsuda, Nucl. Phys. B134, 463 (1978).

(14)   It should be pointed out that the "cutting-rule" for Pomeron as two independent chains is strictly speaking an approximation. Correlation between these two chains are always present, although the effect is small at small tranverse momenta. [See, for instance, Chung-I Tan, Proc. Marseille Conference on "Hadron Physics at High Energies", (1978); P. Aurenche and L.

Gonzales-Mastres, Z. Phys. $\underline{C1}$, 307 (1979)]. It is interesting to note in this connection that a "semi-hard" event in a cylinder topology would lead to the break down of the independent-chain assumption. This suggests a possible perturbative treatment of mini-jets by considering "string kinks" in a topological expansion. Work on this formalism is currectly in progress.

(15) For further detail, see A. Capella in: "Partons in Soft Hadronic Reactions", edited by R. Van de Walle (World Scientific Publishing Co., Singapore, 1981).

(16) A. Capella and J. Tran Thanh Van, Z. Phys. $\underline{C10}$, 249 (1981) and Phys. Lett. $\underline{93B}$, 146 (1980).

(17) A. Krzywicki and B. Petersson Phys. Rev. $\underline{D6}$, 924 (1972); J. Finkelstein and R. D. Peccei, Phys. Rev. $\underline{D6}$, 2606 (1972); R. Feynman and R. Field, Nucl. Phys. $\underline{B136}$, 1 (1978); U. Sukhatme, Phys. Lett $\underline{73B}$, 478 (1978); B. Andersson, G. Gustafson, G. Ingelman, and T. Sjostrand, Phys. Rep. $\underline{97}$, 31 (1983) and references therein.

(18) R. Blankenbecler and S. Brodsky, Phys. Rev. $\underline{D10}$, 242 (1974).

(19) A. Kaidalov, preprint ITEP-116 (1984).

(20) H. Minakata, Phys. Rev. $\underline{D20}$, 1656 (1979); G. Cohen-Tannoudji, El Hassouni, J. Kalinowski, O. Napoli and R. Peschanski, Phys. Rev. $\underline{D21}$, 2689 (1980).

(21) M.G. Ma et al., Z. Phys. $\underline{C30}$, 191 (1986); E. De Wolf et al., Z. Phys. $\underline{C31}$, 13 (1986).

(22) A. Capella, U. Sukhatme and J. Tran Thanh Van, Phys. Lett. 125B, 330 (1983).

(23) A. Pagnamenta and U. Sukhatme, Z. Phys. C14, 79 (1982); K. Fialkowski and A Kotanski, Z. Phys. C20, 1 (1983).

(24) B. Andersson, G. Gustafson and C. Peterson, Phys. Lett. 69B, 221 (1977).

(25) A. Capella, U. Sukhatme and J. Tran Thanh Van, Phys. Lett. 119B, 220 (1982); C. Pajares and A. Varias, Z. Phys. C19, 69 (1983).

(26) A. Capella, U. Sukhatme, C-I Tan and J. Tran Thanh Van, Proceedings of the XVth Multiparticle Dynamics, Lund June 1984, and Phys. Rev. D (in press); A. B. Kaidalov and O. I. Piskunnova, Z. Phys. C30, 145 (1986), and ITEP-157 (1985).

(27) J. Rushbrooke, Proceedings of the International Europhysics Conference on High Energy Physics, Bari, Italy, ed. L. Nitti and G. Preparata (1985).

(28) A. Capella and J. Tran Thanh Van, Phys. Lett. 114B, 450 (1982); P. Aurenche, F. Bopp, Phys. Lett. 114B, 363 (1982).

(29) A. Capella and J. Tran Thanh Van, Z. Phys. C23, 165 (1984); A. Capella, A. Staar and J. Tran Thanh Van, Phys. Rev. D32, 2933 (1985).

(30) A. Kaidalov and K.A. Ter Martyrosian, Phys. Lett. 117B, 247 (1982).

(31) P. Aurenche and F. Bopp and Ranft, Phys. Lett. 147B, 212 (1984).

(32) T. Kanki, Nuc. Phys. B243, 44 (1984).

(33) UA5 collaboration, Phys. Lett. 160B, 193 (1985).

(34) A. Capella and J. Tran Thanh Van, Z. Phys. C18, 85 (1983).

(35) P. Aurenche and F. Bopp and Ranft, Z. Phys. C23, 67 (1983).

(36) A. Capella and K. Krzywicki, Phys. Rev. D18, 4120 (1978).

(37) A. Capella and J. Tran Thanh Van, Proc. Vth Topical Workshop on p̄p Collider Physics, San Vicente (Aosta) 1984, ed. Greco.

(38) K. Böckmann, Physics in Collision 3, ed. G. Bellini, A. Bettini and L. Perasso, (Editions Frontères, 1983).

(39) A. Capella and A. V. Ramallo, Orsay preprint LPTHE 87-08.

(40) E. Berger, Nucl. Phys. B85, 61 (1975).

(41) UA1 Collaboration, C. Ciapetti in The Quark Structure of Matter, ed. M. Jacob and K. Winter, p. 455 (1986); F. Ceradini, Proceedings of the International Europhysics Conference on High Energy Physics, Bari (ed. L. Nitti and G. Preparata, 1985) and references therein.

(42) G. Cohen-Tannoudji, A. Mantrach, H. Navelet, and R. Peschanski, Phys. Rev. D28, 1628 (1983); A. Mueller and H. Navelet, Service de Physique Theorique Report No. SPHT/86-094, 1986 (to be published); L. V. Gribov, E. M. Levin, M. G. Ryskin, Phys. Rep. 100, 1 (1983), and references therein; J. Kwiecinski, Z. Phys.

468

$\underline{C29}$, 561 (1985), preprint IPN Cracow 1338/PH (1986) and Erratum.

(43) T. V. Gaisser and F. Halzen, Phys. Rev. Letters $\underline{54}$, 1754 (1985); G. Pancheri and Y. Srivastava, Phys. Lett. $\underline{159B}$, 69 (1985); G. Pancheri and C. Rubbia, Nuc. Phys. $\underline{A418}$, 117 (1984)

(44) L. Durand and Hong Pi, Phys. Rev. Letters $\underline{58}$, 303 (1987).

(45) A. Capella, J. Tran Thanh Van, and J. Kwiecinski, Phys. Rev. Letters $\underline{58}$, 2015 (1987).

(46) A. D. Martin and C. J. Maxwell, Phys. Lett. $\underline{172B}$, 248 (1986); R. C. Hwa, preprint OITS-354, U. of Oregon.

(47) UA1 Collab., G. Arnison *et al.*, Phys. Lett. $\underline{118B}$, 167 (1982); and G. Giacomelli, Proceedings of the Elastic and Diffractive Scattering, ed. B. Nicolescu and J. Tran Thanh Van, (Editions Frontieres, June 1985).

(48) A. Capella and A. Krzywicki, Phys. Rev. $\underline{D29}$, 1007 (1984).

(49) P. Aurenche, F.W. Bopp and J. Ranft, Phys. Lett. $\underline{147B}$, 212 (1984).

(50) UA4 collab., Proceedings of the Elastic and Diffractive Scattering, ed. B. Nicolescu and J. Tran Thanh Van, (Editions Frontieres, June 1985), and UA4 collab., D. Bernard *et al.*, Phys. Lett. $\underline{166B}$, 459 (1986).

(51) V. Innocente, A. Capella, A.V. Ramallo and J. Tran Thanh Van, Phys. Lett. $\underline{169B}$, 285 (1986).

(52) P. Landshoff and O. Nachtmann, DAMTP preprint 86/27; for a S-matrix derivation, see P. Gauron, B. Nicolescu and S. Ouvry, Phys. Rev. D24, 2501 (1981).

(53) T. Gaisser and C-I Tan, Phys. Rev. D8, 3881 (1973); C-I Tan, Proceedings of IXth Rencontre de Moriond, ed. J. Tran Thanh Van, (Editions Frontieres, March 1974), Proceedings of the Elastic and Diffractive Scattering, ed. B. Nicolescu and J. Tran Thanh Van, (Editions Frontieres, June 1985), and "Increasing Total Cross Sections and Flavoring of Pomeron in QCD", to be published in the Proceedings of the XVIIIth International Symposium on Multiparticle Dynamics, Tashkent, USSR, Sept. 1987.

(54) J. W. Dash, S. T. Jones, and E. Manesis, Phys. Rev. D18, 303 (1978); J. W. Dash and S. T. Jones, Phys. Lett. 157B, 229 (1985), and S. T. Jones and J. W. Dash, preprint UA HEP-869 and CPT-87/PE. 1998.

APPENDIX 1

The expression, Eq. (6), for ρ_k given in Section IV is actually only valid when $x_q = x_{\bar{q}}$ within each pair of sea quark and antiquark. More generally, one obtains[16] the following expression:

$$\rho_k(x_1, x_{2k}; x_2, x_3, \ldots, x_{2k-1}) =$$

$$c_k P \, x_1^{-0.5} (x_2 + x_3)^{-1} x_2^{-0.5} x_3^{-0.5} \cdots x_{2k}^{1.5} \, \delta(1 - x_1 - x_2 - \cdots - x_{2k}). \quad (A1)$$

470

As in Eq. (6), x_1 refers to the valence quark, x_{2k} to the diquark, and x_i ($2 \le i \le 2k-1$), to the sea quarks and antiquarks. Eq. (A1) reduces to Eq. (6) in the limit $x_2 = x_3$, etc. It also tells us that ρ_k has a typical $1/X$ type behavior in the average momentum fraction for each q-$\bar{q}$ pair, e.g., $X = (x_2 + x_3)/2$, for small X. On the other hand, its dependence on the gap between two members of each pair is

$$\rho_k \propto x_2^{-0.5} x_3^{-0.5} = x_2^{-0.5} (X - x_2)^{-0.5} \tag{A2}$$

which is the same as the one governing the rapidity gap between the valence quark and antiquark in a meson, Eq.(4).

The results presented in Sections VI and VII involving multi-chain contributions were obtained without using the Monte-Carlo approach. For simplicity, we have used Eq. (6) and have in addition introduced a cutoff in order to avoid the x^{-1} singularities by using $(x^2 + \mu^2/s)^{1/2}$ in place of x. A value of $\mu \cong 0.1$ GeV was used. However, we have checked that varying μ between 0.05 and 0.3 GeV produces only minor changes in our results.

In a Monte-Carlo approach, it is of course possible to use the exact momentum distribution functions, (A1), and to suppress the cutoff, μ. One generates simultaneously the x_i and x'_i for both incoming hadrons, and keeps only those events where the squared masses, $s_{chain} = x_i x'_i s$, of all chains are above their physical thresholds. In this way one cuts off the x^{-1} singularies in Eq. (A1). A Monte-Carlo program following the procedure outlined above, together with the two-jet Lund Monte-Carlo for the chain fragmentation, has been carried out by the authors in collaboration with X. Artru. The results obtained using this approach are extremely close to the ones reported here.

TABLE 1: A comparison of the data and the dual parton model calculation of the ratio R^h [defined in eq. (7)].

Detected particle h	R^h(experiment)	R^h(DPM)
all (±)	1.00 ± .01	1.02
only positives	0.96 ± .01	0.98
only negatives	1.03 ± .01	1.06
$K^+ + K^-$	0.98 ± .04	1.06
K^+	0.90 ± .05	0.97
K^-	1.08 ± .06	1.17
$p + \bar{p}$	1.05 ± .04	1.06
p	0.93 ± .04	0.92
$\bar{p}$	1.25 ± .07	1.26

472

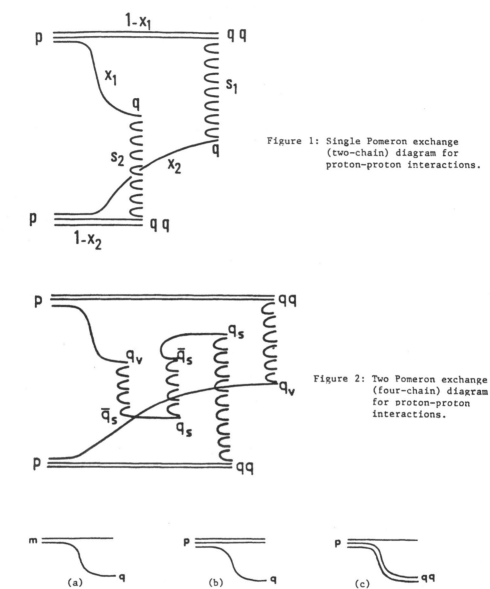

Figure 1: Single Pomeron exchange (two-chain) diagram for proton-proton interactions.

Figure 2: Two Pomeron exchange (four-chain) diagram for proton-proton interactions.

Figure 3: Momentum distributions for (a) quark (antiquark) in a meson ; (b) quark in a proton ; (c) diquark in a proton.

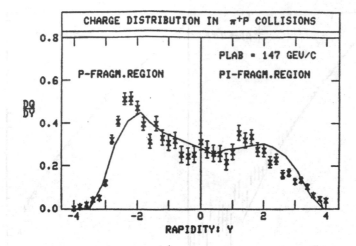

Figure 4: Comparison of the dual parton model (two chain contribution) with charge distribution data from pion – proton collisions.

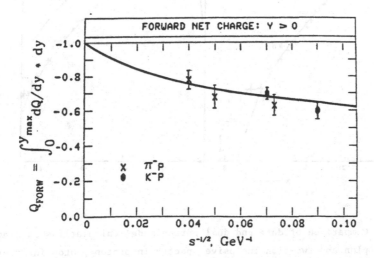

Figure 5: Data on the energy dependence of the net charge in the negative meson hemisphere. The solid line is calculated using the dual parton model.

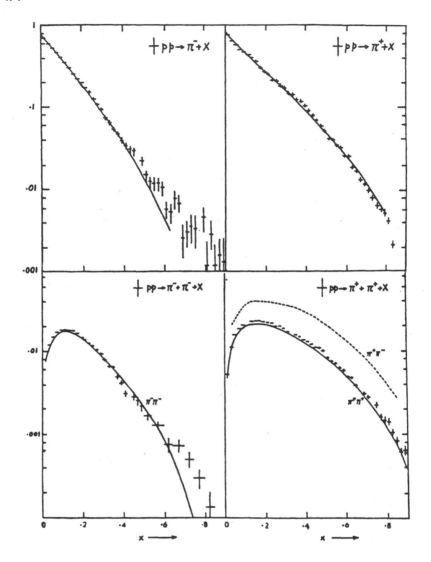

Figure 6: Comparison of data and dual parton model calculations of single
pion and two-pion inclusive spectra in proton-proton interactions.

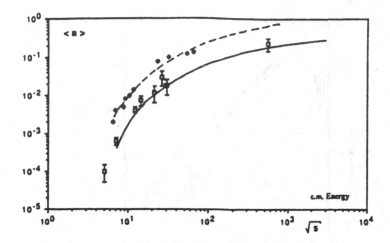

Figure 7: Average number of antiprotons and antilambdas in a pp collision
as a function of c. m. energy (in GeV). Theoretical curves (dashed
for $\bar{p}$ and solid for $\bar{\Lambda}$) are from the multichain dual parton model.

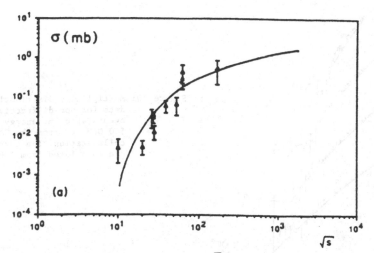

Figure 8: Inclusive cross section for $D\bar{D}$ pair production in pp collisions
versus c. m. energy (in GeV).

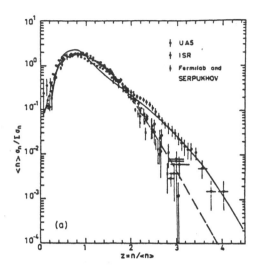

Figure 9: Violation of KNO scaling in the dual parton model.

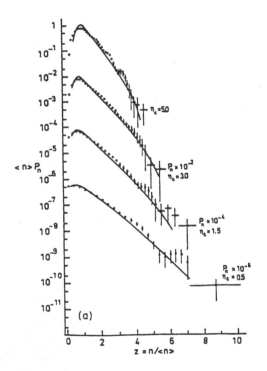

Figure 10: Multiplicity distribution data for non-diffractive events at c. m. energy 540 GeV are from the UA5 collaboration. The curves are calculated from DPM.

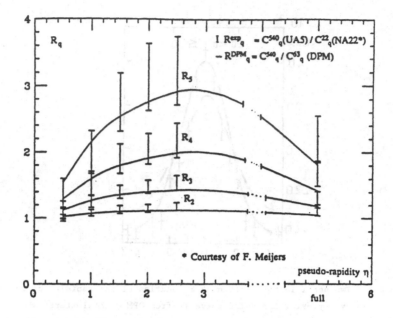

Figure 11: Ratios of various multiplicity moments as a function of the pseudo-rapidity interval.

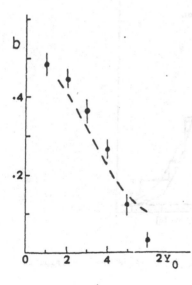

Figure 12: Forward-backward correlation slope $b(y_0)$ as a function of the rapidity gap length $2y_0$.

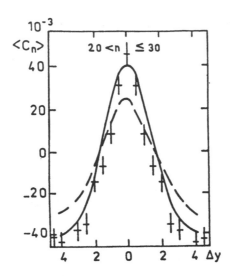

Figure 13: Two particle correlation as a function of the rapidity difference $y = y_1 - y_2$. The solid curve is from DPM with clusters, whereas the dashed curve corresponds to directly produced particles (no clusters or resonances).

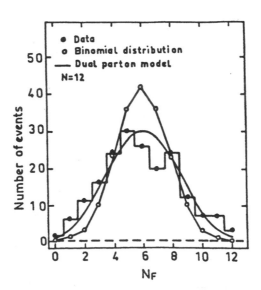

Figure 14: The probability distribution $P(N_F)$ at fixed total multiplicity.

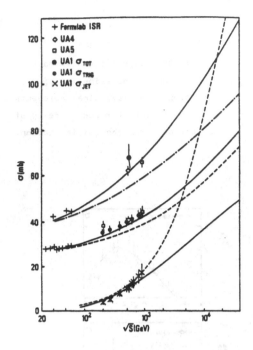

Figure 15: Total, inelastic non-diffractive and minijet cross sections computed in a DPM framework.

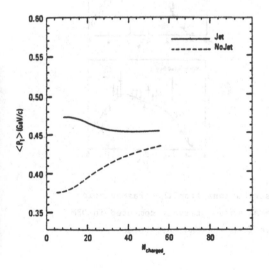

Figure 16: Behavior of the average transverse momentum p_T as a function of the charged particle multiplicity for jet and no-jet samples.

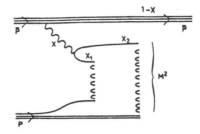

Figure 17: DPM diagram for high mass
diffraction dissociation in pp
collisions. The wavy line represents
a color singlet (Pomeron) made up of
sea partons from the initial hadron.

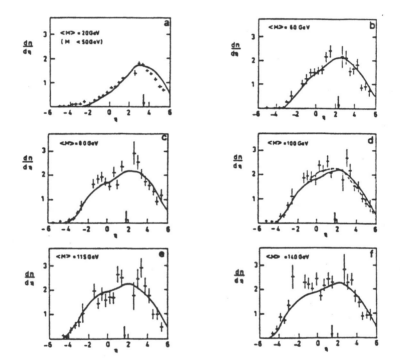

Figure 18: Charged pseudorapidity distributions from the fragmentation of
diffractive states in several mass intervals computed in DPM
(solid lines). The data are from the UA4 collaboration.

APPLICATION OF THE METHODS OF QUANTUM OPTICS TO MULTIHADRON PRODUCTION

G.N. Fowler[1,2] and R.M. Weiner[2]

[1]University of Exeter, England
[2]University of Marburg, F.R.Germany

ABSTRACT

The use of quantum statistical methods applied
with success in quantum optics, is reviewed in
the domain of multiparticle production. The
theoretical formalism is outlined and the conditions under which it can be used in strong
interaction physics are analyzed. Applications
to multiplicity distributions include the rapidity and the energy dependence in hadron and
lepton induced reactions, and the forward-backward correlations in p-p collisions.
The relation between multiplicity distributions
and intensity interferometry is discussed. Further applications include the absence of cascading in hadron-nucleus reactions in terms of
self-induced transparency and Centauros as a
possible consequence of coherent states.

1. Introduction

In high energy processes in which the pion multiplicity is large enough, we may expect the methods of quantum statistics (QS) to be useful. These methods have been applied with great success particularly in quantum optics (QO), superfluidity, superconductivity etc. What distinguishes optical phenomenon from those in particle physics are conservation laws and final state interactions which are present in particle physics. At high energies and high multiplicities the first are unimportant. Neglecting for the moment also the final state interactions, QS reduces then to quantum optics (QO) and we may take over the formalism of QO to interpret the data on multipion production at high energies, provided we consider identical pions. Given the general validity of QS (or QO), it is then clear that any model of multiparticle production must satisfy the laws of quantum statistics and this has far reaching consequences, independent of the particular dynamical mechanism which governs the production process. The first suggestion for this procedure came from Giovannini et al. (1973, 1974) and was further developed by Suzuki (1974) and Namiki (1975). It is the purpose of this paper to review the progress achieved until now in this field.

2. The Theoretical Formalism of QO[*)]

It is convenient to start from the coherent state representation of the statistical operator $\hat{\rho}$ introduced by Glauber (1963). The coherent state $|\alpha_k>$ is defined through

$$\hat{a}_k|\alpha_k> = \alpha_k|\alpha_k> \tag{1}$$

where $\hat{a}_k$ is a pion field annihilation operator referring to the k-th field mode and α_k is a complex eigen-value.

[*)] For a recent general introduction into this subject cf. the excellent textbook by Schubert and Wilhelmi, 1986.

Dropping the mode index we introduce a function $\mathcal{P}(\alpha)$ through

$$\hat{\rho} = \int d^2\alpha \, \mathcal{P}(\alpha) \, \frac{|\alpha\rangle\langle\alpha|}{\pi} \, . \tag{2}$$

In all the cases in which we shall be interested $\mathcal{P}(\alpha)$ in equation (2) is the probability distribution for the stochastic variable α, the field amplitude. In the number representation $|n\rangle$ defined by

$$\hat{n}|n\rangle = n|n\rangle$$

the statistical operator $\hat{\rho}$ is given by

$$\hat{\rho} = \sum_{n,m} \int \mathcal{P}(\alpha) \, \frac{\alpha^n \alpha^{*m} e^{-|\alpha|^2} \, |n\rangle\langle m|}{(n!m!)^{1/2}} \, \left(\frac{d^2\alpha}{\pi}\right) \tag{3}$$

We shall be particularly interested in the diagonal elements

$$\langle n|\hat{\rho}|n\rangle = \int \frac{\mathcal{P}(\alpha) |\alpha|^{2n} e^{-|\alpha|^2}}{n!} \, \frac{d^2\alpha}{\pi} \equiv P(n). \tag{4}$$

$P(n)$ may also be written as a Poisson transform

$$P(n) = \int_0^\infty dx f(x) \, \frac{(x\bar{n})^n}{n!} \, e^{-x\bar{n}} \tag{5}$$

by a suitable redefinition of variables[*]. Here $\bar{n}=\Sigma n P(n)$ is the mean multiplicity.

Certain distributions $P(n)$ have special significance from the QO point of view and are listed in Table 1 along with the corresponding $\mathcal{P}(\alpha)$. The entries in Table 1 correspond in optical language to a pure single mode laser

---

[*] In quantum optics photo-electron counts are measured so that the Poisson transform corresponds to a statistical statement (which depends on the nature of the electron production process) relating the electro-magnetic field amplitude to the electron counts.

with well defined amplitude and phase (6a) (or phase averaged (6b)), a single mode purely chaotic field (7), a k
mode chaotic field (8), a field in which a single mode laser and a chaotic field are superposed (9) (the mode frequencies being the same), and the k mode generalization
thereof (10), hereafter called the partially coherent laser distribution or PCLD for short. It is important to rea-
lize that formulae (8) and (10) have been derived under the
assumption that each mode (or phase space cell) is on the
average equally occupied, $\bar{n}_i = \bar{n}/k$ and that in the appropriate case each cell or mode has the same coherent amplitude $\tilde{\alpha}/\sqrt{k}$.

Table 1

1. Poisson distribution

$$P(n) = \frac{\bar{n}^n e^{-\bar{n}}}{n!}$$

$$\mathcal{P}(\alpha) = \delta(\alpha - \tilde{\alpha}) \tag{6a}$$

$$\mathcal{P}(\alpha) = \delta(|\alpha|^2 - |\tilde{\alpha}|^2) \tag{6b}$$

2. Bose-Einstein distribution

$$P(n) = \frac{\bar{n}^n}{(1+\bar{n})^{n+1}} \tag{7a}$$

$$\mathcal{P}(\alpha) = \frac{e^{-|\alpha|^2/\bar{n}}}{\bar{n}} \tag{7b}$$

3. Negative binomial distribution

$$P(n) = \frac{(n+k-1)!}{(k-1)!n!} \frac{(\bar{n}/k)^n}{(1+\bar{n}/k)^{n+k}} \tag{8a}$$

$$\mathcal{P}(\alpha) = \frac{\exp\{-\sum_{i=1}^{k}|\alpha_i|^2/(\bar{n}/k)\}}{(\bar{n}/k)^k} \tag{8b}$$

4. Glauber-Lachs distribution

$$P(n) = \frac{(p\bar{n})^n}{(1+p\bar{n})^{n+1}} \exp\left(\frac{-|\overset{\sim}{\alpha}|^2}{1+\bar{n}p}\right) L_n\left(\frac{-|\overset{\sim}{\alpha}|^2}{1+\bar{n}p}\right) \tag{9a}$$

$$\mathcal{P}(\alpha) = \frac{e^{|\alpha-\overset{\sim}{\alpha}|^2}/\bar{n}p}{\bar{n}p} \tag{9b}$$

where L_n is the Laguerre polynomial and the quantity p is defined by $p = (\bar{n}-|\overset{\sim}{\alpha}|^2/\bar{n})$.

5. Generalized Glauber-Lachs distribution (PCLD)

$$P(n) = \frac{(p\bar{n}/k)^n}{(1+p\bar{n}/k)^{n+k}} \exp\left\{\frac{-|\overset{\sim}{\alpha}|^2}{1+\bar{n}p/k}\right\} L_n^{k-1}\left(\frac{-k|\overset{\sim}{\alpha}|^2/(p\bar{n})}{1+p\bar{n}/k}\right) \tag{10a}$$

$$\mathcal{P}(\{\alpha_i\}) = \exp\left\{-\sum_{i=1}^{k} \frac{|\alpha_i-\overset{\sim}{\alpha}\sqrt{k}|^2/(p\bar{n}/k)}{(p\bar{n}/k)^n}\right\} \tag{10b}$$

It is convenient to introduce at this point the notation

$$\bar{n}_c = |\overset{\sim}{\alpha}|^2 \text{ and } \bar{n}_{ch} + \bar{n}_c = \bar{n}$$

where $\bar{n}_c$, $\bar{n}_{ch}$ and $\bar{n}$ are the mean numbers of coherent and chaotic quanta respectively and $\bar{n}$ the total mean number. The chaoticity p is now

$$p = \bar{n}_{ch}/\bar{n}.$$

The number of independent field modes, or phase space cells in particle language, may subsequently be modified by a further factor referring to other degrees of freedom, for example different charge states.

Formula (10) has been generalized by Périna and Horak (1969) and applied by Blažek (1984) to include the effect of different frequencies for the chaotic and coherent modes (see also Jaiswal and Mehta (1970)). We refer to this briefly below.

Although from this derivation, k, the number of independent degrees of freedom, is an integer, the formulae can be generalized to non-integral k, as for example in the stochastic cell model of Carruthers and Shih (1984), and through the partial occupancy of the phase space cells according to the following argument.

In QO many modes may be introduced by recognizing that a measurement of counts takes place over a finite time interval, so that what is actually measured is

$$w = \int_0^T I(t)\,dt \tag{11}$$

where I is the field intensity.

It may be shown (cf. Schubert and Wilhelmi, p. 132) that in this case

$$P(n) = \int \mathcal{P}(w)\ e^{-w}\ \frac{w^n}{n!}\ dw \tag{12}$$

and

$$w = \sum_k |\alpha_k|^2 \tag{13}$$

and $\mathcal{P}(w)$ plays the same role as $\mathcal{P}(\alpha)$ in the single mode case.

If the field correlation time γ is introduced through, for example,

$$\langle E(t)\ E^*(t+\tau)\rangle = \Gamma e^{-|\tau|/\gamma}\ e^{-i\omega_0\tau} \tag{14}$$

corresponding to a Lorentz spectrum of central frequency ω_0 (other forms for the spectrum, e.g. Gaussian, may be considered) the multiplicity distribution can be calculated and certain special cases have been examined (see for example Barakat and Blake (1980)). (Note that formula (14) implies that the statistical distribution of the α_i is stationary in time which is equivalent to the assumption of equal average occupancy of the different cells in the particle description.) Of particular interest in this connection are

(i) the purely chaotic case and
(ii) the chaotic-coherent superposition.

(i) Chaotic case

In the limit $T/\gamma \ll 1$ this reduces, in the absence of other degrees of freedom, to the Bose-Einstein formula. In the limit $T/\gamma \gg 1$ one finds the negative binomial distribution (NBD) with $k = T/\gamma$ (Mandel (1959)) so that at least in this limit T/γ represents the number of independent cells in phase space and may be non-integral.

(ii) Chaotic-coherent superposition

Numerical studies of $P(n)$ exist in the optical literature but for our purposes the analytical expressions for the factorial cumulants given by Jaiswal and Mehta (1970) (JM) are the most useful.

To apply the formalism to particle physics we assume that the pion fields are functions of rapidity only, following Betke, Scalapino and Sugar (1974), (it being assumed that the transverse momentum p_T has been averaged over or the restriction is made to $p_T < p_T^0$ where p_T^0 is a conveniently chosen upper limit (a single cell in transverse momentum)) so that T is replaced by the total rapidity and γ becomes a rapidity correlation length ξ. The conjugate variable which replaces frequency is the boost $k_3 = Ez$ (Durand and O'Raifeartaigh 1976) and the spectrum of the optical line is replaced by the boost spectrum which may be infered as in the optical case from the shape of $\bar{n}(y)$.

The factorial cumulants given by JM may now be taken over directly. They are given by

$$\mu_r = \frac{\bar{n}^r}{k^{r-1}} \; (r-1)! \; p^r B_r + r! p^{r-1} \; (1-p)\tilde{B}_r \tag{15}$$

where B_r and $\tilde{B}_r$ are known functions of $\beta \equiv Y/\xi$ and $\Omega \equiv \omega_c - \omega_{ch}$. ω_c and ω_{ch} are the boosts of the coherent and chaotic fields respectively. In particle physics applications one assumes usually, as in QO, that $\Omega = 0$. (Cf. however Blazek (1984).) The factorial cumulants are related to the factorial moments $F_r = \langle n!/(n-r)!\rangle$ by the relation

$$\exp\left(\sum_{n=1}^{\infty} \frac{\mu_n S^n}{n!} \right) = \sum_{k=0}^{\infty} \frac{F_k S^k}{k!}$$

which is to be understood as an identity in S.

Finally ξ is the coherent length of the chaoticity field π defined in analogy to eq. (14) by the relation:

$$\langle \pi(y_1) \; \pi(y_2) \rangle = \bar{n}_{ch} \; e^{-\frac{y_1 - y_2}{\xi}} \tag{16}$$

Before discussing the comparison with experiment of these results, the non-stationarity found already in 1981 (cf. Fowler et al. 1981; see below) and recently reconfirmed by the UA-5 collaboration (J. Alner et al. 1985(a)) should be explained. It is, for example, implied by the observation that the P(n) in different rapidity intervals, which have not the same center, correspond to different values of p, i.e. have different relative widths.

This implies that $\bar{n}_{ch}$ and/or $\bar{n}_c$ are functions of rapidity y. We shall proceed by defining an effective chaoticity p independent of y which represents an average over the p(y).

A different approach has been presented by Fowler et al. (1986) in which the factorial cumulants are derived from a two component model using, instead of a linear superposition of chaotic and coherent fields, a convolution of Bose-Einstein and Poisson distribution in which the mean numbers of the two components are functions of rapidity to be deduced from a fit to data.

3. Comparison with experiment

(I) Introduction

In comparing the theories we have outlined with experiment, a number of theoretical and practical difficulties need to be overcome. In the first place, optical analogues describe identical particles, whereas most experimental data refer to both charges. In addition, incident particles which survive in the final state (generally leading particles) must be excluded and so should diffractive production since it can be argued that these arise from a mechanism different from the one in which we are interested and which populates the central part of the rapidity distribution.

The problem of charges versus identical particles is discussed in section 3(IV). For the corrections due to substraction of leading particles and diffraction we refer to Carruthers and Shih (1987) (leading particle effects have also been included by Blažek (1984) in the model referred to above).

In addition, the available energy for pion production may differ from event to event, which can affect the numerical values of fitted parameters. The experimentally observed distribution $P(n)$ is then related to the intrinsic distribution $P(n|K)$ by the relation

$$P(n) = \int P(n|K) \ \chi(K) dK \tag{17}$$

where $\chi(K)$ is the inelasticity distribution. Models for
$\chi(K)$ have been given by Fowler et al. (1985(a), 1987(b))
and calculations exhibiting the possible effect of $\chi(K)$ on
the multiplicity distribution have been performed by Plümer (1984).

(II) Multiplicity distribution

(a)   Hadron-hadron collisions

The relevant experimental data consist largely of pp
and p$\bar{p}$ events at various energies from approximately 20 GeV
to 540 GeV for various central and non-central rapidity intervals. Fits to the data have been found with NBD for integer k from 2-4, the value depending on rapidity interval
(Carruthers and Shih 1983). Biyajima has fitted the 540 GeV
data with the PCLD with k = 2 and 0.15<p<0.6 (depending on
the rapidity range) and Fowler et al. (1986(a)) have also
found fits with PCLD and k = 1, this time over the entire
energy range referred to, with 0.015<p<0.15.

The most recent comparison of NBD and PCLD fits has been
carried out by Carruthers and Shih (1987) with the substitution n→n-α, 0<α<2. They conclude that the data can as
readily be fitted with NBD, with energy and rapidity dependent k, as by PCLD, with k fixed and p varying with energy
and rapidity. However, as will be shown in section 3(IV),
even the PCLD is a too particular approach to the hadronic
multiplicity distribution, because it ignores the finite
coherence length in rapidity. Therefore the PCLD and, a for-
tiori, the NBD have to be replaced by the more general J.M.
formalism which provides an adequate description of both
the energy and rapidity dependence of P(n) (Fowler et al.
1987).

(b) e^+e^- collisions

The pion multiplicity distribution on e^+e^- collisions

has also been fitted with PCLD (Carruthers and Shih 1984)
and NBD (Derrick et al. 1986), but here the narrowness of
the distribution demands very small values of p, 0<p<0.04
depending on energy in the range 12-35 GeV, and correspon-
dingly large values of k. When the multiplicity distribution in non-central rapidity windows is examined one finds
k>100, p∿0. Of these two alternatives, the PCLD, with its
notion of increasing coherence for energetic pions near
the jet axis, seems the more convincing. The situation is
depicted in fig. 1   (Carruthers and Shih 1986) (note that
$p \cong m^2$). However for this reaction, too, the more general JM
approach appears necessary (Friedlander et al. 1987) (cf.
sec. 3(IV)).

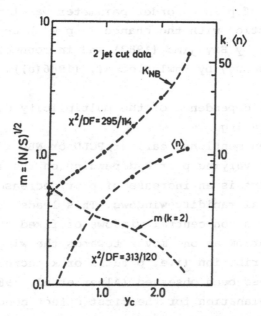

Fig. 1. Best fit parameter to the limited rapidity
($-y_c<0<y_c$) multiplicity distribution for the PCLD
(the left hand coordinate m = N/s)$^{1/2}$, with N = $\bar{n}_{ch}$,
S = $\bar{n}_c$ in our notation) and for the NB (right hand
coordinate k (and $\bar{n}$) (from Carruthers and Shih 1986).

(III) Energy dependence of the multiplicity distribution

Since the distribution over the full rapidity range, available in non-single diffraction events, is known to broaden significantly with energy (no K.N.O scaling), the parameters appearing in the distribution must be energy dependent. Thus for the NBD k must diminish while for the PCLD p must increase. This can be interpreted in the NBD case in terms of a change in size of the contributing clans or clusters (Giovannini and Van Hove 1986, Van Hove and Giovannini 1986) (fewer clusters with larger size), whereas in the PCLD, the change in p can be interpreted as an increase in disorder, implying a transition from an ordered to a disordered state as energy increases. This interpretation of p as an order parameter was first suggested in connection with the change in p with central rapidity interval by Biyajima (1983) and in connection with change with energy by Fowler et al. (1986(a)).

(IV) Rapidity dependence of the multiplicity distribution and β-scaling

As has been mentioned earlier, PCLD or NBD fits are obtainable with varying p or k, depending on the model. The general feature is an increase of p or decrease of k for smaller central rapidity windows. This leads to the suggestion that in non-central windows of fixed size the distributions narrow as one moves towards the wings of the rapidity distribution (i.e. p falls or k increases) and this has indeed been observed (Alner et al. 1985(a)). A natural explanation for the first effect concerned with central windows may be found in the concept of β-scaling to which we now turn.

We follow here very closely the discussion in Fowler et al. (1987(a)) and Friedlander et al. (1987). The main consequence of the J.M. formalism is the concept of

"β-scaling", i.e. the fact that, all other conditions being equal, the set of factorial cumulants (eq. (15)) (in
other words the multiplicity distribution P(n)) measured
in intervals of rapidity of equal width (i.e. having equal
values of β) should be the same, irrespective of the location in rapidity of the interval under consideration or,
conversely, that intervals of different widths Y should
scale with β = Y/ξ. The conditions necessary for testing
this consequence of QO are:

(i)   one must deal with identical bosons;

(ii)   the distributions must be stationary in rapidity;

(iii) for a meaningful comparison of P(n) in two different rapidity intervals, the data obtained from these
intervals should be statistically independent.

In spite of the apparent superabundance of multiplicity distributions on the "high-energy market" a closer look
shows that we are really faced with a scarcity of relevant
data. It would appear at first sight that the ideal place
to look for the effects considered here are the data ga-
thered by the UA-5 collaboration at the CERN p̄p-collider,
especially at √s = 546 GeV (Alner et al. 1985(b)). Indeed,
the mean multiplicity is high, a rather broad pseudo-rapidity plateau extends over better than 6 units and numerous "cuts", i.e. multiplicity distributions recorded in
limited |Y| intervals are available. Unfortunately, the
distributions refer only to all charged particles. This
violates condition (i) above; furthermore the successive
"cuts" contain common information and results derived
from them are strongly correlated; thus condition (iii)
cannot be satisfied. Information is even poorer in results obtained at the ISR.

The only pair of data sets,which comes close to fulfil-
ling the requirement for a meaningful comparison, results

from fixed target experiments of the NA 22 (Adamus et al. 1986) and NA 5 (Dengler et al. 1986) collaborations performed at the CERN SPS at center of mass energies, $\sqrt{s}$ of 20 and 22 GeV respectively, which lie close to the lowest value measured at the ISR (24 GeV). Besides the standard charged particle multiplicities n_{ch}, multiplicities of negative secondaries (n_-) were analyzed. Ideally, explicit values of $P(n_-)$ would be needed to perform the desired comparisons. Instead the published data contain only fits to the (now fashionable, if not necessarily justified) negative binomial distribution in terms of k. For the purpose of a simple test of β-scaling, the k values for the fitted $P(n_-)$ are sufficient. Indeed if for two rapidity intervals of equal width and different locations on the y-axis the $P(n_-)$ are the same, one would expect equality of the k-values. If condition (ii), (i.e. stationarity) is violated, significant differences should appear in the k-values.

Fig. 2 shows as full symbols a plot of (1/k) obtained in the NA 22 experiment (Adamus et al. 1986) for intervals of width Y, centered about y = 0 against the (1/k)-values observed in the NA 5 experiment (Dengler et al. 1986) for intervals of the same width Y, but measured in the forward hemisphere only, i.e. centered about Y/2. The straight line at 45° shows the expectation if β-scaling holds. It is obvious that for Y = 1, which corresponds more or less to the width of the pseudorapidity plateau at this energy, the (1/k)-values agree well with each other, whereas for larger Y significant departures are evident.

It is not necessarily true that β-scaling should hold for all charged secondaries (n_{ch}). One might suspect that if local charge compensation holds, $P(n_{ch})$ should reflect

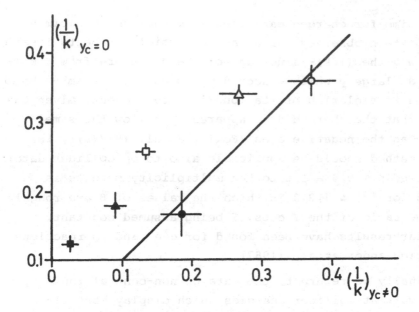

Fig. 2. Plot of the inverse parameter (1/k) of the fitted NBD observed in rapidity intervals Y about y = 0 (from Adamus 1986) against the same quantity observed in one c.m.s. hemisphere only (from Dengler 1986), for different rapidity windows Y; circles Y = 1; triangles Y = 2; squares Y = 3. Full symbols refer to negative secondaries, open symbols to all charged secondaries (from Fowler et al. 1987(a)).

the features of $P(n_-)$, whereas if positive and negative charged pions are emitted independently (if narrow Y intervals are considered) the resulting convolution of $P(n_-)$ and $P(n_+)$ could lead to a violation of β-scaling. As a matter of fact, as will be seen below, this does not happen. Anyway, one might expect that, going from the full rapidity range (where overall charge conservation constrains $P(n_-) = P(n_{ch}/2)$ to narrow Y-windows and where stationarity should be valid) one would find a "washed-out" effect in the test for β-scaling. The open symbols in Fig. 2 show the results of the same analysis as was performed above for n_-, this time for charged particles from the same experiment. The pattern observed for negative particles repeats itself; at low y the (1/k)-values agree; the departure from β-scaling at large y which - according to our formalism - should be due to violation of stationarity, are present. Given the fact that the charged data apparently follow the same trend as the negative ones (Fowler et al. 1987(a)), the J.M. method should be applicable also to $p\bar{p}$ collider data; at fixed p $\sim$ 0.4 a fit to the multiplicity moments is found for $|Y| \leq 1,2,3$ in which the values of β are roughly in the ratio of the Y cuts, ξ being assumed constant. Similar results have been found for e^+e^- and μp reactions by Friedlander et al. (1987).

Finally we return to the data on non-central rapidity intervals at collider energies which display most clearly the dependence of p on rapidity already seen earlier in the data of fig. 2 and the discussion of the e^+e^- data.

This effect may be described rather successfully by the two component model of Fowler et al. (1986) referred to in sec. 2 and in which

$$P(n) = \sum_{n_1+n_2=n} P_1(n_1)\ P_2(n_2) \tag{18}$$

where P_1 is a chaotic distribution (usually chosen to be a Bose-Einstein) and P_2 a Poisson, thought of as arising from a coherent source. The average values $\bar{n}_1$ and $\bar{n}_2$ for each rapidity interval of width 0.4 units are chosen as functions of the center η_c of the pseudorapidity interval so as to fit the experimental data. When the same model is used with $\bar{n}_1$ and $\bar{n}_2$ taken to refer to the entire rapidity range, an average value for the chaoticity may be obtained for different energies. It is found that $\bar{n}_1 \propto W_1^{1/4}$ and $\bar{n}_2 \propto \ln W_2$ consistent with the interpretation of the first source as thermally equilibrated (originating from a quark gluon plasma?) and the second as a coherent bremsstrahlung source. Here W_1 and W_2 are the energies of the two sources respectively, which are related to the total energy $\sqrt{s}$ at a given inelasticity K by

$$K\sqrt{s} = W_1 + W_2$$

(the fluctuations between W_1 and W_2 were neglected).

(V) Forward-backward correlations (F-B)

Further information on the theoretical parameters may be obtained from an analysis of the dependence of the mean number of backward going particles on the fluctuating number of forward going particles, i.e. $\bar{n}_B$ as a function of n_F.

For the NBD and PCLD distributions some assumptions must be made on the probability per event of finding n_1 particles in the forward hemisphere, when n_2 are found in the backward hemisphere. For this a binomial distribution is usually assumed with backward/forward probabilities p,q, whose values depend on particular assumptions about the nature of the sources contributing to the distribution (see for example Carruthers and Shih 1985).

For the NBD it may then be shown that a linear relationship

$$<n_B(n_F)> = a+b\,n_F, \quad a = (1-b)<a>/2,$$

$$b = (\mu_2-\mu_{2F})(\mu_{2F}+<n>/2 \tag{19}$$

(where for clarity we denote average values as <>) is strictly true whereas for the PCLD and JM this is not true. Here μ_2 and μ_{2F} are the second order factorial cumulants for the full rapidity range and for the forward hemisphere, respectively.

Considering, however, the comparison with data in more detail, the NBD predicts a value of k from the backward-forward correlation which differs significantly from that required to fit the multiplicity data assuming $p = q = 1/2$.

To resolve this contradiction Carruthers and Shih (1985) have introduced a cluster production mechanism. (Such a clusterization was used also in a PCLD approach to the F-B problem, where it was found that the amount of coherence is limited (Carruthers and Shih 1985).)

On the other hand the JM formalism can account for both the (approximate) linearity of the F-B correlation, and the value of the slope parameters b without invoking clusters or any new assumptions.

To demonstrate this one assumes a parabolic form for $\langle n_B \rangle$

$$\langle n_B(n_F) \rangle = a + b\, n_F + c\, n_F^2. \tag{20}$$

At given β and p the parameter c can be computed from μ_2 and μ_{2F} (see Fowler et al. 1987). The result is that c is everywhere negative and small enough to have escaped detection. It is however interesting to note that if any tendency to depart from linearity is evident in the data of Uhlig et al. (1978) and Alpgard et al. (1983) it is towards a slight saturation (convex bending) of the $\langle n_B \rangle$ versus n_F curve as expected from a small negative c.

Furthermore the values of the slope parameter b of the F-B correlation are perfectly consistent in the JM formalism with the values of the cumulants of the multiplicity distribution, which we saw is not the case for the NBD. As a matter of fact, the F-B data allow a more precise determination of the parameter p, though still to be taken cum grano salis (see the earlier discussion). The values found are $0.05 - 0.1$ at ISR energies and $0.2 - 0.6$ at collider.

Multiplicity distributions summary:

Three main conclusions can be drawn from the above application of the QO formalism to data:

1. The rapidity dependence of multiplicity distributions in the plateau region, in rapidity ranges which have
a common center, can be understood as due to the scaling of
the factorial cumulants with the width of the rapidity
window $Y = \beta \cdot \xi$ ("beta-scaling").

2. The forward-backward correlation of multiplicities,
in particular its linear behaviour and its energy dependence, are a consequence of beta-scaling without further
assumptions.

3. The chaoticity p shows a marked increase (by a factor of at least 2) in the energy range $\sqrt{s}$ = 24 GeV,
$\sqrt{s}$ = 540 GeV.

It is clear, however, that in the formalism of QO as
described above, this increase of p with energy must stop
at a certain energy, when p reaches its maximum value p=1.
This amounts to the existence of an upper bound for the
(relative) width of the multiplicity distribution (squee-
zed states may, however, change this conclusion (cf. e.g.
Schubert und Wilhelmi 1986; Vourdas and Weiner 1987)).
Furthermore one must recall that the width of P(n) is also
affected by the inelasticity distribution (cf. eq. (17))
and it is conceivable that the observed broadening of P(n)
corresponding to the increase of p, will reverse its trend
just because $\chi(K)$ might narrow with the increase of energy. We refer for a more detailed discussion of this to
Fowler et al. (1986(a)). More recently Sarcevic (1987) has
also suggested the possibility of such a change of trend
in P(n) in a branching model.

Finally one should mention a specific consequence of
the QO method which distinguishes it from all other

500

approaches to multiplicity distributions. This refers to
the fact that the same parameters p, β, etc. which determine P(n) and its cumulants, determine also the differential moments, which are the n-body correlation functions.
Thus in a Hanbury-Brown-Twiss type experiment to which we
turn below the observed increase of p with energy must
find its counterpart in the corresponding change of the
second order correlation functions. This prediction is a
challenge for future experimental work.

Hanbury-Brown and Twiss (H.B.T.) effect[*]

Second order intensity interferometry was used for the
first time by Hanbury-Brown and Twiss (HBT) to measure
stellar radii and to obtain information on the spectral
representation of optical lines through photon coincidence
counting measurements.

In particle physics π meson count rates are measured,
as was first suggested by Goldhaber et al., and when expressed in terms of four momentum difference, information
can be extracted on the size and lifetime of the emitting
source. (For a simple classical account of the phenomenon
see Bowler (1985), for a more detailed account especially
of meson production in nuclei see Gyulassy et al. (1979).)

In order to make contact with the multiplicity data we
have discussed so far, it is natural to choose rapidity as
the relevant variable and consider all p_T limited, i.e. all
p_T less than one coherence length in transverse momentum.
The quantity of interest is then

$$R(y_1,y_2) = \frac{1}{\sigma} \frac{\partial^2\sigma}{\partial y_1 \partial y_2} / (\frac{1}{\sigma}\frac{\partial\sigma}{\partial y_1} \frac{1}{\sigma}\frac{\partial\sigma}{\partial y_2}) \tag{21}$$

[*] (see also W. Zajc, this volume)

which is parametrized with the help of the field correlation function (16)

$$R(y_1 y_2) = 1 + p(2-p) e^{-|y_1 - y_2|/(\xi/2)} \quad . \tag{22}$$

This representation is approximate in that, as shown by Fowler et al. (1979), the effect of coherence (the first factor in R-1) and the dependence on $y_1 - y_2$ do not necessarily factorize (see also Biyajima 1980) but it does have the correct limiting forms when p = 0 and p = 1. The exponential form of the approximate correlation function is a consequence of the choice (14), other choices are possible (see our earlier discussion).

The possibility that the HBT measurements indicated that the pion field could actually be partially coherent, was first suggested by Fowler and Weiner (1977) and independently by Bartnik and Rzążewski (1978).

However, before information on p and ξ can be extracted, various experimental and theoretical effects must be accounted for. In particular final state interactions and ensemble averaging brought about, for example, by correlations with unobserved particles (Fowler et al. 1979, Gyulassy 1982, Suzuki 1987) must be accounted for. (For a review see also Fowler 1984.) On the other hand some qualitative effects, as e.g. the dependence of p and ξ on energy, obtained from the multiplicity distributions, are not expected to be affected by these effects and this makes the simultaneous investigation of P(n) and Bose-Einstein correlations a rewarding task.

Finally it should be appreciated that the effects of non-stationarity, if they are due to changes in the coherent field intensity, may further complicate the HBT analysis.

The analysis can be extended to higher order correlations, all of which should show the effect of partial coherence in a predictable way. If these predictions are fulfilled, the existence of coherent states in the pion field would be established.

4. Other optical phenomena

(I) Self-induced transparency

The particular and controversial feature of the multiplicity distributions, which we have discussed in detail above, is the presence of a coherent state contribution to the pion field, which implies that the statistical operator ρ should have non-vanishing off diagonal matrix elements in the number representation.

Since multiplicity distributions and correlations measure only diagonal elements, it is interesting to look for other processes in which the existence of off diagonal elements is more evident. One obvious source of information is in pA collisions in which the coherent pion field traverses nuclear matter and may possibly excite the nucleus in a coherent manner. Since according to our previous analysis of $p(\bar{p})$-p reactions the field is more coherent in the larger rapidity region, we should look in this region for interesting effects. One such is the possibility that the strong interaction might induce non-linear transparency analogous to self-induced transparency in optics.

This effect is observed when intense beams of light of appropriate frequency pass through matter without attenuation, coherently exciting and de-exciting the material system. Since it is known that in hadron-nucleus collisions, particularly for large forward rapidities, cascading effects are absent, it was suggested by Fowler et al. (1981) that this might indeed be a consequence of self-induced transparency, brought about by non-linear effects enhanced by the strong interaction. This implied that in forward rapidity intervals the multiplicity distribution should be rather Poisson-like, corresponding to a coherent field state. The results from emulsion experiments at 200 GeV are displayed in figures 3 and 4. Here N_h is the number of heavy tracks and is essentially a measure of target size and/or involvement. The quantity Δ should vanish for a Poisson distribution where

$$\Delta \equiv |\ln P_i(0) - \bar{n}_i| \tag{23}$$

and $P_i(0) = \exp -\bar{n}_i$ and $\bar{n}_i$ are the mean multiplicities in four pseudo-rapidity intervals.

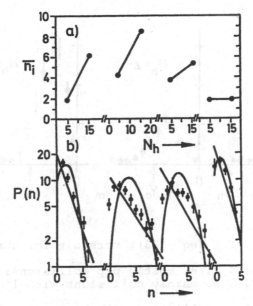

Fig. 3. a) Dependence of the local mean multiplicity $\bar{n}_i$ on N_h in four pseudo-rapidity η intervals, with η increasing from left to right ($\bar{n}_4$ is constant).
b) Local multiplicity distributions for $N_h = 2-8$ in the same η intervals. Curves: Poisson distributions; straight lines: chaotic distributions (from Fowler et al. 1981).

The transparency effect is manifest in the independence of $\bar{n}_i$ on N_h for the largest (pseudo)rapidity interval (fig. 3a). Also evident in fig. 3b is the Poisson-like distribution of $\bar{n}_i$ for the large rapidity interval by comparison with the other $\bar{n}_i$.

The departure from a Poisson distribution, except for the forward interval, is demonstrated in fig. 4 and occurs in all targets including single nucleons (for nucleons this has been confirmed,, as explained earlier, by more recent measurements with different detectors).

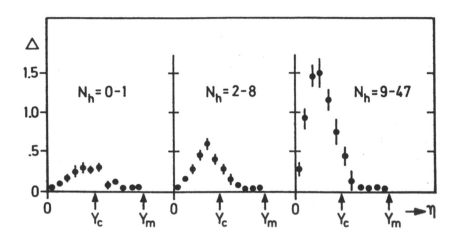

Fig. 4. Plot of Δ (eq. (23)) versus η for three ranges of N_h:
(a) N_h = 0 or 1, i.e. mostly p-p collisions;
(b) N_h = 2-8, i.e. mainly collisions with light (C,N,O) nuclei;
(c) $N_h \geq 9$, i.e. collisions with heavy (Ag, Br) nuclei; (from Fowler et al. 1981).

Centauros as coherent sources

The Centauro phenomenon (the absence of neutral pions in certain very high energy cosmic ray events in apparent contradiction with the charge symmetry of strong interactions) has been the subject of much speculations. We shall discuss it here from an optical point of view as a possible consequence of coherence. This arises from the possibility that at very high energies in nucleus-nucleus (or possibly hadron-nucleus) collisions, excited baryonic states may decay collectively, much as do, in certain circumstances, multiatomic excited states in the optical phenomena of superradiance and superfluorescence. (We leave aside the origin of the coherence we have discussed so far, in hadron-hadron collisions, which presumably lies at a deeper quark gluon level.) Such events produce a pulse of coherent radiation which could account for Centauro type events through a premature decay of the hadronic fireball (Fowler et a. 1982) (see also Wakamatsu 1980 and Fowler et al. 1985).

Another mechanism for Centauro production which arises in hadron-hadron collisions comes about simply through the effect of isospin conservation on a coherent state as shown by Andreev (1981)[*]. This author defines the coherent final state of the pions in the usual way as

$$|f> = e^{-\bar{n}/2} \exp\left\{\int d^3k \sum_i f_i(k) a_i^+(k)\right\} |0> \qquad (24)$$

with

$$\bar{n} = \sum_i \bar{n}_i = \sum_i \int d^3k |f_i(k)|^2 \qquad i = +,-,0.$$

With

$$\underline{f} = f\underline{e}$$

---

[*] We are indebted to Prof. E. Feinberg for drawing our attention to this reference.

506

where $\underline{e}$ is a unit vector in isospin, the state with isospin I and projection I_2 is

$$|f;I,I_2> = e^{-\bar{n}/2} \int d\Omega_e \ Y^A_{I,I_Z}(\theta,\phi) \ \exp\{\int d^3k\underline{f}(k) \ a^+(k)\}0>$$

(25)

From this the probabilities of producing states with different n_i may be evaluated and in the case of small isospin states, which one might expect to dominate in pp collisions, the probability of observing events with no neutral mesons is compatible with the frequency of Centauro events.

Conclusions

The methods of quantum optics could prove to be very powerful tools in the investigation of multiparticle production dynamics. Not only can certain phenomenological features of multiplicity distributions, which demand in other approaches new and sometimes unnatural assumptions, be explained in the QO formalism in a straightforward manner, but connections between apparently unrelated phenomena like multiplicity distributions, HBT interferometry and transparency of nuclear matter become possible. On the other hand the role of certain specific properties of strong interactions like final state interactions and charged states in the applicability of QO to high energy physics is not yet sufficiently understood. The elucidation of these topics is an important task for future investigations.

We are indebted to our colleagues E.M. Friedlander, F.W. Pottag and G. Wilk for the fruitful collaboration in which some of the results presented here are based.

This work has been funded in part by the German Federal Minister for Research und Technology (BMFT) under the contract number 06 MR 777.

REFERENCES

M. Adamus et al., Phys.Lett. 177B, 239 (1986).

G.J. Alner et al., Phys.Lett. 160B, 193 (1985).

G.J. Alner et al., Phys.Rep. 154, 247 (1987).

K. Alpgard et al., Phys.Lett. 123B, 361 (1983).

I.V. Andreev, Sov.Physica JETP Lett. 33, 269 (1981).

R. Barakat and J. Blake, Phys.Lep. C60, 2124 (1980).

E.A. Bartnik and K. Rzążewski, Phys.Rev. D18, 4308 (1978).

N.N. Biswas et al., Phys.Rev.Lett. 37, 175 (1976).

M. Biyajima, Phys.Lett. 92B, 193 (1980).

M. Biyajima, Prog.Theor.Phys. 69, 966 (1983).

M. Biyajima, T. Kawabe and N. Suzuki, Z.Phys.C, Particles and Fields 35, 215 (1987).

J.C. Botke, D.J. Scalapino and R.L. Sugar, Phys.Rev. D9, 813 (1974).

M.G. Bowler, Zeits.f.Phys.C, Particles and Fields 29, 617 (1985).

M. Blazek and T. Blazek, Phys.Lett. 159B, 403 (1985).

A. Breakstone et al., Phys.Rev. D30, 578 (1984).

P. Carruthers and C.C. Shih, Phys.Lett. 127B, 242 (1983).

P. Carruthers and C.C. Shih, in Proc. of the XV. Multiparticle Dynamics Symposium, Lund, eds. G. Gustafson and C. Peterson, World Scientific, Singapore 1984, p. 174.

P. Carruthers and C.C. Shih, Phys.Lett. 165B, 209 (1985).

P. Carruthers and C.C. Shih, Proc. of XVII. Intern.Symp. on Multiparticle Dynamics, Seewinkel, Austria, eds. M. Markytan, W. Majerotto, J. MacNaughton, 1986, p. 653.

P. Carruthers and C.C. Shih, Intern.Journal of Mod.Phys.A, vol. 2, p. 1447 (1987).

F. Dengler et al., Z.Phys.C, Particle and Fields 33, 189 (1986).

M. Derrick et al., Phys.Lett. 168B, 299 (1986).

B. Durand and L. O'Raifeartaigh, Phys.Rev. D13, 99 (1976).

G.N. Fowler and R.M. Weiner, Phys.Lett. 70B, 201 (1977).

G.N. Fowler and R.M. Weiner, Phys.Rev. D17, 3118 (1978).

G.N. Fowler, N. Stelte and R.M. Weiner, Nucl.Phys. A319, 349 (1979).

G.N. Fowler, E.M. Friedlander and R.M. Weiner, Phys.Lett. 104B, 239 (1981).

G.N. Fowler, E.M. Friedlander and R.M. Weiner, Phys.Lett. 116B, 203 (1982).

G.N. Fowler, First Intern. Workshop on Local Equilibrium in Strong Interaction Physics (LESIP I), Bad Honnef 1984 (World Scientific).

G.N. Fowler, R.M. Weiner and G. Wilk, Phys.Rev.Lett. 55, 173 (1985(a)).

G.N. Fowler and R.M. Weiner, Phys.Rev.Lett. 55, 1373 (1985(b)).

G.N. Fowler, E.M. Friedlander, R.M. Weiner and G. Wilk, Phys.Rev.Lett. 56, 15 (1986(a)).

G.N. Fowler, E.M. Friedlander, R.M. Weiner and G. Wilk, Phys.Rev.Lett. 51, 2119 (1986(b)).

G.N. Fowler, E.M. Friedlander, F.W. Pottag, R.M. Weiner, J. Wheeler and G. Wilk, Univ. of Marburg preprint, May 1987(a).

G.N. Fowler, A. Vourdas, R.M. Weiner and G. Wilk, Phys. Rev. D35, 870 (1987(b)).

E.M. Friedlander, F.W. Pottag and R.M. Weiner, to be published in Proc. of 1987 Intern. Symp. on Lepton and Photon Interactions at High Energies, Hamburg 1987.

A. Giovannini, Nuovo Cim. 15A, 543 (1973).

A. Giovannini, Nuovo Cim. 24A, 421 (1974).

A. Giovannini and L. Van Hove, Zs.f.Phys. 30, 391 (1986).

R.J. Glauber, Phys.Rev. 130, 2529 (1963).

G. Goldhaber, S. Goldhaber, W. Lee and A. Pais, Phys.Rev. 120, 300 (1960).

M. Gyulassy, S.K. Kaufman and L.W. Wilsen, Phys.Rev. C20, 2267 (1979).

M. Gyulassy, Phys.Rev.Lett. 48, 454 (1982).

R. Hanbury-Brown and R.O. Twiss, Nature 177, 27 (1956).

R. Hanbury-Brown and R.O. Twiss, Proc.Ray Soc.(Local) A242, 300 (1957).

R. Hanbury-Brown and R.O. Twiss, Proc.Ray Soc. (Local) A243, 291 (1957).

L. Van Hove and A. Giovannini, Proc. XVII. Intern.Symp. on Multiparticle Dynamics, Seewinkel, Austria, eds. M. Markytan, N. Majerotto, J. MacNaughton, World Scientific (1986), p. 56.

A.K. Jaiswal and E.L. Metha, Phys.Rev. A2, 169 (1979).

L. Mandel, Proc.Phys.Soc. 14, 733 (1959).

M. Namiki, I. Okba and N. Suzuki, Prog.Theor.Phys. 53, 775 (1975).

J. Perina and R. Horák, J.Phys. A2, 702 (1969).

M. Plümer, in Proc. of I. Intern.Workshop on Local Equilibrium in Strong Interaction Physics, LESIP I, Bad Honnef, FRG, eds. D.K. Scott and R.M. Weiner, World Scientific, Singapore 1984, p. 185.

I. Sarcevic, Phys.Rev.Lett. 59, 403 (1987).

M. Schubert and B. Wilhelmi, Non-Linear Optics and Quantum Electronics, John Wiley & Sons, New York (1986).

N. Suzuki, Prog.Theor.Phys. 51, 1629 (1974).

N. Suzuki, Phys.Rev. D35, 3359 (1987).

S. Uhlig et al., Nucl.Phys. B132, 15 (1978).

A. Vourdas and R.M. Weiner, Phys.Rev.A (1987), to appear.

M. Wakamatsu, Nuovo Cim. 56A, 336 (1980).

R.M. Weiner, in Proc. of the II. Intern.Workshop on Local Equilibrium in Strong Interaction Physics (LESIP II), eds. P. Carruthers and D. Strottman, 1986, p. 106.

UGA-HE-72
ITP-SB-87-32

A Unified Physical Picture of Multiparticle Emission
in $\bar{p}p$ and e^+e^- Collisions

T. T. Chou

Department of Physics

University of Georgia

Athens, Georgia   30602

Chen Ning Yang

Institute for Theoretical Physics

State University of New York

Stony Brook, New York   11794

Abstract: A unified physical picture for multiparticle emission in
hadron-hadron and e^+e^- collisions is presented. It is based on the wide range
of values of the total angular momentum in hadron-hadron collisions.
Experimental support is reviewed. An extension of the considerations to give
a theory of the momentum distribution of the outgoing particles is compared
with experiments.

Introduction

In a high energy $\bar{p}p$ collision in the 540-GeV $S\bar{p}pS$ Collider at CERN, the total angular momentum of the collision ranges from 0 to $\sim 2000\hbar$ ($\cong 1.5$ fm $\times$ 270 GeV). This wide range of large angular momenta produces in elastic scattering a coherent superposition which manifests itself in the very high forward elastic peak observed in all high energy experiments which have been so far performed. A theory of elastic scattering based on such a concept was advanced in 1967-1968,[1] which was found to be in very good general agreement with experiments.[2] Additional consequences of this theory for diffraction dissociation, limiting fragmentation and other phenomena have been developed in the last twenty years.[3]

In the present paper we summarize our work in the last few years which focuses on the idea that the wide range of angular momenta in high energy hadron-hadron collision is the *cause* of the observed broad multiplicity distribution which has been fitted with KNO-scaling or negative binomials. A natural prediction of our view is that in e^+e^- collisions at high energy where the total angular momentum hardly fluctuates (being 0 or $1\hbar$), the multiplicity distribution is narrow and Poisson-like. If this prediction is right, then one has a *unified physical picture* of multiplicity distribution in hadron-hadron and e^+e^- collisions. We shall present experimental results which strongly support this prediction.

A natural extension of the physical picture mentioned above can be applied to the momentum distribution of emitted particles in the central region[4] (i.e., nonleading particles). Out of such considerations the concept of "partition temperature" T_p was developed. This concept will be discussed together with its application to the angular and momentum distributions of central region particles in both $\bar{p}p$ and e^+e^- collisions.

I. *Multiparticle Production in Hadron-Hadron Collisions*

A. *Multiplicity distribution*

A salient feature for multiparticle particle production processes in the central region in $\bar{p}p$ collisions observed in the early CERN S$\bar{p}p$S Collider experiments at 540 GeV is the large fluctuation[5] in multiplicity distribution P_n, which was shown to satisfy approximately the (KNO)[6] scaling law:

$$\bar{n}P_n = \psi(Z) \tag{I.1}$$

where Z = the KNO-variable $\equiv n/\bar{n}$. More recent measurements and analyses[7] indicate that the multiplicity distributions seem to fit better the negative binomial (NB) form:

$$P_n = \frac{\Gamma(n+k)}{\Gamma(n+1)\Gamma(k)} \left[\frac{\bar{n}}{\bar{n}+k}\right]^n \left[\frac{k}{\bar{n}+k}\right]^k \tag{I.2}$$

where the adjustable parameter k (= 4.6 to 3.2 in the CERN Collider energy range) decreases with $\sqrt{s}$. Such distributions, KNO or NB, imply very large mean value of the fluctuation, $(\Delta n)^2 \equiv \overline{n^2} - \bar{n}^2$, for the observed charged-particle multiplicity: For KNO,

$$(\Delta n/\bar{n})^2 = \text{independent of energy,} \tag{I.3}$$

and for the negative binomial,

$$(\Delta n/\overline{n})^2 = (\overline{n})^{-1} + k^{-1} = \text{increasing with energy.} \tag{I.4}$$

We call these *broad distributions nonstochastic.* In contrast, a *stochastic distribution,* like a Poisson or binomial, gives typically

$$(\Delta n/\overline{n})^2 = 1/\overline{n} \tag{I.5}$$

which is narrow. Furthermore, a stochastic distribution will become infinitely narrow compared to a nonstochastic one as $\overline{n} \to \infty$.

When KNO scaling was first proposed, we were reluctant to accept it, because for a high energy collision with the emission of dozens of particles, it was difficult to believe that the process could be as nonstochastic as the KNO distribution. After experiments at the CERN Collider confirmed in 1981 the broad approximate KNO (i.e., nonstochastic) distribution for the multiplicity, we were forced to *the view that some aspect of the dynamics of hadron-hadron collision is nonstochastic, while other aspects must be stochastic.* The question was then which aspects are stochastic and which are not.

B. *The stochastic and nonstochastic aspects of particle production processes*

   1. *Separation of stochastic from nonstochastic aspects of collision*

   The UA5 Collaboration published[8] in 1983 a probability distribution $P(n_F, n_B)$ of events for 540-GeV $\overline{p}p$ collisions with n_F and n_B denoting the forward and backward charge multiplicities respectively. Their scatter plot is reproduced here in Fig. 1. The exhibited distribution of events is analyzed[8] most conveniently in terms of the variables $n(=n_F+n_B)$ and $z(=n_F-n_B)$. Note that because of symmetry in pp and $\overline{p}p$ collisions, the probability distribution $P(n,z)$ must be even in the variable z.

If one sums the scatter plot over all values of z for a fixed n, and plots the result against n, one gets the experimental multiplicity distribution which is broad and nonstochastic. But if one keeps n constant and plots the probability of events against z, one gets[8,9] a binomial distribution which is, of course, a typical stochastic distribution. To see this, let us evaluate from the data exhibited in Fig. 1 the average value $\overline{z^2}$ for each n. The results fit very well the simple formula

$$[\overline{z^2} \text{ at fixed n}] = 2n \tag{I.6}$$

which is the straight line shown in Fig. 2 against the experimental data points. This very good fit, especially the simple coefficient 2, suggests that the distribution with respect to z, at a fixed n, is a simple one. It turns out that in fact the simple binomial

$$P(n,z) = (\text{function of n}) \ C^{n/2}_{n_F/2}$$
$$= (\text{function of n}) \ C^{n/2}_{(n+z)/4} \tag{I.7}$$

which does satisfy (I.6), gives an excellent fit as exhibited in Fig. 3.

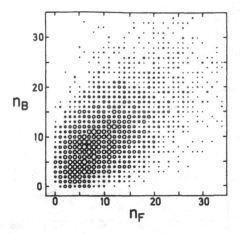

FIG. 1. Scatter plot of backward and forward multiplicities n_B and n_F for the η intervals $4 > \eta_F > 0$ and $0 > \eta_B > -4$ at 540 GeV. For each value of (n_B, n_F) the area of the circle is proportional to the number of events. (Reproduced from Ref. 8.)

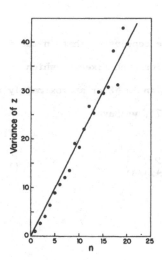

FIG. 2. Plot of the average value of z^2 for fixed n versus n. The straight line represents Eq. (I.6). (Reproduced from Ref. 9.)

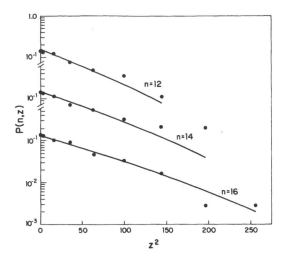

FIG. 3. Plot of the calculated z–distribution with Eq. (I.7) and the experimental data points in log scale for n = 12, 14, and 16. Both the theoretical curves and the experimental data points are normalized. (Reproduced from Ref. 9.)

Through the above analysis, we reached the conclusion that in the n_F—n_B plane, the $z(=n_F-n_B)$ distribution is binomial for each fixed n, which is of course stochastic, but the $n(=n_F+n_B)$ distribution is NB or approximately KNO, hence, nonstochastic. Combining (I.1) and (I.7), we have[9]

$$P(n,z) = \left\{(\bar{n})^{-1}\psi(n/\bar{n})\right\} \left\{[B(n)]^{-1} \ C_{(n+z)/4}^{n/2}\right\} \qquad (I.8)$$

where $\psi(n/\bar{n})$ is the KNO scaling function $[(\bar{n})^{-1}\psi(n/\bar{n})$ may be replaced with the NB-distribution] and $\bar{n}$ is the average total charged multiplicity. The normalization factor $[B(n)]^{-1}$ is inserted so that the expression $C^{n/2}_{(n+z)/4}$ summed over z in steps of z = 2 gives B(n). The values of B(2), B(4), B(6), B(8),....are $2 + (4/\pi)$, $4 + (32/3\pi)$, $8 + (352/15\pi)$, $16 + (1024/21\pi)$,.... Asymptotically, $B(n) = 2^{1+(n/2)}$ for large n. The clear separation exhibited in (I.8) of the stochastic from the nonstochastic aspects of multiparticle production processes gives conceptually a clear and attractive physical picture for high energy hadron-hadron collisions.

2. *Interpretation for the binomial distribution factor in* P(n,z)

The binomial form (I.7) for the multiplicity distribution along a constant $n(=n_F+n_B)$ line in the charge asymmetry variable $z(=n_F-n_B)$ may be realized in many different physical models. But the most intriguing aspect of (I.8) is the factor 2 which came from (I.6). The simplest and most appealing physical picture that will explain this factor is one in which positive and negative charged particles are always produced together in pairs in either the forward or backward hemispheres. The positive charges, n/2 in number, are distributed in the forward and backward hemispheres in a binomial distribution $C^{n/2}_m$, where m is the number of positive charges on one side. The total charge is then assumed to be neutral on each side, so that $n_F = 2m$, and $n+z = 2n_F$. Hence, $m = (n+z)/4$, which yields the z-distribution in (I.7).

If the physical picture described above is approximately correct, then the net charge on either side, forward or backward, must be nearly neutral.

This consideration leads us[10] to the prediction

(root-mean-square net charge in each hemisphere)$/\sqrt{n} \to 0$, as $\bar{n} \to \infty$ (I.9)

which can be experimentally tested.

    In the same spirit that one assumes the total charge to be small in each side, one would also assume the total isospin to be small in each side. The experimental consequence of this latter assumption is that in each hemisphere there will be approximately equal numbers of positive, negative and neutral particles.

3. Cluster models

    The discussion above is one of the general category of cluster models[11] in which clusters are assumed to be emitted in the forward and backward hemispheres, each of which then splitting into k particles within the same hemisphere (or mostly within the same hemisphere). The model of Section I.B.2 corresponds to the case k = 2. Extensive Monte Carlo calculation[11] has been made by the UA5 group to fit $S\bar{p}pS$ data, including a discussion of possible range of values of k. We shall return to the value of k in Part II of this paper when we discuss e^+e^- collisions and compare it with $\bar{p}p$ collisions.

C. Extrapolation of forward-backward multiplicity distribution to very high energies

1. Forward-backward multiplicity correlation

    The analysis in Sections A and B of the probability distribution of events in 540-GeV $\bar{p}p$ collisions led us to the conclusion expressed in (I.8) that, in the n_F—n_B plane, the distribution is nonstochastic (KNO or NB) along the n_F+n_B direction, and the distribution is stochastic (binomial) along the n_F-n_B direction. At 540 GeV the n_F—n_B distribution is in the form of approximately a fat ellipse as shown in Fig. 1.

We have also emphasized that the widths of the two distributions behave very differently as $\bar{n}$ increases. [Compare Eqs. (I.3) and (I.4) with (I.5)]. Thus, if (I.8) should remain valid at higher and higher energies, the population of events would change drastically: Because a stochastic distribution becomes increasingly narrower relative to a nonstochastic distribution when $\bar{n}$ increases, the population in the n_F—n_B plane would become more and more concentrated in an elongated region along the $n_F = n_B$ axis. In other words, the contour lines for $P(n_F, n_B)$ will evolve from (approximately) fat ellipses at 540 GeV to much thinner ellipses as energy increases.

To analyze this problem quantitatively, we introduce a parameter α which measures the elongatedness of the two dimensional n_F—n_B distribution:

$$\alpha \equiv \frac{\text{population average of } (n_F+n_B)^2}{\text{population average of } (n_F-n_B)^2} \tag{I.10}$$

Or $$\alpha \equiv \frac{\text{variance of } n}{\text{variance of } z} \tag{I.11}$$

$\sqrt{\alpha}$ has a simple geometrical meaning: If the contour lines are exactly ellipses, $\sqrt{\alpha}$ is the ratio of the major-axis to the minor-axis. Now, using the relation

$$[\text{variance of } z] = 2\bar{n} \tag{I.12}$$

which is obtained by averaging both sides of (I.6) over all values of n, we get from (I.11),

$$\alpha = [\text{variance of } n]/(2\bar{n})$$
$$= (\Delta n)^2/(2\bar{n})$$
$$\to \infty \quad \text{as } \bar{n} \to \infty \tag{I.13}$$

For the last step, use is made of (I.3) and (I.4).

Equation (I.13) states that, *at very high energies, the population of events will concentrate in an extremely narrow band.*

When that obtains, let us consider a fixed n_F as indicated by the vertical line in Fig. 4. The distribution of n_B along that vertical line would be concentrated near a value of n_B which is close to n_F with a fluctuation of the order of the width of the narrow band, viz., $\sim \sqrt{n}$ or $\sim \sqrt{n_F}$. Explicitly,

$$n_B = n_F + O(n_F^{1/2}). \tag{I.14}$$

Thus, in a high energy collision the forward and backward multiplicities n_F and n_B will exhibit *strong correlation.* Take a case when $\bar{n} = 200$, say. Then $\bar{n}_F = \bar{n}_B = 100$. n_F may fluctuate widely, say, from 50 to 200. But for each n_F, the backward multiplicity,

$$n_B \cong n_F \pm \sqrt{n_F}$$

fluctuates very little.

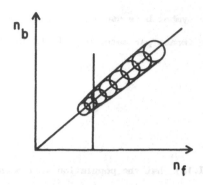

FIG. 4. Multiplicity distribution for hadron-hadron collisions at high energies. n_f and n_b refer respectively to the forward and backward charge multiplicities. We predict that for high energies the distribution becomes a band as shown, which is identical to the prediction that $b_c \to 1$ (see Eq. (I.17). The band can be regarded as a collection of circles each corresponding to one angular momentum. The radius of a circle is $\sim \sqrt{n}$, because the width of the band is $\sim \sqrt{n}$. In other words, the fluctuations of n_f and n_b are $\sim \sqrt{n}$ for each angular momentum. We predict that for each angular momentum the circle represents a product of two Poissons, one in n_f and the other in n_b.

A parameter b_c is commonly used to characterize the correlation of multiplicities in the forward and backward hemispheres, and is defined by

$$
b_c \equiv \frac{d\bar{n}_B}{d\bar{n}_F}
$$

$$
= \frac{\overline{n_F n_B} - \bar{n}_F \bar{n}_B}{\overline{n_F^2} - \bar{n}_F^2}
\qquad (I.15)
$$

(In the literature, the symbol b is usually used. We use b_c here to distinguish it from the impact parameter b.) It is obvious that parameters b_c and α are related:

$$\alpha = \frac{1+b_c}{1-b_c}$$

(I.16)

Thus, our prediction (I.13) that the population will become more and more concentrated in a narrow band as $\bar{n}$ increases means that the correlation parameter b_c for hadron-hadron collisions will increase with energy and approaches unity in the ultrahigh-energy limit:

$$b_c \to 1 \qquad \text{as } \bar{n} \to \infty$$

(I.17)

Is this prediction in agreement with experimental data? We list below some measured b_c values[12] for pp and $\bar{p}p$ collisions over the ISR and $S\bar{p}pS$ energy range:

$$b_c = 0.14 \quad \text{at } \sqrt{s} = 62 \text{ GeV}$$
$$0.44 \qquad \qquad 200 \text{ GeV}$$
$$0.57 \qquad \qquad 546 \text{ GeV}$$
$$0.63 \qquad \qquad 900 \text{ GeV} \qquad\qquad \text{(I.18)}$$

The trend is clearly consistent with our conjecture.

2. Incoherent superposition of collisions at different impact parameters
each with narrow multiplicity fluctuations

Accepting the above conclusion that at ultrahigh energies, both n_F and n_B fluctuate widely, but in a correlated way, with $n_F \cong n_B$, what can be the underlying physical cause for such phenomena?

A natural possibility for such an underlying cause is the wide fluctuation of the total angular momentum, or the impact parameter b, already mentioned in the Introduction above. It is reasonable to assume that for a small impact parameter ($b \cong 0$), i.e., a head-on collision, the multiplicity is large, while for a large impact parameter, say $b > 1.5$ fm, the multiplicity is small.[13] I.e., the multiplicity is assumed to depend on b. Since both sides, forward and backward, *share* the same impact parameter b, this assumption leads naturally to $n_F \cong n_B$. Proceeding along such reasoning, one arrives at the picture illustrated in Fig. 4 in which for each fixed b, the multiplicity fluctuations of both n_F and n_B are stochastic, represented by a circle in Fig. 4. The superposition of such stochastic distribution gives the long narrow band which exhibits wide fluctuations in n ($= n_F + n_B$).

In short, we propose a physical mechanism for approximate KNO scaling in $\bar{p}p$ collisions: *The experimentally observed broad multiplicity distribution (KNO or NB) is the result of an incoherent superposition of collisions at different impact parameters, each of which gives a narrow multiplicity distribution.*

What should be the precise form of the narrow multiplicity distribution for fixed impact parameter which is represented by a circle in Fig. 4? If one assumes this distribution to be a product of two Poissons, one in n_F and one in n_B, both with average $n_0(b)/2$:

$$f(n_0/4, n_F/2) \, f(n_0/4, n_B/2), \tag{I.19}$$

where

$$f(a, m) = \exp(-a) \, a^m/m!, \tag{I.20}$$

then one has a very simple explanation of the binomial factor in (I.7) because

$$f(n_0/4, n_F/2) \, f(n_0/4, n_B/2)$$

$$= \exp(-n_0/2) \, (n_0)^{(n_F+n_B)/2} \, [(n_F+n_B)!]^{-1} \, C_{n_F/2}^{n/2} \tag{I.21}$$

The superposition of distributions (I.21) for different b's, each with its own value for $n_0 = n_0(b)$, gives the n–dependent factor of (I.7). Explorations along this direction is in progress. We remark here that the concept of an average multiplicity n_0 which depends on the impact parameter b has been discussed in the literature before.[13] The assumption that for a fixed b the multiplicity distribution is a product of two Poissons (I.19) is recent.[14]

D.   *Energy partition, partition temperature and single-particle momentum distribution*

  1. *Stochastic energy partition and single-particle spectrum*

     If eventually the population of events clusters into a very narrow band at very high energies and very large average multiplicities $\bar{n}$, the distribution with respect to n_B, for a fixed n_F, would be concentrated near

$n_B \cong n_F$ with a fluctuation of the order of $\sqrt{n_F}$ (Fig. 4). That is, for given n_F, the charge multiplicity distribution in the backward hemisphere would be stochastic.

Accepting this conclusion, we considered it[16] most natural that the energy and momentum distributions of the outgoing central region particles in the backward hemisphere, for this given n_F, would also be stochastic. It would be difficult to concoct a mechanism that could single out multiplicity distribution as the only one that is stochastic.

But the energy-momentum distribution must be subject to certain constraints:

(a) The total energy of the central particles is $E_0 h$ in each hemisphere. While the total energy in the backward direction is E_0 (= $\sqrt{s}/2$ = the c.m. energy of the incoming hadron), only a fraction h of this would be carried by the central region particles in the backward hemisphere. The remaining fraction 1-h would be taken away by the fragmentation (leading) particles in the backward direction. h depends on n_F.

(b) Each emitted particle carries a bremsstrahlung (Bloch–Nordsieck) factor $d^3 p/E$, which was first used[16] in multihadron production in the 1940's.

(c) For each particle, there is a transverse-momentum cutoff factor which we take from experimental results to be

$$g(p_T) = \exp\ (-\alpha p_T) \tag{I.22}$$

where

$$\alpha = 5.25\ (\text{GeV/c})^{-1}\ \text{for 540 GeV collisions.} \tag{I.23}$$

(d) Correlation effects among particles.

526

The simplest stochastic distribution of the particles in the backward
hemisphere, taking into account constraints (a), (b) and (c), but not (d),
gives a probability factor of the form

$$\delta(\sum_i E_i - E_0 h) \; \prod_i (d^3 p_i / E_i) \; g(p_{Ti}) \qquad (I.24)$$

where the summation and product extend over all the particles, charged or
neutral, in the central region in the backward hemisphere. The
single-particle spectrum can now be evaluated[15] from (I.24) by the method of
steepest descent. For large multiplicities, this procedure gives a
single-particle distribution

$$(d^3 p/E) \; g(p_T) \; \exp(-E/T_p) \qquad (I.25)$$

where T_p is a parameter which we have called the partition temperature. (For
an alternative short cut derivation of this result, see Section I.E.1 below.)

Equation (I.25) gives the single-particle distribution along the vertical
line, at a fixed n_F, in Fig. 4. Since the event-populated region is assumed
to be very narrow, as Fig. 4 indicates, fixing n_F is approximately equivalent
to fixing n $(= n_F + n_B)$. Thus, we arrive at the conclusion that for each n
there is a partition temperature T_p.

2. *Comparison with experimental angular distribution*

At 540 GeV for $\bar{p}p$ collisions the $n_F - n_B$ distribution has not yet reached
a narrow band. Nevertheless, we went ahead[15] to test the validity of (I.25)
by evaluating from it the single-particle angular distribution. We write for

the pseudorapidity η: ($\theta \equiv$ production angle in c.m. system)

$$\eta = \cosh^{-1}(1/\sin\theta). \tag{I.26}$$

Now,

$$dn/d\eta = 2\pi \sin^2\theta (dn/d\Omega). \tag{I.27}$$

Thus

$$dn/d\eta = K\ 2\pi\ \sin^2\theta \int_0^{E_0 h} p^2 (dp/E)\ g(p\sin\theta)\ \exp(-E/T_p) \tag{I.28}$$

where K is a normalization constant. Only pions are included in this calculation. (Since the kaon to pion ratio increases with energy, for any accurate fit with experiments, the kaons will have to be included.) It is found that the curve for each multiplicity is well described by (I.28) for one value of T_p. Fig. 5 and Table I summarize these calculations. The good fit to the UA5 data indicates the usefulness of the idea of the partition temperature. It should be emphasized here that there are no adjustable parameters in this computation, the p_T-cutoff factor (I.22) having been taken from experiments. The energy fraction h and normalization constant K are both determined from the curves themselves. If one takes a Gaussian p_T distribution instead of (I.22), the fit to the angular distribution is also good.

E.   *The concept of partition temperature*

  1. *Origin of T_p*

The concept of T_p originates in (i) the δ-function in (I.24) which controls the partition of energy on one side of the collision, and (ii) the method of steepest descent. The *mathematical procedure* used is exactly the same as that for obtaining the canonical ensemble from the microcanonical ensemble in statistical physics as the example below[17] illustrates.

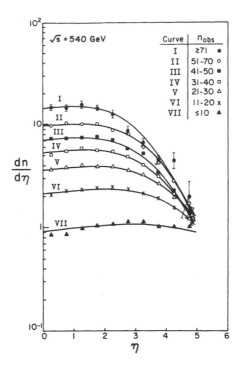

FIG. 5. dn/dη versus η at √s̄ = 540 GeV. Data points are taken from UA5 experiments. (Reproduced from Ref. 15.)

Consider an ideal gas in a three-dimensional (x,y,z) region, x > 0, y > 0, z > 0, bounded by surfaces

$$x = 0, \qquad y = 0, \qquad z = 0, \qquad \text{and}$$

$$z = y \, g(y) \, (m^2 + x^2 + y^2)^{-1/2} \qquad\qquad (I.29)$$

TABLE I. Parameters for multiparticle production processes at $\sqrt{s}$ = 540 GeV. n_{obs}, which labels different charge-multiplicity ranges, is different from the true charge multiplicity by a factor of approximately 1.25 due to experimental corrections. For a given charge multiplicity, the shape of the experimental $dn/d\eta$ curve is well fitted by Eq. (I.28) for one value of T_p. The parameter h and normalization constant K are both determined from the curves themselves. The average values for the impact parameter are rough estimates based on a procedure outlined in T. T. Chou and C. N. Yang, Phys. Lett. 116B, 301 (1982). The total longitudinal momentum transfer is estimated by summing $(e - p_\parallel)$ over all central-region particles (charged and neutral) using Eq. (I.28). Contributions to longitudinal momentum transfer from particles in the fragmentation region are negligible.

n_{obs}	Partition temperature T_p (GeV)	Normalization constant $K(\text{GeV}^{-2})$	n_{cal}	KNO variable $Z=n_{cal}/\sqrt{\bar{n}}$	Energy fraction in central region = h	Average energy per particle in central region (GeV)	Average impact parameter b(fm) (approximate)	Longitudinal momentum transfer (GeV)
≥71	4.4	83	99	3.4	0.45	1.6	0.02	9.2
51-70	6.3	54	73	2.5	0.42	2.1	0.1	6.4
41-50	6.8	40	55	1.9	0.33	2.2	0.3	4.7
31-40	8.8	29	44	1.5	0.32	2.6	0.6	3.6
21-30	14.	20	33	1.1	0.31	3.4	0.9	2.5
11-20	24.	11	21	0.73	0.26	4.4	1.3	1.5
≤10	180.	4.8	11	0.37	0.20	6.6	1.8	0.67

where $g(y) > 0$, and m is a constant. The gas atoms are assumed to have infinite mass M (hence their kinetic energies are negligible,) moving in this region with a potential energy for each particle of

$$V = (m^2 + x^2 + y^2)^{1/2} \qquad (I.30)$$

A microcanonical ensemble for such a gas of N atoms at total energy E_0 has a probability distribution of

$$\delta\left[\sum_i^N V_i - E_0\right] \prod_i^N dx_i dy_i dz_i \qquad (I.31)$$

Integrating over all z_i gives

$$\delta\left[\sum_i (m^2 + x_i^2 + y_i^2)^{1/2} - E_0\right] \prod_i \frac{y_i}{(m^2 + x_i^2 + y_i^2)^{1/2}} \, g(y_i) \, dx_i dy_i \qquad (I.32)$$

If we make the replacements:

$$x_i \rightarrow (p_L)_i$$
$$y_i \rightarrow (p_T)_i \qquad (I.33)$$
$$E_0 \rightarrow E_0 h,$$

(I.32) becomes (I.24).

But this problem has a well known canonical ensemble with the single particle distribution exp(−V/β) dxdydz,

or

$$\frac{y}{(m^2+x^2+y^2)^{1/2}} g(y) \exp[-(m^2+x^2+y^2)^{1/2}(1/\beta)] \, dxdy \qquad (I.34)$$

after the coordinate z is integrated over, where β is the thermodynamical temperature of the gas. Substitution (I.33) together with the replacement $\beta \rightarrow T_p$ converts (I.34) into (I.25).

We note that the number of parameters appearing in (I.25) and (I.34) depends on the number of δ-functions, i.e., constraints or conservation laws, in the problem. Consider the examples: In Planck's black-body radiation formula, there is only one parameter (T), which arises from the conservation of energy. The distribution of a free boson gas has two parameters corresponding to the conservation of energy and of the number of bosons in the system.

2. *Existence of T_p does not imply equilibrium*

Although for the gas model the temperature β is an equilibrium concept, for the high-energy collision problem T_p is just a mathematical parameter that governs the partition of energy in the stochastic process at impact parameter b and neither requires nor implies equilibrium. This point is particularly clear if we concentrate on the factor d^3p/E in (I.25). We want to make[17] the following observations. (a) Replacing d^3p/E with d^3p leads to a single-particle distribution that totally disagrees with the angular distribution given by the UA5 group. (b) In any system in thermal equilibrium, such as for the black-body radiation, it is always the factor d^3p, not d^3p/E, that appears as the density of modes. Does the factor d^3p/E never occur for the problem of black-body radiation? The answer is, yes, it does occur when we discuss certain nonequilibrium phenomena. For example, consider in a block of matter at a finite temperature a cavity which at t=0 is free of atoms and of all radiation. Immediately afterwards, short- and long-wave radiation would begin to fill the cavity. Since the coupling of radiation with matter contains the factor d^3p/E, the long wavelength modes generally couple more strongly with matter and are filled faster. Thus, in such a nonequilibrium situation, the factor d^3p/E does play a role. But in the long run, when equilibrium is established, *the strength of coupling is immaterial*, and only the mode density d^3p would play a role. (c) In a high-energy collision, there is not sufficient time to reach thermodynamical equilibrium in any overlapping part of the colliding hadrons. The Bloch-Nordsieck factor then plays a role and gives rise to the factor d^3p/E.

F. *Discussion of results*

1. *Increase of T_p with b*

   Table I shows that the partition temperature increases with impact distance. This characteristic has a simple qualitative physical interpretation based on the geometrical picture. For a collision with a large impact parameter, the incoming hadrons have only a small overlap and the overlapping regions do not have much chance of exchanging longitudinal momenta. Therefore, they tend to maintain their respective original velocities in the center of mass system and few particles are emitted in the process. One thus expects the average energy per particle that results from these collisions to be large. That is, T_p should be large. On the other hand, for small impact parameters b, the two overlapping parts of the colliding hadrons are expected to exchange more longitudinal momenta, resulting in the emission of many particles with smaller average energy per particle, and therefore, smaller T_p.

2. Inelasticity

We assumed that the leading particles take away a fraction 1-h of the total energy. This fraction according to Table I is larger than 50% for all collisions at 540 GeV. Averaged over all nondiffractive collisions, this fraction is $1-\bar{h} = 72\%$. The remaining fraction $\bar{h} = 28\%$ of the incoming energy represents the average energy that goes into the production of central particles. This is what has been called "inelasticity" by cosmic-ray physicists.[4]

Experimental determination of the value of h for each collision has always in the past come up with an ambiguity, because it is not clear which of the fast particles in the c.m. system should be included in the "central region" and which in the fragmentation region. Since a fast particle contributes a large chunk of energy, this ambiguity has made it very difficult to give any accurate evaluation of the fraction h. With the introduction of the partition temperature T_p and (I.25), this difficulty disappears: Take the 540-GeV $\bar{p}p$ collision. For each multiplicity, T_p is determined from the experimental η distribution, mainly for low values of η, say $0 < \eta < 4$. Having determined T_p, (I.25) then allows for a calculation of the small tail of the curve for $\eta > 5$. The magnitude of this tail then makes possible an accurate evaluation of the probability of emission of a particle with $\eta > 5$ and, thereby, the contribution to h from the $\eta > 5$ part. Thus, this contribution is evaluated not from $\eta > 5$ data which are inaccurate, but from lower-η data together with (I.25).

3. Longitudinal momentum transfer

The value of the total longitudinal momentum transfer for 540-GeV $\bar{p}p$ collisions is listed in the last column of Table I. We see that its value is very small for low-multiplicity events. Even for the high-multiplicity events, its value is only 9.2 GeV/c = 3.4% of the incoming momentum. This presents a picture for small-impact parameter collisions which is not what one would at first anticipate: For such collisions, the incoming hadrons are quite opaque to each other. The survival amplitude $S(b=0) = \exp[-\Omega(0)]$ is[18] only ~ 0.19, so that only ~ 4% of such collisions are elastic. At first sight, this may suggest strong longitudinal momentum transfers for the remaining 96% of the collisions. But in fact, these remaining collisions are very far from a "bang", (i.e., an amalgamation of the two hadrons in one region of space, as in the early Fermi[19] picture). Instead, the longitudinal momentum exchange between the two sides is only ~ 3%. (Of course, there is bound to be a process in which thousands of particles are emitted in the c.m. system, all nearly at rest. Such processes would involve a large longitudinal momentum transfer of ~ 270 GeV/c, and would require h = 1. But that would clearly have an extremely small cross section.)

The smallness of the longitudinal momentum transfer, even for processes involving the production of many particles, indicates what we may call the *persistence of longitudinal momentum* in very–high–energy hadron–hadron collisions. This is a dominating characteristic: For the 540–GeV $\bar{p}p$ collisions, elastic and double–diffraction–dissociation events involve longitudinal momentum transfers of the order of at most a dozen MeV's, while the nondiffractive events involve longitudinal momentum transfers of the order of a few GeV's, not more. Larger longitudinal momentum transfers are *extremely* rare.

4. *Inclusion of particles other than pions*

We speculate that the observed π/K ratio in outgoing particles in $\bar{p}p$ collisions could be just the result of the effect of the masses m_π and m_K in (I.25), when the incoming energy is high enough.

G. *Further predictions*

We present here some additional predictions which are not included in the above discussions.

1. *Spin-spin correlation of emitted particles*

The large angular momentum involved in a high energy collision[20] would lead to transverse-spin-transverse-spin correlations for the outgoing particles which we shall explain as follows with a concrete example.

Consider a $\bar{p}p$ collision at c.m. energy of 540 GeV, producing $\sim$ 21-30 observed charge particles. The average impact parameter[17] involved is $\sim$ 0.9 fm so that the incoming total angular momentum is $\sim 1000\hbar$. The total number of outgoing hadrons[17] is $\sim 33 \times (3/2) = 49$. How do these 49 particles combine to carry away 1000 units of angular momentum?

536

Much of it is certainly carried away as orbital angular momenta by the outgoing particles. But the sum of all these orbital angular momenta is not sufficient to make up the original angular momentum, as the following analysis will demonstrate: Consider all the outgoing pieces that move toward the right in Fig. 6, including the leading particles and those in the central region. Let b_i and p_i^L be the impact parameter[21] and the longitudinal momentum of the i-th such piece. (b_i may be positive or negative according as whether the point of origin P in Fig. 6 is above or below the main axis.) It contributes an orbital angular momentum of $b_i p_i^L$. The total orbital angular momentum of all right-moving outgoing particles is then

$$\sum_R b_i p_i^L = \sum_R b_i E_i - \sum_R b_i (E_i - p_i^L).$$ (I.35)

Now there is little energy transfer in the collision. So E_i is also the energy of the piece of the incoming hadron at point P, which is the same as the longitudinal momentum, $(p_i^L)^{in}$, of that piece of the incoming hadron. Thus,

$$\sum_R b_i p_i^L = \sum_R b_i (p_i^L)^{in} - \sum_R b_i \left[(p_i^L)^{in} - p_i^L \right]$$ (I.36)

Now $\sum_R b_i (p_i^L)^{in}$ is the incoming right-moving hadron's contribution to the incoming angular momentum. Thus, there is a net deficiency of orbital angular momentum in the right-moving outgoing system, by an amount

$$\sum_R b_i \left[(p_i^L)^{in} - p_i^L \right]$$ (I.37)

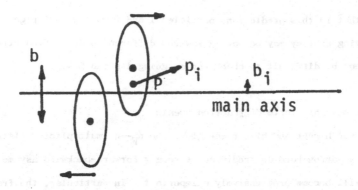

FIG. 6. Impact parameter b_i and momentum p_i of outgoing particle i
originating at point P. (Reproduced from Ref. 20.)

To estimate the magnitude of this deficiency we equate it approximately with

$$\bar{b}_i \sum_R \left[(p_i^L)^{in} - p_i^L \right] \qquad (I.38)$$

Now $\bar{b}_i$ is approximately[17] (1/2)(0.9 fm). The sum in (I.38) is the
longitudinal momentum transfer which was estimated[17] to be 2.5 GeV/c. Thus,
the orbital angular momentum deficiency for the right moving particles
is ~ (1/2)(0.9 fm)(2.5 GeV/c) $\cong 6\hbar$.

In other words, the 49 outgoing particles must contribute ~ 12ℏ of spin
angular momentum in order to make up the total angular momentum. There is
thus a strong necessity for the spins of the outgoing particles to line up
parallel to each other in the transverse direction:

$$(\sigma_1^T \cdot \sigma_2^T)_{average} > 0. \qquad (I.39)$$

Notice that in this prediction, particles 1 and 2 may be both right moving, left moving or they may be moving towards different sides. Prediction (I.39) should not be difficult to check for outgoing Λ's and Σ's.

2. Decrease of single-diffraction events

If for higher and higher energies, the n_F—n_B multiplicity distribution becomes a narrow band as predicted, strongly forward-backward asymmetrical events will become progressively unimportant. In particular, the fraction of single-diffraction events, events populating near the n_F- and n_B- axis in Fig. 1 for which there are only a few diffractive fast particles on one side and many particles emitted on the other, will become rare. This is consistent with the recent UA4 observation[22] that the ratio of single-diffraction to total cross sections decreases with increasing energy.

H. Extrapolation to higher and lower energies

We have made[17] extrapolations of the angular distribution to $\sqrt{s} = 53$ GeV (CERN-ISR), 2 TeV (Fermilab Tevatron) and 40 TeV (SSC) with the following assumptions:

(i)   The p_T cutoff factor $g(p_T)$ is taken to be $g(p_T) = \exp(-\alpha p_T)$, where we take

$\alpha = 5.8$ (GeV/c)$^{-1}$ for $\sqrt{s} = 53$ GeV,

$\alpha = 5.0$ (GeV/c)$^{-1}$ for $\sqrt{s} = 2$ TeV,

$\alpha = 4.4$ (GeV/c)$^{-1}$ for $\sqrt{s} = 40$ TeV.

(ii)   The value of $\bar{n}_{ch}$ for these energies are taken to be 13, 41 and 78, respectively.

(iii)   The parameter h is taken to be a function only of the KNO variable $Z = n/\bar{n}$.

The results are presented in Fig. 7 and Table II.

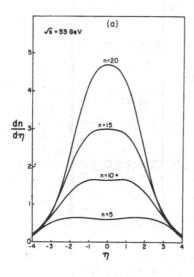

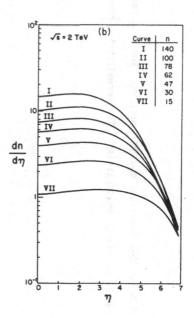

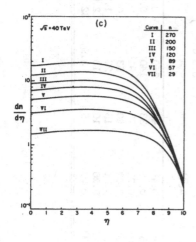

FIG. 7. Calculated dn/dη versus η at (a) √s = 53 GeV, (b) √s = 2 TeV, and (c) √s = 40 TeV. (Reproduced from Ref. 17.)

TABLE II. Parameters for single-particle distribution at ISR, Tevatron, and SSC energies. The relation between α and $\langle p_T \rangle$ is given by $\alpha = 2\langle p_T \rangle^{-1}$. n is the total charge multiplicity. The normalization constant K is in GeV^{-2} and partition temperature T_p in GeV.

Z	h(Z)	$\sqrt{s}$ = 53 GeV (α = 5.8 GeV^{-1})			$\sqrt{s}$ = 2 TeV (α = 5.0 GeV^{-1})			$\sqrt{s}$ = 40 TeV (α = 4.4 GeV^{-1})		
		n	K	T_p	n	K	T_p	n	K	T_p
3.4	0.45	44	350	0.25	140	71	17	270	62	300
2.5	0.42	33	140	0.43	100	48	23	200	44	400
1.9	0.33	25	91	0.49	78	36	25	150	33	420
1.5	0.32	20	53	0.70	62	27	31	120	26	520
1.2	0.31	15	27	1.2	47	19	43	89	18	700
0.73	0.26	10	13	2.0	30	11	61	57	11	950
0.37	0.20	5	3.9	32.	15	5.1	130	29	5.3	1700

II. *Multiparticle production in two-jet events in* e^+e^- *collision*

A. *Does* e^+e^- *collision give stochastic (i.e., narrow) or nonstochastic (i.e., wide) multiplicity distribution?*

The crucial conclusion of Part I above is that the wide multiplicity distribution in $\bar{p}p$ collisions is due to the large fluctuation of the impact parameter, or angular momentum, of the collision. If this conclusion is correct, then it follows that for e^+e^- collision, where the total angular momentum is 0 or 1, (which is the angular momentum of the virtual intermediate photon), there should be no wide multiplicity fluctuation. In fact, the natural assumption is that the multiplicity distribution in e^+e^- collisions is a Poisson distribution. For those events which are classified as two-jet events, the natural assumption is that the distribution $P(n_F, n_B)$ is a product of two Poissons, one in n_F, one in n_B. These assumptions imply that the multiplicity distributions are stochastic (narrow). We made[14,23] these conjectures in 1985 in an effort to gain a unified understanding of multiplicity distributions in $\bar{p}p$ and e^+e^- collisions. In particular, we conjectured the following:

(i) The distribution of the forward, backward charged multiplicities $P(n_F, n_B)$ is a product distribution:

$$P(n_F, n_B) = g(n_F) \, h(n_B) \qquad\qquad (II.1)$$

(ii) The separate distribution on each side is Poisson with a "cluster" size k:

$$P(n_F, n_B) = f(a, n_F/k) \, f(a, n_B/k) \qquad\qquad (II.2)$$

where

$$f(a, n_F/k) = (\text{constant})\ a^{(n_F/k)}/(n_F/k)! \qquad \text{(II.3)}$$

(iii) We conjectured further that $k = 2$ in analogy with the distribution[8,9] with respect to $z = n_F - n_B$ for $\bar{p}p$ collisions.

B.   *Comparison with experiments*

   1. *Multiplicity distribution curves*

   The most important qualitative difference between a KNO or negative binomial distribution and a Poisson distribution is

$$\Delta n/\bar{n} = (\text{constant})(\bar{n})^{-1/2} \qquad \text{(Poisson)} \qquad \text{(II.4)}$$

$$\Delta n/\bar{n} > (\text{constant}) \qquad \text{(KNO or negative binomial)} \qquad \text{(II.5)}$$

This difference in principle could resolve the question whether in e^+e^- collisions a Poisson distribution obtains. Unfortunately, existing data for e^+e^- collisions cover too narrow a range of values for $\bar{n}$ and are insufficient to resolve this question.

But the shapes of the multiplicity distribution for $\bar{p}p$ and e^+e^- collisions are distinctively different (Fig. 8) with the latter case fitting Poisson distribution well. In the recent experiment of Derrick et al.[24] at 29 GeV for e^+e^- collisions, the fit with a Poisson is even better (Fig. 9): We emphasize that with a Poisson distribution, there is to the left of the peak a region of the curve where the second derivative is positive. As $\bar{n}$

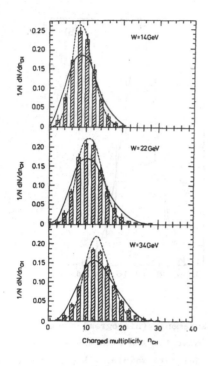
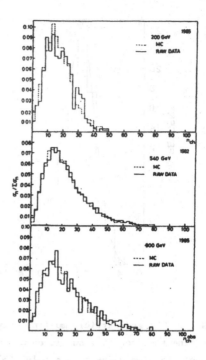

FIG. 8. Comparison of charged-particle multiplicity distribution for e^+e^- annihilation and $\bar{p}p$ collisions. The e^+e^- curves (left column) are reproduced from a TASSO publication, Ref. 27. No corrections for gluon bremsstrahlung effects have been made. The $\bar{p}p$ curves (right column) are reproduced from a UA5 report, Ref. 12. For details see these original papers.

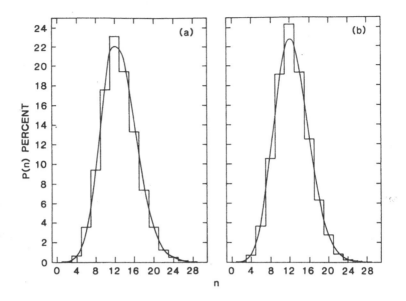

FIG. 9. Charged-particle multiplicity measurements (histogram) compared to the line that connects points on a Poisson distribution with the same mean value: (a) inclusive data sample, (b) two-jet data sample. (Reproduced from Ref. 24.)

increases, this region characteristically widens. Both of these phenomena are apparent in the e^+e^- curves in Fig. 8, but are absent in the $\bar{p}p$ curves. Furthermore, e^+e^- curves are quite symmetrical on the two sides of the peak, which is a characteristic also of Poisson distributions but not of the $\bar{p}p$ curves.

2. *Consequence of recent results of Derrick et al.*

We shall demonstrate[25] below that the recent careful analysis of Derrick et al.[24] of their e^+e^- 29-GeV data strongly indicates that conjectures (i) and (ii) above are correct; but conjecture (iii) is in error, since their result indicates that k = 1 instead of k = 2.

Derrick et al.[24] found that for fixed n_F, the average $\langle n_B \rangle$ is essentially independent of n_F. The average slope was given as

$$b = -\,0.001 \pm 0.015 \qquad\qquad (II.6)$$

which is also consistent with earlier TASSO data.[26] Although these results do not prove the product form (II.1) of the n_F—n_B distribution, it is strongly suggestive of it. Assuming (II.1) we use the condition

$$\mathbb{P}(n_F,\ n_B) = \mathbb{P}(n_B,\ n_F)$$

which follows from charge conjugation invariance of the strong and electromagnetic interactions. Thus

$$g(x)\ h(y) = g(y)\ h(x)$$
i.e., $\qquad g(x)/h(x) = g(y)/h(y)$

Since x and y are independent, this shows that $g(x)/h(x) = $ constant. By normalizing g and h, we see that they are identical and we arrive that

$$\mathbb{P}(n_F,\ n_B) = g(n_F)\ g(n_B) \qquad\qquad (II.7)$$

Now Derrick et al.[24] found that the distribution of $n = n_F + n_B$ fits beautifully a Poisson. In the language of (II.7), this means

$$\sum_{n_F} g(n_F)\ g(n-n_F) = \text{(constant)}\ a^n/n! \qquad (n = \text{even}) \qquad (II.8)$$

Equation (II.8) can be solved with the method of generating functions as follows. Define

$$G(x) \equiv \sum_{m=0}^{\infty} g(m) \, x^{m} \qquad (II.9)$$

Then

$$[G(x)]^{2} = \sum_{m_1, m_2} x^{m_1 + m_2} g(m_1) \, g(m_2)$$

which on account of (II.8) becomes

$$[G(x)]^{2} = (constant) \sum_{n} a^{n} x^{n}/n! = (constant) e^{ax}. \qquad (II.10)$$

Thus

$$G(x) = (constant) e^{ax/2}$$

which leads to the explicit expression for $g(m)$ as a Poisson

$$g(m) = (constant) \, (a/2)^{m}/m! \qquad (II.11)$$

Comparison with (II.3) shows that the cluster size is $k = 1$ (identifying $n_F = m$), and not $k = 2$. Thus, we conclude that the results of Derrick et al.[24] strongly suggest the validity of conjecture (i) and (ii), but give $k = 1$ instead of $k = 2$.

The above derivation is not exact because $n = n_F + n_B$ must be even. Thus, the n_F, n_B values form a checkerboard pattern and (II.7) is an interpolative equation. In such an interpolation, the value of $\sum_{n_F} g(n_F)g(n-n_F)$ for n = odd is also approximately given by (II.8). Thus, (II.10) is approximately, but not exactly correct. Therefore, (II.11) is also approximately but not exactly correct.

Confirmation of conjectures (i) and (ii) gives strong support to our unified physical picture[14,23] that the multiplicity distribution of e^+e^- is narrow and that of $\bar{p}p$ is broad because of the wide range of angular momentum available in the latter case. But why should the cluster size be $k = 2$ for $\bar{p}p$ and $k = 1$ for e^+e^-? We believe the answer perhaps resides in the same mechanism that produces[14] another difference between the two cases: The different average charged multiplicities when one concentrates on those events with zero impact parameter in $\bar{p}p$ collisions and compares them with two-jet e^+e^- events at a proper energy (Fig. 10). The quark-gluon system in the two cases start from very different[14] initial conditions, even though with the same total energy. The resultant average cluster size and average multiplicity may, therefore, be different.

3. Partition temperature T_p

The same argument that we gave in Section I.D for the single-particle momentum distribution in $\bar{p}p$ collisions applies also to e^+e^- collisions, the only difference being that for e^+e^- collisions there is only one partition temperature because there is only one impact parameter $b = 0$ for a given $W = \sqrt{s}$. We have fitted[14] the TASSO data[27] for the momentum distribution of

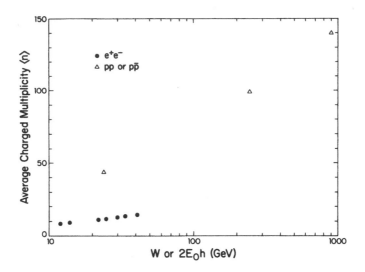

FIG. 10. Plot of average charged multiplicity versus total energy W for e^+e^- annihilation or central energy $2E_0h$ for $\bar{p}p$ collisions at approximately zero impact parameter. The average multiplicities for $\bar{p}p$ collisions are crude estimates subject to large uncertainties. (Reproduced from Ref. 14.)

single charged particles and found that

$$
\begin{array}{lll}
\text{at } W = 14 \text{ GeV} & T_p = 1.6 \text{ GeV} & \\
\text{at } W = 34 \text{ GeV} & T_p = 3.3 \text{ GeV} & (\text{II.12})
\end{array}
$$

The fits are good and are reproduced in Figs. 11 and 12.

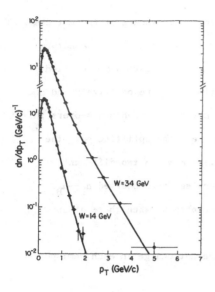

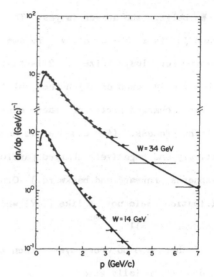

FIG. 11. The calculated dn/dp$_T$ distributions for e$^+$e$^-$ annihilation at W=14 and 34 GeV. The experimental data points are taken from Ref. 27. (Reproduced from Ref. 14.)

FIG. 12. The calculated charged-particle momentum spectrum dn/dp for e$^+$e$^-$ annihilation at W=14 and 34 GeV. The experimental data points are taken from Ref. 27. (Reproduced from Ref. 14.)

4. Root-mean-square net charge in each jet

Although the mechanism for the different values of the cluster size, k = 1 for e$^+$e$^-$ and k = 2 for $\bar{p}$p collisions, is not clear, we observe they are based on the very good fits in Fig. 9 (for e$^+$e$^-$) and Fig. 3 (for $\bar{p}$p), both of which are parameterless (except for k). Without understanding the underlying mechanism, are there nevertheless other experimentally measureable quantities that may be correlated with the cluster size?

We believe the answer is yes. The root-mean-square of the net charge Q in each jet is such a quantity. To see this, we observe that a natural mechanism for cluster size k = 2 is that neutral clusters are emitted stochastically, each of which then splitting into one positively and one negatively charged particle. There must not be too much cross-overs in this splitting process. (A cross-over is one where the splitting puts the positively and negatively charged particles into the two different hemispheres, forward and backward.) Otherwise, the observed n_F—n_B distribution would not be like (I.7) which obtains when there are no cross-overs at all.

If there are no cross-overs, then the net charge in each hemisphere should be essentially zero:

$$\overline{Q^2} \cong 0. \tag{II.13}$$

On the other hand, for cluster size k = 1, the natural mechanism is that the particles are stochastically emitted, so that there is equal forward and backward probability for each outgoing particle. By the central limit theorem the root-mean-square net charge in each hemisphere should then be

$$(\overline{Q^2})^{1/2} = (\text{constant}) \, n^{1/2}. \tag{II.14}$$

We have studied[23] a model like this and found

$$(\overline{Q^2})^{1/2} = (1/2) \, n^{1/2}. \tag{II.15}$$

The details are as follows:

Let the number of particles with charge +1 or −1 be both equal to $n/2$ and let each particle be randomly assigned to either the forward or backward hemisphere. We can find the distribution of the net charge Q in the forward hemisphere by the method of generating functions: The probability that there are Q_+ particles of charge +1 in the forward hemisphere is the coefficient of x^{Q_+} in

$$[(1/2)(1 + x)]^{n/2}.$$

Similarly, the probability that there are Q_- particles of charge −1 in the forward hemisphere is the coefficient of $x^{-(Q_-)}$ in

$$[(1/2)(1 + x^{-1})]^{n/2}.$$

The net charge in the forward hemisphere $Q = Q_+ - Q_-$. Thus, the probability for Q is the coefficient of x^Q in

$$[(1/2)(1 + x) \, (1/2)(1 + x^{-1})]^{n/2}. \tag{II.16}$$

552

It is now easy to evaluate for this model the moments of Q:

$$\overline{Q} = 0, \qquad \overline{Q^2} = n/4. \qquad\qquad\qquad (II.17)$$

I.e., $(\overline{Q^2})^{1/2} = (1/2) \, n^{1/2}.$ \qquad\qquad\qquad (II.15)

There is not much data with which to check (II.14). TASSO[28] has reported that $\overline{Q^2} = 1.53$ at $W = 33$ GeV, at which $\overline{n} = 13$.

5. Lack of $\sigma_1^T \cdot \sigma_2^T$ correlation

The prediction about spin correlations in (I.39) for $\overline{p}p$ collisions was based on the large average transverse angular momentum of $\overline{p}p$ events. Since the average transverse angular momentum is small for e^+e^-, we predict that for e^+e^- two-jet events

$$(\sigma_1^T \cdot \sigma_2^T) \text{ average} = 0. \qquad\qquad\qquad (II.18)$$

Acknowledgments: This work is supported in part by the U. S. Department of Energy under grant No. DE-FG09-84ER40160 and by the National Science Foundation under grant No. PHY 85-07627.

REFERENCES

1. T. T. Chou and C. N. Yang, in *High-Energy Physics and Nuclear Structure*, edited by G. Alexander (North—Holland Publishing Co., Amsterdam, 1967), pp. 348–359; Phys. Rev. 170, 1591 (1968); Phys. Rev. Lett. 20, 1213 (1968); Phys. Rev. 175, 1832 (1968).

2. Further theoretical work on elastic scattering can be found in A. W. Chao and C. N. Yang, Phys. Rev. D8, 2063 (1973); F. Hayot and U. P. Sukhatme, Phys. Rev. D10, 2183 (1974); T. T. Chou, Phys. Rev. D11, 3145 (1975) and D19, 3327 (1979); T. T. Chou and C. N. Yang, Phys. Rev. D17, 1889 (1978) and D19, 3268 (1979); Phys. Rev. Lett. 46, 764 and 1350(E) (1981); T. T. Chou, T. T. Wu and C. N. Yang, Phys. Rev. D26, 328 (1982); T. T. Chou and C. N. Yang, Phys. Lett. 128B, 457 (1983). For experimental support, see A. Bohm et al., Phys. Lett. 49B, 491 (1974); N. Kwak et al., Phys. Lett. 58B, 233 (1975); E. Nagy et al., Nucl. Phys. B150, 221 (1979); UA4 Collaboration, M. Bozzo et al., Phys. Lett. 147B, 392 (1984); S. R. Amendolia et al., Phys. Lett. 146B, 116 (1984) and 178B, 435 (1986).

3. J. Benecke, T. T. Chou, C. N. Yang and E. Yen, Phys., Rev. 188, 2159 (1969); T. T. Chou and C. N. Yang, Phys. Rev. Lett. 25, 1072 (1970); T. T. Chou, Phys. Rev. Lett. 27, 1247 (1971); T. T. Chou and Chen Ning Yang, Phys. Rev. D7, 1425 (1973); T. T. Chou and Chen Ning Yang, Nucl. Phys. B107, 1 (1976); T. T. Chou and Chen Ning Yang, Phys. Rev. D22, 610 (1980).

4. "Leading particle" is a term used by cosmic ray physicists to characterize outgoing particles in the very forward and backward directions with large energies. If we write the total energy carried away by such particles divided by $\sqrt{s}$ as 1-h, then h was called the inelasticity. See G. Cocconi, Phys. Rev. 111, 1699 (1958). The nonleading particles are called central region particles.

5. UA5 Collaboration, K. Alpgård et al., Phys. Lett. 107B, 310 and 315 (1981); and UA1 Collaboration, G. Arnison et al., Phys. Lett. 107B, 320 (1981).

6. Z. Koba, H. B. Nielsen and P. Olesen, Nucl. Phys. B40, 317 (1972).

7. UA5 Collaboration, G. J. Alner et al., Phys. Lett. 160B, 193 and 199 (1985). This is the latest experimental analysis. The introduction of negative binomial distributions into the analysis of multiplicity data involves many papers and authors. They will not be listed here.

8. UA5 Collaboration, K. Alpgård et al., Phys. Lett. 123B, 361 (1983).

9. T. T. Chou and Chen Ning Yang, Phys. Lett. 135B, 175 (1984).

10. A model which illustrates Eq. (I.9) can be found in T. T. Chou and Chen Ning Yang, Phys. Lett. 167B, 453 (1986).

11. UA5 Collaboration, G. J. Alner et al., Nucl. Instr. and Methods (to be published).

12. J. G. Rushbrooke, in Multiparticle Dynamics 1985, proceedings of the XVI Symposium, Kiryat Anavim, Israel, 1985, edited by J. Grunhaus (World Scientific, Singapore, 1985), p. 289.

13. This assumption is very natural in the geometrical picture. It is discussed in the literature by S. Barshay, Phys. Lett. 42B, 457 (1972); H. B. Nielsen and P. Olesen, Phys. Lett. 43B, 37 (1973); A. Bialas and E. Bialas, Acta Physica Polonica B5, 373 (1974); S. Barshay and Y. Yamaguchi, Phys. Lett. 51B, 376 (1974); T. T. Chou and Chen Ning Yang, Phys. Lett. 116B, 301 (1982).

14. T. T. Chou and Chen Ning Yang, Phys. Rev. Lett. 55, 1359 (1985).

15. T. T. Chou, Chen Ning Yang and E. Yen, Phys. Rev. Lett. 54, 510 (1985).

16. F. Bloch and A. Nordsieck, Phys. Rev. 52, 54 (1937); H. W. Lewis, J. R. Oppenheimer and S. A. Wouthuysen, Phys. Rev. 73, 127 (1948).

17. T. T. Chou and Chen Ning Yang, Phys. Rev. D32, 1692 (1985).

18. T. T. Chou and Chen Ning Yang, Phys. Lett. 128B, 457 (1983).

19. E. Fermi, Progr. Theor. Phys. 5, 570 (1950); Phys. Rev. 81, 683 (1951).

20. Chen Ning Yang and T. T. Chou, J. Phys. Soc. Jpn. 55, 53 (1986) Suppl.

21. The impact parameter is a two-dimensional quantity. b_i is the projection of the impact parameter into the impact plane. Similarly for p_i^L.

22. UA4 Collaboration, D. Bernard et al., Phys. Lett. 186B, 227 (1987).

23. T. T. Chou and Chen Ning Yang, Phys. Lett. 167B, 453 and 171B, 486(E) (1986).

24. M. Derrick *et al.*, Phys. Rev. D$\underline{34}$, 3304 (1986).

25. T. T. Chou and Chen Ning Yang, Phys. Lett. $\underline{B}$ (in press).

26. TASSO Collaboration, M. Althoff *et al.*, Z. Phys. C $\underline{29}$, 347 (1985).

27. TASSO Collaboration, M. Althoff *et al.*, Z. Phys. C $\underline{22}$, 307 (1984).

28. TASSO Collaboration, R. Brandelik *et al.*, Phys. Lett. $\underline{100B}$, 357 (1981).

BRANCHING PROCESSES IN MULTIPARTICLE PRODUCTION

Rudolph C. Hwa

Institute of Theoretical Science and Department of Physics

University of Oregon, Eugene, OR 97403, USA

Abstract

A brief review of the stochastic methods is given, and various evolution equations of common interest are discussed. The relevance of branching processes to particle physics is then considered. First, the connection between branching and perturbation theory (in ϕ^3 and QCD) is reviewed. Then, the application of branching to soft hadronic production is discussed in the framework of the geometrical branching model. An attempt is made to show that Furry branching is a generally valid description of all multiparticle production processes, hard or soft.

CONTENT

1. INTRODUCTION

The study of multiparticle production in hadronic collision processes, though shunned by many for lack of a workable theory or because of the unlikelihood of any spectacular experimental discoveries, is actually a fascinating subject to those who have probed deeply enough to uncover some underlying physical features that are simple and interesting. Admittedly, there exist many approaches to the subject, whose relationships with one another are in some instances quite obscure. Such a phenomenon need not, however, be regarded as a sign of malaise. It merely reflects the complexity of the system under investigation. Many roads leading to the ultimate description are possible, each revealing a partial picture of the whole.

An incontroversial nature about this subject is that it involves many aspects of modern physics, not limited to particle physics. Hadrons have sizes, so they exhibit diffractive features in scattering, but their fragmentations have non-trivial energy dependences that cannot be simply related to the existence of one intrinsic scale. Hadron constituents are point-like, but except for hard scattering perturbation theory is ineffective. To treat the many-body problem of the simultaneous interaction of all constituents, statistical methods have been used. Such an approach stands in apparent opposition to a dynamical approach, exemplified by the study of the time evolution of stretched color strings. No doubt each point of view has some domain of validity that can claim some support from phenomenology. Successful avenues into the future are likely to be found in attempts to relate and incorporate the various approaches and to discover the underlying unity.

Our aim in this review is to show that stochastic branching seems to be a feature shared by a number of processes involving multiparticle production. In hard processes where detailed calculations can be done in the perturbation theory for specific interactions, branching is found to be relevant. This is quite amazing in view of the scarcity of dynamical details in stochastic methods. In soft processes, though one does not have any proof as rigorous as one can provide in perturbation theory, there are also evidences that branching is relevant. Putting the two together, we find the situation very hopeful that perhaps stochastic branching is the unifying feature that underlies all multiparticle production processes, when the number of particles is large enough. The fact that the evolution equation for Furry branching is independent of the strength of the basic coupling gives specific inducement to regard Furry branching as the mechanism that can interpolate between hard and soft processes. To be sure, much further work needs to be done. But with the possibility of that unity lying on the horizon, it seems worthwhile to review the subject at this point.

We start from the very beginning of stochastic methods on the assumption that it is an unfamiliar subject to most particle physicists. We shall discuss the evolution equations of some most common types and present them in a coherent way. Calculations in perturbation theory are then reviewed for ϕ^3 and QCD interactions, and the results are discussed in the context of branching. Finally, the branching mechanism is applied to the description of soft hadronic production with emphasis on conceptual issues rather than technical details.

2. STOCHASTIC PROCESSES

We review in this section the basic ideas in treating stochastic processes. Many books have been written on the subject, a few of which are listed in Refs. 1-6. We give here only the elements that form the background for the material to be discussed in the following sections.

2.1 Chapman-Kolmogorov Equation

A quantity of fundamental importance in stochastic processes is the conditional probability $P(x_2,t_2|x_1,t_1)$ defined to be the probability density that a random variable X has the value x_2 at time t_2, given that its value at an earlier time t_1 is x_1. It is normalized so that

$$\int P(x_2,t_2|x_1,t_1)dx_2 = 1 \tag{2.1.1}$$

The conditional probability $P(x_n,t_n|x_1,t_1;...;x_{n-1},t_{n-1})$ depends on the variable X having values $x_1,\cdots,x_{n-1}$ at times $t_1,\cdots,t_{n-1}$, where $t_1 < t_2 <...< t_{n-1} < t_n$. A Markov process is one for which the above conditional probability depends only on the variable X at t_{n-1}, i.e.

$$P(x_n,t_n|x_1,t_1;\cdots;x_{n-1},t_{n-1}) = P(x_n,t_n|x_{n-1},t_{n-1}) \qquad (2.1.2)$$

This hierarchical structure immediately leads to the chain relationship, which is called the Chapman-Kolmogorov equation,

$$P(x_3,t_3|x_1,t_1) = \int P(x_3,t_3|x_2,t_2) \, P(x_2,t_2|x_1,t_1) \, dx_2 \qquad (2.1.3)$$

for any t_2 in the interval (t_1,t_3). Note that there is no integration over t_2 in (2.1.3), which looks deceptively similar to the integral equation of the kernel in Feynman's path-integral formulation of quantum mechanics. The essential difference is that (2.1.3) refers only to probability functions and therefore has no complications arising from phases of amplitudes.

Defining the transition rate from x_1 to x_2 to be $W(x_2|x_1)$, we then have for incremental time change ε

$$P(x_2,t_1+\varepsilon|x_1,t_1) = \varepsilon \, W(x_2|x_1)$$
$$+ \delta(x_1 - x_2) \, [\, 1 - \varepsilon \int W(x'_2|x_1) \, dx'_2] + O(\varepsilon^2) \qquad (2.1.4)$$

where the quantity inside the square bracket stands for the probability that no transition takes place during ε. The fact that the r.h.s. of (2.1.4) is independent of t_1 is a property of the Markov process. Applying (2.1.4) to the transition between t_2 and t_3 in (2.1.3), and using the abbreviated notation $P(x,t) = P(x,t|x_1,t_1)$ for fixed x_1 and t_1, one gets

$$\frac{\partial}{\partial t}P(x,t) = \int dx' \, [\, W(x|x') \, P(x',t) - W(x'|x) \, P(x,t)] \qquad (2.1.5)$$

This is the differential form of the Chapman-Kolmogorov equation, and is usually referred to simply as the master equation. In fact, it is also called the forward equation, to be distinguished from the backward equation, which can similarly be derived by substituting (2.1.4) into the part in (2.1.3) that involves the t_1 - to - t_2 transition:

$$\frac{\partial}{\partial t_1} P(x,t|x_1,t_1) = \int dx' [(P(x,t|x',t_1) W(x'|x_1) - P(x,t|x_1,t_1)W(x'|x_1)]$$

$$(2.1.6)$$

For the most part, we shall be working from the forward equation (2.1.5), whose interpretation is physically more transparent: the first term on the right-hand side represents gain from x' and the second term loss from x.

2.2 Fokker-Planck Equation

From (2.1.5) we can derive the Fokker-Planck equation under certain assumptions. Firstly, the transition rate $W(x|x')$ is a sharply peaked function of $z \equiv x - x'$, centered at $z = 0$. Secondly, both $W(x|x')$ and $P(x'|t)$ are slowly varying functions of x'. Thus keeping z fixed, we may expand in x' the first term on the right-hand side of (2.1.5) around x (i.e. regarding $x' = x - z$):

$$W(x|x') P(x'|t) = W(x + z|x) P(x,t) - z\frac{\partial}{\partial x}[W(x + z|x) P(x,t)]$$

$$+ \frac{z^2}{2} \frac{\partial^2}{\partial x^2}[(W(x + z|x) P(x,t)] \qquad (2.2.1)$$

The termination of the expansion at the second order can be justified under certain limiting processes.[6] Substituting (2.2.1) into (2.1.5) yields the Fokker-Planck equation

$$\frac{\partial}{\partial t}P(x,t) = -\frac{\partial}{\partial x}[A_1(x) P(x,t)] + \frac{1}{2} \frac{\partial^2}{\partial x^2}[A_2(x) P(x,t)] \qquad (2.2.2)$$

where

$$A_j(x) = \int z^j W(x + z|x)\, dz \qquad (2.2.3)$$

The middle term in (2.2.2) is usually referred to as the drift term, while the last term gives rise to diffusion in x.

We shall come back to consider the Fokker-Planck equation in Sec. 5.4 in connection with the Poisson transform of scaling functions.

2.3 Branching Processes

If the random variable is discrete, like the multiplicity n of electrons in a cosmic-ray shower, and if the conditional probability $P(n_2,t_2|n_1,t_1)$ is zero for n_2 less than n_1 at any time t_2 greater than t_1, then we say that the process involves only branching, and no recombination. A branching process need not be Markovian, but it must have the property that the n particles at any given time do not interact with one another — they can only create more particles separately and independently. There can be many species of particles, like electrons, positrons and photons, each capable of initiating a family (or jet) of more particles in the same stochastic manner. The use of quarks and gluons as illustrative examples is deliberately avoided here because I want to side-step the question of what t stands for in a quark jet, say, at this point.

A central property of the branching process, according to the above description, is that

$$P(n,t|2,0) = \sum_{n_1 n_2} P(n_1,t|1,0)\, P(n_2,t|1,0)\, \delta_{n,n_1+n_2} \qquad (2.3.1)$$

i.e., each of the two initial particles can start its own branch of the family tree. Such a convolution can more simply be expressed in terms of the generating functions. Define

$$G_l(s,t) = \sum_n s^n P(n,t|l,0) \tag{2.3.2}$$

Then from (2.3.1) we have

$$G_2(s,t) = [G(s,t)]^2 \tag{2.3.3}$$

where we have used the notation $G(s,t) \equiv G_1(s,t)$. It is straightforward to generalize (2.3.3) to

$$G_l(s,t) = [G(s,t)]^l \tag{2.3.4}$$

Let us now restrict the branching process to be Markovian, and use the invariance property under time translation, i.e.,

$$P(n_2,t_2|n_1,t_1) = P(n_2,t_2 - t_1|n_1,0) \tag{2.3.5}$$

to write the above simply as $P(n_2,n_1;t_2 - t_1)$. The Chapman-Kolmogorov equation, (2.1.3), now takes the form

$$P(n,1;t+t') = \sum_l P(n,l;t') P(l,1;t) \tag{2.3.6}$$

Using (2.3.2) and (2.3.4), we obtain[7]

$$G(s,t+t') = G[G(s,t'),t] \tag{2.3.7}$$

This is a useful relation that is true only for branching processes.

Adapting (2.1.6) to the present problem, we have

$$P(n,1;\varepsilon) = \varepsilon W_{n1} + \delta_{n,1}(1 - \varepsilon\lambda) \tag{2.3.8}$$

where $W_{nn'}$ denotes $W(x|x')$ for the discrete case, and

$$\lambda = \sum_{n'=2}^{\infty} W_{n'1} \tag{2.3.9}$$

Defining

$$g(s) = \sum_{n=2}^{\infty} s^n W_{n1} \tag{2.3.10}$$

[and so $\lambda = g(1)$], we obtain from (2.3.8)

$$G(s,\varepsilon) = s + \varepsilon[g(s) - \lambda s] \tag{2.3.11}$$

The substitution of this into (2.3.7) with $t' = \varepsilon$ yields, upon Taylor expansion in ε,

$$G(s,t + \varepsilon) = G(s,t) + \varepsilon[g(s) - \lambda s]\frac{\partial}{\partial s}G(s,t) \tag{2.3.12}$$

whereupon we get

$$\frac{\partial G(s,t)}{\partial t} = [g(s) - \lambda s]\frac{\partial G(s,t)}{\partial s} \tag{2.3.13}$$

This is an alternate form of the forward equation, (2.1.5), specialized to the branching processes. Note that because it is homogeneous in the derivatives, $G_l(s,t)$ is also a solution of the same equation.

The backward equation can also be derived if we let t in (2.3.7) to be the infinitesimal time interval, and by using (2.3.11) we have

$$G(s,\varepsilon + t) = G(s,t) + \varepsilon[g(G(s,t)) - \lambda G(s,t)] \tag{2.3.14}$$

It then follows that

$$\frac{\partial G(s,t)}{\partial t} = g[G(s,t)] - \lambda G(s,t) \qquad (2.3.15)$$

This is related to the backward equation, (2.1.6). It is actually simpler to solve than (2.3.13) in practical cases.

3. EVOLUTION EQUATIONS AND THEIR SOLUTIONS

We describe in this section some typical evolution equations involving discrete transitions due to the creation and annihilation of particles. The variety of solutions span the spectrum of the most commonly used distribution functions applied to the analyses of particle multiplicities in recent years. Although the branching process is only one of these, we include the other stochastic processes in this discussion in order to contrast the various approaches.

Rewriting the forward master equation (2.1.5) for discrete variables, we have

$$\frac{d}{dt}P_n(t) = \sum_{n'} W_{nn'} P_{n'}(t) - \sum_{n'} W_{n'n} P_n(t) \qquad (3.0.1)$$

where $P_n(t)$ will hereafter be referred to as the multiplicity distribution. For simplicity we have restricted our attention to the case of one species only. We consider here only two basic types for the transition rate $W_{nn'}$:

creation: $\qquad W_{nn'} = c_{n'} \delta_{n',n-1} \qquad\qquad (3.0.2)$

annihilation: $W_{nn'} = a_{n'} \delta_{n',n+1} \qquad\qquad (3.0.3)$

recalling that n' is the multiplicity at an early time than n. Eq.(3.0.2) can describe birth, emission, fragmentation, or branching, while (3.0.3) can represent death, absorption, recombination, or decay. The basic evolution equation is therefore[3]

$$\frac{d}{dt} P_n(t) = a_{n+1} P_{n+1} + c_{n-1} P_{n-1} - (a_n + c_n) P_n \tag{3.0.4}$$

We consider below four special cases of this equation.

3.1 Poisson Distribution

In the case of independent emission, we put

$$a_n = 0, \qquad c_n = \beta \tag{3.1.1}$$

so that (3.0.4) becomes

$$\frac{d}{dt} P_n = \beta (P_{n-1} - P_n) \tag{3.1.2}$$

Its solution is the Poisson distribution

$$P_n = \frac{(\beta t)^n}{n!} e^{-\beta t}, \quad \bar{n} = \beta t \tag{3.1.3}$$

It is a purely coherent distribution, and is the narrowest one consistent with quantum statistics. Multiplicity distributions associated with a single multiperipheral chain[8] or the simple fragmentation of a color string are of this type. In those cases t corresponds to the total rapidity.

The generating function G(s), defined by

$$G(s) = \sum_n s^n P_n , \tag{3.1.4}$$

for the Poisson distribution is

$$G(s) = \exp\left[(s-1)\bar{n}\right] \tag{3.1.5}$$

The factorial moments are

$$\overline{n(n-1)...(n-r+1)} = \frac{d^r}{ds^r}G(s)\,|_{s=1} \,, \tag{3.1.6}$$

which in this case is simply $\bar{n}^r$. In particular, the dispersion d, defined by

$$d^2 = \overline{n^2} - \bar{n}^2 \,, \tag{3.1.7}$$

for the Poisson distribution is

$$d^2_{Poisson} = \bar{n} \,. \tag{3.1.8}$$

Evidently, the normalized dispersion

$$\gamma_2 = d^2/\bar{n}^2 = 1/\bar{n} \quad \text{(Poisson)} \tag{3.1.9}$$

becomes infinitesimal, as $\bar{n}$ becomes very large.

3.2 Bose-Einstein Distribution

For a system of identical bosons the properties of the creation and annihilation operators imply the following dependences on n:

$$a_n = \alpha n, \quad c_n = \beta(n+1) \tag{3.2.1}$$

The corresponding evolution equation is therefore

$$\frac{d}{dt}P_n = \alpha(n+1)P_{n+1} + \beta n P_{n-1} - [(\alpha+\beta)n + \beta]P_n \tag{3.2.2}$$

The solution can be put in the form

$$P_n(t) = p(t)q(t)^n, \quad p = 1-q, \tag{3.2.3}$$

where

$$q(t) = \frac{1 - (\beta/\alpha) \, C \, \exp[(\alpha - \beta)t]}{1 - C \, \exp[(\alpha - \beta)t]} \tag{3.2.4}$$

C is an integration constant that can be fixed by the initial condition. The average multiplicity is $q/(1 - q)$; hence,

$$\bar{n} = \frac{\beta}{\alpha - \beta} [1 - \frac{\alpha}{\beta C} e^{(\beta - \alpha)t}] \tag{3.2.5}$$

If $\alpha > \beta$, then asymptotically, in the limit $t \to \infty$, we have

$$q = \frac{\beta}{\alpha}, \qquad \bar{n} = \left(\frac{\alpha}{\beta} - 1\right)^{-1} \tag{3.2.6}$$

$$P_n = \bar{n}^n / (1 + \bar{n})^{n+1} \tag{3.2.7}$$

This is the Bose-Einstein distribution. If the stationary state is in thermal equilibrium at temperature T, then $P_n \propto e^{-nE/T}$ where E is the boson energy, and $\alpha / \beta = e^{E/T}$; hence, the familiar formula $\bar{n} = (e^{E/T} - 1)^{-1}$ follows.

3.3 Negative Binomial Distribution

A generalization of the Bose-Einstein distribution can be achieved by generalizing (3.2.1) to:

$$a_n = \alpha n, \qquad c_n = \beta n + \gamma \tag{3.3.1}$$

The resultant evolution equation is

$$\frac{d}{dt} P_n = \alpha(n + 1) P_{n+1} - (\beta n + \gamma) P_n + (\beta n - \beta + \gamma) P_{n-1} - \alpha n \, P_n \tag{3.3.2}$$

which can with a bit more effort be solved as above. However, since we are actually only interested in the stationary solution, we can obtain that far more easily by recognizing that the two lines of (3.3.2) should separately vanish when $dP_n / dt = 0$. This is because the first line contains the loss and gain terms of the $n + 1$ state; those transitions should be independent of the loss and gain terms of the $n - 1$ state, which comprise the second line of (3.3.2). Define $q = \beta/\alpha$ and $k = \gamma/\beta$, then the conditions for $dP_n / dt = 0$ are

$$(n + 1)P_{n+1} = (n + k) qP_n ; \quad (n + k - 1) qP_{n-1} = nP_n \qquad (3.3.3)$$

They admit the joint solution

$$P_n = P_n^k \equiv \frac{\Gamma(n + k)}{\Gamma(n + 1)\, \Gamma(k)} p^k q^n \qquad (3.3.4)$$

where $p = 1 - q$. This is the Pascal or negative binomial distribution [1] for integer k. It is also called the Polya distribution [1,5] when k is non-integer.

The generating function for the B.E. distribution, $P_n = pq^n$, is

$$G_{BE}(s) = p/(1 - qs), \qquad (3.3.5)$$

while that for the negative binomial distribution is

$$G_{NB}(s) = [p / (1 - qs)]^k \qquad (3.3.6)$$

In this sense P_n^k in (3.3.4) is a generalized Bose-Einstein distribution.

From (3.1.6) and (3.3.6) we obtain

$$\bar{n} = \frac{qk}{1 - q} = \frac{\gamma}{\alpha - \beta} \qquad (3.3.7)$$

$$\gamma_2 = (\overline{n^2}/\overline{n}^2) - 1 = \frac{1}{k} + \frac{1}{\overline{n}} \qquad (3.3.8)$$

The first equality of (3.3.7) gives

$$q = \frac{\overline{n}/k}{1 + \overline{n}/k}, \qquad p = \frac{1}{1 + \overline{n}/k} \qquad (3.3.9)$$

Substituting these into (3.3.4) facilitates the interpretation of the generalized B.E. distribution as composing of k cells of equal average occupancy $\overline{n}/k$.[9] Carruthers has assembled a large number of other interpretations of P_n^k in a variety of physical and mathematical contexts.[10] One can add to the list laser amplification by identifying γ in (3.3.7) with the term due to spontaneous emmision and $(\alpha-\beta)^{-1}$ as the amplification factor due to stimulated emission.[11]

It has been suggested[12] that (3.3.4) can be the solution of the equation

$$\frac{d}{dt}P_n^k = -\alpha(n+1)P_{n+1}^k + \alpha n \ P_n^k \qquad (3.3.10)$$

for which $\overline{n} = ke^{\alpha t}$. Hence, for non-vanishing $\overline{n}$ as $t \to \infty$, it is necessary that $\alpha > 0$. However, for positive α, (3.3.10) is not an evolution equation of the gain-loss type described by (3.0.4). The physical meaning of (3.3.10) in that case is therefore not clear.

3.4 Furry Distribution

In studying the multiplicities of electrons in cosmic ray showers, Furry[13,14] considered the evolution equation for

$$a_n = 0, \quad c_n = \beta n \tag{3.4.1}$$

so that

$$\frac{d}{dt} P_n = \beta(n-1) P_{n-1} - \beta n P_n \tag{3.4.2}$$

Since this is the prime example of a branching process, we examine its property and derive its solution in some detail.

Using (3.4.1) and (3.0.2) in (2.3.9) and (2.3.10), we have $\lambda = \beta$ and $g(s) = \beta s^2$. We thus obtain from (2.3.13) and (2.3.15) the forward and backward equations for the generating function, $G(s,t) = \sum_n s^n P_n(t)$,

$$\frac{\partial G(s,t)}{\beta \, \partial t} = s(s-1) \frac{\partial G(s,t)}{\partial s} \tag{3.4.3}$$

$$\frac{\partial G(s,t)}{\beta \, \partial t} = G^2(s,t) - G(s,t) \tag{3.4.4}$$

the former being directly derivable from (3.4.2). Equation (3.4.4) can be simply integrated, giving

$$G(s,t) = [1 + \sigma(s) \, e^{\beta t}]^{-1} \tag{3.4.5}$$

Substituting this into (3.4.3) leads to an equation on $\sigma(s)$ whose solution is $\sigma(s) = \sigma_0(1 - s^{-1})$. The constant σ_0 can be determined by the initial condition. If at $t = 0$, $P_n(0) = \delta_{n1}$, then $G(s,0) = s$ and so $\sigma_0 = -1$. Hence, we get

$$G(s,t) = \frac{p(t)\ s}{1 - q(t)\ s} \qquad (3.4.6)$$

where

$$p(t) = e^{-\beta t}, \qquad q(t) = 1 - e^{-\beta t}. \qquad (3.4.7)$$

It then follows from (3.4.6) by expanding the inverse power that

$$P_n(t) = p(t)\ [q(t)]^{n-1}. \qquad (3.4.8)$$

If initially there are k particles, i.e., $P_n(0) = \delta_{nk}$, then we have $G_k(s,0)$ = s^k. Instead of finding the corresponding σ_0, it is more convenient to make use of (2.3.4), or equivalently to recall the comment that follows (2.3.13); we obtain

$$G_k(s,t) = [G(s,t)]^k = \left(\frac{ps}{1 - qs}\right)^k \qquad (3.4.9)$$

Now, by negative binomial expansion[1]

$$(1 - qs)^{-k} = \sum_r \binom{-k}{r}(-qs)^r = \sum_r \binom{k+r-1}{r}(qs)^r \qquad (3.4.10)$$

we can determine P_n from (3.4.9). Let us use the notation F_n^k for the solution; thus we have

$$G_k(s,t) = \sum_{n=k}^{\infty} s^n\ F_n^k(t) \qquad (3.4.11)$$

$$F_n^k(t) = \binom{n-1}{n-k}\ [p(t)]^k\ [q(t)]^{n-k} \qquad (3.4.12)$$

We shall refer to F_n^k as the Furry distribution. Note that it is related to the negative binomial distribution in (3.3.4) by

$$F_n^k(p) = P_{n-k}^k(p), \quad q = 1 - p. \tag{3.4.13}$$

This is not surprising in view of the similarity in their generating functions, c.f. (3.3.6) and (3.4.9).

Although $\bar{n}$ and $\bar{n^2}$ can be obtained from the evolution equation directly without the benefit of the explicit solution we give here their dependences on t from the direct manipulations with $G_k(s,t)$:

$$\bar{n} = \frac{\partial}{\partial s} G_k(s,t)|_{s=1} = \frac{k}{p(t)} \tag{3.4.14}$$

$$\overline{n(n-1)} = \frac{\partial^2}{\partial s^2} G_k(s,t)|_{s=1} = \frac{k}{p}[-1 + k + (\frac{1}{p} - 1)(k + 1)]$$

$$\gamma_2 = (\bar{n^2} - \bar{n}^2)/\bar{n}^2 = \frac{1}{k} - \frac{1}{\bar{n}} \tag{3.4.15}$$

From (3.4.7) we thus have for the evolution parameter w

$$w = \bar{n}(t) / k = e^{\beta t}. \tag{3.4.16}$$

The Furry distribution, when expressed in terms of w, now takes the form

$$F_n^k(w) = \frac{\Gamma(n)}{\Gamma(k)\,\Gamma(n-k+1)} (\frac{1}{w})^k (1 - \frac{1}{w})^{n-k} \tag{3.4.17}$$

which may be continued to non-integral k.

A useful formula can be derived from the above. Consider

$$\sum_{n=k}^{\infty} \frac{\Gamma(n+m)}{\Gamma(n)} F_n^k(w) = w^{-k} \sum_{n=k}^{\infty} \frac{\Gamma(n+m)}{\Gamma(k)\Gamma(n-k+1)} (1 - \frac{1}{w})^{n-k}$$

$$= \frac{\Gamma(k+m)}{w^k \Gamma(k)} \sum_{p=0}^{\infty} \frac{\Gamma(p+k+m)}{\Gamma(k+m)\Gamma(p+1)} \left(1 - \frac{1}{w}\right)^p$$

The last summation is the negative binomial expansion of $[1 - (1-1/w)]^{-(k+m)}$.
Consequently, we have

$$\sum_{n=k}^{\infty} \frac{\Gamma(n+m)}{\Gamma(n)} F_n^k(w) = \frac{\Gamma(k+m)}{\Gamma(k)} w^m \qquad (3.4.18)$$

From this we get, for $m = 1$ and 2, and in light of (3.4.16),

$$\sum_{n=k}^{\infty} n \, F_n^k = kw = \bar{n} \qquad (3.4.19)$$

$$\sum_{n=k}^{\infty} (n+1)n \, F_n^k = \bar{n}^2 + \bar{n}w \qquad (3.4.20)$$

The latter is consistent with (3.4.15). Higher moments can similarly be obtained from considering higher values of m.

4. BRANCHING AND PERTURBATION THEORY

Having discussed the stochastic origin of branching in the previous two sections, we now want to establish its connection with particle physics. The relevance of stochastic equations to multiparticle production at high energies was pointed out by Giovannini [15] as far back as 1972. The connection was made more precise in perturbation theory by Konishi, Ukawa and Veneziano. [16] In the ϕ^3 theory in six dimensions the multiplicity distribution is shown to satisfy the Furry branching equation, but in QCD there are complications due to infrared divergence. To have a connection between branching and perturbation theory is important, because in order to apply stochastic methods to soft hadronic processes we need an anchoring assertion that in the limit of weak coupling due to asymptotic freedom the branching distribution agrees with the result in perturbation theory. When the coupling is strong, perturbation method fails, but branching may still be a valid description of soft processes, since its evolution equation does not rely on the coupling constant being small. Because of its importance we shall describe the connection in some detail.

4.1 Cross Sections, Distributions and their Moments

Before getting into perturbation theory, we first need to review some well-established notions about cross sections and distributions, at least for the purpose of defining our notations. For simplicity we shall consider the production of n identical bosons.

Denoting the momentum fraction of the i-th particle in the c.m. frame by x_i, we define

$$\sigma_n(x_1,...,x_n) = \frac{d\sigma(A+B \to 1+...+n)}{dx_1 dx_2...dx_n} \tag{4.1.1}$$

to be the exclusive cross section for detecting n particles with momenta $x_1,...,x_n$. The semi-inclusive cross section for observing k of those n particles having momenta $x_1,...,x_k$ is

$$\sigma_n(x_1,...,x_k) = \frac{1}{(n-k)!} \int \sigma_n(x_1,...,x_n) \prod_{j=k+1}^{n} dx_j \tag{4.1.2}$$

The k-particle inclusive distribution allows any number of particles to accompany the k particles, so it is

$$D(x_1,...,x_k) = \frac{1}{\sigma} \sum_{n=k}^{\infty} \sigma_n(x_1,...,x_k) \tag{4.1.3}$$

576

where σ is the total inelastic cross section. On the other hand, the n-particle topological cross section counts the number of particles without specifying their momenta, so it is given by

$$\sigma_n = \frac{1}{n!} \int \sigma_n (x_1,...,x_n) \prod_{j=1}^{n} dx_j \qquad (4.1.4)$$

The corresponding multiplicity distribution is therefore

$$P_n = \sigma_n / \sigma \qquad (4.1.5)$$

whose sum over all n is 1, leaving out the elastic cross section. In phenomenology sometimes the diffractive cross sections (single and/or double) are also left out in both the numerator and denominator in (4.1.5).

The inclusive distribution $D(x_1,...,x_k)$ and the multipliticy distribution P_n are related as follows.

$$\int D(x_1,...,x_k) \prod_{j=1}^{k} dx_j = \frac{1}{\sigma} \sum_{n=k}^{\infty} \int \sigma_n(x_1,...,x_k) \prod_{j=1}^{k} dx_j$$

$$= \frac{1}{\sigma} \sum_{n=k}^{\infty} \frac{1}{(n-k)!} \int \sigma_n(x_1,...,x_n) \prod_{j=1}^{n} dx_j$$

$$= \sum_{n=k}^{\infty} \frac{n!}{(n-k)!} P_n = <n(n-1)...(n-k+1)> \equiv \phi_k . \qquad (4.1.6)$$

ϕ_k defined above is called the factorial moment.[3,17] The moments of the inclusive distribution are defined by

$$M(n_1,...,n_k) = \int_0^1 D(x_1,...,x_k) \prod_{j=1}^{k} (x_j^{n_j-1} dx_j) \qquad (4.1.7)$$

where the integrals are from 0 to 1 for every one of the variables. Clearly, we have

$$\phi_k = M(n_1=1,...,n_k=1) \tag{4.1.8}$$

Let us consider two generating functions defined as follows:

$$F(v) = \sum_{k=1}^{\infty} \frac{v^k}{k!} \phi_k \tag{4.1.9}$$

$$G(s) = \sum_{n=1}^{\infty} s^n P_n \tag{4.1.10}$$

They are related by simple manipulation:

$$G(s) = \sum_{n=1}^{\infty} [1 + \sum_{k=1}^{n} \binom{n}{k}(s-1)^k] P_n$$

$$= 1 + \sum_{k=1}^{\infty} \frac{(s-1)^k}{k!} \sum_{n=k}^{\infty} \frac{n!}{(n-k)!} P_n$$

yielding

$$G(s) = 1 + F(s-1) \tag{4.1.11}$$

This equation will prove to be useful below.

4.2 Fragmentation Function in the Tree Approximation.[16]

We summarize here some of the essential properties of an asymptotically free theory in its description of the fragmentation function for the decay of a parton of high virtuality Q^2. For notational simplicity let us consider only one species of partons. That would be sufficient when we apply the formalism to the ϕ_6^3 theory in the following subsection. For QCD we need only generalize the equations to a two-channel problem involving quarks and gluons, with the basic structure of the tree diagrams unchanged.

Let D(x,t) denote the inclusive distribution for finding a parton with momentum fraction x in the fragmentation of a parent parton in a hard process characterized by a large scale Q^2, in terms of which the evolution parameter t is defined:

$$t = \frac{1}{2\pi b} \ln [\alpha_s(Q_0^2)/\alpha_s(Q^2)] \qquad (4.2.1)$$

In QCD the constant b is $(11-2N_f / 3) / 4\pi$, and $\alpha_s(Q^2) = (b \ln Q^2/\Lambda^2)^{-1}$. The fragmentation function satisfies the Altarelli-Parisi evolution equation[18]

$$\frac{d}{dt}D(x,t) = \int_x^1 \frac{dz}{z} \, P(z) \, D(x/z,t) \qquad (4.2.2)$$

where P(z) is the splitting function that specifies the elementary parton decay process (i → j+k) with the j parton having momentum fraction z, k parton 1 – z. The solution of this equation can be represented by the diagram in Fig. 1(a), where the dot signifies the degradation of virtuality through the emission of partons (not exhibited in the diagram).

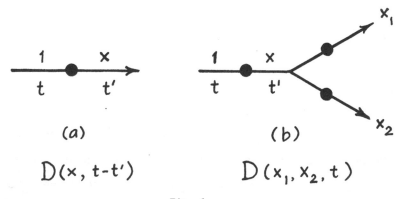

(a)

$D(x, t-t')$

(b)

$D(x_1, x_2, t)$

Fig. 1

In the leading log approximation and in the (axial) gauge so that no interference terms are allowed when squaring amplitudes, the tree diagrams that add up to $D(x,t)$ for the one-parton inclusive distribution can be organized to give the two-parton distribution $D(x_1, x_2, t)$ that has the diagrammatic representation shown in Fig. 1(b). The essence of this diagram is that in a tree diagram there is one vertex at which a parton splits into two branches, in each of which is to be found the specified final partons, all other partons being integrated over. Thus $D(x_1, x_2, t)$ can be expressed in terms of three single-parton distributions and one splitting function as follows:

$$D(x_1, x_2, t) = \int_0^t dt' \int_0^1 dx\, dz\, dy_1\, dy_2\, D(x, t-t')\, \hat{P}(z)$$
$$D(y_1, t')\, D(y_2, t')\, \delta(x_1 - xzy_1)\, \delta(x_2 - x(1-z)y_2) \quad (4.2.3)$$

The splitting function $\hat{P}(z)$ differs from $P(z)$ in (4.2.2) in the absence of the $\delta(1-z)$ term, which corresponds to a virtual process that regularizes $\int P(z)\, z^{-1}\, dz$, while the diagram in Fig. 1(b) describes splitting into two real branches.

If we take the moments of (4.2.3) according to (4.1.7), we get

$$M(n_1, n_2, t) = \int_0^t dt'\, M(n_1 + n_2 - 1, t-t')\, P_{n_1 n_2}\, M(n_1, t')\, M(n_2, t')$$
$$(4.2.4)$$

where

$$P_{n_1 n_2} = \int_0^1 dz\, z^{n_1-1}\, (1-z)^{n_2-1}\, \hat{P}(z) \quad (4.2.5)$$

Of particular interest to us is (4.2.4) for $n_1 = n_2 = 1$, in which case we have

$$\phi_2(t) = \int_0^t dt' \, \phi_1(t-t') \, P_{11} \, \phi_1^2(t') \tag{4.2.6}$$

This equation can be generalized to the case of k-parton inclusive distribution, for which the schematic tree diagram is shown in Fig.2 for a particular partition of k into two k_1 and k_2 clusters. The sum of all tree diagrams for a k-inclusive process can be put in the form of Fig. 2, provided we sum over all possible (k_1, k_2) partitions. Note that the two bubbles in Fig. 2 leading to the k_1 and k_2 partons can each in themselves be described by diagrams like Fig.2. In other words we have a system of non-linear integral equations. It is

$$\phi_k(t) = \int_0^t dt' \, \phi_1(t-t') \, P_{11} \, \frac{1}{2} \sum_{k_1=1}^{k} \binom{k}{k_1} \phi_{k_1}(t') \, \phi_{k-k_1}(t') \tag{4.2.7}$$

The one-half factor is necessary to compensate for the over-counting in the sum due to the symmetry in k_1 and $k_2 = k - k_1$ interchange, as can be verified for the case $k = 2$.

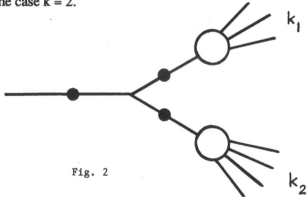

Fig. 2

Substituting (4.2.7) into (4.1.9), we have

$$F(v,t) = v \, \phi_1(t) + \frac{1}{2} P_{11} \int_0^t dt' \, \phi_1(t-t') \, F^2(v,t') \tag{4.2.8}$$

This is the integral equation that we have sought to derive. It clearly reveals the recursive nature of the tree diagrams, and expresses the essence of branching. The first term on the right side corresponds to the 1-parton inclusive process shown in Fig. 1(a), while the second term corresponds to k-parton inclusives, $k \geq 2$, depicted in Fig. 2.

4.3 ϕ_6^3 Theory and Furry Branching

To proceed further from (4.2.8) we need the explicit forms for $\phi_1(t)$ and P_{11}, which specify the basic nature of the branching. $\phi_1(t)$ can be easily obtained by taking the first moment of (4.2.2), whereupon by trivial integration, one gets

$$\phi_1(t) = \exp(A_1 t) \tag{4.3.1}$$

where

$$A_1 = \int_0^1 dz\, P(z) \tag{4.3.2}$$

which is proportional to the anomalous dimension for $n = 1$. On the other hand, P_{11} is, from (4.2.5),

$$P_{11} = \int_0^1 dz\, \hat{P}(z) \tag{4.3.3}$$

Since $P(z)$ and $\hat{P}(z)$ differ by a δ-function, $P(z) = \hat{P}(z) - c\delta(1 - z)$, we have $A_1 = P_{11} - c$.

In the ϕ^3 theory in six dimensions, which was considered by KUV[16] and Taylor[19] in their studies of the jet physics, there is no soft infrared divergence, although there is collinear divergence. In that theory $\hat{P}(z) = z(1 - z)$, and c is determined by the requirement

$$\int_0^1 dz\, z\, P(z) = 0 \qquad (4.3.4)$$

which results in c = 1/12. [The zdz in (4.3.4) is due to d = 6 in z^{d-5} dz.] Indeed, one has

$$A_1 = \frac{1}{2} P_{11} \qquad (4.3.5)$$

This relation actually follows directly from the invariance of $\hat{P}(z)$ under the interchange of z with $1 - z$, which is an inherent property of a ϕ^3 theory. It should be pointed out that the quantities in QCD corresponding to A_1 and P_{11} are divergent because of infrared singularities. It is due to the absence of such complications in the ϕ_6^3 theory that the consideration here is so simple.

To apply these results for $\phi_1(t)$ and P_{11} to the branching problem, let us first take the derivative of (4.2.8) and obtain with the help of (4.3.1)

$$\frac{\partial}{\partial t} F(v,t) = A_1 F(v,t) + \frac{1}{2} P_{11} F(v,t)^2 \qquad (4.3.6)$$

The boundary condition is $F(v,0) = v$. If we now apply (4.3.5) and $F = G - 1$ from (4.1.11), we get

$$\frac{\partial}{\partial t} G(s,t) = A_1 [G^2(s,t) - G(s,t)] \qquad (4.3.7)$$

and $G(s,0) = s$. This is the final result that we have aimed for. It coincides with the backward evolution equation, (3.4.4), for Furry branching. The solution of (4.3.7) in the form (3.4.5) is a result obtained by Taylor[19] without the use of tree diagrams. Thus the connection between Furry branching and the perturbation theory in the leading log approximation (for ϕ_6^3 coupling) is established.

4.4. Hinz-Lam Model [20]

Instead of the unphysical world of $d = 6$, Hinz and Lam considered a model in which the dimension is essentially two. This is more physical for low-p_T physics because for limited transverse momenta, the phase space of the produced particles at very high energy has effectively only one spatial (longitudinal) dimension. Strictly, the phase-space factor $\rho(i \rightarrow a+b)$ for the decay of $i \rightarrow a+b$ in two dimensions is

$$\rho^{(2)} = \frac{1}{2}[(m_i^2 - m_a^2 - m_b^2)^2 - 4\,m_a^2\,m_b^2]^{-1/2} \tag{4.4.1}$$

where m_i etc. are the masses. When m_i is near the threshold of $m_a + m_b$, $\rho^{(2)}$ diverges; however, in that region the transverse directions cannot be ignored, and $\rho(i \rightarrow a+b)$ should behave more like $\rho^{(4)}$ than $\rho^{(2)}$. Hinz and Lam made the simplifying assumption that

$$\rho^{(2)}(m_i) = \frac{1}{2m_i^2}\,\theta(m_i - m_a - m_b) \tag{4.4.2}$$

for all m_i. This expression for $\rho^{(2)}$ is very good at high m_i, and is a reasonable approximation at low m_i.

The Hinz-Lam model then proceeds by considering all tree diagrams with ϕ^3 coupling. Every propagator then ends at a vertex; the associated wave-function factor and the two-body phase-space factor are summarized by a (vertex) function $v(m_i^2)$. The integration over m_i of every internal line is then replaced by an integration over $dt_i = v(m_i^2)\,dm_i^2$ with trivial integrand.

By studying all tree diagrams for a particular n-body production process, and summing over all contributions with symmetry factors carefully taken into account, it is found that the corresponding multiplicity distribution $\breve{P}_n$ satisfies

$$\frac{d\,\breve{P}_n}{dt} = \frac{1}{2}(n-1)\,\breve{P}_{n-1} \tag{4.4.3}$$

This equation differs from (3.4.2) by the absence of the loss term. Thus $\breve{P}_n$ is not the Furry distribution F_n^k. In fact, it fails to satisfy unitarity, since (4.4.3) implies that $\Sigma\,\breve{P}_n \neq 1$. As an ununitarized version of F_n^k, it may be regarded as the Born term in a unitarity equation. In studying this model, one takes advantage of the explicit knowledge of the exclusive distributions, and hopes that unitarity correction does not totally remove some of the features associated with $\breve{P}_n$.

As an application of the model, one can determine the exclusive rapidity distribution for every tree diagram.[21] From that result one can construct a t-channel representation for every s-channel tree diagram. The duality thus inferred is useful in relating branching processes to other approaches, in particular, that of the dual parton model.[22] Another application of the model is that it allows one to examine in detail the effects of finite-energy correction. Since stochastic equations, such as (3.0.4) and (3.4.2), involve no scale, it is usually quite unclear how a branching equation should be modified at finite energy when the particle masses are not neglected. In the context of the Hinz-Lam model, it turns out that the finite-energy modification to the multiplicity distribution is unexpectedly small owing to an intricate structure of the branching chains.[23]

4.5   QCD Branching

The derivation of Furry branching in the ϕ_6^3 theory is relatively simple because we only have to deal with summing leading collinear singularities. In QCD the essential complication is the presence of infrared (IR) singularities. The quantity A_1 in (4.3.2) first of all has to be generalized to a matrix to account for the quark and gluon channels, but more significantly certain matrix elements are divergent. This is due to the soft gluon contribution, as can be seen explicitly in, for example, the splitting function for $g \rightarrow g + g$

$$P_{gg}(x) = 6\left[\frac{x}{(1-x)_+} + \frac{1-x}{x} + x(1-x) + (\frac{11}{12} - \frac{N_f}{18})\, \delta(x-1)\right] \qquad (4.5.1)$$

The integral of this over x diverges at $x = 0$. Since the parton multiplicity depends critically on the behavior near $x = 0$, the IR problem must be treated carefullly, if perturbative QCD is to be applied to jet fragmentation.

The effect of the IR divergence on the anomalous dimension is that it introduces a pole at $N = 1$ in every finite order in α_s. Consider, for example, the singlet anomalous dimension, which in the leading order is the Nth moment of P_{gg}

$$\gamma_N^s = \frac{\alpha_s}{2\pi} \int dx\, x^{N-1}\, P_{gg}(x) \qquad (4.5.2)$$

since the splitting function for $g \rightarrow q\bar{q}$ has no singularity at $x = 0$. To second order in α_s it behaves near $N = 1$ as

$$\gamma_N^s = \frac{\bar{\alpha}_s}{N-1}\, [1 - \frac{2\,\bar{\alpha}_s}{(N-1)^2} + ...], \quad \bar{\alpha}_s = 3\alpha_s/\pi \qquad (4.5.3)$$

The leading divergent terms can actually be summed, yielding a branch-point singularity at $N = 1$:[24]

$$\gamma_N^s = - \frac{N-1}{2} + [(\frac{N-1}{2})^2 + \bar{\alpha}_s]^{1/2} \qquad (4.5.4)$$

A quick way of seeing this result is to examine the integral equation for the ladder diagram

$$D(x,q^2) = \int^{q^2} \frac{dk^2}{k^2} \frac{\alpha_s}{2\pi} \int_x^1 \frac{dz}{z} P(z) \, D(\frac{x}{z}, zk^2) \qquad (4.5.5)$$

If $P(z)$ were such that the integral over z converges at $z = 0$, and if α_s is a running coupling constant, then by differentiation with respect to $-\ln \alpha_s(q^2)$ one obtains the Altarelli-Parisi equation, (4.2.2), in which the dependence on $\ln zq^2$ of D in the integrand is written as t because the $z \rightarrow 0$ limit is not an important kinematical region. Now, consider the IR singular case where P_{gg} is dominated by $6/z$. If we assume α_s not to run, we get

$$\frac{d}{d \ln q^2} D(x,q^2) = \bar{\alpha}_s \int_x^1 dz \, z^{-2} D(\frac{x}{z}, zq^2) \qquad (4.5.6)$$

The Nth moment of $D(x,q^2)$, according to (4.1.7), then satisfies [25]

$$\frac{d}{d \ln q^2} M(N,q^2) = \bar{\alpha}_s \int_0^1 dz \, z^{N-2} M(N, zq^2) \qquad (4.5.7)$$

This equation admits the solution $M(N,q^2) \sim (q^2)^{\gamma_N^s}$, which requires

$$\gamma_N^s = \bar{\alpha}_s / (N-1+\gamma_N^s) \qquad (4.5.8)$$

whose solution is (4.5.4). What this means is that the double logarithms arising from collinear and IR singularities can be summed in leading order to yield a finite anomalous dimension. It is then possible to generalize the jet calculus[16] and show that the parton distribution can be described by a branching process similar to that discussed in Sec. 4.3.[24]

The above result is, however, incorrect because it takes into account ladder diagrams only. It was found[26] that in the small x region the non-ladder diagrams can contribute in the leading order and interfere with the ladder diagram contributions. In fact, the large-angle emission of soft gluons gets cancelled out, thereby reducing the anomalous dimension to [27]

$$\gamma_N^s = -\frac{N-1}{4} + [(\frac{N-1}{4})^2 + \frac{1}{2}\bar{\alpha}_s]^{1/2} \qquad (4.5.9)$$

in double log accuracy. This corresponds to the solution of a modified (4.5.8), in which the denominator is replaced by $N - 1 + 2\gamma_N^s$. That in turn corresponds to a modification of the phase space of integration in (4.5.5) such that the virtuality of the M-function in the integrand in (4.5.7) is z^2q^2, instead of zq^2. The implication is that a larger range of the soft gluons effectively do not contribute. This has an important effect on the nature of the final state.

One may represent the calculation of the n-parton distribution P_n by the schematic bubble diagrams shown in Fig. 3. At small x the crossed diagrams in the second term on the right-hand side cancel the large angle part of the soft gluons in the first term. Thus in leading double log one recovers a branching tree diagram without interference such as that in Fig. 2, except that one must be carefull about angular ordering of the emitted gluons.

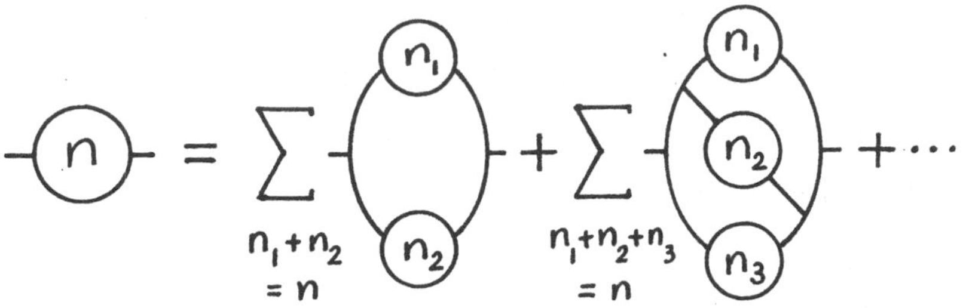

Fig. 3

Angular ordering can be described in terms of the angular variable [24]

$$\zeta = q_1 \cdot q_2 / q_1^0 q_2^0 \tag{4.5.10}$$

where q_1 and q_2 are the momenta of the two gluons that a parent gluon (of momentum q) splits into: $q \rightarrow q_1 + q_2$, as shown in Fig. 4. In a frame in which q_1^0 and q_2^0 are large compared to the virtualities of the produced gluons, ζ becomes $1 - \cos\theta$, where θ is the angle between $\vec{q}_1$ and $\vec{q}_2$; consequently, it follows from

$$q^2 = q_1^2 + q_2^2 + 2 q_1^0 q_2^0 \zeta \tag{4.5.11}$$

that Q, defined as $| q^2 |^{1/2}$, is proportional to θ, when θ is small. Thus when neither one of the gluons is soft, an ordering in the virtual masses in the ladder diagrams implies a similar ordering in the opening angles of successive gluons. Interference is therefore not possible unless one of the emitted gluon is soft. For the crossed diagram in Fig. 3 it is precisely in the phase space region where the angles are not ordered that the contribution is as important as that of the ladder diagram, so that a cancellation can take place .

If ζ_1 and ζ_2 are the angular variables for the splitting of the daughter gluons with q_1 and q_2, respectively, as indicated in Fig. 4, then the effect of the interference of soft gluons is taken into account to leading order by ordering the ζ variables according to

$$\zeta_1, \zeta_2 < \zeta < \zeta_{max} \qquad\qquad (4.5.12)$$

where ζ_{max} is associated with the initial vertex of production. The ladder diagrams for the branching process thus regain their validity, when (4.5.12) is imposed successively.

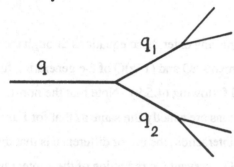

Fig. 4

A Monte Carlo calculation of the jet evolution with angular ordering has been carried out by Marchesini and Webber.[28] Explicit event structures showing the momenta of partons in the branching product can be exhibited in such a scheme. Indeed, the single-gluon inclusive distribution has a pronounced dip at small x, a result of the depletion of soft gluons due to the interference effects. The multiplicity distribution exhibits KNO scaling.

The moments of the multiplicity distribution can be calculated analytically from the evolution equations of the generating functions. With $G_g(s)$ and $G_q(s)$ defined as in (4.1.10) for the gluon and quark jets, (s being

590

the parameter of the generating function , having nothing to do with energy), their angular-ordered evolution equations at scales greater than some IR cutoff Q_0 are [25,29]

$$Q\frac{\partial}{\partial Q}G_g(s,Q) = \int_{Q_0/Q}^{1-Q_0/Q} dz \, (\alpha_s/\pi) \, \{P_{gg}(z) \, [G_g(s,zQ) \, G_g(s,(1-z)Q) - G_g(s,Q)]$$

$$+ P_{qg}(z) \, [G_q(s,zQ) \, G_g(s,(1-z)Q) - G_g(s,Q)]\} \qquad (4.5.13)$$

$$Q\frac{\partial}{\partial Q}G_g(s,Q) = \int_{Q_0/Q}^{1-Q_0/Q} dz \, (\alpha_s/\pi) \, P_{qq}(z) \, [G_q(s,zQ) \, G_g(s,(1-z)Q) - G_q(s,Q)]$$

$$(4.5.14)$$

The effects of angular ordering enter these equations through the limits of integration and the arguments zQ and $(1-z)Q$ of the generating functions in the integrands, as discussed following (4.5.9). Note that the non-linear structure of these evolution equations are exactly the same as that for Furry branching, shown in (4.3.7) for ϕ_6^3 interaction; the major difference is that the z dependences of G_g and G_q prevent the reduction of the z integrations into multiplicative factors like A_1. That, of course, is the mechanism that turns the pole at $N = 1$ in each order into a branch point, as we have already discussed. Thus QCD branching is strictly not of the type of stochastic processes discussed in Sec. 3. This is a consequence of the fact that angular ordering implies some memory of past branchings in a chain, so it cannot be a purely Markovian process. However, apart from those details the general structure of the evolution is primarily that of a branching process.

The multiplicity moments, $R_k \equiv \phi_k / \phi_1^k$, can be determined from (4.5.13) and (4.5.14). It is found[29] that they are close to those of a negative binomial distribution. Although the method includes next-to-leading order corrections, the distribution calculated is broader than the observed data in e^+e^- annihilation, unless the QCD scale Λ is adjusted for every moment. Also, the ratio of the charge particle multiplicities in a gluon jet to that in a quark jet, $<n>_g / <n>_q$, measured in e^+e^- annihilation experiment at 29 GeV,[30] has been found to be 1.29, which is significantly lower than the value of 9/4 expected in the leading order consideration.

Recently, an integral equation method for calculating multiplicity distribution has been developed,[31,32] bringing considerable simplification to the problem, especially when dealing with lower multiplicity moments. The method provides a systematic and tractable procedure to extend the analysis to include higher-order corrections.[33]

Before the IR problem was carefully investigated, the interest in knowing the nature of QCD branching as a stochastic process led Giovannini to consider an arbitrary cut-off in the soft gluons and to write down a set of coupled evolution equations, similar to (4.5.13) and (4.5.14), except that the coupling coefficients are effective constants.[7] The problem is then reduced to one having the characteristics discussed in Sec.3, the only complication being that it involves quark and gluon channels. The evolution equations are difficult to solve in general. But in special limiting cases, the solutions can be readily obtained. The result for a gluon jet without quarks is the Furry distribution, while that for a quark jet without $g \rightarrow q+\bar{q}$ is the Polya distribution. For reasons to be given in the next subsection, Giovannini's evolution equations should strictly not be regarded as an appropriate description of QCD branching.

More recently, the problem was reconsidered by B. Durand and Sarcevic[34] under certain approximations for mathematical simplicity, which facilitates an explicit examination of the energy dependences of the solutions.

Earlier, Biyajima and Suzuki [35] discovered the connection between Giovannini's equations and the Fokker-Planck description of stochastic evolution, a subject to which we shall return in Sec. 5.4. None of these later investigations addressed the problem of branching in perturbative QCD, so we shall not discuss them further here.

4.6 Summary and Comments

We have reviewed three ways to relate branching processes to perturbative calculations: the ϕ_6^3 theory, Hinz-Lam model with ϕ^3 coupling, and QCD. Of these, only the ϕ_6^3 theory gives a clean connection with Furry branching. The Hinz-Lam model gives only a part of the evolution equation for Furry branching but provides an explicit algebraic description of each exclusive tree diagram. Perturbative QCD is the most complicated one of the three due to its IR singularity. The high-Q^2 jets in leading double-log approximation can be described by a quasi-Markovian branching process of the Furry type; no analytic expression for the multiplicity distribution can be given.

Because of the fundamental importance of QCD, more work needs to be done to fully elucidate the properties of jets. As it stands, one can do Monte Carlo simulation of jet fragmentation and describe the branching process step-by-step in double-log accuracy. The leading order result unfortunately does not agree with the present data. Thus until the e^+e^- annihilation experiments can be done at higher energies, one must go to non-leading order calculations which are extremely complicated. One does not expect the result of such calculations to correspond to branching processes because of important interference effects. The work of Tesima appears to be very promising in that direction.[33]

The basic complication of QCD is side-stepped when the branching equations for quarks and gluons are approximated by constant coupling coefficients.[7] While it is of some interest to determine the branching properties of a system of quarks and gluons with simplified couplings so that the evolution equations can be solved analytically, it is not clear how one should interpret the solutions obtained. More specifically, it is unclear whether the partons have been given some effective mass to regularize the IR singularity, or the interaction modified so as to render the branching process Markovian. In either case since the precisely known QCD dynamics has been altered, the problem is presumably not intended to describe high-Q^2 jets, for which perturbative QCD needs no modification, only the non-leading order contributions calculated. One then may speculate that the simplified evolution equation might simulate low-p_T hadronic processes. But since the dynamics of quark and gluon interactions has been altered in an unspecified way, one is not even certain whether the system consists of fermionic quarks and bosonic vector gluons with the usual colors and flavors. In that circumstance the motivation for studying a system of quarks and gluons seems to be lost. Thus it is not clear in what specific way the study of such a system provides more reliable information than the stochastic evolution of a simple system with ϕ^3 coupling.

Our purpose for demonstrating a connection between perturbation theory and branching processes is primarily to establish the latter as a bridge between perturbative and non-perturbative processes in hadronic interactions. The relevance of branching to soft hadronic processes cannot be investigated with the same degree of precision as we have with hard processes in this section. That is the subject of the next section. Although the description of jets in QCD has its mathematical complications, branching as a physical picture of how partons are produced in the process of degradation of the virtualities of the partons has certainly been established here as a relevant and useful concept, at least at high enough Q^2 to justify leading order results. We shall appeal to this same picture when we discuss soft production, except that the virtuality will be replaced by some other comparable quantity. In this sense we regard the branching mechanism not only as a bridge but also as a unifying characteristics of all multiparticle production processes, when the number of particles produced is large.

5. BRANCHING IN SOFT HADRONIC PROCESSES

Multiparticle production in soft hadronic processes is perhaps the most controversial subject in modern physics.[36] The diverse views represented by the various articles in this review volume reveal the lack of unanimity, which has arisen out of the chronic absence of a workable theory. I discuss in this section the reasons and evidences why I believe that Furry branching is relevant to particle production at low p_T. The views expressed will necessarily be rather subjective. At the end a brief summary of the Fokker-Planck approach to the evolution equation will be given in Sec. 5.4.

5.1   Multiparticle Production at Low p_T

The fact that the average transverse momentum $<p_T>$ of the produced particles for $\sqrt{s} < 100$ GeV is constant at around 350 MeV indicates not only that the geometrical size of the hadrons provides the only relevant scale in the problem, but also that the corresponding length scale is too coarse to reveal the point-like structure of the hadron constituents and their interactions. For this reason and in recognizing that soft production is basically a many-body problem, I believe that QCD, perturbative or non-perturbative, is operationally irrelevant. The use of quarks and gluons as the basic quanta in low-p_T physics is therefore inappropriate. In e^+e^- annihilation at high energy the string picture of a quark and an antiquark stretching a color flux tube is sensible because the ends of the string are well defined in this highly virtual process, and their momenta are precisely known. In hadron-hadron collisions, the picture is far more complex. The string model is ambiguous because the ends of the strings are not well defined and their momenta not incontrovertible. I mention this here mainly to register the note that for central production the use of the quark picture and associated strings is not inevitable, despite the apparent popularity of the modified Lund model [37] and the dual parton model.[38] The quark picture can be relevant in the fragmentation region because the hadron structure affects the inclusive cross sections there, and hadron structure can only be described in terms of its constituents.[39] Our concern in this section will only be in the central region. The remarks made above are not meant so much to discredit QCD as to state that an unreliable application of QCD is not any better than omitting QCD altogether. Just as the many-body problem in condensed-matter physics is not built on QED, so also the many-body problem in strong-interaction physics needs a theoretical description that is more efficient than QCD.

The formulation of a problem that involves many particles requires the language designed for treating many-body systems. When the basic dynamics responsible for multiparticle production is unclear, or is believed to be less important that the statistics of the problem, then statistical or stochastic methods may be more appropriate. Whether or not this attitude is correct cannot be determined a priori; it is only by phenomenology can one learn whether a statistical approach can capture the main features of multiparticle production. A number of attempts have been initiated in recent years to describe the soft production processes by statistical methods.[40-43] An important feature of the data is that the KNO scaling of multiplicity distribution is broken by the results of the UA5 group experimenting at the CERN-Sp$\bar{p}$S.[44] Generally speaking, it is hard for a purely statistical model to explain such an energy dependence. Thus, in some models a new component must be introduced. The analysis of data by the negative binomial distribution results in an energy dependence of the k parameter [45] that is physically unacceptable. Phenomenology therefore led some to a more general picture in which there is a mixture of chaos and coherence.[41-43] Whatever the interpretation may be, what is not controversial is that soft hadronic processes involve many strands of physics; thus it is not wise at an exploratory stage to deal with too many of them at once.

The discussion below focuses on one issue: why is there KNO scaling in the CERN-ISR energy range? We avoid the issue of the violation of KNO scaling here for the reason just mentioned above, not because we cannot handle it. In fact, that problem has already been successfully treated in the same framework but with the inclusion of jet production also.[46,47] As a first step in the construction of that framework, we have shown that Furry branching together with impact-parameter smearing can give rise to an excellent description of KNO scaling,[48] which is the topic we now consider. Our emphasis will be on the conceptual basis and some details not fully discussed in the literature, so that this review serves to supplement the published papers, to which the reader is directed for details.

5.2 Furry Branching in Central Production.

Before we address the question of what does the branching, if not quarks and gluons, we first ask: why Furry branching, and not other types of stochastic processes? Although the answer should ultimately rely on what works, we can indicate some hints without elaborate phenomenology. In order to have KNO scaling the normalized dispersion γ_2 of the multiplicity distribution $P_n(\bar{n})$ should not change with energy $\sqrt{s}$, and therefore $\bar{n}(s)$. Let us compare three commonly discussed distributions: Poisson, Polya, and Furry. (It is interesting to note that all three have been considered in detail in connection with the study of cosmic radiation.[14]) Their normalized dispersions are

$$\text{(a)} \quad \text{Poisson:} \qquad \gamma_2 = 1/\bar{n}, \qquad\qquad\qquad (3.1.9)$$

$$\text{(b)} \quad \text{Polya:} \qquad \gamma_2 = 1/k + 1/\bar{n}, \qquad\qquad (3.3.8)$$

$$\text{(c)} \quad \text{Furry:} \qquad \gamma_2 = 1/k - 1/\bar{n}. \qquad\qquad (3.4.15)$$

Poisson's γ_2 decreases with increasing s. For Polya (or negative binomial) distribution γ_2 can be constant only if k decreases, a behavior which has been regarded as being unphysical. Only for Furry distribution can γ_2 have a reasonable chance of being constant, provided that k increases as $\bar{n}$ does, a behavior that has to be given physical interpretation. The s dependence of γ_2 in (3.4.15) was first noted by B. Durand and S. Ellis [12] in the context that with fixed k the increase of γ_2 might befit an explanation for the KNO scaling violation. However, the problem of scaling violation is far more complex than that observation alone can solve, since apart from the fact that the higher multiplicity moments increase dramatically with s in ways unaccountable by the simple Furry distribution,[49,50] the strong increase in the forward-backward multiplicity correlation at the CERN collider precludes any naive identification of the observed $P_n(\bar{n})$ with either the Polya or the Furry distribution.[46] Thus again, we return to our point of view that any discussion about the KNO scaling violation is premature until one first understands the mechanism for KNO scaling itself.

So far we have only indicated the motivation for considering Furry branching on the grounds that γ_2 has a chance of being constant . That alone is, of course, not sufficient. A number of questions must be answered, if one is to take the branching process seriously. First, what is the "time" t in the evolution equation (3.4.2)? Second, what _does_ the branching , and what is k? Third, the fluctuation in multiplicities should surely be influenced by the uncertainties in the impact parameters of the collisions. How would that affect branching?

A stochastic evolution equation such as (3.4.2) involves real time, which has no comparable place in particle physics. Combining (3.4.7) and (3.4.14), we have

$$\beta t = \ln (\bar{n}/k) \qquad\qquad (5.2.1)$$

which relates t to the energy $\sqrt{s}$ through $\bar{n}(s)$. We may therefore interpret the evolution equation as follows.[51] If the multiplicity of a hadronic reaction is n at s, then at s + Δs there is one more particle produced, which is related to the n particles at s by being the branching product of one of the n particles. Stochasticity enters the problem when we allow any one of the n particles to have an equal chance in branching. At t = 0, the Furry distribution has the initial condition $F_n^k(0) = \delta_{nk}$. A strict application of the above interpretation would mean that at some low s (the lowest value at which we can still use Furry branching) there are exactly k particles produced. That is a very special initial condition, too restrictive to be physically applicable except in particular models such as the fireball model.[40] It is therefore more useful if we relax the above interpretation and emphasize more the analogy with QCD branching. Let us then interpret Furry branching for soft processes as follows.

At some fixed energy s, an average of $\bar{n}(s)$ particles are produced. During hadronization there is an evolution process leading from k high-mass clusters to the n final particles. The branching in each one of those clusters may be viewed in much the same way as in the degradation of virtuality in the branching of a jet, as discussed in the previous section. In this picture s plays the same role as the virtuality Q^2 that initiates a hard process. Note the similarity between (5.2.1) and (4.2.1). One may then imagine an evolution parameter (call it τ) that increases along the branches as the cluster masses decrease; the value of τ is determined by the average multiplicity (of branching clusters and/or particles) at each stage of the evolution in accordance to (5.2.1). Thus τ need not be thought of as time; it is a parameter that orders the particle production process, just as the variable ln ln q^2 does in a tree diagram in jet physics. In this interpretation a line in the interior of a tree diagram need not be identified with a field off-shell. Instead, it represents an on-shell cluster which can subdivide. Thus quarks and gluons have no role to play at this level, consistent with our view expressed earlier; the ϕ^3 tree diagrams are used mainly as a convenient device for graphical organization.

The branching process can only describe what happens in going from k clusters to n particles. The value of k must be specified from other considerations. Naively, one would expect k to be an integer. But there are two physical reasons why it need not be so. One is that the Furry distribution F_n^k should be used only in a statistical sense. A possible point of view is to attribute a distribution in k to the "initial condition", so that the effective k value, k_{eff}, can be non-integral. A brief discussion of this is given in Sec.

5.4. In fact, a Poisson distribution in k has been tried in Ref.49 to fit ISR
and S$\bar{p}$pS data using a modified Furry distribution in n. However, as
mentioned earlier, our concern here is not KNO scaling violation. The other
reason for us to consider non-integral k is related to the impact parameter b.
The number of inital clusters must depend on b in some way. Since b is a
continuous variable that must be integrated over, it is necessary to regard k(b)
as a smooth function of b and to give its non-integral value a probabilistic
interpretation.

In dealing with b-smearing we shall have to specify k(b), which is to
be related to the eikonal function Ω(b). One of the central problems to
understand in soft hadronic processes is, I believe, the connection between
k(b) and Ω(b). Even in an approach that does not involve Furry branching,
there should be some quantity that describes the productivity of particles at b.
How that is related to the opacity at b is at the heart of the problem of treating
multiparticle production in a geometrical picture.

Finally, we remark that there is no deductive way to argue in support
of using Furry branching in soft hadronic processes. Lacking a workable
theory, we can only examine all the salient issues and rely on phenomenology
to affirm its applicability. The one guide post we have is that in hard
processes perturbation method has shown us that the production mechanism
is bascially Furry branching. But Furry branching does not depend on the
smallness of the coupling constants. Thus it is hoped that in the limit of low
momentum transfers, for which the coupling constants become so large as to
render the perturbation method useless, Furry branching may still be a valid
approach to describe the multiplicity distribution.

5.3 Geometrical Branching Model

Having discussed the conceptual issues at length in the above subsection, we now briefly discuss the geometrical branching model (GBM), which to my knowledge is the only model, built in the geometrical framework and endowed with Furry branching, that gives a quantitatively successful description of the multiplicity distribution ,[48] forward-backward multiplicity correlation,[46] and rapidity-interval dependence [52] in the ISR energy region. By being geometrical, the model possesses also the virtues of the geometrical description of elastic scattering.[53] We shall focus only on the KNO scaling feature of the model, since it relies on branching in a crucial way.

The key point in the GBM is that branching takes place in every collision at any impact parameter. Thus in a geometrical scaling description we have

$$P_n = \int_0^\infty dR^2 \, [1-e^{-2\Omega(R)}] \, F_n^{k(R)}(w) \qquad (5.3.1)$$

where R is the scaled impact parameter, $b/b_0(s)$, and $\Omega(R)$ the eikonal function, which is known from elastic differential cross section.[53] The crucial link between the geometrical structure of hadrons and the branching mechanism of particle production is the quantity $k(R)$. If we denote

$$<n> = \sum_n nP_n, \qquad \bar{n}(R) = \sum_n nF_n^{k(R)} \qquad (5.3.2)$$

then $<n> = \{\bar{n}(R)\}$ where the curly brackets stand for $\int dR^2[1-\exp-2\Omega(R)]$. Now define

$$<k> = \{k(R)\} \qquad (5.3.3)$$

so that we can write

$$k(R) = <k> h(R) \qquad (5.3.4)$$

where $h(R)$ is a normalized distribution in R, satisfying $\{h(R)\} = 1$. In principle, all three quantities in (5.3.4) can depend on s; however, we shall regard $h(R)$ as being essentially independent of s. This assumption fits in conveniently with the properties of geometrical scaling; indeed, in the absence of any knowledge about $h(R)$, we further assume that it has the form

$$h(R) = h_0 \, \Omega^{\gamma}(R) \qquad (5.3.5)$$

where γ is an adjustable parameter. This equation relates the relative number of initial clusters at R to the opacity of hadrons at R. It was found [48] for $\gamma = 0.3 \pm 0.05$ that not only does P_n exhibit KNO scaling behavior, but also that the KNO scaling curve for pp collision at ISR is well-fitted. For such a small value of γ, $h(R)$ falls off slowly with R^2. At $R^2 = 2$, $h(R^2)$ is about half of $h(0)$, while $\Omega(R^2)$ is roughly only 5% of $\Omega(0)$. Thus we learn that it is rather easy to produce particles even at large R^2 where the opacity is small. This is in accord with the expectation that collisions at large R^2 should join smoothly with diffractive processes, and that there can be a large number of particles produced in diffractive dissociation. Another way of stating the result is that the KNO distribution is broad due to impact parameter smearing, in which a wide range of the impact parameter contributes to the multiplicities measured.

To get the energy independence of the KNO distribution, we also need to give $<k>$ an s dependence so that C_2 is constant, where $C_p = <n^p> / <n>^p$. [The constancy of C_p for $3 \le p \le 5$ follows in an intricate way from the chosen value of γ.] It is straightward to show that [48]

$$<k> = [<n>^{-1} + C_2 - \{h^2\}]^{-1} \qquad\qquad (5.3.6)$$

Thus $<k>$ increases with $<n>$, but saturates asymptotically. This is physically very reasonable, since, as s increases, the average number of initial clusters should increase when more and more of the hadronic overlap is excited to dense matter distributed in rapidity (before branching), until the whole system in a particular overlap is excited. The average number of initial clusters may saturate, but the excitation may reach higher and higher cluster masses, so that more branching can take place within each cluster, and $<n>$ can increase without limit. Our phenomenological result indicates that $<k>$ is roughly between 4 and 4.5 in the ISR range. For $R^2 = 2.5$, we have $h(R^2) \approx 0.5$, so $k(s,2.5)$ is about two clusters. This matches very well with diffractive dissociation. For $R^2 = 0$, we have $h(0) \approx 1.5$ and $k(s,0)$ is about 6 to 7, which is a large number covering a total rapidity range of about 5. Thus after branching the produced particles from neighboring clusters are expected to overlap in rapidity, leading to high multiplicity correlation between adjacent rapidity windows, compared to the correlation between windows that are separated by a gap. This is, of course, what has been observed in the correlation experiments.[46,54]

The GBM has been extended to higher s by the inclusion of jet production in a way that is free from the arbitrariness in the definition of jets.[46,47] Using the data on σ_{el} and σ_{tot} at one energy ($\sqrt{s} = 540$ GeV) to determine the magnitude of the component that has hard scattering, we have been able to provide a coherent description of a number of observed phenomena thoughout the CERN collider energies, including the violation of geometrical scaling and KNO scaling, and the increase of forward-backward multiplicity correlation. Taken all together, we believe that this is the best phenomenological evidence that one can find to support the use of Furry branching in multiparticle production.

5.4 The Fokker-Planck Approach

In the foregoing subsections the emphasis has been on the physics issues of soft hadronic production. As a final topic of review, we discuss now a more mathematical development of stochastic evolution, which may have relevance to soft processes. It involves the use of the Fokker-Planck equation, initiated by Biyajima [55,35] and later extended by Shih.[56] Although those authors claim contact with QCD branching, inappropriately named for Giovannini's evolution equations ,[7] the proper significance of the approach is the interesting mathematical relationship between the Fokker-Planck equation, (2.2.2), and the general evolution equation, (3.0.4), involving both creation and annihilation transitions.

The key device of this approach is the Poisson transform,[57,58,17] often used in quantum optics,[59,60]

$$P_n(t) = \int_0^\infty dx\ f(x,t)\ p_n(x\bar{n}) \tag{5.4.1}$$

where $p_n(x\bar{n})$ is the Poisson distribution (3.1.3)

$$p_n(x\bar{n}) = e^{-x\bar{n}}\ (x\bar{n})^n/n! \tag{5.4.2}$$

If we denote (5.4.1) by $P_n(t) = \langle p_n(x\bar{n})\rangle_t$, then we have

$$P_{n-1}(t) = \langle p_n(x\bar{n})\ n\ /x\bar{n}\ \rangle_t \tag{5.4.3}$$

$$P_{n+1}(t) = <p_n(x\bar{n}) \; x\bar{n} / n+1) >_t \qquad (5.4.4)$$

Now, let $f(x,t)$ satisfy the Fokker-Planck equation, (2.2.2),

$$\frac{\partial}{\partial t} f(x,t) = -\frac{\partial}{\partial x} [A_1(x)f(x,t)] + \frac{1}{2} \frac{\partial^2}{\partial x^2} [A_2(x)f(x,t)]. \qquad (5.4.5)$$

Then from (5.4.1) and by integration by parts, we have

$$\frac{\partial}{\partial t} P_n(t) = <p_n(x\bar{n}) \{A_1(x)(-\bar{n} + n/x) + \frac{1}{2}A_2(x) [-\frac{n}{x^2} + (-\bar{n} + \frac{n}{x})^2]\}>_t \quad (5.4.6)$$

Let

$$A_1(x) = \mu_1 - \mu_2 x, \qquad A_2(x) = 2\,\mu_3 x. \qquad (5.4.7)$$

With the help of (5.4.3) and (5.4.4), we thus obtain (3.0.4), rewritten here for convenience:

$$\frac{d}{dt} P_n(t) = a_{n+1}P_{n+1}(t) + c_{n-1} P_{n-1}(t) - (a_n+c_n) P_n(t) \qquad (5.4.8)$$

where

$$a_n = (\mu_2 + \mu_3 \bar{n})n, \qquad c_n = (\mu_1 + \mu_3 n)\bar{n} \qquad (5.4.9)$$

This establishes the connection between the two evolution equations, (5.4.5) and (5.4.8), for general μ_1, μ_2 and μ_3.

We have seen in Sec. 3.3 that the negative binomial distribution is the stationary solution of (5.4.8) for $a_n = \alpha n$ and $c_n = \beta n + \gamma$, specified in (3.3.1). Comparison with (5.4.9) yields $\alpha - \beta = \mu_2$ and $\gamma = \mu_1 \bar{n}$. It then follows from (3.3.7) that $\mu_1 = \mu_2$. Furthermore, one also gets $k = \gamma/\beta =$

μ_1/μ_3. Using the freedom in renormalizing t in (5.4.5), we may choose $\mu_1 = \mu_2 = 1$. Thus, in summary, we have in this case

$$A_1(x) = 1 - x, \qquad \tfrac{1}{2}A_2(x) = x/k \tag{5.4.10}$$

The solution of (5.4.5) with (5.4.10) has been discussed by Biyajima, [55] along the line first developed in quantum optics. The stationary solution is

$$f(x,t) \to \psi_k(x) = \Gamma(k)^{-1} k^k x^{k-1} e^{-kx} \tag{5.4.11}$$

which is the well-known form (gamma distribution) necessary to give the negative binomial distribution P_n^k through Poisson transform, and is also the scaling form of $\bar{n}P_n^k$ in the limit of $\bar{n}/k \gg 1$, $x = n/\bar{n}$.[61] The resurgence of interest in the use of the negative binomial distribution to describe the multiplicity distributions in hadronic collisions was spurred by Carruthers and Shih[41] with surprising longevity despite the rise and fall of the KNO scaling property. Since the subject is well covered by their recent review article, [17] no more will be added here.

We turn next to Furry branching, for which $a_n = 0$ and $c_n = \beta n$, as required in (3.4.1). From (5.4.9) we therefore obtain $\mu_1 = 0$, $\mu_2 = -\mu_3\bar{n} = -\beta$. It then follows from (5.4.7) that

$$A_1(x) = \beta x, \qquad A_2(x) = 2\beta x / \bar{n} \tag{5.4.12}$$

Although this establishes the equivalence of (5.4.5) and (3.4.2), unfortunately the Furry distribution $F_n^k(t)$, given in (3.4.12), does not correspond to any particular solution f(x,t) of the Fokker-Planck equation. The reason is that the initial condition $F_n^k(0) = \delta_{nk}$ cannot be specified by any corresponding condition on f(x,0) through the Poisson transform. However, Furry branching can still be a sensible physical process, if we relax the stringent initial condition to a Poisson distribution, i.e.

$$F_n^k(0) = p_n(k) \qquad (5.4.13)$$

Whether this is reasonable or not depends on the physics problem at hand. Mathematically, (5.4.13) allows the Fokker-Planck approach to become immediately available with $f(x,0) = \delta(x - k\bar{n})$. The t dependence of such a solution and the phenomenological application of the solution (without the concern for any other physics issues) have been considered by Biyajima and coworkers.[49]

6. CONCLUSION

We have shown how the production of many particles in both hard and soft processes can be described by branching. The situation is, however, not totally satisfactory. On one hand, in hard processes the connection with branching is established through well-defined perturbation theories, which unfortunately cannot produce results that agree with the present data. On the other hand, in soft processes the connection is made only in the framework of a phenomenological model, and there is no solid theoretical basis for it, despite the good description of the data. Nevertheless, the status of our collective understanding of multiparticle production has never been better.

The suggestion made in this article that the branching process is the basic mechanism underlying all multiparticle production processes, if proven to be correct, provides some real hope for an eventual theory, or at least the direction in which such a theory is to be found. For the first time we are beginning to see grounds for unification. Perturbative high-Q^2 processes and non-perturbative low-p_T reactions can now be joined through Furry branching. Dynamical and statistical approaches share a common ground in the mathematical description. The s-channel tree diagrams can be related to the t-channel dual diagrams so that the possibility of incorporating the string approach is now open. Since Furry branching involves the production of clusters which hadronize through branching, contact with the cluster models should not be hard to establish. Finally, being geometrical in its formulation, the GBM certainly is intimately related to a whole area of work based principally on the geometrical approach.

On the experimental side, it is unfortunate that at the highest energies now available at the hadron colliders the production of jets has caused a great deal of confusion and controversy among those working on minimum-bias physics. In principle, high-p_T jet fragmentation and initial-state bremsstrahlung are also branching processes. But to demonstrate experimentally and convincingly that the production mechanism for the low-p_T particles is indeed of the Furry type, it is necessary to separate out the hard component. Such a separation affects the soft component due to the non-linearity of unitarity, so the geometrical scaling component cannot be isolated without introducing s dependence. In other words, the no-jet component, measured by, say, the UA1 group, is not the scaling component. Nevertheless, in the GBM it is that component whose multiplicity distribution is the impact-parameter smearing of the Furry distribution. A direct check of that at the high-energy colliders is conceivable, though rather difficult.

Phenomenological investigations of the effects of branching at all energies are only in the beginning stage. The theoretical problems of relating

the branching process to the other approaches mentioned above essentially
have not yet begun. There is now some hope for a broad understanding of
the physics of hadrons in high-energy collisions. Thus at this point we
have good reasons for optimism.

ACKNOWLEDGMENT

I am grateful to many physicists who have helped me at various stages
to understand the subject discussed here, or with whom I have collaborated.
They are, among others, Wei Chen, Harry Lam, Bob Mazo and Kiusau
Tesima. This work is supported in part by the U.S. Department of Energy
under Grant no. DE-FG06-85ER40224-A001.

REFERENCES

1. W. Feller, An Introduction to Probability Theory and its Applications,
    (Hohn Wiley, N.Y. 1950).
2. E. Parzen, Stochastic Processes, (Holden-Day, San Francisco, 1962).
3. N.G. van Kampen, Stochastic Processes in Physics and Chemistry,
    (North-Holland, Amsterdam, 1981).
4. S.K. Srinivasan, Stochastic Theory and Cascade Processes, (Elsevier,
    N.Y., 1969).
5. A. Ramakrishnan, in Handbuck der Physik, ed. S. Flugge, vol. III/2,
    (Springer-Verlag, Berlin, 1959), p.524.
6. C.W. Gardiner, Handbook of Stochastic Methods, (Springer-Verlag,
    Berlin, 1983).
7. A. Giovannini, Nucl. Phys. B 161, 429 (1979).
8. D. Amati, S. Fubini, and A. Stangellini, Nuovo Cimento 26, 826
    (1962).

9.   W. Knox, Phys. Rev. D$\underline{10}$, 65 (1974).

10. P. Carruthers, in Strong Interactions and Gauge Theories, Proc. of the 21st Rencontre de Moriond, Les Ares, 1986, (Editions Frontieres, 1986), p.229.

11. K. Shimoda, H. Takahashi, and C.H. Townes, J. Phys. Soc. Japan $\underline{12}$, 686 (1957).

12. B. Durand and S.D. Ellis, in Design and Utilization of the SSC, Snowmass 1984, Proc. of the 1984 Division of Particles and Fields Summer Study, Snowmass, Colorado, 1984 edited by R. Donaldson and J.G. Morfin (Fermilab, Batavia, Illinois, 1985), p.234.

13. W.H. Furry, Phys. Rev. $\underline{52}$, 569 (1937).

14. N. Arley, Stochastic Processes and Cosmic Radiation, (John Wiley, New York, 1948), p.92.

15. A. Giovannini, Nuovo Cimento $\underline{10}$A, 713 (1972).

16. K. Konishi, A. Ukawa, and G. Veneziano, Nucl. Phys. B$\underline{157}$, 45 (1979).

17. P. Carruthers and C.C. Shih, Int. J. Mod. Phys. A$\underline{2}$, 1447 (1987).

18. G. Altarelli and G. Parisi, Nucl. Phys. B$\underline{126}$, 298 (1977).

19. J.C. Taylor, Phys. Lett. $\underline{73}$ B, 85 (1978).

20. D.C. Hinz and C.S. Lam, Phys. Rev. D$\underline{33}$, 3256 (1986).

21. R.C. Hwa and C.S. Lam, Mod. Phys. Lett. A$\underline{1}$, 303 (1986).

22. A. Capella, U. Sukhatme, C.I. Tan, J. Tran Thanh Van, Phys. Lett. $\underline{81}$B, 68 (1979); A.Capella, U. Sukhatme, J. Tran Thanh Van, Z. Physik C$\underline{3}$, 329 (1980); G. Cohen-Tannoudji, A. El-Hassouni, J. Kalinowski, O. Napoly and R. Peschanski, Phys. Rev. D$\underline{21}$, 2689 (1980); H. Minakata, Phys. Rev. D$\underline{20}$, 1656 (1979).

23. R.C. Hwa and C.S. Lam, Phys. Rev. D$\underline{34}$, 3370 (1986).

24.   A. Bassetto, M. Ciafaloni, and G. Marchesini, Nucl. Phys. B163, 477 (1980).

25.   A. Bassetto, M. Ciafaloni, and G. Marchesini, Phys. Rep. 100, 201 (1983).

26.   A.H. Mueller, Phys. Lett. 104 B, 161 (1981); B.I. Ermolaev and V.S. Fadin, JETP Lett. 33, 269 (1981).

27.   A. Bassetto, M. Ciafaloni, G. Marchesini, and A.H. Mueller, Nucl. Phys. B207, 189 (1982).

28.   G. Marchesini and B.R. Webber, Nucl. Phys. B238, 1 (1984).

29.   E.D. Malaza and B.R. Webber, Nucl. Phys. B267, 702 (1986).

30.   M. Derrick et al. Phys. Lett. 165B,449 (1985).

31.   K. Tesima, Phys. Lett. 133 B, 115 (1983).

32.   T. Munehisa and K. Tesima, Nucl. Phys. B277, 575 (1986).

33.   K. Tesima, OITS-367 (1987).

34.   B. Durand and I. Sarcevic, Phys. Rev. D36, 2693 (1987).

35.   M. Biyajima and N. Suzuki, Phys. Lett. 143B, 463 (1984).

36.   For a comparative survey, see R.C. Hwa, in Proc. of the Shandong Workshop on Multiparticle Production, Jinan, China, 1987, edited by R.C. Hwa and Xie Q.-B. (World Scientific, Singapore, 1988).

37.   B. Andersson, in Proc. of the Shandong Workshop, loc. cit.; see also his article in this volume.

38.   A. Capella, in Proc. of the Shandong Workshop, loc. cit.; see also his article in this volume.

39.   R.C. Hwa, in Partons in Soft-Hadronic Processes, edited by R.T. Van de Walle (World Scientific, Singapore, 1981), p.137; and in Short-Distance Phenomena in Nuclear Physics, NATO ASI Series, vol. 104,

edited by D.H. Boal and R.M. Woloshyn (Plenum Press, N.Y., 1983), p.79.

40. Liu L.-S. and Meng T.-C., Phys. Rev. D27, 2640 (1983); Chou K.-C., Liu L.-S., and Meng T.-C., ibid. D28, 1080 (1983); Cai X., Liu L.-S., and Meng T.-C., ibid. D29, 869 (1984); Cai X., Chao W.-Q., Meng T.-C., and Huang C.-S., ibid. D33, 1287 (1986).

41. P. Carruthers and C.C. Shih, Phys. Lett. 127B, 242 (1983); 137B, 425 (1984); P. Carruthers, in Proc.of the Workshop on Local Equilibrium in Strong Interaction Physics, Bad Honnef, W. Germany, 1984, edited by D. Scott and R. Weiner (World Scientific, Singapore, 1985), p.390.

42. G.N. Fowler and R.M. Weiner, Phys. Rev. D17, 3118 (1978); G.N. Fowler, E. Friedlander, R.M. Weiner and G. Wilk, Phys. Rev. Lett. 56, 14 (1986); R.M. Weiner, in Hadronic Matter in Collision, Proc. of the LESIP II, Santa Fe, N.M., 1986, edited by P. Carruthers and D. Strottman (World Scientific, Singapore, 1986), p.106.

43. M. Biyajima, Prog. Theor. Phys. 69, 966 (1983); ibid. 70, 1468 (1983).

44. G.J. Alner et al., Phys. Lett. 138B, 304 (1985); 167B, 476 (1986).

45. G.J. Alner et al., Phys. Lett. 160B, 193 (1985); 167B, 476 (1986).

46. R.C. Hwa, OITS-354 (1987), to be published in the Phys. Rev.

47. W.R. Chen and R.C. Hwa, OITS-372 (1987) to be published in the Phys. Rev.

48. W.R. Chen and R.C. Hwa, Phys. Rev. D36, 760 (1987).

49. M. Biyajima, T. Kawabe, and N. Suzuki (to be published); se also M. Biyajima in Proc. of the Shandong Workshop, loc. cit

50. C. Maxwell, Phys. Lett. 185B, 192 (1987).

51.   C.S. Lam and M.A. Walton, Phys. Lett. 140B, 246 (1984).

52.   R.C. Hwa, OITS-369 (1987).

53.   T.T. Chou and C.N. Yang, Phys. Rev. 170, 1591 (1968); A.W.
       Chao and C.N. Yang, Phys. Rev. D8, 2063 (1973).

54.   S. Uhlig et al., Nucl. Phys. B132, 15 (1978).

55.   M. Biyajima, Phys. Lett. 137B, 225 (1984); 139B, 93 (1984).

56.   C.C. Shih, Phys. Rev. D33, 3391 (1986).

57.   P.L. Kelley and W.H. Kleiner, Phys. Rev. 136, A316 (1964).

58.   A. Giovannini, Nuovo Cimento 15A, 543 (1973); M. LeBellac, J.L.
       Meunier, and G. Paut, Nucl. Phys. B62, 350 (1973).

59.   R.J. Glauber, Phys. Rev. 131, 2766 (1963); in Physics of Quantum
       Electronics, edited by P.L. Kelley, B. Lax, and P.E. Tannenwald
       (McGraw Hill, N.Y., 1966).

60.   J.R. Klauder and E.C.G. Sudarshan, Fundamentals of Quantum
       Optics (W.A. Benjamin, N.Y. 1968).

61.   L. Mandel, Proc. Phys. Soc (London), 74, 233 (1959).

Multiparticle Production at High Energy and Superclustering Processes

Chia Chang Shih

Department of Physics and Astronomy

University of Tennessee

Knoxville, TN 37996-1200

Abstract

We have investigated multiplicity distribution as a supercluster in
hadronic production processes. This process creates unstable clusters
at intermediate stages and hadrons at the final stage. It includes
Poisson-weighted distributions and negative binomial distributions as the
special cases. It also includes modified gamma distribution for intermediate neutral clusters decay, impact parameter smearing, and multiplicity projection to a rapidity window. While a discussion of the
conservation of energy is presented through a model of bremsstrahlung
analogy, the implication of local charge conservation is examined via a
classical clustering process. Aspects of quantum statistics are formulated as superclustering distributions in the quantum mechanical functional space.

Section I.   Introduction

The past decade has seen an explosive increase in the energy available involving particle collisions. Developments in theory and experimentation have led to universal acceptance of quantum chromodynamics as the basic strong interaction mechanism. Accompanying this increasing collision energy is an ever-increasing production of soft secondary particles. Despite the existence of many extensive experiments. a comprehensive understanding of the dynamics of the soft hadron systems is however still lacking.[1]

A wide range of phenomenological formulations currently exist for soft hadron processes. However, many tend to differ substantially in the interpretation of the dynamical contents of the experimental information. We believe much of the success of these phenomenological formulations stems from the stochastic nature of the strongly interacting production processes, rather independent of the details of the individual representation. It is therefore constructive to investigate common features shared by different representations. In this way we hope to screen out uninteresting statistical backgrounds and eventually focus on intrinsic collective behaviors of the many-body system.

In this article we concentrate on a technical aspect of the multi-particle distribution.[2] We review the concept of superclustering in relation to hadronic multiplicity distributions at high energies. The word superclustering is used to distinguish this phenomenon from clustering, because the latter has rather different meaning to different

616

people. We define superclustering as a two- or multistage processes for global multiplicity distributions. At the first stage of the production process, individual clusters are produced according to a given statistical law. For example, clustering may be described by partially coherent (or even sub-Poissonian) distributions. At the second stage, the clusters are considered as the sources of the particle production. The corresponding distributions may then be as general as the clustering distribution just mentioned. Hadronic distribution is therefore a multistage clustering phenomenon.

The concept of clustering is not at all new among multiparticle production theories.[3,4] Quite often an investigation of the short-range two-body correlations between the produced particles also leads to the construction of local neutral clusters that are intimately related to the clustering properties that we are interested in.[5] However, we would like to emphasize global and semi-global characteristics of production processes, including three-body and higher order correlations.

In this paper we shall restrict ourselves to models that can essentially be calculated analytically. There are many interesting dynamical models such as the Lund Model[6], the dual parton model[7-8], quark combination models[9], and thermodynmical QCD "gas" models[10] that contain features of superclustering properties. Because a considerable amount of numerical analysis is needed for the discussion of these models, we shall not review them in this paper. We hope that by evaluating and reviewing simple models in the following sections, we may still obtain a fairly accurate intuition concerning the importance and even the necessity of incorporating many aspects of superclustering distribution in a realistic model of the production processes.

In section II we shall briefly review formulation concerning super-clustering probability distribution, its moments and its KNO scaling[1,2]. In Section III we shall focus on the superposition of the Poisson distribution as a special case of superclustering distributions[2,11-12]. This leads to a Poisson-transform representation of the overall distribution and is very useful for the study of multi-plicity distributions.[1,2] In Section IV we shall discuss the negative binomial distribution as a superclustering distribution[13]. We shall demonstrate the intrinsic freedom and ambiguity at the level of individual subprocesses in constructing a negative binomial distribution for the overall distribution.[2] In Section V, we review a recent formulation of a modified gamma distribution[14], which includes analytic expressions obtained here enable us to examine explicitly properties of the interme-diate cluster distributions without numerical analysis. It raises some doubt as to the interpretation of the phenomenological results. In Sec-tion VI, we present a simplified impact parameter smearing model[12,15-16] and compare it with previous analyses. This comparison demonstrates the ease of constructing phenomenological models that are qualitatively satisfactory. In Section VII, we review the probability distribution of the multiplicity distribution in a rapidity window.[17-19] This allows us to examine the compact properties of binomial distribution as a statistical projection of the probability distribution.[20-21] In section VIII, we use a simplified version of bremsstrahlung analogy to demonstrate a simple example on the constraint

618

due to energy conservation.[22] In section IX, we concentrate on the consequences of global charge conservation.[23] In Section X, we shall reformulate quantum stochastic processes as superclustering distributions[1,24]. We shall conclude with remarks on further investigations along this direction.

Section II. Formulation

A. Probability distributions

We shall first define the probability of having c clusters as p_c with $\Sigma\, p_c = 1$. Once a cluster is created, its subsequent evolution into hadrons is, by definition, independent of the other clusters. The probability of creating n_j particles in the j-th cluster is given by $f_j(n_j)$ with $\Sigma f_j(n_j) = 1$. In a total multiplicity measurement, neither p_c nor f_j are observed directly. What can be measured is the sum total of all the particles produced by all the clusters. We need therefore to relabel the particles in terms of an overall index N and evaluate

$$P_N = \Sigma\, p_c \,\Sigma\, f_1(n_1)\ldots f_c(n_c)\; \delta(n_1+\ldots+n_c-N) \tag{1}$$

However, for most of the general cases, analytic calculations are tedious and cumbersome. There is no obvious short cut for analysis other than numerical evaluations. Since we are interested in the general properties of the overall distribution P_N for semi-quantitative studies, we shall assume that the same distribution governs the evolution of each cluster, i.e.,

$$f_j(n) = f(n), \qquad j = 1, \ldots, c. \tag{2}$$

The relationships between the f_j, P_c, and P_N can now be expressed indirectly through the generating function

$$G(x) = \Sigma \, x^N \, P_N = g_c(y), \tag{3}$$

where

$$g_c(y) = \Sigma \, y^c \, P_c, \tag{4}$$

and

$$y = g_f(x) = \Sigma \, x^n \, f(n), \tag{5}$$

Even for simple f_c and $f(n)$, the associated P_N are often rather complicated. It is sometime easier to work on the various statistical moments.

B. Moments

Moments of the distribution functions can be calculated from Eqs.(3) to (5) through relationships such as

$$\begin{aligned}
\xi_M &= \langle\langle N(N-1)\ldots(N-M+1)\rangle\rangle \\
&= \partial^M \, G(x) / \, \partial x^M \big|_{x=1}
\end{aligned} \tag{6}$$

For example,

$$<<N>> = <c> <f> \tag{7}$$

Throughout this paper, we shall designate $<<...>>$ and capital letters to refer to an overall superclustering system. we shall also express the relationships between the moments in terms of the normalized cumulant moments Γ_j. These Γ_j's are most sensitive to the shape of the overall distributions in terms of the scaled variables and are useful in comparing different probability distributions at different $<<N>>$. For example,

$$\Gamma_2 = \gamma_{2c} + \gamma_{2f}/<c> \tag{8}$$

$$\Gamma_3 = \gamma_{3c} + 3\gamma_{2c}\gamma_{2f}/<c> + \gamma_{3f}/<c>^2 \tag{9}$$

$$\Gamma_4 = \gamma_{4c} + 6\gamma_{3c}\gamma_{2f}/<c> + 3\gamma_{2c}\,\gamma_{2f}^2/<c>^2 + 4\gamma_{2c}\gamma_{3f}/<c>^2 + \gamma_{4f}/<c>^3 \tag{10}$$

Here the lower cases γ_{jc} and γ_{jf} refer to the normalized cumulant moments corresponding to p_c and $f(n)$. Moments of higher orders can be calculated in a straightforward fashion and are not included here. It is already clear that the overall distribution P_N is always broader than the individual distributions of p_c or $f(n)$ as a scaling distribution. This can be reflected in inequalities relationships among various cumulant moments such as

$$\Gamma_j < \gamma_{cj}, \ \gamma_f/<c>^{j-1} \tag{11}$$

C. Scaling limit

Upon considering the scaling distribution $\Psi = \langle\langle N\rangle\rangle P_N$, different probability distributions can be compared as a function of $Z = N/\langle\langle N\rangle\rangle$. $\Psi(Z,\langle\langle N\rangle\rangle)$ is directly related to the generating function $G(x)$ through

$$G(1-x/\langle\langle N\rangle\rangle) = \int e^{-x}\Psi(Z)dZ \tag{12}$$

Thus Ψ, the inverse Laplace transform of $G(-\langle\langle N\rangle\rangle(1-x))$, is a function of Z and $\langle\langle N\rangle\rangle$. Whether or not the KNO limit exists, (Ψ independent of $\langle\langle N\rangle\rangle$, when $\langle\langle N\rangle\rangle = \infty$ with constant Z), Eq(1) predicts the violation of the KNO limit at finite $\langle\langle N\rangle\rangle$. There is no need to interpolate Ψ for P_N at finite $\langle\langle N\rangle\rangle$.[25] Notice also that a KNO limit does not require both $\langle c\rangle$ and $\langle f\rangle$ to be ∞. Thus the scaling behavior of p_c or $f(j)$ can be very different from P_N.

In the following, we shall discuss many different forms of p_c and f_j. They are motived by various physical arguments and conservation require-ments.

Section III. Poisson superposition

A. Poisson superposition and Poisson transform

In stochastic processes, Poisson distribution occupies a very unique position. A subprocess can be Poissonian if the individual occurrence satisfies certain criteria such as independency, stationarity and linearity. A Poisson distribution can also be identified with the diagonal matrix element of a coherent state in its number space. Thus the $f_j(n)$ distribution may be a Poissonian distribution in many circumstances. Furthermore, convolution of Poisson distributions is a Poisson distribution, whether or not $<n_j>$'s are independent of j. If the f_n's are Poissonian, we can always write

$$P_N = \Sigma \; p_c \; e^{c<n>} \; (c<n>)^N/N! \tag{13}$$

And in the case of a continuous c,

$$P_N = \int dx \; F(x) \; e^{-x<N>} \; (x<N>)^N/N! \tag{14}$$

The above expression has been referred to as the Poisson-transform distribution and has been used extensively both in the phenomenology of hadronic multiplicity distributions of strong interaction and in photon distributions of quantum optics. In the KNO scaling limit, $\psi(Z)$ is exactly $F(Z)$. The asymptotic KNO scaling function is therefore also the kernel of the Poisson transform representation. In this type of distribution the asymptotic KNO function $\psi(Z)$ leads to a precise definition of P_N at finite $<<N>>$ through Eq(13). Since there is no ambiguity in defining P_N, there is no need of further guessing the P_N at finite $<<N>>$. In other words, KNO scaling violation is automatically built to in Eq(13) and (14).

An important example of this group is the negative binomial (NB). Using F(x) as a simple exponential

$$\int F(x) \ dx = \int dx_1 \ldots x_k \ e^{-(x_1 + \ldots + x_k)} \tag{15}$$

we obtain

$$P_N = K^K / \Gamma(K)(<<N>>/K)^N / (1 + <<N>>/K)^{(K+N+1)} \tag{16}$$

a negative binomial of order K. The above very natural way of introducing a negative binomial is however not unique. Using a more general expression for $f(n)$, we can also obtain negative binomial distributions without using the Poisson distribution as the fundamental subprocess. Because negative binomials have received much attention as a useful and convenient way in representing experimental data, we shall focus on them in the next section.

Section IV. Negative binomial and superclustering

Recent successes of negative binomial distributions have encouraged renewed interests in the origin of this distribution. For example, Giovannini and Van Hove have emphasized the equivalence between the NB and a special class of superclustering distribution with a logarithmic $f(n)$.[3] The possibility of logarithmic behavior for the individual clusters has subsequently been investigated [26-27].

In order to have a negative binomial, its generating function must be in the form

$$G(x) = (1 + <<N>>(1-x)/K \ )^{-K} \tag{17}$$

624

If we assume that p_c is Poissonian, we also have

$$g_c(x) = \exp(-<c>(1-x)) \tag{18}$$

Eq.(3) then determines uniquely the generating function of $f(n)$ in the form

$$g_f(y) = 1-K(\ln(1+<<N>>(1-y)/K)/<c> \tag{19}$$

leading to the expression for $f(n)$

$$f(n)= K/<c>(<<N>>/(K+<<N>>))^n /n \ , \ n\neq 0 \tag{20}$$

and

$$f(0) = 1-K\ln(1+<<N>>/K)/<c> \tag{21}$$

Even with this strong assumption on the form of $f(n)$, it is not possible to relate $<c>$ directly to $<<N>>$. This is achieved through the additional requirement that $f(0) =0$, leading to

$$<c> = K \ \ln(1+<<N>>/K) \tag{22}$$

and

$$<n> = <<N>>/(K\ln(1+<<N>>/K)) \tag{23}$$

Even though the above choice allows us to eliminate one free parameter, the physical reason for such a choice is rather ambiguous. Certainly the possibility exists of an individual cluster decaying completely through neutral particles. Whether or not the probability distribution of $f(n)$ takes on a simple form after excluding the $n=0$ component should be determined phenomenologically.[27]

   The non-uniqueness of the construction of the NB in this approach is actually more serious than what is indicated above. There is in fact an infinite number of choices of the functional form of $f(n)$, all leading to a NB for P_N. For example, takings a $k = 1$ negative binomial for g_c,

$$g_c(y) = (1+<c>(1-y))^{-1} \tag{24}$$

we get

$$g_f(y) = 1 + K(1-(1+<<N>>/K(1-y))^K)/<c> \tag{25}$$

A choice of $f(0) = 0$ leads to

$$<c> = (1+<<N>>/K)^K -1 \tag{26}$$

which does not behave logarithmically. As long as p_c is narrower than the required distribution of P_N, a distribution of $f(n)$ can be found.

Section V. Modified gamma distribution and neutral clusters

Recently a modified gamma distribution in conjunction with binomial distribution has been proposed by Goulianos for the hadronic multiplicity distributions[14]. The numerical analysis in this work is extensive and successful. Since this model is a good example of superclustering distributions, we shall include more specific formulation in the following.

Let us first assume that for total multiplicity distributions the value of the allowed N is even only ($N = 2N'$). Furthermore, each pair of the charged particles is the decay product of c neutral clusters according to a binomial distribution, instead of the modified Gamma distribution in Ref.14. That is,

$$P_N = \Sigma\, p_c\, B(c|N';q)\, \delta(N-2N') \tag{27}$$

where $B(c|N';q)$ is the binomial distribution given explicitly by

$$B(c|N';q) = c!/(N'!(c-N')!)\, q^{N'}(1-q)^{c-N'} \tag{28}$$

Explicit relations can be established through the generating functions of P_N and p_c. I.e.,

$$G_e(x) = \Sigma\, P_{2N}'\, x^{2N'} \;=\; \Sigma\, p_c \Sigma\, B(c|N',q)x^{2N'}.$$

$$= g_c(y) \tag{29}$$

where $y = x^2 q + 1 - q$.

In terms of the moments, we get

$$<<N>> = 2q <c> \ ,$$
(30)

$$<N(N-1)>> = 4 \ q^2<c(c-1)> + 2q<c>.$$
(31)

This leads to

$$\Gamma_2 = \gamma_{2c} + 2(1-q)/<<N>>$$
(32)

Higher moments can be worked out in a similar fashion. But the second cumulant Γ_2 is sufficient for our discussion. Since Γ_2 is always larger than γ_{2c}, the KNO distribution of $<<N>>P_N$ is always wider than the corresponding $<c>p_c$. At large $<<N>>$, the asymptotic KNO distribution of $<<N>>P_N$ is controlled by $<c>P_c$. As was noticed in Ref.[14] , with $q = 1/2$, the second term gives $1/<<n>>$, similar to that of a negative binomial. However, at finite $<<N>>$, the scaling violation contribution from the first term of Eq(32) is not zero. Notice that whenever $<<N>>$ is finite, the value of $<c>$ is also finite (unless q tends to 0). There is an important contribution due to the finite scaling violation of $<c>P_c$.[2] Empirically it is inconsistent to use a negative binomial distribution for $<c>p_c$.

One way to avoid this difficulty is to assume an approximate scaling form for the clustering distribution $<c>P_c$. In Ref.14, Goulianos used the Gamma distribution for $<c>P_c$ evaluated at discrete points of z. Thus

$$<c>p_c \sim \mu^\mu/\Gamma(\mu)z^{\mu-1}e^{-\mu z} \tag{33}$$

Here p_c is almost normalized to 1 for the values of $<n>$ of interest.

Section VI. Impact parameter smearing and superclustering

Multiplicity distribution can also be investigated through geometrical scaling.[9,15-16]   In this formulation an eikonal function $\Omega(s,b)$ $=b/b_0(s)$ is introduced so that

$$\sigma_{in} = \pi \int db^2 \ (1-\exp(-2\Omega(s,b)) \tag{34}$$

If the multiplicity distribution at a fixed s and b is given by $f_n(s,b)$, the overall distribution P_N is

$$P_N = \int dR^2 \ p_R \ f_N(s,b) \tag{35}$$

with $p_R = 1 - \exp(-2\Omega(R))$.

Impact parameter smearing (IPS) is reflected in the fact that the shape of $f_N(s,R)$ is a function of s and R, so that

$$<N(s,R)> = \Sigma N f_N(s,b) \neq <<N>> \tag{36}$$

The remaining task is to visualize a proper choice of f_N. In this spirit Hwa et al. have used a Furry process for the underlying dynamics of

f_N[15]. Here, we shall use a Poisson process instead.[19] Geometrical aspects of the impact parameter formulation is however kept as close as possible to the parametrization of [18] for the sake of comparison. We shall use

$$\Omega = \Omega_0 \exp(-\beta R^2) \tag{37}$$

with $\Omega_0 = 1.4$ and $\beta = 1.624$. This leads to

$$P_N = 1/\beta \int_0^{\Omega_0} d\Omega/\Omega(1-\exp(-2\Omega)) <N(s,R)>^N/N! \ e^{-<N(s,R)>} \tag{38}$$

In terms of the moments, we get

$$\xi_M = <<N(N-1)...(N-M+1)>> = 1/\beta \int_0^{\Omega_0} d\Omega/\Omega p_R <N(b,R)^M> \tag{39}$$

Again using the parametrization of [15]

$$<N(s,R)> = h_0 \ \Omega^\gamma(R) \tag{40}$$

we get

$$\xi_M = <h^M> = 1/\beta \ h_0^M \int d\Omega \ p_R \ \Omega^{\gamma M-1} \tag{41}$$

The above $<h^M>$ are also evaluated in [15], but their relationships to the overall moments of P_N are different.

In order to produce a similar overall P_N distribution between a Furry process and a Poisson process, we need a slightly stronger smearing fluctuation. This is because a Poisson distribution is narrower than the Furry distribution. Indeed, a value of $\gamma = 0.47 \pm 0.3$ instead of 0.45 of Ref.[15] leads to values of ξ_M rather close to experimental data up to the IRS energies. When these values are translated to the moments $C_M = <<N^M>>/<<N>>^M$, the numerical values for M ≤5 are similar to the values obtained in [15]. The apparent agreement reflects the fact that IPS leads to qualitatively satisfactory agreement of ξ_M between ISP models and the data up to IRS energies. On the other hand, we would like to caution that a comparison between the moments can be deceptive, since the C-moments are typically not very insensitive to differences in models. The best test is to compute P_N directly. Short of that, we should like to choose moments such as the normalized cumulant moments Γ_M, which are somewhat more sensitive than the C_M.

The situation is similar at 540 Gev of the collider energy. In Ref.[16] a slightly different parametrization of Ω is used, with the subprocess $f_n(s,b)$ also changed from a Furry distribution to a negative binomial. There the numerical fitting to the C-moments are quite reasonable, but the resultant fits of P_N are too low in the peak region of P_N. In the work of Ref.[12] yet another parametrization of Ω is adopted in conjunction with a Poisson subprocess. The fitting of the P_N at 540 Gev is better than that of Ref.[15]. Similar fits at 200 and 900 Gev are also quite reasonable. Further investigation along this direction for KNO scaling violation[1] should be very desirable.

Section VII. Conditional distribution and rapidity window

Given a set of total multiplicity distributions, various selection
rules can be given to examine the correlation and fluctuation of the
multiplicity distribution of a subset of data. The net effect of this
selection is often a statistical projection of the original distribution,
which is in itself a statistical essembly. It is unavoidable that within
the window of the subset, the multiplicity distribution is a superclustering distribution.

One of the simplest statistical projections of the distributions is
the binomial distribution. This distribution is essentially observed
phenomenologically for e+e- production processes[28]. We shall now
identify p_c as the distribution $P_T(N_T)$ of the total event, and P_N as the
distribution of the number of particles N_W in the subspace window. Thus

$$P_W(N_W) = \Sigma\ P_T(N_T)\ B(N_T|N_W;q) \tag{42}$$

Here $B(N_T|N_W;q)$ is the binomial distribution characterized by a parameter
q. I.e.,

$$B(N_T|N_W;q) = N_T!/(N_W!(N_T-N_W)!)q^{N_W}(1-q)^{(N-N_W)} \tag{43}$$

Expressed in terms of the corresponding generating function, we get

$$G_W(x) = \Sigma\ P_W(N_W)\ x^{N_W} \tag{44}$$

where

$$G_T(x) = \Sigma \, P_T(N_T) \, x^{N_T} \tag{45}$$

Thus the binomial projection often leads to a distribution of N_W identical to that of N_T (form invariant)[18]. For example, if P_N is a k-th NB, we get for the generating function

$$G_W(x) = (1 + N_W(1-x)/k)^{-k} \tag{46}$$

leading to a k-th NB for the N_W also. There are a variety of statistical projections similar to the binomial relationship above. In the work of Mong et al.[17] $B(N_T|N_W;q)$ is replaced by $B(N_T/2|N_W/2;q)$. Eq.(46) can be generalized to clusters of size K. [19,21]. It is similar in structure to the isospin projection of Eq(28) and the forward backward hemisphere projection of Ref.[20,29] $B(N_T|N_W;q)$ can also be generalized to $B(N_{T+}|N_{W+N};q_+) \, B(N_{T-}|N_{W-};q_-)$ for a separate projection among the + and - charges.[18,30] It would then correspond to the weakest form of global charge correlation.

The main feature of the statistical binomial projection of Eq(42) can be summarized through the moment Γ_2,

$$\Gamma_2 = 1/k + 1/N_W \tag{47}$$

Thus the KNO distribution is broader for N_W than for N_T. The above analysis can be further generalized to trinomial distributions.[15,31] Experimentally it is known that as the narrower rapidity window is narrowed, the $<N_W>P_W(N_W)$ scaling distribution becomes progressively wider than the prediction of Eq(46) and (47).[25,32-33] A simple binomial projection is therefore not very satisfactory. Eq(43) needs to be further convoluted. We shall refer to Ref.[15] for some typical analysis in this direction.

Section VIII. Bremsstrahlung analogy and energy conservation

Recently Bialas et al. have emphasized the effect of energy conservation to the multiplicity distribution[19]. Because for hadronic processes the leading particles and the neutrals always carry a finite amount of energy, soft hadronic multiplicity distribution may be considered as the superclustering distribution compounding the distribution of the effect energy W_{ef} and the hadronic distribution with a given effective energy W_{ef}. We shall now identify the cluster index c with the continuous variable W_f, and discuss the distribution of N for a given W_{ef}. To demonstrate this effect, we shall use a model of longitudinal phase space, in the same spirit of the bremsstrahlung analogy of Bialas et al. [22], and the partition temperature of Chou and Yang [34]. For a given W_{ef}, we shall assume that the probability of producing N charged hadrons is given by

$$P_N(W_{ef}) = c \; \Sigma \int_{|p_m|} \Pi \; \beta \; dp_j/e_j \; \delta(\Sigma p_j - W_{ef})/N! \tag{48}$$

Here we have assumed a symmetrical interval of p_j, so that total momentum $\Sigma p_j \neq 0$. In order to proceed analytically, we shall further approximate p_j by e_j, and use $e_j = x_j^2$. Then

$$P_N(W_{ef}) = c \ (2\beta)^N \int \Pi \ dx_j/x_j \ \delta(\Sigma x_j^2 - W_{ef})/N! \qquad (49)$$

The above expression can be integrated explicitly in terms of generalized spherical coordinates in the limit of large W_{ef}, we then obtain

$$P_N(W_{ef}) = c' \ a^N/N! \qquad (50)$$

with

$$a \sim 2\beta \ln(W_{ef})$$

Because $P_N(W_{ef})$ is normalized to 1, $c' = e^{-a}$. We therefore obtained a Poisson distribution. Superposition and convolution of $P_N(W_{ef})$ follows exactly the same formulation as that of the previous sections of this article. The only difference is that p_c is now visualized as the distribution for the overall energy of the charged articles.

We would like to emphasize that the resultant Poisson distribution of eq(48) is a reflection of a crucial N! factor, which was essentially put in by hand. Had we counted the states differently and excluded the N! factor in Eq.(49), Eq(50) would then be replaced by a negative binomial.[35]

In the analysis [22] a factor p_j/e_j was not replaced by 1 as in Eq.(49). Because $e_j > p_j$, the net effect is to suppress slightly the high N events, leading to a slightly subPoissonian distribution for $P_N(W_{ef})$.

The analysis of this section can be extended to include a QCD estimate of the inelasticity effect. In this content the factors that compound and convolute the $P_N(W_{ef})$ is given by the gluon distribution from perturbative QCD. Since numerical calculation is needed, we shall refer the work of Weiner et al. in Ref.[36]

Section IX. Clusters and global charge conservation

In this section we shall restrict ourselves to a classical analysis of charge conservation and its implication on the charge content of intermediate charged clusters. The easiest process, which is consistent with both stochasticity and charge conservation, is the stochastic production of intermediate neutral clusters, that branch into two or more charged pairs and neutrals.[14,23] The fluctuation of n_{ch} then reflects both the intrinsic fluctuation of the number of neutral clusters n_{cl} and the decay processes. Given the probability of producing a cluster in an interval dy as $dp_{cl} = \lambda \, dy$, and with $n_{cl} = 0$ at $y = 0$, we get the well-known Poisson distribution in n_{cl},

$$P(n_{cl}) = exp(-\lambda y)(\lambda y)^{n_{cl}}/n_{cl}!$$

This may be taken as the fundamental subprocess of the hadronic collision processes, but it is not a likely candidate for the underlying mechanism of e+e- processes. This is because it predicts somewhat too large a fluctuation in n_{ch} than a Poisson distribution. For example, the moment γ_2 is given by

$$\gamma_2 = 1/ <n_{cl}> = 2 / <n_{ch}> > \gamma_2(\exp) \sim 1/ <nch>. \tag{51}$$

Alternatively, we may first relax the constraint due to charge conservation. We shall assume that the probability of producing a + or − charge in an interval dy is independently $dp(\pm) = \lambda(\pm) dy = \lambda dy$. This leads to the branching equations for the probability $P(n_+, n_-)$ of producing n_+ positive and n_- negative charges

$$dP(n_+, n_-)/d(\lambda y)= P(n_+-1, n_-) + P(n_+, n_--1) -2 P(n_+, n_-) \tag{52}$$

With $P(0,0) = 1$ at $y=0$, we obtain a joint Poisson distribution

$$P(n_+, n_-)= e^{-\lambda y} (\lambda y)^{n_+}/n_+! e^{-\lambda y}(\lambda y)^{n_-}/n_-! \tag{53}$$

Since Poisson distributions are form invariant under convolution, the distribution in n_{ch} is also Poissonian. i.e.,

$$P_{ch}(n_{ch}) = e^{-2\lambda y}(2\lambda y)^{n_{ch}}/n_{ch}! \tag{54}$$

The values of $P_{ch}(n_{ch})$ are therefore quite close to the experimentally observed values in e+e- systems. On the other hand, there is a finite probability P_Q for the net charge $Q =n_+ - n_-$ to be nonzero. The conditional probability $P(n_{ch}|Q)$ for fixed Q is given by

$$P(n_{ch}|Q) = P_Q^{-1} \Sigma\ e^{-2\lambda y}(\lambda y)^{n_{ch}}/((n_{ch}+Q)/2)!/(n_{ch}-Q)/2)!) \qquad (55)$$

At this point, we may be tempted to simply ignore the other $Q \neq 0$ terms as being unphysical. But all $Q \neq 0$ states are actually coupled to the $Q = 0$ states through the branching relationships represented in Eq.(52). Their suppression would also invalidate the prediction in $P(n_{ch}|Q=0)$. Difficulty of this kind is common in many phenomenological analyses of the previous sections.[37]

Since charge is conserved event by event, any unbalanced charge Q has to be associated with the source of hadronization within an event. To count the total number of charges, we need to include the charge content of the source as well. A charge state (n_+,n_-) should then be assigned to a $(n>,n>)$ state, where $n> = max(n_+,n_-) = n_{ch}/2 = n_{cl}$. The probability of $P_{ch}(n_{ch})$ is now given by

$$P_{ch}(n_{ch})= P(n>,n>) + 2\ \Sigma\ P(n>,n_-) \sim e^{-\lambda y}\ (\lambda y)^{n_{cl}}\ /n_{cl}! \qquad (56)$$

Notice that Eq(56) is essentially a Poisson distribution in $n_{ch}/2 = n_{cl}$, and is somewhat too broad for the e+e- systems. We may consider the above type of branching equations Eq(2) to be applied to individual components (such as jets) of the total process. If charge compensation can be achieved between net charge of the components, the multiplicity distribution of n_{ch} may remain similar to Eq(56). Otherwise, we would expect an even broader distribution than Eq(56).

The above argument suggests that correction to the random walk of the charge imbalance Q is nontrivial, and constraints on Q should lead to rather different results in $P_{ch}(n_{ch})$. Two considerations are needed.

We should modify the structure of the fundamental subprocesses, such as the Poisson distribution. We should also allow the possibility of clustering of the sub-processes. To demonstrate this more concretely, we shall examine the case where the emitting source is allowed to be singly charged only. Random walks between n_+ and n_- are therefore restricted to ± 1. Eq.(4) is now replaced by .

$$dP(n,n)/d(\lambda y) = -2P(n,n) + P(n-1,n) + P(n,n-1) \tag{57}$$
$$dP(n+1,n)/d(\lambda y) = -P(n+1,n) + P(n,n)$$
$$dP(n,n+1)/d(\lambda y) = -P(n,n+1) + P(n,n)$$

With $P(0,0) = 1$, we get $P(n+1,n) = P(n,n+1)$. The solution of Eq(7) can be expressed in terms of the generating functions

$$Q_E(x,y) = \Sigma\, x^{2n}\, P(n,n) = ((s_+ + \lambda)e^{s_+ y} - (s_- + \lambda)e^{s_- y}) /(s_+ - s_-) \tag{58}$$
and
$$Q_0(x,y) = \Sigma\, x^{2n+1}\, (P(n+1,n)+P(n,n+1)) = 2\lambda x(e^{s_+ y} - e^{s_- y})/(s_+ - s_-)$$

where
$$s_{1,2} = 3\lambda(3 \pm (1+8\lambda x^2)^{1/2})/2$$

Since the net charge of the source is ± 1, it is more sensible to re-assign both $(n+1,n)$ and $(n,n+1)$ state to $(n+1,n+1)$ state. We then get

$$Q(x,y) = \Sigma\, x^{2n}\, P_{ch}(n_{ch}) = Q_E(x,y) + x\, Q_0(x,y) \tag{59}$$

characterized by

$$\langle n_{ch} \rangle = 4\lambda y/3 + 8(1-e^{-3\lambda y})/9 \sim 4\lambda y/3 \qquad (60)$$

$$\gamma_2 \sim 1/\langle n_{ch} \rangle + 1/(9\langle n_{ch} \rangle)$$

The distribution represented by Eq(59) is almost as narrow as a Poisson distribution, and is rather close to what is observed experimentally in e+e- processes

Since $Q(x/\langle n_{ch} \rangle) \sim \exp(-x)$, $\langle n_{ch} \rangle P_{ch}(n_{ch})$ shows an asymptotic delta function dependence on $z_{ch} = n_{ch}/\langle n_{ch} \rangle$, and does not satisfy KNO scaling. It is interesting to observe that the requirements of a nearly neutral source have led the distribution of the individual charged sectors $P_+(n_+)$ = $P_-(n_-) = P_{ch}(n_{ch})$ to be subPoissonian.[38] Eq.(57) can be generalized to allow a distribution of charged states for the radiation source. A more complicated supercluster distribution follows. On the other hand, classical branching processes involving more complicated birth and death subprocesses usually lead to n_{ch} fluctuations wider than a Poisson process.[23,39]

Section X. Quantum statistics and superclustering

Considering the stochastic process of hadronic production as a quantum process, the statistics of multiplicity distribution would in general be superclustering distribution in our formulation. This is be-cause any stochastically produced entity would also display quantum fluctuation. We shall now consider two examples, both of which have been successfully applied in photon statistics of quantum optics[40-42] and the hadron statistics of strong interaction.

A. Partial coherent laser distribution

Consider first the secondary hadrons interacting to an stationary source, we may construct an effective Hamiltonian of the quantum system as a forced harmonic oscillator.[1,43] Furthermore, if the coupling is linear, the resultant excited state of the hadronic matter is a coherent state $|\alpha>$, with $a|\alpha> = \alpha|\alpha>$, where a is the destruction operator of the hadron. Since the state $|\alpha>$ is not an eigen state in the hadronic number state $|n>$, we expect fluctuation n. For example, the distribution of n is related to the diagonal matrix element $<n|\alpha><\alpha|n>$ and is a Poisson distribution. In a more realistic system, the strength of the source is also fluctuating. This may be expressed through a density matrice &ho.(α) in $|\alpha>$. Thus

$$P_n = \int \rho(\alpha) \, d^2\alpha \, <n|\alpha><\alpha|n> \tag{61}$$

$$= \int \rho(\alpha) \, d^2\alpha \, \alpha^n/n!e^{-\alpha} \tag{62}$$

It is clear that in this example, the cluster distribution p_c can be identified as a distribution in the complex plane of α. For example, if the phase information is not important, and ρ is a Gaussian

$$\rho(\alpha) = \exp(-|\alpha|^2)$$

we get the Bose-Einsten distribution (k = 1 NB),

$$P_n = <n>^n/(1+<n>)^{-n-1}, \quad <n> = |\alpha|^2. \tag{63}$$

On the other hand, if the phase information is not completely destroyed, so that

$$\rho(\alpha) = \exp(-|\alpha-\beta|^2/N)/\pi N \tag{64}$$

we would get a partial coherent laser distribution,(PCLD)[1,40-41] with

$$P_n = <n>^n/(1+<n>)^{n+1}\exp(-S/(1+N))L_n(-S/((1+N)) \tag{65}$$

where $S = |\beta|^2$ $<n> = S + N$, and L_n is the Laguerre polynomial. Here S and N are the signal and noise content of the quantum mechanical system. The above expression is a very convenient interpolation between the Poisson formula as $S \to 0$ and the Bose-Einstein formula (k =1 negative binomial)as $N \to 0$. The above distribution can also be convoluted to a k cell formula.

Technically, if we are confined to the information on the total multiplicity distribution, it is rather difficult to distinguish the PCLD distribution with other more classical distributions,

B. Jaiswal and Mehta formulation

Jaiswal and Mehta formulation(JM) is a natural generalization of the classical photon production process.[42] It was introduced to soft pion production by Fowler and Weiner.[44] With a given strength of radiation field coupled to the production of the secondaries, the distribution of the secondaries is assumed to be Poissonian. On the other hand, the field of radiation contains in itself both a coherent and a chaotic component extending throughout the whole functional space. It is therefore necessary to average over the functional space of the entire radiation field. We get

$$P_n = <<P(W)exp(-W)dW/n!>>_W \tag{66}$$

$$W = \alpha \int_0^T |V(t)|^2 \, dt \tag{67}$$

Here the average over the radiation field is expressed through two component fields

$$V(t) = V_S(t) + V_N(t) \tag{68}$$

where V_S stands for a coherent source field and V_N stands for a chaotic noise field, which satisfies

$$<V_N(t)V_S(t')> = \Gamma(|t-t'|) \tag{69}$$

The above expression is rather difficult to evaluate analytically for P_n, and is analyzed in terms of the factorial moments $\mu_M = <<N(N-1)...(N-M+1)>>$. We shall refer to Ref.[44] for more details of the expressions on the μ_M.

Only in the case of a totally chaotic source, $(V_S(t)= 0)$, with a Lorentz spectrum,

$$\Gamma(|t-t'| = \Gamma_0 \, e^{-\gamma|t-t'|} \tag{70}$$

an analytic expression of the generating function G(x) is available in
the literature,[45]

$$G(x) = e^{\gamma}/(\cosh(z)+(\gamma T/z+z/\gamma T)\sinh(z)/2), \tag{71}$$

where

$$z= (2\gamma T<n>(1-x) +\gamma^2 T^2)^{1/2}$$

For total multiplicity distributions, JM distribution leads to reasonable representation of the data similar to the situation with the PCLD
distributions. On the other hand, JM formulation contains in itself the
richness of the quantum mechanical two-body correlation structure. Further investigation is needed to determine the applicability of this unique
quantum mechanical model of the production process.

Section XI. Conclusion

In this paper we have examined Basic aspects and formulations of
superclustering distribution for the hadronic multiplicity distributions
at high energies. The examples we presented here cover a wide range of
simple phenomenological models.

Superclustering distribution is clearly an unavoidable feature of the
strongly interacting many-body hadronic system. Even though the formu-

lation tends to be somewhat complicated, it permits vest amount of freedom in constructing a reasonable superclustering distribution. In fact the superclustering formulation is so flexible, it becomes difficult to determine the representation uniquely without exploring correlations at the semi-global level.

Consideration of conservation requirements such as conservation of energy and conservation of charge remains one of outstanding challenges of any stochastic formulation. In a sense, the separation between a dynamical and a stochastic input becomes less obvious. There is also the outstanding question of the importance of the quantum mechanical effects in a stochastic production process. We hope the investigation of super-cluster processes eventually leads to a better understanding of the strong interacting many-body systems.

Acknowledgments

This research was supported in part by National Science Foundation Grant NO. NSF phy-8417526 and by The Science Alliance , a Stat of Tennessess Center of excellence , through the University of Tennessee, Knoxville. The author has benefited greatly from discussions with Drs. P. Carruthers, R. M. Weiner, L. Van Hove, R. C. Hwa, C. S. Lam, T. Meng, M. Biyajima and K. Goulianos.

Reference

(1) P. Carruthers and C. C. Shih, Modern Phys. A2 (1987) 1544.

Modern Phys. Letts. A 2 (1987) 89; Phys. Lett. 127B (1983) 242.

(2) C. C. Shih, Phys. Rev. D34 (1986) 2710.

(3) S. Pokorski and L. Van Hove, Acta Phys. Polo. B5 (1974) 229.

(4) P.N. Dremin and E. L. Feinberg, Sov. J. Part. Nucl. 10 (1979) 394.

(5) G. Giacomelli and M. Jacob, Phys. Rep. 55 (1979) 1.

(6) B. Andersson and W. Hofmann Phy. Lett 169B (1986) 364.

B. Anersson et al. Phys. Rep 97 (1983) 31.

(7) P. Aurenche and F. W. Bopp, Zeit. fur Phys.C13 (1982) 459.

(8) A. Capella, J. kwiecinski and J. Tran Than Van, Phy.Rev.Lett 58 (1987)2015

(9) Qu-bing Xie, Wen-chuan Mo and Yu-fa Li Shandong Univ. preprint (1987)

(10) P. Carruthers and I. Sarceivic, Phys. Lett.192B (1987) 321.

(11) J. Finkelstein, UCSJ, preprint (1987).

(12) D. Na. Huang and E. Yen, National Tsin Hua Univ., preprint (1987).

(13) A. Giovannini and L. Van Hove, Z. Phys. C30 (1986) 391.

(14) K. Goulianos Phys. Lett. 193B, (1987) 151.

(15) W. R. Chen and R. C. Hwa, Phys. Rev. D36 (1987) 760.

(16) C. K. Chew, S. Date' and D. Kiang Modern Phy. Lett. A1(1986) 553.

(17) Chao Wei-qin, gao Chong-shou, Meng Ta-chung and Pan Ji-cai, Fachbereich

Physik preprint, FUB-HEP/87-1

(18) C. C. Shih, Univ. of Tenn., preprint (1987). window

(19) D. W. Huang National Tsing Hua Univ. preprint (1987).

(20) T. T. Chou and C. N. Yang, Phys. Lett. 135 (1984) 175.

(21) Chao Wei-qin, Gao Chong-shou, Meng Ta-chung and Pan Ji-cai,

646

Fachbereich Physik preprint (1987).

(22) A. Bialas, E. H. Groot and T. W. Ruijgrok Phys. Rev. D36 (1987) 752.

(23) C. C. Shih, Univ. of Tenn., preprint (1987).

(24) G. F. Fowler, E. M. Friedlander, R.M. Weiner, G. Wilk, Phys. Rev. Lett.56(1986)14.

(25) R. Szwed and G. Wrochna, Z. Phys. C29 (1985) 255.

(26) R. Szwed, G. Wrochna, and A. K. Wroblewski, Univ.of Warsaw preprint,(1987)

(27) M. Adamus et al. NA22 Cloob, Phy. Lett 177B (1986) 239.

(28) M. Derrick, et al,Phys.Lett.168B (1986) 299; Phys.Rev D34,(1986) 3304.

(29) P. Carruthers and C. C. Shih, Phy. Lett. 165B (1985) 209.

(30) C C Shih and P. Carruthers preprint (1987).

(31) Cai Xu, Chao Wei-qin, Meng Ta-Chung and Huang Chao-shang,Phy.Rev.D33 (1986)1287, W Chao, T. Meng, and J. Pan Phys. Lett.B176,(1987) 211.

(32) J. Alner et al. (UA5 Collaboration) Phy. Lett. 151B (1985) 309.

(33) M. Adamus et al. (NA22 Collaboration) preprint (1987) HEN-284.

(34) T. T. Chou, C. N. Yang, and E. Yen Phy.Rev.Lett.54 (1985) 510; Phys. Rev. Letts 167B(1986) 453.

(35) C. S. Lam, Phys. Rev. D28 (1983) 1228.

(36) G. F. Fowler, A. Vourdas, and R. W. Weiner Phy.Rev.D35 (1987)870.

(37) C. C. Shih, p. 141 in "Hadronic Matter in Collision", Proc. of the 2nd. Intern.Workshop on Local Equilibrium in Strong Interaction physics. ed. P. Carruthers, and D. Strottman (World Scientific, Singapore, 1986).

(38) C. C. Shih, Phys. Rev. D34 (1986) 2710.

(39) M. biyajima and N. Suzuki Prog. Theo.Phys. 73 (1985) 918.

(40) R. J. Glaiber, p.788 in "physics of Quantum Electronics" ed. P. L. Kelly, B. Lax and P. E. Tamnwald (mcGraw-Hill, NY 1966).

(41) G. Lachs, Phy. Rev. 138 (1965) 1012.

(42) A. K. Jaiswal and C. L. Mehta, Phys. Rev. A2, (1970) 168.

(43) P Carruthesr and C. C. Shih Phy. Lett. 137B (1984) 425.

(44) G. N. Fowler and R.M. Weiner, Phys. Lett. 70B,(1977) 201;D17,(1978)3118.

(45) G. Bedard, Phys. Rev. 151 (1966) 1038.

FUB-HEP/87-13

"Fireball" Models and Statistical Methods in Multiparticle Production Processes

Meng Ta-chung

Fachbereich Physik der FU Berlin

Berlin, Germany

* Supported in part by Deutsche Forschungsgemeinschaft (DFG: Me 470/5-1, 5-2)

1. Introduction

Before the large accelerators were built, multihadron production processes could only be observed in cosmic-ray experiments. It is therefore not surprising that many of the concepts, ideas, and methods used in describing such processes have their origins in cosmic-ray physics. In particular fireball models (also known as two center models) have already been suggested and extensively discussed in this connection by Cocconi[1], Takagi[2] Niu[3] and Ciok et al.[4,5] and many others[6] in the 1950's.

Section 2 of the present paper is a survey of the fireball idea. It contains a brief discussion on the experimental and theoretical background of this idea — including a discussion on the question: "What is the essence of the fireball models?". Section 3 is a report on a data analysis. Here, the stochastic aspect of multiparticle production processes is discussed; and its importance is pointed out. It is shown in particular that standard statistical methods can be used to describe the rapidity and azimuthal angle dependence of multiplicity and those of transverse energy distributions. Section 4 is a summary of the statistical model proposed by the Berliner group. Some concluding remarks are made in Section 5.

2. "Fireballs"

The word "fireball" in connection with multiparticle production processes was first used by Cocconi [1] in his 1958 article on "Empirical model for ultrarelativistic nucleon-nucleon collisions". It is now used not only by comic-ray physicists, but also by particle physicists and by nuclear physicists, although its meaning is, unfortunately, model-dependent.

This term is usually associated with a specific physical picture for the final state, and/or with a given scenario for the process in which multiparticle production takes place. For example, according to the "tentitative definition" given by Miesowicz in his 1971 review article[5]:
" A fireball is a group of mesons, which have been produced in the high-energy interaction of hadrons. The momentum distribution of the constituent mesons in the rest system of this group

corresponds to an (isotropic) phase space distribution, the average energy of the mesons being about 0.5 GeV ". It is not difficult to imagine that this definition can be, and in fact has been, modified in various ways in order to describe other (in particular, new) experimental facts and/or to incorporate with different theoretical models, especially with statistical-hydrodynamical models[7]. For example: In order to take the observed short-range rapidity-correlations into account, one replaces the word "mesons" by "hadronic clusters which decay into mesons and/or baryon-antibaryon pairs". In order to fit the multiplicity distribution, the rapidity-distribution, the correlation, and other data, one proposes suitable distributions for the hadronic clusters in phase space, and/or modify the original two-fireball- or "two-center"- picture by changing the number of fireballs in such processes[8]. (We note that the word "fireball" has a number of synonyms, for example "nova", "conglomerate", "cluster", "super cluster" , "clan", etc. .) Here, we do not discuss in detail the definitions and models which have been proposed in the past. The interested readers are referred to the papers in Refs. 1-8 and the papers cited therein. In our discussion, we concentrate our attention on the question: What are the characteristic features of the fireball models which have stood the test of time?

A careful examination shows that the answer to this question should be the following. Multiparticle production processes in high-energy collisions can be envisaged to take place in two stages: the formation stage, and the decay stage of excited systems — "fireballs". The two stages can be considered as independent processes, in the sense that the decay depends only on the properties of the excited systems but not on the specific way in which the systems are formed.

The striking similarity between the characteristic features of the fireballs and those of the compound nuclei, suggests that multiparticle production processes in high-energy collisions are predominantly statistical processes.

3. Data analyses and standard statistical methods

We show in this section that a number of recently observed striking phenomena in multiparticle production processes are pure statistical effects.

First example:

Rapidity-dependence of multiplicity distributions in high-energy collisions have been studies by UA5 Collaboration[9] and other experimental groups[10–13]. They measured the distributions $P_w(n_w; s)$ of charge multiplicity n_w in different rapidity windows w for hadron-hadron[9,10], lepton-lepton[11], lepton-hadron[12] and hadron-nucleus[13] reactions at various total cms energies $\sqrt{s}$. They found in particular that, for every one of these reactions at a given $\sqrt{s}$, the distribution for every rapidity window w can be described by a two-parameter fit (negative binomial fit with parameters $\langle n_w \rangle$ and k_w for every w) . Many models[14] have been proposed to describe the data and/or to interpret the corresponding parameters. It is observed that, although the dynamical details of these models are very much different, all the results seem to agree well with the data.

Chao, Pan and myself carried out a systematic analysis[15] of the above-mentioned multiplicity distributions data. The purpose of the analysis was to see, how much dynamical details is really needed, in order to understand the observed rapidity-dependence. The method we used to approach this problem is a very simple one:

Suppose we know the multiplicity distribution P(n;s) of a given hadron emitting system, and the probability $q_w(s)$ of observing one (charged, nor negatively charged) hadron in the rapidity window w is $q_w = \langle n_w \rangle / \langle n \rangle$, where $\langle n \rangle$ is the average multiplicity of the emitting system and $\langle n_w \rangle$ is the corresponding value in the window w. If the dynamical details do not play an important role in understanding the rapidity-dependence of the multiplicity distributions, then, $P_w(n_w; s)$ should be calculable from $P(n; s)$ and $q_w(s)$ by using standard statistical methods. The result of our analysis shows that it is indeed the case. The binomial distribution

$$P_w(n_w; s) = \sum_n P(n; s)\binom{n}{n_w} q_w^{n_w}(1 - q_w)^{n-n_w} , \qquad (1)$$

652

and its generalization for cases in which more emitting systems contribute indeed give a good description of the data. Some of the results are shown in Figs.1-3. Further results and detailed calculations can be found in Ref.15 . The following empirical fact should also be mentioned in this connection. The multiplicity distributions in limited rapidity intervals at ISR energies have been presented in a very recent paper by Giacomelli[16] (see Fig.4). It is very interesting to see that the distributions for the intervals $2.5 < |\eta| < 3.0$, $1.5 < |\eta| < 2.0$, and $0.5 < |\eta| < 1,0$ fall on one curve. In our opinion, this is simply due to the fact that the rapidity distribution of the observed charged hadrons is more or less flat in this region $|\eta| < 3$, and hence the q_w-values [in Eq.(1)] should be the same when the corresponding rapidity intervals are of equal size.

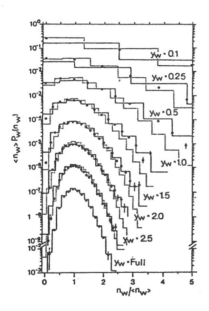

Fig.1: Multiplicity (n_w) distributions for charged hadrons observed in the rapidity window w: $|y| < y_w$ plotted in the variables $\langle n_w \rangle P_w(n_w)$ vs $z_w = n_w/\langle n_w \rangle$. Here, $\langle n_w \rangle$ is the average value of n_w, and $P_w(n_w)$ is the probability of observing n_w charged hadrons inside the *symmetric* window $|y| < y_w$ plot for e^+e^- annihilation processes at $\sqrt{s} = 29$ GeV. The data are taken from Ref.11. The solid lines are the calculated result based on the assumption that the observed hadrons are produced via decay of charge-neutral clusters. The dashed lines correspond to the result obtained under the assumption that the observed charged hadrons are produced independently.

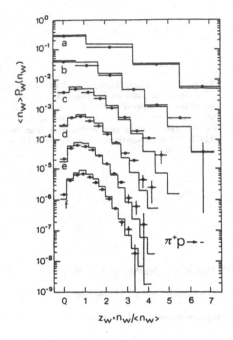

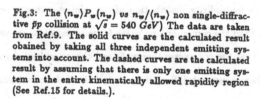

Fig.2: The $\langle n_w \rangle P_w(n_w)$ vs $n_w/\langle n_w \rangle$ plot for the reaction $\pi^+ p \rightarrow$ negatively charged hadron + anything, at $\sqrt{s} = 22\ GeV$. The data are taken from Ref.10. The size of the rapidity windows $(-y_w < y < y_w)$ are indicated by a, b, c, d, e, and f which stands for $y_w = 0.5, 1.0, 1.5, 2.0, 2.5,$ and 3.0, respectively. The histograms are the calculated result (see text).

Fig.3: The $\langle n_w \rangle P_w(n_w)$ vs $n_w/\langle n_w \rangle$ non single-diffractive $\bar{p}p$ collision at $\sqrt{s} = 540\ GeV$) The data are taken from Ref.9. The solid curves are the calculated result obained by taking all three independent emitting systems into account. The dashed curves are the calculated result by assuming that there is only one emitting system in the entire kinematically allowed rapidity region (See Ref.15 for details.).

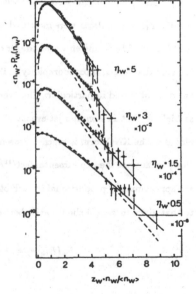

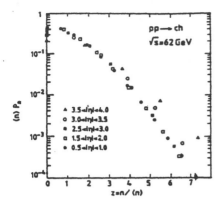

Fig.4: Charged multiplicity distributions in the KNO form for different intervals of the modulus of the c.m. pseudorapidity variable for minimum bias events in $\bar{p}p$ interactions at $\sqrt{s} = 62\ GeV$ (preliminary uncorrected data from the ABCDHW collaboration). The figure is taken from Ref.16.

Second example:

It has been observed by UA1 Collaboration[17] that a considerable fraction of "minijet-events" exist in the data sample of minimum-bias events. ("Minijets" are also known as "low-transverse-energy jets".) Such events were found when the charge multiplicity (n) distributions and transverse energy (E_t) distributions were measured in the rapidity region $|\eta| < 2.5$ under the condition that at least 5 GeV of E_t should enter the trigger cone of radius $R = [(\Delta\phi)^2 + (\Delta\eta)^2]^{\frac{1}{2}} = 1$ in the $\phi-\eta$ space (ϕ:azimuthal angle, η:pseudorapidity). The minijet data sample have very striking properties. It has been observed in particular: The average multiplicity of minijet-events is approximately twice as high as that in non-minijet-events. The multiplicity distribution of the minijet-events, when ploted in the KNO form is much narrower than the corresponding curve for minimum-bias events. Now, because of the experimental fact[18] that the transverse energy E_t and charge-multiplicity n are approximately proportional to each other, it was not difficult to guess[19] that relations similar to that given in Eq.(1) should exist between E_t-distributions $P(E_t; s)$ and $P_w(E_{tw}; s)$:

$$P_w(E_{tw}; s) = \int dE_t/\varepsilon\, P(E_t; s) B(E_t, E_{tw}; q_w; \varepsilon). \qquad (2)$$

Here, P_w is the distribution of transverse energy (E_{tw}) observed in the window w, $q_w(s) = \langle E_{tw} \rangle / \langle E_t \rangle$ is the average probability for a unit of transverse energy to be inside the given rapidity-window w (The energy scale is fixed by the parameter ε); $\langle E_{tw} \rangle$ and $\langle E_t \rangle$ are respectively the average value of E_{tw} and E_t. $B(E_t, E_{tw}; q_w; \varepsilon)$ is the generalized binomial distribution for continuous random variables E_t/ε and E_{tw}/ε:

$$B(E_t, E_{tw}; q_w; \varepsilon) = \frac{\Gamma(E_t/\varepsilon + 1)}{\Gamma(E_{tw}/\varepsilon + 1)\Gamma(E_t/\varepsilon - E_{tw}/\varepsilon + 1)} q_w^{E_{tw}/\varepsilon} (1 - q_w)^{E_t/\varepsilon - E_{tw}/\varepsilon}. \tag{3}$$

Since the minijet-trigger can be viewed as a moving window with an energy-cut, the probability density for a given amount of transverse energy E_{tw}^j (j for jet) to be inside w and for a part of this (E_{tw}^c) to enter the trigger cone c with threshold energy $E_{to}(= 5 \; GeV)$ can be written as:

$$P_w^j(E_{tw}^j; E_{to}) = \int_0^{E_{tw}^j/2} dE_{tw}^c/\varepsilon P_w(E_{tw}^j; s) B(E_{tw}^j/2, E_{tw}^c; q_c; \varepsilon) \theta(E_{tw}^c - E_{to}). \tag{4}$$

Note that the trigger cone receives contributions from the "jet side" only, while those on the "away side" which compensate the momentum of the jet are not included. We calculated the transverse energy distribution for minijet events from Eq.(4), where we used the empirical value for $q_c = 2\langle E_{tw}^c \rangle / \langle E_{tw}^j \rangle$ and an empirical fit for $P_w(E_{tw}; s)$. The result for $\langle E_{tw}^j \rangle P_w^j$, when ploted as a function of $Z_w = E_{tw}^j / \langle E_{tw}^j \rangle$ is shown in Fig.5, where the data for the corresponding multiplicity distribution $\langle n_w^j \rangle P_w^j$ as a function of $n_w^j / \langle n_w^j \rangle$ are given. The calculated result, as well as qualitative arguments (given in Ref.19) show that the narrowness of the distribution for the minijet sample is attributable to the method of event selection. We also calculated other quantities measured by UA1 Collaboration in this connection: the multiplicity distribution for non-minijet events (shown in Fig.6), the average multiplicities and the average transverse energy per particle both for minijet- and for non-minijet-events, and the relative occurrence of minijet for different s-values. The obtained results are in agreement with the data[17]. It should be mentioned that all these quantities depend significantly on the threshold energy E_{to} of the trigger cone. This dependence is illustrated in Fig. 7.

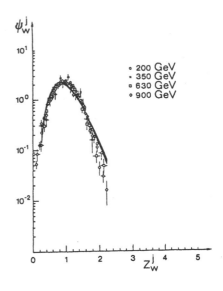

Fig.5: $\psi_w^i = \langle E_{tw}^j \rangle P_w^j (E_{tw}^j)$ vs $z_w^j = E_{tw}^j / \langle E_{tw}^i \rangle$, KNO-plot for low-$E_t$-jets (minijet) events for $\sqrt{s} = 200, 350, 630$, and 900 GeV. The results for different energies are very close to one another. All of them are inside the narrow band of the given curve. The corresponding multiplicity distribution data (see text) are taken from UA1 experiments (Ref.17).

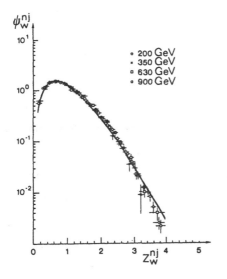

Fig.6: ψ^{nj} (definition similar to that for ψ^j) for non-low-E_t-jet (non-minijet) events, for same values of $\sqrt{s}$ as in Fig.5.

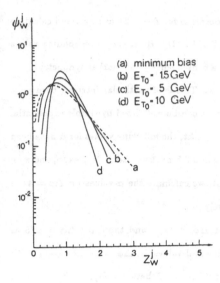

ψ_w^j

10^1

10^0

10^{-1}

10^{-2}

(a) minimum bias
(b) E_{T_0} = 15 GeV
(c) E_{T_0} = 5 GeV
(d) E_{T_0} = 10 GeV

c b
d a

0   1   2   3   4   5
z_w^j

Fig.7: Threshold-dependence of KNO-plots for low-E_t-jet (minijet) events. The functions ψ^j and z^j are defined in the same way as in Fig.5 and in the text. E_{t_o} is the threshold energy in Eq.(4).

Third example:

In order to see, in general, whether it is useful to differentiate between events which have low-, medium- or high-transverse-energy E_t in limited $\eta-$ and/or $\phi-$windows, and if it is the case, where and how the corresponding E_t-cuts should be made, a systematic data analysis[20] has been carried out by the Berliner group. The result of this analysis can be summarized as follows: The transverse energy spectra in pseudorapidity and/or azimuthal angle windows are related to one another in an extremely simply way. In fact, the relationship is nothing else but that given in Eq.(2), where q_w can in general be calculated by the geometry of the experimental apparatus. Exceptional behaviours appear only in events associated with very large transverse energies $E_t \geq 40\ GeV$. The validity of Eq.(2) has been tested; and the result of some of the tests are given below:

Test A: We consider the experiment of B. Brown et al.[21], in which high-transverse-energy events produced in proton-proton collisions at 400 GeV/c were studied using a large-acceptance

658

multiparticle spectrometer. We use the cross section obtained for $\Delta\phi = 2\pi$ as input and calculate

that for $\Delta\phi = 4\pi/5$ and that for $\Delta\phi = \pi/5$ (See Fig. 2 of Ref.21). Here, the corresponding values

for q_w are respectively 2/5 and 1/10, which is simply a consequence of azimuthal symmetry.

As a first step, we calculate the slopes of the curves. This is of particular interest, because it

has been shown empirically[21] that all three sets of data can be well fitted by simple exponential

functions $d\sigma/dE_t \sim exp(-\alpha E_t)$ in the large-E_t region. In fact, the following values for α have been

given by Brown et al.[21] : $\alpha = 0.84 \pm 0.02$, 1.25 ± 0.05 and 2.5 ± 0.1 for $\Delta\phi = 2\pi$, $4\pi/5$, and $\pi/5$,

respectively. By using $\alpha = 0.84$ for $\Delta\phi = 2\pi$ as input, we calculate the α-values for $\Delta\phi = 4\pi/5$

and $\Delta\phi = \pi/5$. The results are: 1.3 and 2.4 respectively.

Then, we calculate the cross-sections $d\sigma/dE_t$ for $\Delta\phi = 4\pi/5$ and those for $\Delta\phi = \pi/5$ as

functions of E_t by inserting the corresponding experimental values for $\Delta\phi = 2\pi$ into Eq.(2). We

use the value for $E_t = 0$ to determine the proportionality constant between $d\sigma/dE_t$ and $P(E_t; S)$

and thus obtain the absolute $d\sigma/dE_t$ values for $\Delta\phi = 4\pi/5$ and $\pi/5$. These are shown as solid

curves in Fig.8.

Fig.8: Cross section obtained with three different azimuthal acceptances as a function of transverse-energy contained with the trigger modules. Data are taken from Ref.21. The solid lines for the $\Delta\phi = 4\pi/5$ and the $\Delta\phi = \pi/5$ data set are the calculated rsult of Test A. The dashed line is explained in Ref.20.

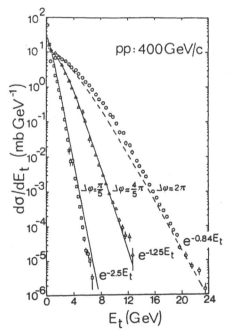

Test B: We consider the proton-proton data in the experiment of C. De Marzo et al[22], in which large transverse-energy cross sections of 300 GeV/c pions and protons on hydrogen have been measured with a segmented calorimeter covering the central rapidity region — $0.88 < y < 0.67$ and 2π in azimuth. Similar to Test A, here we take $d\sigma/dE_T$ for $\Delta\phi = 2\pi$, insert it as input in Eq.(2) and calculate that for $\Delta\phi = \pi$ and that for $\Delta\phi = \pi/2$. Here the corresponding values for q_w are obviously 1/2 and 1/4 respectively. Because of the uncertainties in extrapolating the data in the low-E_T region, we determine the proportionality constant between $d\sigma/dE_T$ and $P(E_T; s)$ at $E_T = 3\ GeV$. The calculated result is shown, together with the data of Ref.22 in Fig.9.

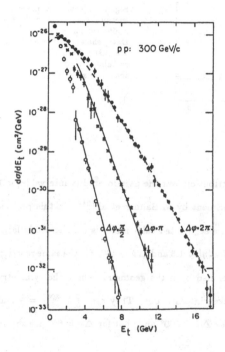

Fig.9: Cross sections versus transverse-energy obtained in regions of various azimuthal acceptances. Data are taken in from Ref.22 second paper. The solid lines for $\Delta\phi = \pi$ and $\pi/2$ are the calculated result of Test B. The dashed line for $\Delta\phi = 2\pi$ is explained in Ref. 20.

660

Test C: We consider the total transverse energy distribution in the UA1-experiment of G. Arnison et al.[23] at the CERN proton-antiproton collider. The data[23] are for $\sqrt{s} = 540\ GeV$, and pseudo-rapidity region $|\eta| < 1.5$. Here we use exactly the same input as that in the second example (about minijets), which is an empirical E_T-distribution for $|\eta| < 2.5$ at $\sqrt{s} = 540\ GeV$. Because of the ϕ-symmetry and the flatness of the η-distribution in the central rapidity region, q_w can be obtained also in this case from the geometry: $q_w = 3/5$. The calculated result is shown, together with the data[23], in Fig.10.

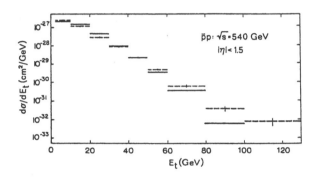

Fig.10: Total transverse-energy distribution in the rapidity region $|\eta| < 1.5\ GeV$. Data, shown as dashed lines, are taken from Ref.23. The calculated result of Test C is shown as solid lines.

Test D: We consider the transverse-energy distribution over the pseudorapidity interval $|\eta| < 1$ an azimuthal range $\triangle\phi = 300°$ in the UA2 experiment by M.Banner et al.[24]. In this proton-antiproton collision experiment at $\sqrt{s} = 540\ GeV$, a segmented calorimeter is used to study large $E_T - jets$. Here, we use as input the UA1 data[23] for $|\eta| < 1.5$ and $\triangle\phi = 360°$ for the same reaction at the same energy. The values for q_w are determined from the geometry, where the symmetry in $\dot\phi$ and the flatness in the η-distribution are taken into account. They are: $\frac{2}{3} \times \frac{300°}{360°} = \frac{5}{9}$ and $\frac{2}{3} \times \frac{60°}{360°} = \frac{1}{9}$ respectively. The calculated result for $\triangle\phi = 300°$ and that for $\triangle\phi = 60°$ are shown in Fig.11 together with the UA2 data[24]

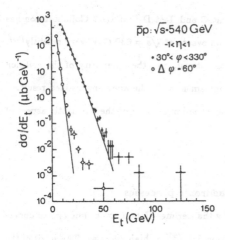

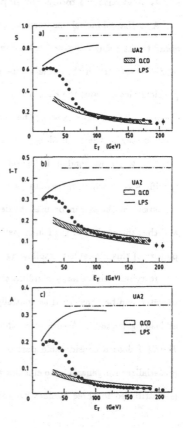

Fig.11: Transverse energy distributions over two restricted azimuthal regions. Data are taken from Ref.24. The solid curves are the calculated result of Test D.

Fig.12 (a-c): Mean values of sphericity (a) 1-thrust (b) and aplanarity (c) as a function of total transverse energy E_t. The solid curve and the dashed band represent the predictions of the LPS model and of the QCD jet model including detector effects, as discussed in Ref.25. The range of values representing the QCD simulations includes the effects of different initial state parton subprocesses and of the entire variation of the weight factors entering the underlying event contribution. The dashed line shows the limiting value that the variables can attain when all the calorimeter cells are given the same energy deposition. The results are taken from Ref.25.

It should be pointed out in connection with Test C and Test D that UA2 Collaboration has published the result of a shape analysis of $\bar{p}p$ collision events at $\sqrt{s} = 630\ GeV$ very recently[25]. In this experiment, they measured the sphericity, the thrust, and the aplanarity of the collision events. The result, shown in Fig.12, is in excellent agreement with the above-mentioned result.

The examples given in this section explicitly show the importance of the stochastic aspect of multiparticle production processes.

4. A statistical model for nondiffractive hadronic processes

A statistical model has been suggested[26-29] by the Berliner Group some time ago to understand the non-single-diffractive hadron-hadron collision data[30] at high energies. The spirit of the model is very much the same as the conventional fireball models. The purpose, the underlying physical picture, and the obtained result have, however, little in common with those discussed in the past. To be more precise, the multiparticle production processes are assumed to take place via two stages: the formation- and the decay-stage of excited systems; and these two stages are assumed to be independent in the sense that the observed final state hadrons do not depend on the specific way in which the excited systems are formed. The purpose of this model is to show that statistical methods can be used to describe the formation and the decay of the excited systems, and that the applicability of such methods can be checked experimentally. The underlying physical picture of this model is the following.

In typical high-energy nondiffractive hadron-hadron collision events, the projectile P and the target T can be considered as spatially extended objects which go through each other and appear as leading particles. While going through and interacting with each other, the colliding objects P and T lose a considerable part of their energies and momenta. A part of this "lost energy" materializes (in general, into clusters which subsequently decay into hadrons). This part of energy is distributed in three distinct excited systems C^*, P^*, and T^* which are characterized by their locations in rapidity space. (Here C stands for the central rapidity region, and P and T stands for the projectile- and target-fragmentation regions, respectively.)

The existence *of three excited systems* as intermediate state in nondiffractive hadron-hadron collisions is a postulate. It is motivated by the experimentally observed "fragmentation products" and "central plateau" in such processes. It is also a natural consequence of the usual quark-gluon picture[31]. In terms of this picture, in which hadrons are made out of valence quarks and gluons, and on the average about 35-50% of the momentum of a high-energy hadron is carried by its valence quarks and the rest by the gluons (and/or seaquark pairs), the hadronic matter created in system C^* is due to the interaction between the gluons (and/or seaquark pairs) in the projectile and their counterparts in the target. That is, the clusters in system C^* obtain their energies and momenta from *two* sources, namely, the gluon part (gluons and/or sea-quark pairs) of the projectile and that of the target. Similarly, the materialization/excitation energy in the system $P^*(T^*)$ is also due to two energy-momentum sources, namely, the energy-momentum of the quark part and that of the gluon part of the projectile (target). That is to say, the quark-gluon picture implies that the number of such two-source systems in high-energy nondiffractive hadron-hadron collision processes should be *three*.

Each of the three excited systems $(C^* . P^*, T^*)$ is assumed to have the following property. The amount of energy contributed by each of the two sources to the materialization/excitation energy of the system is a *random* variable. After their formation, the excited system *"forgets its history"* in the sense that the probability of forming such a system characterized by the total amount of materialization energy does not depend on the specific way of its formation. That is, it "does not remember" how much of this energy is due to the one, and how much of it is due to the other source.

This implies in particular that the materialization/excitation energy E_i^* of the system $i(i = C^*, P^*, T^*)$ is distributed in the following way (we omit the asterisks on the subscripts; that is, we write E_C^* instead of $E_{C^*}^*$, etc.):

$$\langle E_i^* \rangle P(E_i^*) = 4 \frac{E_i^*}{\langle E_i^* \rangle} \exp\left[-2\frac{E_i^*}{\langle E_i^* \rangle}\right], \qquad i = C^*, P^*, T^*, \tag{5}$$

where $\langle E_i^* \rangle$ is the average value of E_i^*. This means, the distribution of the total internal energy of the created clusters in any one of the systems C^*, P^*, and T^* can be determined, provided that the corresponding average value is known.

In order to see that the excitation energy distribution is indeed given by Eq.(5),. we consider any one of the three systems $i(i = C^*, P^*, T^*)$. Let us denote, for the sake of simplicity, the excitation energy E_i^* of this system by E, the two energy-momentum sources of this system by 1 and 2. Let $f_1(E_1)$ be the probability for the system i to obtain an amount E_1 from source 1, and $f_2(E_2)$ that for the system i to obtain E_2 from source 2. Then, since the two sources 1 and 2 are independent from each other, the probability $f(E_1, E_2)$ for i to obtain E_1 from source 1 and E_2 from source 2 is simply the product $f_1(E_1)f_2(E_2)$. Now, if this probability $f(E_1, E_2)$ depends only on E, the sum of E_1 and E_2, but independent of E_1 and E_2, we have

$$\frac{d}{dE_1}[f_1(E_1)f_2(E - E_1)] = 0. \tag{6}$$

Hence

$$\frac{d}{dE_1}[ln f_1(E_1)] - \frac{d}{d(E - E_1)}[ln f_2(E - E_1)] = 0, \tag{7}$$

which gives

$$f_1(E_1) = A_1 \exp(-B E_1), \tag{8}$$

$$f_2(E_2) = A_2 \exp(-B E_2), \tag{9}$$

where A_1, A_2 and B are positive real numbers. (We recall that f_1 and f_2 are probabilities.) Therefore, the probability for the system i to be in the state characterized by the total materialization energy E is

$$
\begin{aligned}
P(E) &= \int dE_1 dE_2 \delta(E - E_1 - E_2) f_1(E_1) f_2(E_2) \\
&= \int_0^E dE_1 A_1 A_2 \exp(-BE) \\
&= A_1 A_2 E \exp(-BE).
\end{aligned}
\tag{10}
$$

Taken together with the normalization condition

$$
\int P(E) dE = 1,
\tag{11}
$$

and the definition for the average value of E,

$$
\int E P(E) dE = \langle E \rangle
\tag{12}
$$

we obtain from Eq.(10) the distribution given in Eq. (5) when we replace E by E_i^*. $(i = C^*, P^*, T^*)$.

We note that Eq.(5) is the key result of the proposed model . Its validity can be, and has already been, tested in various ways:

First test:

We consider the distribution of the transverse energy flow (E_{tc}, i.e. the sum of the energy of every produced particle times the sine of its production angle) in the central rapidity region[18] (Here, the index t stands for transverse, and c stands for central.). Since the particles produced in the central rapidity region are the decay products of the clusters of the excited system C^*, and the sum of the masses of the clusters is the excitation energy E_c^*, it is clear that E_{tc} should be proportional to E_c^*. Hence we should see

$$
\langle E_{tc} \rangle P(E_{tc}) = 4 \frac{E_{tc}}{\langle E_{tc} \rangle} \exp\left(-2 \frac{E_{tc}}{\langle E_{tc} \rangle}\right),
\tag{13}
$$

666

where $P(E_{tc})$ is the probability of observing an amount of E_{tc} transverse energy in the central rapidity region, and $\langle E_{tc} \rangle$ is its mean value. Such a behaviour has indeed been observed experimentally[18].

Second test:

We consider the multiplicity (n_c) distributions of charged hadrons in the central rapidity region[30]. Taken together with the empirical fact[18] that the transverse energy and the multiplicity of the charged hadrons in the central rapidity region is proportional to each other, we obtain from Eq.(13):

$$\langle n_c \rangle P(n_c) = 4 \frac{n_c}{\langle n_c \rangle} \exp\left(-2 \frac{n_c}{\langle n_c \rangle}\right). \tag{14}$$

Also this is in good agreement with the data[30]. See Fig.13 and Fig.14

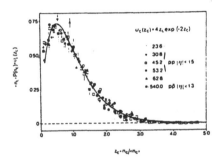

Fig.13: The scaled multiplicity distribution in the central rapidity region. The experimental data are taken from Ref.30. The curve is obtained from Eq.(14).

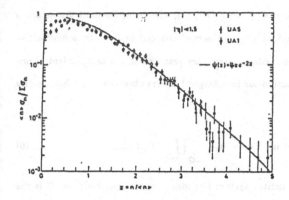

Fig.14: The scaled multiplicity distribution in the central rapidity region, shown as a log-plot. The curve is obtained from Eq.(14). This figure is taken from the second paper of Ref.30.

It should be pointed out in this connection that the size $|2y_c|$ of the central rapidity region $y \leq |y_c|$ depends on the total c.m.s. energy $\sqrt{s}$ of the colliding system. The value for $|y_c|$ is approximately 1.3 to 1.5 in the CERN ISR energy range ($23 \ GeV \leq \sqrt{s} \leq 63 \ GEV$). It increases logarithmically with increasing s, and is about 2 to 2.5 at $\sqrt{s} = 540 \ GeV$. Hence, it is expected that at higher and higher energies, the transverse energy and the multiplicity distributions will have the forms given by Eqs.(13) and (14) in larger and larger rapidity intervals around (the center of the c.m.s. frame) $y = 0$.

This means, in particular, for a given rapidity window, we expect to see that KNO scaling is only approximately valid, even when the given rapidity window is inside all the corresponding central rapidity regions (i.e. inside the central rapidity regions for processes at all possible s-values. The multiplicity distributions in limited rapidity windows should be calculated with the method given in Section 3.

Third test:

We consider charge distributions in the full rapidity space. Since all three excited systems contribute, we have

$$n_{nd} = n_C + n_P + n_T \tag{15}$$

where n_i is the contribution of the system i, $(i = C^*, P^*, T^*)$; also in n_i we omit the asterisks on the subscripts and n_{nd} is the multiplicity of the observed charged hadrons in nondiffractive hadron-hadron collisions. Since all three excited systems are assumed to the independent of one another, we can write the multiplicity distribution for charged hadrons observed in the full rapidity region in non-diffractive processes as

$$P(n_{nd}) = \int 2\delta(n_C + n_T + n_P - n_{nd}) \prod_{i=C^*, P^*, T^*} P(n_i) dn_i. \tag{16}$$

Here, $P(n_i)$ is the probability that the excited system i produces n_i charged hadrons. It is the expression given in Eq.(5), provided that the relationship between the multiplicity of the produced charged hadrons and the excitation energy in the system i are proportional to each other, that is;

$$\frac{n_i}{\langle n_i \rangle} = \frac{E_i^*}{\langle E_i^* \rangle}, \qquad i = C^*, P^*, T^*. \tag{17}$$

The relationship between $\langle n_C \rangle$ and $\langle n_{nd} \rangle$, the average value of n_{nd}, is not known a priori. In order to see whether the proposed model is consistent with the existing data[30], we write

$$\langle n_C \rangle = \alpha \langle n_{nd} \rangle \tag{18}$$

where α is a s-dependent parameter. It follows from Eq.(15) and the $P^* - T^*$ symmetry (We recall that the leading particles are not included in the excited systems P^* and T^*) that

$$\langle n_P \rangle = \langle n_T \rangle = \frac{1-\alpha}{2} \langle n_{nd} \rangle. \tag{19}$$

This means, $\langle n_i \rangle$ $(i = C^*, P^*, T^*)$ can be determined from the experimental value for $\langle n_{nd} \rangle$, provided that $\alpha = \alpha(s)$ is known.

Comparison with experiments shows that α increases monotonically with increasing energy. Its value is between 0.15 and 0.5 for $14\, GeV \leq \sqrt{s} \leq 63\, GeV$, and becomes 0.75 for $\sqrt{s} = 540\, GeV$.

The result shows that the model is indeed consistent with the existing data. The following remarks may be helpful in understanding the details.

(i) The scaling behaviour of $\langle n_{nd} \rangle P_{nd}$ with respect to $n_{nd}/\langle n_{nd} \rangle$ observed in the energy range $14 \ GeV \leq \sqrt{s} \leq 63 \ GeV$ is due to the fact that the expression obtained from Eqs.(16), (18) and (19) is insentitive to the variation of α for $0.15 < \alpha < 0.5$.

(ii) By using the values $\alpha = 0.5$ for $\sqrt{s} = 63$, $\alpha = 0.75$ for $\sqrt{s} = 540 \ GeV$ and experimental values for $\langle n_{nd} \rangle$, the product $(1-\alpha)\langle n_{nd} \rangle$ calculated at these two energies turns out to be the same ($= 6.8$) . That is, the average multiplicity of the P^{*}- and that of the T^{*}-system are the same for $\sqrt{s} = 63$ and $540 \ GeV$. This result is in agreement with the hypothesis of limiting fragmentation.

(iii) Based on the observation mentioned in (ii) we assume:

$$\{1 - \alpha(s)\}\langle n_{nd} \rangle = 6.8 \qquad \text{for all } \sqrt{s} > 63 \ GeV. \tag{20}$$

This can be used to determine the s-dependence of α provided that of $\langle n_{nd} \rangle$ is known. Hence, by extrapolating the empirical formula (e.g. $\langle n_{nd} \rangle = a + b\ell ns + c(\ell ns)^2$ where a, b and c are constants), we are able to calculate, in this model, multiplicity distributions at higher energies. Our predictions for energies up to the next generation accelerators are given in Ref.28. (See Fig.15.)

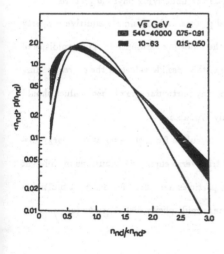

Fig.15: The scaled multiplicity distribution for nondiffractive hadron-hadron collisions at different total c.m.s. energy ranges.

(iv) In terms of the quark-gluon picture[31], the excited system C^* obtains its excitation energy from the gluon part of the projectile- and that of the target-hadron. Hence, the observed increase of $\alpha(s)$ with increasing s can be interpreted as the increasing importance of gluon-gluon interaction at higher energies.

These tests show that the key result, as given by Eq.(5), is indeed in good agreement with the data. It should be pointed out, however, that the distribution of excitation energies alone is not sufficient to describe experimental results related to rapidity distributions of the final state hadron. On the other hand, we note that in the proposed model, nothing has yet been said about the rapidity distributions of the constituents of the excited systems — neither in their formation nor in their decay stage.

Various possibilities have been discussed in the past[1-8,26-28] The location(s) of the excited system(s) in rapidity space are usually different in different fireball and/or statistical models. If there is one (and only one) such excited system, as it is for example the case in Fermi's statistical model, the location of the "center" of the system has to coincide with the total center of mass of the colliding objects. It is also trivial if there are only two (including the leading particles) systems, as in the case of diffractive processes. But, in general, not only the number of excited systems, but also their location in rapidity space depend very much on the underlying physical picture.

In our model[29], the characteristics of the physical picture and the simple additive property of rapidity under Lorenz transformation have let us to the following assumption. The rapidities of the clusters in a given excited system may have any one of the possible values in the corresponding kinematical region. Since there is no a priori reason why any particular one of these values should be prefered, we assume that their occurences are equally probable.

Taken together with this assumption, one can now calculate from the excitation energy distribution given in Eq.(5) multiplicity distributions and transverse energy distributions in different rapidity regions, rapidity distributions for all produced particles and those for different multiplicity ranges, short-range rapidity correlations and long-range multiplicity distributions for presently available, as well as for future accelerator energies etc. . We do not discuss these calculations in this paper. They can be found in Refs.29 .

The proposed model for nondiffractive hadron-hadron collisions can be, and has already been generalized to describe multiparticle production processes in hadron-nucleus and nucleus-nucleus collisions at comparable energies.

In fact, a multisource model for particle production in high-energy hadron-nucleus collisions has been proposed[32]. Here we note that in hadron-nucleus (h-A) collisions, the incident hadron in general hits more than one nucleons inside the target nucleus (of atomic number A); and the final states depend on *how many nucleons* the incident hadron meets and *how it interacts* with these nucleons. In this connections it is convenient to use the concept of effective target ET[33] which is the group of nucleons inside the target nucleus hit by the incident hadron as it "goes through" the target.

Suppose ν of the ν_{ET} nucleus inside the ET interact with the incident hadron h nondiffractively and every one of the ν nondiffractive collisions can be described by the statistical model for nondiffractive hadron-hadron collisions[26-29]. Then, the multiparticle production process in such a $h - ET$ collision can be envisaged to take place as follows.

After the ν nondiffractive collisions between the incident hadron h and ν of the nucleons in the ET, one P^*-system, one C^*-system and ν T^*-systems will be formed. The formation and the decay of the P^*- and the T^*-systems are exactly the same as those in hadron-hadron collisions. That is, the materialization energy of each system is randomly obtained from two energy sources. Hence, the distribution for the materialization energy and (taken together with the assumption that the multiplicity of the produced charged hadrons and the materialization energy are proportional to each other) the distribution for the multiplicity of the charged hadrons is also given by Eqs. (5) and (17), where in this case i stands only for P^*- and for any one of the ν T^*-systems. Since the C^*-system obtains its materialization energy randomly from $\nu + 1$ sources, the corresponding multiplicity distribution for charged hadrons is:

$$P_{C\nu}(n_C|1+\nu) = \frac{(\nu+1)^{\nu+1}}{\nu!} \; \frac{1}{\langle n_C \rangle_\nu} \left[\frac{n_C}{\langle n_C \rangle_\nu}\right]^\nu \exp\left[-(\nu+1)\frac{n_C}{\langle n_C \rangle_\nu}\right] \qquad (21)$$

Furthermore, since the P^*-, the C^*- and the $\nu\, T^*$.systens are independent from one another, the probability for observing n charged particles in such an event is

$$P(n|\nu) = \Sigma' P_P(n_P) P_{C\nu}(n_C|1+\nu) P_T(n_{T1}) P_T(n_{T2})...P_T(n_{T\nu}), \qquad (22)$$

where the dash on the summation sign means that the sum over $n_P, n_C, n_{T1},\ n_{T2}, ..., n_{T\nu}$ should be taken such that the following condition is satisfied.

$$n = n_P + n_C + n_{T1} + n_{T2} + ... + n_{T\nu}, \qquad (23)$$

Eqs. (21) and (22) are the key results of the multisource model for high-energy hadron-nucleus collisions. Hence, it is useful to see first whether these results are consistent with the experimental findings. Such tests are indeed possible.

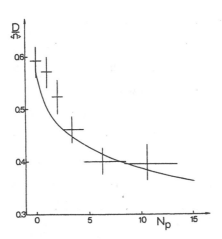

Fig.16: The ratio of the dispersion D to the mean multiplicity $\langle n \rangle$ as a function of N_P, the number of identified protons. The data are from Ref.34. The curve is the calculated result given in Ref.32.

In order to do this, we recall that two of the most striking new features which have been observed in recent hadron-nucleus collision experiments[34,35] are the following. (a) The ratio $D/\langle n \rangle$ between the dispersion $D = [(n-\langle n \rangle)^2]^{\frac{1}{2}}$ and the average multiplicity $\langle n \rangle$ is a function of the number N_p of identified protons which correspond to the grey measured in emulsion experiments. For small N_p this ratio is close to the value for proton-proton reactions, but it decreases with increasing N_p. The data points[34] are shown in Fig.16

(b) Mid-rapidity transverse energy spectra have been measured[35] in 200 GeV $p - Au$ and $\pi^+ - Au$ collisions. The obtained spectra are the *same* which shows that the mid-rapidity spectrum is independent of the quantum numbers of the incident hadron. The data can be well-described by the following empirical formula[35] [$d\sigma/dE_t$ given in b/GeV].

$$d\sigma/dE_t = 0.173\, E_t^{2.36} \exp(-0.727\, E_t). \qquad (24)$$

It can be readily shown[32,36] that both of these properties are natural consequences of the multisource model. We do not discuss the details here, and refer the interested reader to the results given in Refs.32 and 36 . It should be mentioned: (i) The multisource model can be generalized to include nucleus-nucleus collisions. This result is given in Ref.36. (ii) Both in hadron-nucleus and nucleus-nucleus collision, geometrical average has to be made in order to compare with experiments.

5 Concluding remarks

We have seen in Section 2: The essential features of the fireball models discussed in the past suggest that multiparticle production processes in high-energy collisions are predominantly statistical processes.

We have seen in Section 3: A number of striking phenomena observed in recent multiparticle production experiments[9-13,16-18,21-25] are pure statistical effects. It may seem disappointing that these phenomena can be understood without introducing new dynamics. But, we think, we did gain new insight into the reaction mechanism of multiparticle production processes through these experiments and the corresponding analyses[15,19,20]: They explicitly show what a dominating role the stochastic aspect play in multiparticle production processes; and they clearly demonstated how useful it is to differentiate between pure statistical from dynamical effects in understanding the multiparticle production experiments.

We have seen in Section 4: Statistical methods can be used to construct a model[26-29,32,36] which gives an adequate description of the multiparticle production data in nondiffractive hadronic processes.

The usefulness of statistical methods in multiparticle production studies has become increasingly clear. This is, in fact, the reason why we not only have carried out the analyses mentioned in Section 3, but also have introduced[37] new concepts and methods to study hadronic cluster problems (which we did not have a chance to discuss in this review). This is also the reason why we not only proposed a statistical model for nondiffractive hadron-hadron collisions[26-29], but also generalized it to include hadron-nucleus[32] and nucleus-nucleus collisions[36] (which we could not discuss in detail in this paper).

It should be pointed out however, while statistical methods are extremely useful in analysing and describing various multiparticle production phenomena, the final goal of our research is of course to understand the *dynamics* of multiparticle productions processes. The result of the tests mentioned in this paper gives us confidence that we are on the right track.

Acknowledgements

The author thanks all his collaborators, especially Cai Xu, Chaò Wei-qin, Chou Kuang-chao, and Liu Lian-sou for their valuable contributions to the multiparticle prodution research program in Berlin. He thanks Renate Rothe for her marvelous secretarial work. This program has been supported in part by Deutsche Forschungsgemeinschaft (DFG: Me 470/4-3, 5-1 and 5-2) and by Max Planck Gesellschaft.

References

1 G. Cocconi, Phys. Rev. **111**, 1699 (1958)

2 S. Takagi, Prog. Theor. Phys. **7**, 123 (1952)

3 K. Nu, Nuovo Cimento, **10**, 994 (1958)

4 P.Ciok, T. Coghen, J. Gierula, R. Holyński, A. Jurak, M. Miesowicz, T. Saniewsak, O. Stanisz and J. Pernegr, Nuovo Cimento 8,166; **10**, 741 (1958)

5 M. Miesowicz, in Progress in Elementary Particle and Cosmic Ray Physics, vol. X ed. by J.G. Wilson and S.A. Wouthhuysen, North Holland, (1971), p.101

6 See for example, Ref. 1-5 and the papers cited therein

7 See for example, the following review articles and the references given therein:

E.L. Feinberg, in Proc. of the 2nd International Conference on Elementary Particles, 1-12 Sept. 1973 in Aix-en-Provence, France; J. de Physique Coll. C-1, Suppl.10, (1973) p.356

P. Carruthers, in this review volume

8 See, for example, Ref. 1-7 and the papers cited therein

9 G.J. Alner et al. (UA5 Coll.) Phys. Lett. **160B**,193,199 (1985)

10 M. Adamus et al. (NA22 Coll.) Phys. Lett. **177B**, 239 (1986)

11 M. Derrick et al., Phys. Lett. **B168**, 299 (1986)

12 M. Arneodo et al., (EMC Coll.) Z. Phys. **C31**, 1 (1986)

13 F. Dengler et.al. Z. Phys. **C33**,187 (1986)

14 See, for example, the models discussed in the following review article

P. Carlson, in Proceedings of the 23rd International Conference on High-Energy Physics, 16-23 July 1986, Berkeley, Ca. USA, ed. S.C. Loken, World Scientific, p.1346

W. Kittle, in Proceedings of the 18th International Symposium on Multiparticle Dynamics 8-12, Sept. 1987, Tashkent, USSR, ed. I.M. Dremin, World Scientific (in press)

and the references given therein

15 Chao Wei-qin, Meng Ta-chung and Pan Ji-cai, Phys. Lett. **176B**, 211 (1986); Phys. Rev. **D35**, 152 (1987)

16 G. Giacomelli, Talk given at Les Rencontres de Physique de la Vallee d'Aoste, La Thuile, Aosta, Valley, Italy 1987, Bologna-preprint DFUB 87/13

17 G. Ciapetti (UA1Coll), in Proc. 5th Topical Workshop on Proton-Antiproton Collider Physics, Saint Vincent, Italy, 1985, edited by M. Greco, Work Scientific Singapore (1985), p. 488

F. Ceradi (UA1-Coll.) in Proceedings of the 16th International Symposium on Multiparticle Dynamics, June 9-14, 1985, Jerusalem, Israel, ed. J. Grunhaus, Editions Frontiers, World Scientific (1985), p. 531

18 See, for example, S. Geer, (UA1 Collaboration), An.Fis. SerA **79**, 94 (1983) and the references given therein

19 Chao Wei-qin, Meng Ta-chung and Pan Ji-cai, Phys. Rev. Lett. **85** 1399 (1987)

20 Meng Ta-chung and Pan Ji-cai, Phys. Rev. D (in press)

21 B. Brown et al., Phys. Rev. Lett. **49**, 711 (1982)

22 C. De Marzo et al., Phys. Lett. **112B**, 173 (1982); and Nucl. Phys. **B212**, 375 (1983)

23 G. Arnison et al. (UA1-Collaboration), Phys. Lett. **123B**, 115 (1983)

More recent references on UA1 jet experiments can for example be found in:

F. Ceradini, in Proceedings of the 23rd International Conference on High-Energy Physics, Berkeley, CA (1986)p. 1051

24 M. Banner et al. (UA2-Collaboration), Phys. Lett. **118B**, 203 (1982)

More recent references can for example be found in:

A. Roussrie, in Proceedings of the 10th Warsaw Symposium on Elementary Particle Physics, May 25-29, 1987, Kasimierz, Poland, ed. Z. Ajduk, (in press)

25 R. Ansari et al. (UA2-Coll.) Z. Phys. **C36**, 175 (1987)

26 Liu Lian-sou and Meng Ta-chung, Phys. Rev. **D27**, 2640 (1983)

27 Chou Kuan-chao, Liu Lian-sou and Meng Ta-chung, Phys. Rev.**D28**, 1080 (1983)

28 Cai Xu, Liu Lian-sou and Meng Ta-chung, Comm. Theory. Phys. **4**, 847 (1985)

29 Cai Xu, Chao Wei-qin, Huang Chao-shang and Meng Ta-chung, Phys. Rev. **D33**, 1287 (1986).

30 See, for example, the following papers and the references given therein

W. Thome et al., Nucl. Phys. **B129**, 365 (1977)

F.G. Rushbrooke, in Proceedings of the DPF Workshop on $\bar{p}p$ Options for the Supercollider, Chicago, 1984, edited by J. Pilcher and A.R. White (Physics Department, University of Chicago, (1984)

G. Ekspong in Proceedings of the 16th International Symposium on Multiparticle Dynamics, June 9-14, 1985, Jerusalem, Israel, Ed. J. Grunhaus, Editions Frontiers, World Scientific, p. 309

P. Carlson, Ref. 14 and papers in this review volume

W. Kittel, Ref. 14

31 S. Pokorski and L. van Hove, Acta Phys. Pol. **B5**, 229 (1974)

P. Carruthers and Ming Duong-Van, Phys. Rev. **D28**, 130 (1983)

32 Cai Xu, Chao Wei-qin and Meng Ta-chung, Phys. Rev. **D36**, 2009 (1987)

33 Meng Ta-chung, Phys. Rev. **D15**, 197 (1977)

34 C. de Marzo et al., Phys. Rev. **D26**, 1019 (1982)

35 A. Bamberger et al. (NA35 Coll.), Phys. Lett. **B184**, 271 (1987)

T.H. Humanic (NA35 Coll.) in Proceedings of the International Europhysics Conference on High-Energy Physics, June 25-July 1, 1987 in Uppsala, Sweden (in press)

36 Liu Lian-sou, Meng Ta-chung and Peng Hong-an, Phys. Rev. **D** (in press)

37 Chao Wei-qin, Gao Chong-shou, Meng Ta-chung and Pan Ji-cai, Phys. Rev. **D36**, 2702 (1987)

TOWARD A STATISTICAL DESCRIPTION OF DECONFINEMENT TRANSITION

Peter Carruthers
Physics Department
University of Arizona
Tucson, Arizona 87521
and
Theoretical Division
Los Alamos National Laboratory
Los Alamos, New Mexico 87545

and

Ina Sarcevic
Theoretical Division
Los Alamos National Laboratory
Mail Stop B285
Los Alamos, New Mexico 87545

ABSTRACT

We review a statistical model for the multiparticle production in hadron-hadron collision based on analogy with Feynman-Wilson gas. In particular, we consider the approach introduced by Scalapino and Sugar in which the probability functional has a Ginzburg-Landau form. We show that in the constant field approximation, there is KNO scaling at very large energies. When we include the fluctuations around the mean field, we find that there is a crossover at the point at which correlation length becomes comparable with the size of the system and the probability distribution obeys KNO scaling. We give theoretical predictions for the observable quantities such as multiplicities, moments and two-particle correlations at arbitrary temperature point [1]. We compare our theoretical predictions with experimental data and find that at present energies the hadronic system is not close to the crossover point. We argue that onset of KNO scaling is an indication of a phase transition of a hadronic matter to a quark-gluon plasma.

1. INTRODUCTION

Recently, there has been intense experimental and theoretical effort to study the various aspects of high energy hadron-hadron collisions [2]. The possibility of investigating the hadrons formation in these collisions can give a deeper insight into strong interaction physics. One of the most interesting features of these collisions, in which this problem can be studied, is the multiparticle production in hadron-hadron collision and in particular KNO scaling and its violation.

In 1972, Koba, Nielsen and Olesen [3] proposed at sufficiently high energies, a scaling law for the probability distribution

$$\bar{n}\, p_n = \Psi\left(\frac{n}{\bar{n}}\right) \tag{1}$$

where p_n is the probability distribution of getting n particles with mean multiplicity $\bar{n}$ and $\Psi\left(\frac{n}{\bar{n}}\right)$ is energy independent function. Their prediction was based on assumption of validity of Feynman scaling for the many body inclusive cross section. The KNO scaling also implies that moments defined as

$$C_q = \frac{\overline{n^q}}{\bar{n}^q} \tag{2}$$

are energy independent. This scaling law was found to be approximately valid up to ISR energies [4] but its violations have been observed at CERN collision energies [5]. Here we investigate KNO scaling in the context of the statistical model based on the analogy with a Feynman-Wilson gas.

First we note that the number of hadrons produced in high energy hadron-hadron collision is large enough to be described with statistical theory. Therefore it is not suprising that in recent years the methods of statistical physics have been increasingly successful in describing the empirical features of multihadron production experiments [6]. Here we mention the success of the negative binomial probability distribution and its generalizations in describing the energy dependence of the KNO plot. However these successes have been mostly limited to global data or other grossly averaged measures of the S Matrix such as the forward-backward correlation. We extend these techniques to the detailed n particle inclusive cross section following a suggestive framework proposed some time ago by Scalapino, Sugar and their collaborators [7]. The idea is to introduce a field $\phi(y)$ as a random variable (y being the rapidity; as customary we neglect the dependence on the transverse momentum) whose probability functional $\Phi(y)$ is given in Ginzburg-Landau form (here $Y = \ln(s/m^2)$ is the maximum rapidity in the laboratory frame)

$$- \ln \Phi \equiv F[\phi(y)] = \int_0^Y dy[a\phi^2(y) + b\phi^4(y) + c(d\phi(y)/dy)^2] \tag{3}$$

For example the inclusive cross sections are given by

$$\frac{1}{\sigma} \frac{d\sigma}{dy_n \cdots dy_n} = \frac{1}{Z} \int \delta\phi e^{-F[\phi]} \phi^2(y_1) \cdots \phi^2(y_n) \tag{4}$$

where the "partition function" of the one dimensional Feynman-Wilson "gas" [8] is

$$Z = \int \delta\phi e^{-F[\phi]} \tag{5}$$

We imagine that the effective field ϕ represents the quark "condensate" $\langle\bar{q}q\rangle$. For parameters of the model leading to $\phi = 0$ we shall speak of the "unconfined phase", confident that QCD technology will eventually support this language.

Our statistical theory contains only the macroscopic observables and the microscopic degrees of freedom are integrated out. It is worth pointing out that the phenomenological Ginzburg-Landau theory of superconductivity was followed by its BCS [8] microscopic theory. In principle, one would eventually hope to be able to start from some basic theory such a QCD and in the same way as the BCS theory established the nature of the confinement in analogy with the superconducting ground state as a coherent plasma of Cooper pairs. The analogous to the Ginzburg-Landau "order parameter" ϕ in particle physics is the Higgs field. The two-dimensional Gross-Neveu model [9] seem to exhibit very similar properties as the superconductor. Namely, after the symmetry breaking, the composite fields ($\bar{\Psi}\Psi$ and $\Psi\gamma_5\Psi$) develop a nonvanishing vacuum expectation values. This approach deserves further study and is presently under investigation [10].

The resemblance of these formulas to statistical mechanics suggests the exploitation of well-known techniques to explore the possible phase structure of hadronic matter concealed in existing or future data.

The usual procedure is to note the formal resemblance of the generating function $G(z)$ to the grand partition function; recall that G is related to the probability $P_n = \sigma_n/\sigma_{in}$ and the factorial moments by

$$G(z) = \sum_{m=0}^{\infty} z^m P_m = \sum_{m=0}^{\infty} \frac{(z-1)^m}{m!} K_m \tag{6}$$

The parameter z plays the role of the fugacity, which is in turn related to the quasi-chemical potential μ by $z = e^{\beta\mu}$ where β is $1/kT$. Since the factorial moments

$$K_m = \frac{\overline{n(n-1)\ldots(n-m+1)}}{\bar{n}^m} \tag{7}$$

are related to the inclusives by

$$K_m = \frac{1}{\sigma} \int dy_1 \ldots dy_m \frac{d\sigma}{dy_1 \ldots dy_m}$$

$$= \frac{1}{z} \int \delta\phi \, e^{-F[\phi]} \int dy_1 \ldots dy_m \, \phi^2(y_1) \ldots \phi^2(y_m) \tag{8}$$

we easily verify the expression:

$$G(z) = \frac{1}{z} \int \delta\phi e^{-F[\phi]} e^{(z-1)\int_0^Y \phi^2(y)dy} \tag{9}$$

The probability functional e^{-F} describes the distribution of (non-interacting) particles at the "moment" of hadronization or more accurately, on the phase space hypersurface on which hadronization occurs. The form of Eq. (3) is to be regarded as an effective free energy derived from an understanding of the collective behavior of the microscopic degrees of freedom (here quarks and glue). Thus the observed distributions give one a glimpse of the behavior of the hadronic system just prior to its separation into free asymptotic states.

In hadronic collision experiments the only controllable variable is the initial kinetic energy of the colliding particles. In order to relate the changes of observable quantities to the model we need to allow for the possibility that the parameters a,b and c, being related to averages over the microscopic motions of the system, can change slowly as the c.m. energy changes. For simplicity we shall assume these parameters to depend on a single parameter T, which is not necessarily a thermodynamic temperature. We shall use thermodynamic language, but insist that the approach has greater validity than that of equilibrium thermodynamics. When we speak of "critical temperature" T_c we simply mean that value of the parameter (on which a(T), b(T) and c(T) depend) for which a (T)/b(T) goes through zero. Likewise, we shall choose the parametrization so that "low temperature" $T < T_c$ corresponds to that domain where a < 0, b > 0 (implying a nonzero external order parameter $\phi_0 \neq 0$) and "high temperature" $T < T_c$ to a > 0, b > 0, i.e. zero order parameter. Since ϕ is the effective meson field of the problem, and since ϕ_0 would be expected to "dissolve" (i.e. $\phi_0 = 0$) on going to an unconfined phase, it is interesting to explore the qualitative behavior of the system near the hypothetical phase transition (in one dimensional system it corresponds to crossover). That the quasi-temperature T should parameterize an out-of-equilibrium phase transition (or crossover) does not lessen the interest of the phenomenon.

From this generating functional we can derive most quantities of interest: correlation functions, probability distributions, moments, multiplicities to the extent that the functional integral (9) can be evaluated. In this paper we are particularly interested in the behavior of these physical quantities when the relevant parameter space (a,b,c) is such that a phase transition inhabits the system.[†]

Our work [1] is complementary to that of previous authors who have recognized the potential importance of a phase transition for hadronic phenomena. Antoniou et al [11] have employed the Scalapino-Sugar pseudo-statistical mechanical system to relate KNO scaling to a phase transition in the Feynman-Wilson gas. Fowler, Weiner and collaborators have emphasized a possible onset of a predominant quark-gluon plasma in a real-space thermo-dynamic system [12].

In recent years the measurement of hadronic multiplicities by the UA5 group [13] at 200, 540 and 900 GeV c.m. energies has proven that the KNO "plot" of $\bar{n}$ Pn vs n/$\bar{n}$ does not remove all of the energy dependence of the multiplicity distributions. Consequent theoretical discussions have not succeeded in clarifying why or under what circumstances such scaling can be expected. One under-utilized mechanism is in our view the possibility that the system contains a phase transition or crossover point. Experience shows (especially from Wilson's renormalization group analysis of critical point phenomena) that some quantitative results can be obtained without a precise knowledge of dynamics, provided one is suitably close to the critical point.

[†] We note that, as ponted out by Scalapino and Sugar in Ref. 3, Eq. (2) has a simple formulation in terms of the diagonal representation of the density operator in quantum optics. In this case the field ϕ is the complex analytic signal representing the real meson field in terms of which we formulate the model prolem. We find this formulation appealing in part since it is so close to the representation we have previously advocated [2] for the best description of the multiplicity distributions.

2. CONSTANT FIELD APPROXIMATION

First we consider the simplest case when ϕ is an arbitrary constant not necessarily equal to the mean field value. Since ϕ^2 is a density of pions it is easy to see that $\phi^2 = n/Y$, where n is the number of pions. The generating function can be explicitly evaluated in the continuum aprroximation

$$G(z) = \frac{1}{G(z=1)} \int_0^\infty \frac{1}{\sqrt{n}} d(\sqrt{n}) e^{-(a+1-z)n} e^{-bn^2/Y} \tag{10}$$

Using standard formulae [14] we get for $a \geq 0$

$$G(z) = e^{(a+1-z)^2 Y/8b - a^2 Y/8b} \frac{D_{-1/2}((a+1-z)\sqrt{Y/2b})}{D_{-1/2}(a\sqrt{Y/2b})} \tag{11}$$

The corresponding multiplicities can be found by taking the first derivative of the generating function and setting $z = 1$. We obtain the following expression

$$\bar{n} = \sqrt{Y/2b} \frac{\Gamma(3/2)}{\Gamma(1/2)} \frac{D_{-3/2}(a\sqrt{Y/2b})}{D_{-1/2}(a\sqrt{Y/2b})} \tag{12}$$

As shown in [11], KNO scaling requires that $G(z)$ be a function of $\bar{n}\ln z$ near $z = 1$. Since the generating function (11) cannot be written as the function of $\bar{n}\ln z$, there is no KNO scaling for constant ϕ and finite energies. However if we consider a limit of very large energies ($Y \gg (2b/a^2)$) when the chemical potential approaches zero ($z \to 1$) the generating function becomes

$$G(z)_{Y large} \to [1-2\bar{n}\ln z]^{-1/2} \tag{13}$$

where $\bar{n} = 1/2a$. This implies that the factorial moments defined by Eq. (7) are energy independent and that the probability distribution obeys KNO scaling.

Another possibility is that the crossover occurs for ϕ = const.. If we take the limit $a \to 0$ in the expression for the generating function given by Eq. (10) we get

$$G(z) = 2^{3/4} \frac{\Gamma[1/2]}{\Gamma[1/4]} e^{(1-z)^2 Y/8b} D_{-1/2}[(1-z)\sqrt{Y/2b}] \tag{14}$$

The corresponding multiplicites are

$$\bar{n} = \sqrt{Y/b} \frac{\Gamma[3/4]}{\Gamma[1/4]} \sim (\ln s)^{1/2} \tag{15}$$

Clearly, $G(z)$ is a function of $\bar{n}(1-z)$(or $\bar{n}\ln z$ when $z \to 1$) which shows that there is KNO scaling. Using the expression for $G(z)$ we note that the only possibility of having the crossover is at very large energies. Since the prediction for the multiplicities given by Eq. (15) does not agree with the experimental data we proceed with looking at the effect of the fluctuation about the specific constant field, namely the mean field value.

3.   GAUSSIAN APPROXIMATION

In this section we include the fluctuations around the mean field value of the field. By expanding $\phi(y)$ in the complete set of functions

$$\phi(y) = \frac{1}{\sqrt{Y}} \sum_{\text{all } k}^{\infty} a_k e^{i2\pi ky/Y} \tag{16}$$

and keeping only terms up to second order in fluctuations we get the following expression for the free energy

$$F[\phi] = \sum_{k \neq 0}^{\infty} a_k^2 [a+6b\phi_{MF}^2 + c(2k\pi/Y)^2] + aY\phi_{MF}^2 + bY\phi_{MF}^4 \tag{17}$$

where ϕ_{MF} is the mean field value of the field. Since $\phi^2(y)$ is a density of pions there is a condition imposed on the coefficients a_k's, namely $\sum_{k=0}^{\infty} a_k^2 = n$. The grand partition function is given by

$$G(z) = \sum_k \frac{d^2 a_k}{\bar{n}_k} e^{\frac{-a_k^2}{\bar{n}_k} [1+(1-z)\bar{n}_k]} {}^{(a+z-1)Y\phi_{MF}^2 + bY\phi_{MF}^4} \tag{18}$$

where

$$\bar{n}_k = \frac{1}{a+6b\phi_{MF}^2+c(2ky\pi/Y)^2} \tag{19}$$

After doing the integral the generating function becomes

$$G(z) = \sum_k [1+(1-z)\bar{n}_k]^{-1} e^{[a+z-1]Y\phi_{MF}^2 + bY\phi_{MF}^4} = e^{\frac{Y}{2\sqrt{\beta}}(\sqrt{\alpha}-\sqrt{\alpha'})} e^{[a+z-1]Y\phi_{MF}^2 + bY\phi_{MF}^4} \tag{20}$$

where $\alpha = a+6b\phi_{MF}^2$ and $\alpha' = \alpha - z+1$. The corresponding multiplicities are

$$\bar{n} = \frac{Y}{2\pi} \int \frac{dk}{\alpha+\beta k^2} + Y\phi_{MF}^2 = \frac{Y}{2\sqrt{\beta\alpha}} + Y\phi_{MF}^2 \tag{21}$$

where $\beta = c(2\pi/Y)^2$. To determine the nature of the phase transition (or crossover) we consider the specific heat of the gas near the critical point. Specific heat is defined as

$$C_p \sim \frac{d^2 f}{dT^2} \quad \text{where}$$

$$f(T) = \ln \quad e^{-F[\phi]+(z-1)\int_0^Y \phi^2 dy} \quad \delta\phi \tag{22}$$

Using the expression for $F[\phi]$ given by Eq. (17) we get

$$C_p \sim \frac{Y}{2\pi} \int \frac{dk}{(a+\beta k^2+1-z)^2} \left|\frac{d\alpha}{dT}\right|^2 \tag{23}$$

Up to now we have considered the most general case when the gas is at some finite arbitrary temperature. Since the expression for the free energy is different for the temperature above the critical value, then for the temperature below we now proceed to discuss these two cases separately.

a)   FEYNMAN-WILSON GAS AT $T > T_c$

When $T > T_c$, $\alpha = a, a > 0$ and the mean field solution is $\phi_{MF} = 0$. The grand partition function given by Eq. (20) becomes

$$G(z) = e^{\frac{Y(\sqrt{a}-\sqrt{a'})}{\sqrt{\beta}}} \tag{24}$$

where $a' = a-z+1$. The corresponding multiplicities and moments C_2, C_3, C_4 and C_5 are

$$\bar{n} = \frac{Y}{4\pi\sqrt{ac}}$$

$$C_2 = 1 + \frac{1}{\bar{n}} + \frac{1}{k_0}$$

$$C_3 = 1 + \frac{1}{\bar{n}^2} + \frac{3}{\bar{n}} + \frac{3}{k_0}\left(1+\frac{1}{\bar{n}}\right) + \frac{3}{k_0^2} \tag{25}$$

$$C_4 = 1 + \frac{6}{\bar{n}} + \frac{7}{\bar{n}^2} + \frac{1}{\bar{n}^3} + \frac{6}{k_0}\left(1+\frac{3}{\bar{n}}\right) + \frac{7}{k_0\bar{n}^2} + \frac{18}{k_0^2\bar{n}} + \frac{15}{k_0^2} + \frac{15}{k_0^3}$$

$$C_5 = 1 + \frac{10}{\bar{n}} + \frac{25}{\bar{n}^2} + \frac{15}{\bar{n}^3} + \frac{1}{\bar{n}^4} + \frac{10}{k_0} + \frac{45}{k_0^2} + \frac{105}{k_0^3}\left(1+\frac{1}{k_0}\right) + \frac{60}{k_0\bar{n}}$$

$$+ \frac{150}{k_0^2\bar{n}}\left(1+\frac{1}{k_0}\right) + \frac{75}{k_0\bar{n}^2}\left(1+\frac{1}{k_0}\right) + \frac{15}{k_0\bar{n}^3}$$

where $k_0 = Y/\zeta$ and $\zeta = \sqrt{\beta/\alpha}$.

When $z \to 1$, Eq. (24) shows that if $a \neq 0$ $G(z)_{z \to 1} \to [1-\ln z\bar{n}]^{-1}$. This corresponds to the system far from the crossover point. We note that coefficient a for $T > T_c$ can be written as $a = \gamma(T - T_c)$, $\gamma > 0$. Then the specific heat given by Eq. (23) becomes

$$C_p = \frac{\gamma^2}{2Y\pi\sqrt{\beta}} [\gamma(T-T_c) + 1 - z]^{-3/2} \tag{26}$$

When $z \to 1$

$$C_p \to \frac{\sqrt{\gamma}}{2Y\pi\sqrt{\beta}} (T - T_c)^{-3/2} \tag{27}$$

We note that when temperature $T \to T_c$, the specific heat (27) is discontinuous as expected at the crossover point.

b)   FEYNMAN-WILSON GAS AT $T < T_c$

Now we consider a gas of pions at temperature below critical. In this case $\phi_{MF}^2 \neq 0$, $\alpha = -2a$ and $a < 0$. The free energy has in addition a contribution from the mean field value. Therefore, the generating function given by Eq. (20) becomes

$$G(z) = e^{\frac{Y}{\sqrt{\beta}}(\sqrt{\alpha}-\sqrt{\alpha'})} \; e^{\frac{aY(1-z)}{2b}} \tag{28}$$

where $\alpha' = \alpha + 1 - z$. The corresponding multiplicities and moments C_2 C_3, C_4 and C_5 are

$$\bar{n} = \frac{Y}{2\sqrt{-2a\beta}} - \frac{aY}{2b} = \frac{Y^2}{4\pi\sqrt{-2ac}} - \frac{aY}{2b}$$

$$C_2 = 1 + \frac{1}{\bar{n}} + \frac{1}{k'_0}$$

$$C_3 = 1 + \frac{1}{\bar{n}^2} + \frac{3}{\bar{n}} + \frac{3}{k'_0}\left(1 + \frac{1}{\bar{n}}\right) + \frac{3}{k'^2_0}\left(1 - \frac{a\sqrt{\alpha\beta}}{Yb}\right)$$

$$C_4 = 1 + \frac{6}{\bar{n}} + \frac{7}{\bar{n}^2} + \frac{1}{\bar{n}^3} + \frac{6}{k'_0}\left(1 + \frac{3}{\bar{n}}\right) + \frac{7}{k'_0\bar{n}^2} + \frac{3}{k'^2_0} + \frac{18}{k'^2_{0-}}\frac{1}{(\bar{n}+\alpha)}$$

$$+ \frac{12}{k'^2_0}\left(1 - \frac{\alpha}{\bar{n}+\alpha}\right) + \frac{15}{k'^3_0}\;1 - \frac{\alpha(\alpha+2\bar{n})}{(\alpha+\bar{n})} \tag{29}$$

$$C_5 = 1 + \frac{10}{\bar{n}} + \frac{25}{\bar{n}^2} + \frac{15}{\bar{n}^3} + \frac{1}{\bar{n}^4} + \frac{10}{k'_0} + \frac{45}{k'^2_0} - \frac{30}{k'^2_0}\frac{\alpha}{\bar{n}+\alpha} + \frac{105}{k'^3_0}$$

$$- \frac{(75\alpha^2+150\alpha\bar{n}+30\alpha)}{k'^3_0\;(\alpha+\bar{n})} + \frac{105}{k'^4_0}\frac{1}{(1+\frac{\alpha}{\bar{n}})^3} + \frac{60}{k'_0\bar{n}}(1+\alpha) + \frac{150}{k'^2_0\bar{n}} - \frac{120\alpha}{k^2_0(\bar{n}+\alpha)}$$

$$+ \frac{75}{k'_0\bar{n}^2} + \frac{15}{k'_0\bar{n}^3} + \frac{150}{k'^3_0\bar{n}}\frac{1}{\left(1+\frac{\alpha}{\bar{n}}\right)^2} + \frac{75}{k'^2_0\bar{n}^2}\frac{1}{\left(1+\frac{\alpha}{\bar{n}}\right)^2} + \frac{15}{k'_0\bar{n}^3}$$

where $k'_0 = k_0 + 4a^2Y(1 + Y/16b)/b$.

In the limit when $z \to 1$ (chemical potential $\mu \to 0$) the grand partition function is

$$G(z) \rightarrow e^{\ln z \bar{n}} \sim [1 - \bar{n}\ln z]^{-1} \tag{30}$$

We recall that $\phi_{MF}^2 = -a/2b$ for $T < T_c$. Therefore $\alpha = -2a = 2\gamma(T_c - T)$, $\gamma > 0$. Then the specific heat is given by

$$C_p = \frac{4\gamma^2}{\gamma\sqrt{\beta}2\pi}[\gamma(T_c - T) + 1 - z]^{-3/2} \tag{31}$$

When $z \rightarrow 1$

$$C_p = \frac{4\sqrt{\gamma}}{\gamma\sqrt{\beta}2\pi} (T_c - T)^{-3/2} \tag{32}$$

Note that the coefficient of the $(T_c-T)^{-3/2}$ differs if T_c is approached from above or below (cf. Eq. (27)) which indicates discontinuity of the specific heat.

4. TWO-PARTICLE CORRELATIONS IN RAPIDITY SPACE

Experimental data on two-particle correlations provide additional tests for production models. We derive results for the two particle field and intensity correlation using the same (Gaussian) approximation employed above. Assuming for simplicity translation invariance of the field correlation functions we can write

$$\langle\phi(y+r) \phi(y)\rangle = \frac{1}{2\pi} \int dk \bar{n}_k e^{ikr} \tag{33}$$

where $\bar{n}_k$ is given by Eq. (19). If we define $\kappa = \frac{\sqrt{\alpha}}{\sqrt{\beta}}r$ we get

$$\langle\phi(y+r)\phi(y)\rangle = \frac{1}{2\sqrt{2\pi\alpha\beta}} \sqrt{\kappa}K_{1/2}[\kappa] \tag{34}$$

In the limit when $\alpha \to 0$, $\kappa \to 0$. If the energy is finite (Y finite) we can use the expression for the asymptotic behavior or the modified Bessel function $K_{1/2}(z \to 0) \to 1/2\Gamma[1/2]\sqrt{2/z}$. However if we consider the large energy limit with α finite we get $\langle \phi(y+r)\phi(y) \rangle_{\kappa large} \to \frac{1}{\sqrt{\alpha\beta}} e^{-\kappa}$. When $\kappa \to \infty$ and the system is far from the critical point the two particle correlation function vanishes.

The correlation function $C(y_1, y_2)$ defined as $\langle \phi^2(y_1)\phi^2(y_2) \rangle - \langle \phi^2(y_1) \rangle \langle \phi^2(y_2) \rangle$ has the following behavior

$$C(y_1, y_2) \sim e^{-|y_1 - y_2|/\zeta} \tag{35}$$

where $\zeta = \sqrt{\beta/\alpha}$. This agrees qualitatively with UA5 experimental data [15].

5. SUMMARY AND CONCLUSION

We have explored a suggestive statistical model for multiparticle production. In particular we have investigated the critical behavior of a one dimensional Feynman-Wilson "gas". We have shown that in the constant field approximation this gas approaches the crossover point at very large energies. This implies KNO scaling, although with a wrong variation multiplicity ($\bar{n} \sim \sqrt{\ln s}$). When we include the fluctuations around the mean field value the phase transition at finite temperature and finite energies implies KNO scaling with a definite prediction for the experimentally measurable quantities such as probability distribution, multiplicities and moments. All these observable quantities have different energy dependence for $T > T_c$ and $T < T_c$. For example, multiplicities increase as $\ln^2 s$ for $T > T_c$ while for $T < T_c$ they increase as A lns + B $\ln^2 s$. If we compare this prediction with the experimental data we conclude that Feynman-Wilson "gas" is still below the "critical temperature." Moments C_2 have the same form as for the negative binomial distribution with $k_0 = Y/\zeta$. If we are considering limiting rapidity space than $k_0 = y/\zeta$. Therefore we can "explain" the increase of the parameter k with rapidity y (i.e KNO scaling function becomes narrow as rapidity is increasing).

Also, the parameter k_0 is different below and above the critical temperature (See Eq. (25) and Eq. (29)). This confirms our previous statement that the "gas" has not reached the critical point yet.

Moments C_3 differ from the same for the negative binomial distribution by $1/k_0^2$ for $T > T_c$ and by $1/k'_0{}^2(1-3a\sqrt{\alpha\beta}/Yb)$ for $T < T_c$. We note that for large k_0 and k'_0, moments C_3 are the same as for the negative binomial distribution.

Another interesting observation is that in the limit of very large energies <u>and</u> far from the critical point we get the Poisson distribution which does not obey KNO scaling (this could be important for e^+e^- data).

The specific heat of the gas is discontinuous at the critical point. We have obtained the two-particle correlations for arbitrary energies and finite temperature which agree qualitatively with the experimental data. We have shown that at the critical point the correlation fun-ction diverges if the energy is finite and vanishes if the energy is infinite.

The behavior of the correlation function $C(y_1,y_2)$ given by Eq. (34) agrees qualitatively with experimental data (Ref. 8). Another indication of approaching the point of crossover would be observation of the increasing of $C(0.0)$ with energy.

694

These results are summarized in systematic form in Table I. Theoretical values for multiplicities and moments are compared with experimental data in Table II. Our model could also be applied to nucleon-nucleus, nucleus-nucleus, and heavy-ion collisions.

Acknowledgements

We are indebted to D. Scalapino for many helpful comments.

Table I. Theoretical prediction for the observable quantities such as multiplicities, moments and two-particle correlation functions. The parameter α is $\alpha = -2a$ for $T < T_c$ and $\alpha = a$ for $T > T_c$.

	$T < T_c \ (a<0)$	$T > T_c \ (a > 0)$
$\bar{n}$	$\dfrac{Y^2}{4\pi\sqrt{-2ac}} - \dfrac{aY}{2b}$	$\dfrac{Y^2}{4\pi\sqrt{ac}}$
k	$\dfrac{Y}{\xi_1} + 4\dfrac{a^2 Y}{b}\left(1+\dfrac{Y}{16b}\right)$	$\dfrac{Y}{\xi_2}$
ξ	$\xi_1 = \sqrt{\dfrac{\beta}{(-2a)}}$	$\xi_2 = \sqrt{\dfrac{\beta}{(a)}}$
C_2	$1 + \dfrac{1}{n} + \dfrac{1}{k}$	$1 + \dfrac{1}{n} + \dfrac{1}{k}$
C_3	$1 + \dfrac{1}{n^2} + \dfrac{3}{n} + \dfrac{3}{k}\left(1+\dfrac{1}{n}\right) + \dfrac{3}{k^2}\left(1 - \dfrac{2\pi a\sqrt{-2ac}}{Y^2 b}\right)$	$1 + \dfrac{1}{n^2} + \dfrac{3}{n} + \dfrac{3}{k}\left(1+\dfrac{1}{n}\right) + \dfrac{3}{k^2}$
C_4	$1 + \dfrac{6}{n} + \dfrac{7}{n^2} + \dfrac{1}{n^3} + \dfrac{6}{k_0'}\left(1+\dfrac{3}{n}\right) + \dfrac{7}{k_0'^2} + \dfrac{3}{k_0'^2} + \dfrac{18}{k_0'^2}\dfrac{1}{n+\alpha}$ $+ \dfrac{12}{k_0'^2}\left(1 - \dfrac{\alpha}{n+\alpha}\right) + \dfrac{15}{k_0'^3}\left(1 - \dfrac{\alpha(\alpha+2n)}{(\alpha+n)}\right)$	$C_4 = 1 + \dfrac{6}{n} + \dfrac{7}{n^2} + \dfrac{1}{n^3} + \dfrac{6}{k_0}\left(1+\dfrac{3}{n}\right)$ $+ \dfrac{7}{k_0^2 n} + \dfrac{18}{k_0^2 n} + \dfrac{15}{k_0^2 k_0} + \dfrac{15}{k_0^3}$

TABLE I. (Cont.)

	$T < T_c\,(a<0)$	$T > T_c\,(a > 0)$
c_5	$1 + \dfrac{10}{\bar{n}} + \dfrac{25}{\bar{n}^2} + \dfrac{15}{\bar{n}^3} + \dfrac{1}{\bar{n}^4} + \dfrac{10}{k_0'} + \dfrac{45}{k_0'^2} - \dfrac{30}{k_0'^2}\,\dfrac{\alpha}{\bar{n}+\alpha} + \dfrac{105}{k_0'^3}$ $- \dfrac{(75\,\alpha^2 + 150\alpha\bar{n} + 30\alpha)}{k_0'^3\,(\alpha + \bar{n})} + \dfrac{105}{k_0'^4}\,\dfrac{1}{(1+\frac{\alpha}{\bar{n}})^3} + \dfrac{60}{k_0'\bar{n}}\,(1+\alpha) + \dfrac{150}{k_0'^2\bar{n}}$ $+ \dfrac{120\,\alpha}{k_0'^2(\bar{n}+\alpha)} + \dfrac{75}{k_0'^2\bar{n}^2} + \dfrac{15}{k_0'^2\bar{n}^3} + \dfrac{150}{k_0'^3\bar{n}}\,\dfrac{1}{(1+\frac{\alpha}{\bar{n}})^2} + \dfrac{75}{k_0'^2\bar{n}^2}\,(1+\dfrac{\alpha}{\bar{n}}) + \dfrac{15}{k_0'^2\bar{n}^3}$	$c_5 = 1 + \dfrac{10}{\bar{n}} + \dfrac{25}{\bar{n}^2} + \dfrac{15}{\bar{n}^3} + \dfrac{1}{\bar{n}^4} + \dfrac{10}{k_0} + \dfrac{45}{k_0^2}$ $+ \dfrac{105}{k_0^3}\left(1+\dfrac{1}{k_0}\right) + \dfrac{60}{k_0\bar{n}} + \dfrac{150}{k_0^2\bar{n}}\left(1+\dfrac{1}{k_0}\right)$ $+ \dfrac{75}{k_0^2\bar{n}^2}\left(1+\dfrac{1}{k_0}\right) + \dfrac{15}{k_0^2\bar{n}^3}$

696

Table II.

Theoretical results for $\bar{n}$, c_2 and c_3 are given for the energy range 11.5 GeV $\leq \sqrt{s} \leq 900$ GeV. These values were obtained using the theoretical predictions for $\bar{n}$, c_2, c_3, c_4 and c_5 for $T < T$ given by Eq. (29). The experimental data for $\bar{n}$, c_2, c_3, and c_4 and c_5 (in the brackets) are taken from Ref. 16.

$\sqrt{s}$	$\bar{n}$	c_2	c_3	c_4	c_5	ξ	$-a$	c	$0.1b^{-1}$
11.5	8.59 (6.23±0.04)	1.169 (1.21±0.01)	1.582 (1.70±0.02)	2.254 (2.68±0.06)	4.6179 (4.6±0.2)	0.46	0.63	0.50	0.01
30.4	12.44 (10.07±0.1)	1.171 (1.18±0.01)	1.576 (1.60±0.02)	2.298 (2.43±0.06)	3.852 (4.1±0.2)	0.97	0.24	1.30	0.03
62.2	15.77 (13.6±0.1)	1.186 (1.20±0.01)	1.631 (1.65±0.02)	2.436 (2.56±0.06)	4.523 (4.4±0.2)	1.47	0.14	2.25	0.06
200	22.06 (21.6±0.5)	1.263 (1.26±0.03)	1.963 (1.91±0.12)	3.310 (3.3±0.3)	7.386 (6.6±0.9)	3.13	0.05	5.72	0.14
540	28.27 (28.3±0.2)	1.306 (1.31±0.03)	2.170 (2.12±0.11)	3.904 (3.98±0.09)	9.275 (8.8±1.0)	4.44	0.03	9.23	0.23
900	31.78 (35.1±0.6)	1.344 (1.34±0.03)	2.356 (2.22±0.13)	4.127 (4.3±0.4)	11.0 (9.4±1.1)	5.44	0.03	12.01	0.30

References

[1] P. Carruthers and I. Sarcevic, Phys. Lett. 189B (1987) 442.

[2] For review see P. Carruthers and C. C. Shih, LA-UR-87-1655 to be published in International Journal of Modern Physics A.

[3] Z. Koba, H. B. Nielsen and P. Olesen, Nucl. Phys. B40 (1974) 317.

[4] W. Tome et al., Nucl. Phys. B129 (1972) 365; J. Firestone et al., Phys. Rev. D14 (1976) 2902.

[5] UA5 Collaboration, G. J. Alner et al., Phys. Lett. 138B (1984) 304; ibid, Phys. Lett. 160B (1985) 193.

[6] P. Carruthers, LA-UR-86-1540, in Proceedings of the XXIst Recontre de Moriond, Les Arcs, Savoie, France.

[7] D. J. Scalapino and R. L. Sugar, Phys. Rev. D8 (1973) 2284; J. C. Botke, D. J. Scalapino and R. L. Sugar, Phys. Rev. D9 (1974) 813; ibid. Phys. Rev. D10 (1974) 1604.

[8] K. Wilson, Cornell Report No. CLNS-131, 1970, in Proceedings of the Fourteenth Scottish Universities Summer School in Physics, 1973, edited by R. L. Crawford and R. Jennings (Academic, London, 1974).

[9] D. Gross and A. Neveu, Phys. Rev. D10 (1974) 3235.

[10] P. Carruthers and I. Sarcevic, in preparation.

[11] N. G. Antoniou, A. I. Karanikas and S. D. P. Vlassopulos Phys. Rev. D29 (1984) 1470; ibid Phys. Rev. D14 (1976) 3578.

[12] G. N. Fowler, R. M. Weiner and G. Wilk, Phys. Rev. Lett. 55 (1985) 173.

[13] UA5 Collaboration, G. J. Alner et al., Phys. Lett. 138B (1984) 304; ibid Phys. Lett. 160B (1985) 193.

[14] Handbook of Mathematical Functions, edited by M. Abramowitz and I. A. Stegun (Dover, New York, 1965).

[15] UA5 Collaboration, p. 155 in Proceedings of the XV International Symposium on Multiparticle Dynamics, 1984, edited by G. Gustafson and C. Peterson (World Scientific), 1984).

[16] UA5 Collaration, G. J. Alner et al., Phys. Lett. 160B (1985) 199.